Student Solutions Manual

Chemistry & Chemical Reactivity

EIGHTH EDITION

Alton J. Banks
North Carolina State University

John C. Kotz
SUNY Oneonta

Paul Treichel
University of Wisconsin, Madison

John Townsend
West Chester University of Pennsylvania

BROOKS/COLE
CENGAGE Learning

Australia • Brazil • Japan • Korea • Mexico • Singapore • Spain • United Kingdom • United States

For product information and technology assistance, contact us at **Cengage Learning Customer & Sales Support, 1-800-354-9706**

For permission to use material from this text or product, submit all requests online at **www.cengage.com/permissions** Further permissions questions can be emailed to **permissionrequest@cengage.com**

ISBN-13: 978-1-111-42698-9
ISBN-10: 1-111-42698-8

Brooks/Cole
20 Davis Drive
Belmont, CA 94002-3098
USA

Cengage Learning is a leading provider of customized learning solutions with office locations around the globe, including Singapore, the United Kingdom, Australia, Mexico, Brazil, and Japan. Locate your local office at: **www.cengage.com/global**

Cengage Learning products are represented in Canada by Nelson Education, Ltd.

To learn more about Brooks/Cole, visit **www.cengage.com/brookscole**

Purchase any of our products at your local college store or at our preferred online store **www.cengagebrain.com**

Printed in the United States of America
1 2 3 4 5 6 7 15 14 13 12 11

Table of Contents

Chapter 1
Basic Concepts of Chemistry

PRACTICING SKILLS

Matter: Elements and Atoms, Compounds and Molecules

1. The name of each of the elements:

(a)	C	carbon	(c)	Cl	chlorine	(e)	Mg	magnesium—typically confused with manganese (Mn)
(b)	K	potassium—from *Latin*, Kalium	(d)	P	phosphorus—frequently confused with Potassium	(f)	Ni	nickel

3. The symbol for each of the elements:

(a)	barium	Ba	(d)	lead	Pb
(b)	titanium	Ti	(e)	arsenic	As
(c)	chromium	Cr	(f)	zinc	Zn

5. In each of the pairs, decide which is an element and which is a compound:

(a) Na and NaCl—Sodium(Na) is an element and Sodium chloride(NaCl) is a compound.

(b) Sugar and carbon—Sugar($C_xH_yO_x$) is a compound, and carbon(C) is an element.

(c) Gold and gold chloride—Gold(Au) is an element, and gold chloride ($AuCl_x$) is a compound.

Physical and Chemical Properties

7. Determine if the property is a physical or chemical property for the following:

(a)	color	a physical property
(b)	transformed into rust	a chemical property
(c)	explode	a chemical property
(d)	density	a physical property
(e)	melts	a physical property
(f)	green	a physical property (as in (a))

Physical properties are those that can be observed or measured without changing the composition of the substance. Exploding or transforming into rust results in substances that are **different** from the original substances—and represent chemical properties.

9. Descriptors of physical versus chemical properties:
   (a) Color and physical state are physical properties (colorless, liquid) while **burning** reflects a chemical property.
   (b) Shiny, metal, orange, and liquid are physical properties while **reacts readily** describes a chemical property.

Energy

11. To move the lever, one uses mechanical energy. The energy resulting is manifest in electrical energy (which produces light); thermal energy would be released as the bulb in the flashlight glows.

13. Which represents potential energy and which represents kinetic energy:
    (a) thermal energy represents matter in motion--kinetic
    (b) gravitational energy represents the attraction of the earth for an object—and therefore energy due to position--potential
    (c) chemical energy represents the energy stored in fuels--potential
    (d) electrostatic energy represents the energy of separated charges—and therefore potential energy.

GENERAL QUESTIONS

15. For the gemstone turquoise:
    (a) Qualitative: blue-green color   Quantitative: density; mass
    (b) Extensive: Mass   Intensive: Density; Color
    (c) Volume: $\dfrac{2.5 \text{ g}}{1} \cdot \dfrac{1 \text{ cm}^3}{2.65 \text{ g}} = 0.94 \text{cm}^3$

17. Of the observations below, those which identify chemical properties:
    Chemical properties, in general, are those observed during a chemical change—as opposed to during a physical change.
    (a) Sugar soluble in water--Physical
    (b) Water boils at 100°C--Physical
    (c) UV light converts O_3 to O_2--**Chemical**
    (d) Ice is less dense than water--Physical

2

19. (a) The symbols for the elements in fluorite: Ca(calcium) and F(fluorine);

(b) Shape of the crystals: cubic

Arrangement of atoms in the crystal: indicates that the fluorine atoms are arranged around the calcium atoms in the lattice in such a way as to form a cubic lattice.

21. In the photo, one sees obvious regions that are darker (iron) and those that are lighter (sand). This mixture is **heterogeneous**. Separating the iron from the sand can be done easily, remembering that the components of a mixture retain their properties—so the iron, attracted to a magnet—can be separated from the sand by dragging a magnet through the mixture.

23. Identify physical or chemical changes:

(a) As there is no change in the composition of the carbon dioxide in the sublimation process, this represents a physical change.

(b) A change in density as a function of temperature does not reflect a change in the composition of the substance (mercury), so this phenomenon represents a physical change.

(c) The combustion of methane represents a change in the substance present as methane is converted to the oxides of hydrogen and carbon that we call water and carbon dioxide—a chemical change.

(d) Dissolving NaCl in water represents a physical change as the solid NaCl ion pairs are separated by the solvent, water. This same phenomenon, the separation of ions, also occurs during melting.

25. A segment of Figure 1.2 is shown here:

The macroscopic view is the large crystal in the lower left of the figure, and the particulate view is the representation in the upper right. If one imagines reproducing the particulate (sometimes called submicroscopic) in all three dimensions—imagine a molecular duplicating machine—the macroscopic view results.

27. A substance will float in any liquid whose density is greater than its own, and sink in any liquid whose density is less than the substance's. The piece of plastic soda bottle

(d =1.37 g/cm³) will float in liquid CCl_4 and the piece of aluminum (d = 2.70 g/cm³) will sink in the liquid CCl_4.

29. Categorize each as an element, a compound, or a mixture:

   (a) Sterling silver is a mixture—an alloy—of silver and other metals, to improve the mechanical properties. Silver is a very soft metal, so it is alloyed with copper (frequently) to produce a material with better "handling" characteristics.

   (b) Carbonated mineral water is a mixture. It certainly contains the compound water AND carbon dioxide. The term "mineral" implies that other dissolved materials are present.

   (c) tungsten—an element

   (d) aspirin—a compound, with formula $C_9H_8O_4$

31. Indicate the relative arrangements of the particles in each of the following:

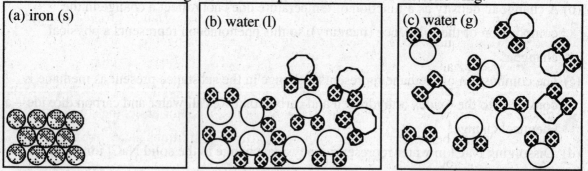

(a) iron (s)   (b) water (l)   (c) water (g)

33. When the three liquids are placed into the graduated cylinder, they will "assemble" in layers with increasingly smaller densities (from the bottom to the top) in the cylinder.

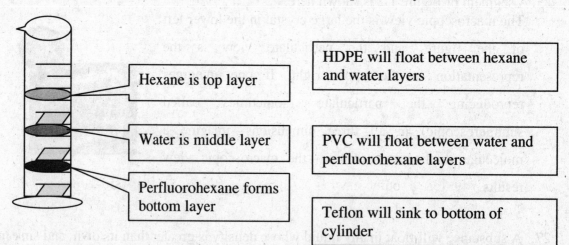

Hexane is top layer

Water is middle layer

Perfluorohexane forms bottom layer

HDPE will float between hexane and water layers

PVC will float between water and perfluorohexane layers

Teflon will sink to bottom of cylinder

[Shadings are added to the top of each layer to provide clarity only (NOT to indicate the color of the liquids). Similarly the parallelogram symbols indicating the plastic samples are shaded—only to provide clarity in locating them—and not to imply any specific colors.]

35. HDPE with a density of 0.97 g/mL will float in any liquid whose density is greater than 0.97 g/mL and sink in any liquid whose density is less than that of HDPE. Of the liquids listed, HDPE should float in ethylene glycol, water, acetic acid, and glycerol.

37. To confirm that a sample of silvery meal is silver, one could determine the density of the metal.

39. If one excretes too much sugar, the concentration of the sugar "solution" in the body would decrease, resulting in urine with a higher density. If one excretes too much water, the concentration of the sugar "solution" in the body would increase, with a concomitant decrease in density of the urine.

41. For the reaction of elemental potassium reacting with water:
    (a) States of matter involved: **Solid** potassium reacts with **liquid** water to produce **gaseous** hydrogen and aqueous potassium hydroxide solution (a homogenous mixture).
    (b) The observed change is chemical. The products (hydrogen and potassium hydroxide) are quite different from elemental potassium and water. Litmus paper would also provide the information that while the original water was neither acidic nor basic, the solution produced would be basic. (It would change the color of red litmus paper to blue.)
    (c) The reactants: potassium and water
        The products: hydrogen, potassium hydroxide solution, heat, and light
    (d) Potassium reacts **vigorously** with water. Potassium is less dense than water, and floats atop the surface of the water. The reaction produces enough heat to ignite the hydrogen gas evolved. The flame observed is typically violet-purple in color. The potassium hydroxide formed is soluble in water (and therefore not visible).

43. Since gases rise to an area with a similar density as their own, balloons with helium and neon—with densities less than that of dry air--will float (and, if untethered, float away), while the balloons containing argon and krypton will "sink"—to the lowest nearby surface.

45. The dissolution of iodine in ethanol (to make a solution) is a **physical** change, with iodine being the solute and ethanol the solvent.

Chapter 1
Let's Review

PRACTICING SKILLS

Temperature Scales

1. Express 25 °C in kelvins:

$$K = (25 \text{ °C} + 273) \text{ or } 298 \text{ K}$$

3. Make the following temperature conversions:

°C	K
(a) 16	$16 + 273.15 = 289$
(b) 370 - 273 or 97	370
(c) 40	$40 + 273.15 = 310$

Note no decimal point after 40

Length, Volume, Mass, and Density

5. The distance of a marathon (42.195 km) in meters; in miles:

$$\frac{42.195 \text{ km}}{1} \bullet \frac{1000 \text{ m}}{1 \text{ km}} = 42195 \text{ m}$$

$$\frac{42.195 \text{ km}}{1} \bullet \frac{0.62137 \text{ miles}}{1 \text{ km}} = 26.219 \text{ miles}$$

The factor (0.62137 mi/km) is found inside the back cover of the text.

7. Express the area of a 2.5 cm x 2.1 cm stamp in cm^2 ; in m^2 :

$$2.5 \text{ cm} \bullet 2.1 \text{ cm} = 5.3 \text{ cm}^2$$

$$5.3 \text{ cm}^2 \bullet \left(\frac{1 \text{ m}}{100 \text{ cm}}\right)^2 = 5.3 \times 10^{-4} \text{ m}^2$$

9. Express volume of 250. mL beaker in cm^3; in liters (L); in m^3 ; in dm^3:

$$\frac{250. \text{ mL}}{1 \text{ beaker}} \bullet \frac{1 \text{cm}^3}{1 \text{mL}} = \frac{250. \text{ cm}^3}{1 \text{ beaker}}$$

$$\frac{250. \text{ cm}^3}{1 \text{ beaker}} \bullet \frac{1 \text{L}}{1000 \text{cm}^3} = \frac{0.250 \text{ L}}{1 \text{ beaker}}$$

$$\frac{250.\ cm^3}{1\ beaker} \cdot \frac{1m^3}{1x10^6 cm^3} = \frac{2.50\ x\ 10^{-4}\ m^3}{1\ beaker}$$

$$\frac{250.\ cm^3}{1\ beaker} \cdot \frac{1L}{1000 cm^3} \cdot \frac{1\ dm^3}{1\ L} = \frac{0.250\ dm^3}{1\ beaker}$$

11. Convert book's mass of 2.52 kg into grams:

$$\frac{2.52\ kg}{1\ book} \cdot \frac{1\ x\ 10^3 g}{1\ kg} = \frac{2.52\ x\ 10^3\ g}{book}$$

13. What mass of ethylene glycol (in grams) possesses a volume of 500. mL of the liquid?

$$\frac{500.\ mL}{1} \cdot \frac{1\ cm^3}{1\ mL} \cdot \frac{1.11\ g}{cm^3} = 555\ g$$

15. To determine the density, given the data, one must first convert each length to units of cm:

$$\left[\frac{1.05\ mm}{1} \cdot \frac{1\ cm}{10\ mm}\right] \cdot \frac{2.35\ cm}{1} \cdot \frac{1.34\ cm}{1} = 0.3306\ cm^3\ (\ 0.331\ to\ 3sf)$$

The density is calculated Mass/Volume or $\frac{2.361\ g}{0.3306\ cm^3} = 7.14 \frac{g}{cm^3}$. Given the selection of

metals, the identity of the metal is **zinc**.

Energy Units

17. Express the energy of a 1200 Calories/day diet in joules:

$$\frac{1200\ Cal}{1\ day} \cdot \frac{1000\ calorie}{1\ Cal} \cdot \frac{4.184\ J}{1\ cal} = 5.0\ x\ 10^6\ Joules/day$$

19. Compare 170 kcal/serving and 280 kJ/serving.

$$\frac{170\ kcal}{1\ serving} \cdot \frac{1000\ calorie}{1\ kcal} \cdot \frac{4.184\ J}{1\ cal} \cdot \frac{1\ kJ}{1000\ J} = 710\ kJoules/serving$$

So 170 kcal/serving has a greater energy content.

Accuracy, Precision, Error, and Standard Deviation

21. Using the data provided, the averages and their deviations are as follows:

Data point	Method A	deviation	Method B	deviation
1	2.2	0.2	2.703	0.777
2	2.3	0.1	2.701	0.779
3	2.7	0.3	2.705	0.775
4	2.4	0.0	5.811	2.331
Averages:	2.4	0.2	3.480	1.166

Note that the deviations for both methods are calculated by first determining the average of the four data points, and then subtracting the individual data points from the average (without regard to sign).

(a) The average density for method A is 2.4 ± 0.2 grams while the average density for method B is 3.480 ± 1.166 grams—if one includes all the data points. *Data point 4* in *Method B* has a large deviation, and *should probably be excluded* from the calculation. If one omits data point 4, Method B gives a density of 2.703 ± 0.001 g.

(b) The percent error for each method:

Error = experimental value - accepted value

From Method A   error = $(2.4 - 2.702) = 0.3$

From Method B   error = $(2.703 - 2.702) = 0.001$ (omitting data point 4)

error = $(3.480 - 2.702) = 0.778$ (including all data points)

and the percent error is then:

$$(\text{Method A}) = \frac{0.3}{2.702} \cdot \frac{100}{1} = \text{ about 10\% to 1 s.f.}$$

$$(\text{Method B}) = \frac{0.001}{2.702} \cdot \frac{100}{1} = \text{ about 0.04\% to 1 s.f.})$$

(c) The standard deviation for each method:

$$\text{Method A: } \sqrt{\frac{(0.2)^2 + (0.1)^2 + (0.3)^2 + (0.0)^2}{3}} = \sqrt{\frac{0.14}{3}} = 0.216 \text{ or } 0.2 \text{ (to 1 s.f.)}$$

$$\text{and for Method B: } \sqrt{\frac{(0.777)^2 + (0.779)^2 + (0.775)^2 + (2.331)^2}{3}} =$$

$$\sqrt{\frac{7.244}{3}} = 1.554 \text{ or } 1.55 \text{ (to 3 s.f.)}$$

(d) If one counts all data points, the deviations **for all data points** of Method A are less than those for **the data points of** Method B, Method A offers *better precision*. On the other hand, omitting data point 4, Method B offers both *better accuracy* (average closer to the accepted value) and *better precision* (since the value is known to a greater number of significant figures).

Exponential Notation and Significant Figures

23. Express the following numbers in exponential (or scientific) notation:

(a) $0.054 = 5.4 \times 10^{-2}$ To locate the decimal behind the first non-zero digit, we move the decimal place to the right by 2 spaces (-2); 2 significant figures

(b) $5462 = 5.462 \times 10^3$ To locate the decimal behind the first non-zero digit, we move the decimal place to the left by 3 spaces (+3); 4 significant figures

(c) $0.000792 = 7.92 \times 10^{-4}$ To locate the decimal behind the first non-zero digit, we move the decimal place to the right by 4 spaces (-4); 3 significant figures

(d) $1600 = 1.6 \times 10^{3}$ To locate the decimal behind the first non-zero digit, we move the decimal place to the left by 3 spaces (+3); 2 significant figures

25. Perform operations and report answers to proper number of s.f.:

(a) $(1.52)(6.21 \times 10^{-3}) = 9.44 \times 10^{-3}$ (3 sf since each term in the product has 3)

(b) $(6.217 \times 10^{3}) - (5.23 \times 10^{2}) = 5.694 \times 10^{3}$ [Convert 5.23×10^{2} to 0.523×10^{3} and subtract, leaving 5.694×10^{3}. With 3 decimal places to the right of the decimal place in both numbers, we can express the difference with 3 decimal places.

(c) $(6.217 \times 10^{3}) \div (5.23 \times 10^{2}) = 11.887$ or 11.9 (3 s.f.)

Recall that in multiplication and division, the result should have the same number of significant figures as the **term** with the fewest significant figures (3 in this case).

(d) $(0.0546)(16.0000)\left[\dfrac{7.779}{55.85}\right] = 0.121678$ or 0.122 (3 s.f.)

the same rule applies, as in part (c) above: The first term has 3 s.f., the second term 6 s.f.—yes the zeroes count, and the third term 4 s.f. Dividing the two terms in the last quotient gives an answer with 4 s.f. So with 3, 6, and 4 s.f. in the terms, the answer should have **no more than** 3 s.f.

Graphing

27. Plot the data for number of kernels of popcorn versus mass (in grams):

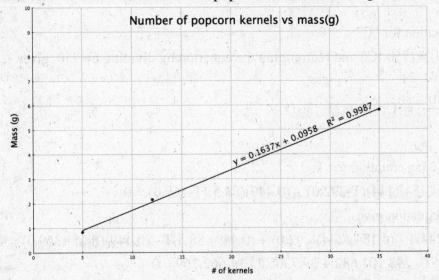

The best straight line has the equation, $y = 0.1637x + 0.0958$, with a slope of 0.1637.

This slope indicates that the mass increases by a factor of 0.1637grams with each kernel of popcorn. The mass of 20 kernels would be: mass = (0.1637)(20) + 0.0958 or 3.3698 grams. To determine the number of kernels (x) with a mass of 20.88 grams, substitute 20.88 for mass (i.e. y) and solve for the number of kernels.

20.88 g = 0.1637(x) + 0.0958; (20.88 – 0.0958) = 0.1637x and dividing by the slope:

$$\frac{[20.88 - 0.0958]}{0.1637} = x \text{ or } 126.96 \text{ kernels—approximately } 127 \text{ kernels.}$$

29. Using the graph shown, determine the values of the equation of the line:

(a) Using the first and last data points, we can calculate the slope (rise/run):

$$\frac{20.00 - 0.00}{0.00 - 5.00} = -4.00$$

The intercept (b) is the y value when the x value is zero (0). Substituting into the equation for the line:
Y = (-4.00)(0) + b

We can read this value from the graph (20.00). The equation for the line is:

Y = -4.00x + 20.00

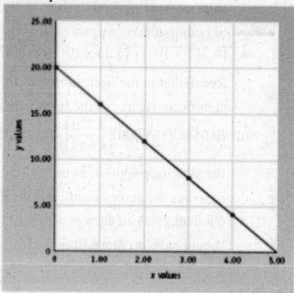

(b) The value of y when x = 6.0 is:

Y = (-4.00)(6.00) + 20.00 or -4.00.

Solving Equations

31. Solving the equation for "C":

(0.502)(123) = (750.)C   and rearranging the equation by dividing by 750. gives

$$\frac{(0.502)(123)}{750.} = C = 0.0823 \ (3 \text{ sf})$$

33. Solve the following equation for T:

The equation: (4.184)(244)(T-292.0) + (0.449)(88.5)(T-369.0) = 0

Expanding the equation gives:

(4.184)(244)T - (4.184)(244)(292.0) + (0.449)(88.5)T - (0.449)(88.5)(369.0) = 0

1,020.896T - 298,101.632 + 39.7365T - 14,662.769 = 0

grouping the terms (1,020.896T + 39.7365T) gives 1,060.6325T likewise

(- 298,101.632 + - 14,662.769) gives -312,764.401.

10

The equation now is equivalent to 1,060.6325T -312,764.401= 0. If we add the negative term to both sides:

1,060.6325T = 312,764.401 and T = 312,764.401/1,060.6325 or 294.884797986.

This is clearly too many significant figures. Rounding to 3sf, gives a value of 295.

General Questions

35. Express the length 1.97 Angstroms in nanometers? In picometers?

$$\frac{1.97 \text{ Angstrom}}{1} \bullet \frac{1 \times 10^{-10} \text{m}}{1 \text{ Angstrom}} \bullet \frac{1 \times 10^{9} \text{ nm}}{1 \text{ m}} = 0.197 \text{ nm}$$

$$\frac{1.97 \text{ Angstrom}}{1} \bullet \frac{1 \times 10^{-10} \text{m}}{1 \text{ Angstrom}} \bullet \frac{1 \times 10^{12} \text{pm}}{1\text{m}} = 197 \text{ pm}$$

37. Diameter of red blood cell = 7.5 μm

(a) In meters: $\dfrac{7.5 \text{ μm}}{1} \bullet \dfrac{1 \text{ m}}{1 \times 10^{6} \text{μm}} = 7.5 \times 10^{-6} \text{m}$

(b) In nanometers: $\dfrac{7.5 \text{ μm}}{1} \bullet \dfrac{1 \text{ m}}{1 \times 10^{6} \text{μm}} \bullet \dfrac{1 \times 10^{9} \text{nm}}{1\text{m}} = 7.5 \times 10^{3} \text{nm}$

(c) In picometers: $\dfrac{7.5 \text{ μm}}{1} \bullet \dfrac{1 \text{ m}}{1 \times 10^{6} \text{μm}} \bullet \dfrac{1 \times 10^{12} \text{pm}}{1\text{m}} = 7.5 \times 10^{6} \text{ pm}$

39. Mass of procaine hydrochloride (in mg) in 0.50 mL of solution

$$\frac{0.50 \text{ mL}}{1} \bullet \frac{1.0 \text{ g}}{1 \text{ mL}} \bullet \frac{10. \text{ g procaine HCl}}{100 \text{ g solution}} \bullet \frac{1 \times 10^{3} \text{ mg procaine HCl}}{1 \text{ g procaine HCl}} = 50. \text{ mg procaine HCl}$$

41. The volume of the marbles is the initial question, since its volume will add to the volume of water initally present (61 mL). The final volume is about 99 mL, so the volume occupied by the marbles is 99 ml – 61 ml, or 38mL. Knowing that marbles have a collective mass of 95.2 g, the density is : $D = \dfrac{M}{V} = \dfrac{95.2 \text{ g}}{38 \text{ mL}} = 2.5 \text{ g/mL}$ (to 2 sf).

43. For the sodium chloride unit cell:

(a) The volume of the unit cell is the (edge length)^3
With an edge length of 0.563 nm, the volume is (0.563 nm)^3 or 0.178 nm^3 The volume in cubic centimeters is calculated by first expressing the edge length in cm:

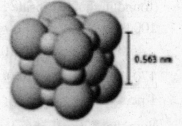

0.563 nm

sodium chloride, NaCl

$$\frac{0.563 \text{ nm}}{1} \bullet \frac{1 \times 10^{2} \text{cm}}{1 \times 10^{9} \text{nm}} = 5.63 \times 10^{-8} \text{cm}$$

The volume is (5.63 × 10^{-8} cm)^3 or 1.78 × 10^{-22} cm^3

11

(b) The mass of the unit cell is:

$M = V \cdot D = 1.78 \times 10^{-22} cm^3 \cdot 2.17 \ g/cm^3 = 3.86 \times 10^{-22} g$

(c) Given the unit cell contains 4 NaCl "molecules", the mass of one "molecule is:

$$\frac{3.86 \times 10^{-22} \ g \ for \ 4 \ NaCl \ pairs}{4 \ NaCl \ pairs} = 9.66 \times 10^{-23} \ g/ion \ pair$$

45. The accepted value for a normal human temperature is 98.6 °F. On the Celsius scale this corresponds to:

$$°C = \frac{5}{9}(98.6 - 32) = 37 \ °C$$

Since the melting point of gallium is 29.8 °C, the gallium should melt in your hand.

47. The heating of popcorn causes the loss of water.

(a) The percentage of mass lost upon popping:
$$\frac{(0.125g - 0.106g)}{0.125g} \times 100 = 15\%$$

(b) With an average mass of 0.125 g, the number of kernels in a pound of popcorn:
$$\frac{1 \ kernel}{0.125g} \cdot \frac{453.6g}{1 \ lb} = 3628.8 \ kernels \ or \ 3630 \ (to \ 3 \ s.f.)$$

49. The mass of NaF needed for 150,000 people for a year:

This problem can be done many different ways. One way is to begin with a factor that contains the units of the "answer" (mass NaF in kg). Since NaF is 45% fluoride (and 55 %Na), we can write the factor: $\frac{100.0 \ kg \ NaF}{45.0 \ kg \ F^-}$. Note that the expression of kg/kg has the same value of g/g—and provides the "desired" units of our answer. A concentration of 1ppm can be expressed as: $\frac{1.00 \ kg \ F^-}{1.00 \times 10^6 \ kg \ H_2O}$ [We could use the fraction with the masses expressed in **grams**, but we would have to convert grams to kg. Note this factor can be derived from the factor using grams if you multiply BOTH numerator and denominator by 1000.]. Using the data provided in the problem, plus conversion factors (found in the rear inside cover of your textbook)

$$\frac{100.0 \ kg \ NaF}{45.0 \ kg \ F^-} \cdot \frac{1.00 \ kg \ F^-}{1.00 \times 10^6 \ kg \ H_2O} \cdot \frac{1 \ kg \ H_2O}{1 \times 10^3 \ cm^3 H_2O} \cdot \frac{1 \times 10^3 \ cm^3 H_2O}{1.0567 \ qt \ H_2O} \cdot \frac{4.00 \ qt \ H_2O}{1 \ gal \ H_2O} \cdot$$

$$\frac{170 \ gal \ H_2O}{1 \ person-day} \cdot \frac{1.50 \times 10^5 person}{1} \cdot \frac{365 \ day}{1 \ year} = 8.0 \times 10^4 \ kg \ NaF/year$$

Note that 170 gal of water per day limits the answer to 2 s.f.

51. Mass of sulfuric acid in 500. mL (or 500. cm^3)solution.

$$\frac{38.08 \text{ g sulfuric acid}}{100.00 \text{ g solution}} \cdot \frac{1.285 \text{ g solution}}{1.000 \text{ cm}^3 \text{ solution}} \cdot \frac{500.\text{cm}^3 \text{solution}}{1} = 244.664 \text{ g sulfuric acid}$$

or 245 g sulfuric acid (3 sf—note 500. has 3 sf)

53.(a) Volume of solid water at –10°C when a 250. mL can is filled with liquid water at 25°C:

The volume of liquid water at 25 degrees is 250. mL (or cm^3). The mass of that water is:

$$\frac{0.997 \text{ g water}}{1.000 \text{ cm}^3 \text{ water}} \cdot \frac{250.\text{ cm}^3}{1} = 249.25 \text{ g water (249 to 3 s.f.)}$$

That mass of water at the lower temperature will occupy:

$$\frac{1.000 \text{ cm}^3 \text{ water}}{0.917 \text{ g water}} \cdot \frac{249.25 \text{ g water}}{1} = 271.81 \text{ (or 272 cm}^3 \text{ to 3 sf)}$$

(b) With the can being filled to 250. mL at room temperature, the expansion (an additional 22 mL) can not be contained in the can. (Get out the sponge—there's a mess to clean up.)

55. Calculate the density of steel if a steel sphere of diameter 9.40 mm has a mass of 3.475 g:

The radius of the sphere is 1/2(9.40mm) or 4.70mm. Since density is usually expressed in cm^3, express the radius in cm (0.470cm) and substitute into the volume equation:

$$V = \frac{4}{3}\Pi r^3 = \frac{4}{3}(3.1416)(0.470\text{cm})^3 = 0.435 \text{ cm}^3 \text{(to 3 s.f.)}$$

The density is 3.475 g /0.435 cm^3 = 7.99 g/cm^3.

57. (a) Calculate the density of an irregularly shaped piece of metal:

$$D = \frac{M}{V} = \frac{74.122 \text{ g}}{(36.7 \text{ cm}^3 - 28.2 \text{ cm}^3)} = \frac{74.122 \text{ g}}{8.5 \text{ cm}^3} = 8.7 \text{ g/cm}^3$$

Note that the subtraction of volumes leaves only 2 sf, limiting the density to 2 sf

(b) From the list of metals provided, one would surmise that the metal is **cadmium**. Since the major uncertainty is in the volume, one can substitute 8.4 and 8.6 cm^3 as the volume, and calculate the density (resulting in 8.82 and 8.62 g/cm^3 respectively). The hypothesis that the metal is cadmium is reasonably sound.

59. Mass of Hg in the capillary:

Mass of capillary with Hg	3.416 g
Mass of capillary without Hg	3.263 g
Mass of Hg	0.153 g

To determine the volume of the capillary, calculate the volume of Hg that is filling it.

$$\frac{0.153 \text{ g Hg}}{1} \cdot \frac{1 \text{ cm}^3}{13.546 \text{ g Hg}} = 1.13 \times 10^{-2} \text{ cm}^3 (3 \text{ sf})$$

Now that we know the volume of the capillary, and the length of the tubing (given as 16.75 mm—or 1.675 cm), we can calculate the radius of the capillary using the equation:

Volume = $\pi r^2 l$.

1.13×10^{-2} cm^3 = (3.1416)r^2(1.675 cm), and solving for r^2:

$$\frac{1.13 \times 10^{-2} \text{ cm}^3}{(3.1416)(1.675 \text{ cm})} = 2.15 \times 10^{-3} \text{ cm}^2 = r^2$$

So **r** is the square root of (2.15 x 10^{-3} cm^2) or 4.63 x 10^{-2} cm. The diameter would then be twice this value or 9.27 x 10^{-2} cm.

Copper

61.(a) The number of Cu atoms in a cube whose mass is 0.1206g.

$$\frac{0.1206 \text{ g}}{\text{cube}} \cdot \frac{1 \text{ atom Cu}}{1.055 \times 10^{-22} \text{g}} = 1.143 \times 10^{21} \text{ atoms Cu}$$

Fraction of the lattice that contains Cu atoms:

Given the radius of a Cu atom to be 128 pm, and the number of Cu atoms, the total volume occupied by the Cu atoms is the volume occupied by ONE atom (4/3 Πr^3) multiplied by the total number of atoms:

Volume of one atom: $4/3 \cdot 3.1416 \cdot (128 \text{pm})^3 = 8.78 \times 10^6$ pm^3

Total volume: (8.78 x 10^6 pm^3/Cu atom)(1.143 x 10^{21} atoms Cu) = 1.00 x 10^{28} pm^3

The lattice cube has a volume of (0.236 cm)^3 or (2.36 x 10^{9}pm)^3 or 1.31 x 10^{28} pm^3

The fraction occupied is: total volume of Cu atoms/total volume of lattice cube:

$$\frac{1.00 \times 10^{28} \text{ pm}^3}{1.31 \times 10^{28} \text{ pm}^3} = 0.763 \text{ or } 76\% \text{ occupied (to 2 sf)}$$

The empty space in a lattice is due to the inability of spherical atoms to totally fill a given volume. A macroscopic example of this phenomenon is visible if you place four marbles in a square arrangement. At the center of the square there are voids. In a cube, there are obviously repeating incidents.

(b) Estimate the number of Cu atoms in the smallest repeating unit:

Since we know the length of the smallest repeating unit (the unit cell), let's calculate the volume (first converting the length to units of centimeters:

$$L = 361.47 \text{ pm so } L = 361.47 \text{ pm} \cdot \frac{1 \times 10^2 \text{ cm}}{1 \times 10^{12} \text{ pm}} = 361.47 \times 10^{-10} \text{cm}$$

$$V = L^3 = (3.6147 \times 10^{-8})^3 = 4.723 \times 10^{-23} \text{ cm}^3$$

Since we know the density, we can calculate the mass of one unit cell:

$$D \times V = 8.960 \text{ g/ cm}^3 \times 4.723 \times 10^{-23} \text{ cm}^3 = 4.23 \times 10^{-22} \text{ g}$$

Knowing the mass of one copper atom (1.055×10^{-22} g) we can calculate the number of

Cu atoms in that mass: $\dfrac{4.23 \times 10^{-22} \text{ g}}{1.055 \times 10^{-22} \text{ g/Cu atom}} = 4.0 \text{ Cu atoms}$

As you will learn later, the number of atoms for a face-centered cubic lattice is 4.

IN THE LABORATORY

63. The metal will displace a volume of water that is equal to the volume of the metal.

The difference in volumes of water (20.2-6.9) corresponds to the volume of metal. Since 1 mL = 1 cm^3, the density of the metal is then:

$$\frac{\text{Mass}}{\text{Volume}} = \frac{37.5 \text{ g}}{13.3 \text{ cm}^3} = \text{or } 2.82 \frac{\text{g}}{\text{cm}^3}$$

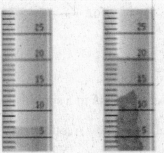
Graduated cylinders with unknown metal (right)

From the list of metals provided, the metal with a density closest to this is **Aluminum**.

65. The plotted data result in the graph below:
Using Cricket Graph™ to plot the "best straight line", one gets the equation:
$y = 248.4x + 0.0022$.
The concentration when Absorbance = 0.635 is (from the graph) 2.55×10^{-3} g/L
The slope of the line is the coefficient of x or 248.4.

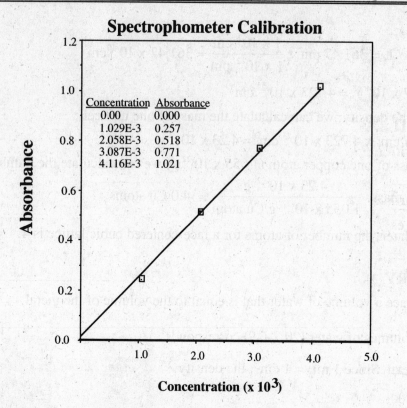

Spectrophotometer Calibration

Concentration	Absorbance
0.00	0.000
1.029E-3	0.257
2.058E-3	0.518
3.087E-3	0.771
4.116E-3	1.021

67. Insert the data in a spreadsheet (here, Excel is used), to obtain the results:

Student	% Acetic Acid
1	5.22
2	5.28
3	5.22
4	5.30
5	5.19
6	5.23
7	5.33
8	5.26
9	5.15
10	5.22
Average	5.24
St.Dev	0.05

Only the values 5.30%, 5.33%, and 5.15% fall outside the Average ± St.Dev.

Chapter 2
Atoms, Molecules and Ions

PRACTICING SKILLS

Atoms: Their Composition and Structure

1.

Fundamental Particles	Protons	Electrons	Neutrons
Electrical Charges	+1	-1	0
Present in nucleus	Yes	No	Yes
Least Massive	1.007 u	**0.00055 u**	1.009 u

3. Begin by expressing the diameter of the nucleus and the electron cloud in the same units.

2mm (diameter of nucleus) = 2×10^{-3} m (since 1 m = 10^3 mm, 1 mm = 10^{-3} m)

The ratio of diameters: $\dfrac{\text{Electron cloud}}{\text{nucleus}} = \dfrac{200\text{m}}{2 \times 10^{-3}\text{m}}$. So we set the actual diameters in the

same ratio: $\dfrac{200\text{m}}{2 \times 10^{-3}\text{m}} = \dfrac{1 \times 10^{-8}\text{cm}}{\text{x}}$ and solving for x:

$200 \text{ x} = (2 \times 10^{-3})(1 \times 10^{-8}) = 2 \times 10^{-11}$ and x = 1×10^{-13} cm. Note that we left the actual diameter of the electron cloud in units of centimeters, so the **ratio** would be the same as if we had changed the units to **meters**. Leaving the actual diameter in units of centimeters means that our "diameter of the nucleus" will have units of centimeters.

5. Isotopic symbol for:
   (a) Mg (at. no. 12) with 15 neutrons : 27 $^{27}_{12}Mg$
   (b) Ti (at. no. 22) with 26 neutrons : 48 $^{48}_{22}Ti$
   (c) Zn (at. no. 30) with 32 neutrons : 62 $^{62}_{30}Zn$

The mass number represents the SUM of the protons + neutrons in the nucleus of an atom.

The atomic number represents the # of protons, so (atomic no. + # neutrons)=mass number

7.

substance	protons	neutrons	electrons
(a) magnesium-24	12	12	12
(b) tin-119	50	69	50
(c) thorium-232	90	142	90
(d) carbon-13	6	7	6
(e) copper-63	29	34	29
(f) bismuth-205	83	122	83

Note that the number of protons and electrons are **equal** for any **neutral atom**. The number of protons is **always** equal to the atomic number. The mass number equals the sum of the numbers of protons and neutrons.

Isotopes

9. The mass of a ^{16}O atom is 15.995 u. The mass relative to the mass of an atom of ^{12}C, which has a mass of 12.000 u will be 15.995/12.000 or 1.3329 (5 significant figures or 5 sf).

11. Isotopes of cobalt (atomic number 27) with 30, 31, and 33 neutrons:

would have symbols of $^{57}_{27}Co$, $^{58}_{27}Co$, and $^{60}_{27}Co$ respectively.

13. Hydrogen has three isotopes:

Name	# Protons	# Neutrons	# Electrons
Protium	1	0	1
Deuterium	1	1	1
Tritium	1	2	1

The ONLY difference between the isotopes of an element is in the **number** of neutrons.

Isotope Abundance and Atomic Weight

15. Thallium has two stable isotopes ^{203}Tl and ^{205}Tl. The more abundant isotope is:___?___ The atomic weight of thallium is 204.4 u. The fact that this weight is closer to 205 than 203 indicates that the **205 isotope is the more abundant** isotope. Recall that the atomic weight is the "weighted average" of all the isotopes of each element. Hence the more abundant isotope will have a "greater contribution" to the atomic weight than the less abundant one.

17. The atomic mass of lithium is: (0.0750)(6.015121) + (0.9250)(7.016003) $=$ 6.94 u Recall that the atomic mass is a weighted average of all isotopes of an element, and is obtained by **adding** the *product* of (relative abundance x mass) for all isotopes.

19. The two stable isotopes of silver are Ag-107 and Ag-109. The masses of the isotopes are respectively: 106.9051 and 108.9047. The atomic weight of Ag on the periodic table is 107.868. Since this weight is a weighted average, a 50:50 mixture of the two would have an atomic weight exactly mid-way between the two isotopic masses. Adding the masses of these two isotopes yields (108.9047 + 106.9051) = 215.8098. One-half this value is 107.9049. Given the proximity of this number to that of the published atomic weight indicates that the two stable isotopes of silver exist in a 50:50 mix.

21. The average atomic weight of gallium is 69.723 (from the periodic table). If we let **x** represent the abundance of the lighter isotope, and (**1-x**) the abundance of the heavier isotope, the expression to calculate the atomic weight of gallium may be written:

$$(x)(68.9257) + (1 - x)(70.9249) = 69.723$$

[Note that the sum of all the isotopic abundances must add to 100% -- or 1 (in decimal notation).] Simplifying the equation gives:

$$68.9257 \text{ u } x + 70.9249 \text{ u} - 70.9249 \text{ u } x = 69.723 \text{ u}$$
$$-1.9992 \text{ u } x = (69.723 \text{ u} - 70.9249)$$
$$-1.9992 \text{ u } x = -1.202 \text{ u}$$
$$x = 0.6012$$

So the relative abundance of isotope 69 is 60.12 % and that of isotope 71 is 39.88 %.

The Periodic Table

23. Comparison of Titanium and Thallium:

Name	Symbol	Atomic #	Atomic Weight	Group #	Period #	Metal, Metalloid, or nonmetal
Titanium	Ti	22	47.867	4B (4)	4	Metal
Thallium	Tl	81	204.3833	3A (13)	6	Metal

25. Periods with 8 elements: **2**; Periods 2 (at.no. 3-10) and 3 (at.no. 11-18)
    Periods with 18 elements: **2**; Periods 4 (at.no 19-36) and 5 (at.no. 37-54)
    Periods with 32 elements: **1**; Period 6 (at.no. 55-86)

27. Elements fitting the following descriptions:

	Description	Elements
(a)	Nonmetals	C, Cl
(b)	Main group elements	C, Ca, Cl, Cs
(c)	Lanthanides	Ce
(d)	Transition elements	Cr, Co, Cd, Cu, Ce, Cf, Cm
(e)	Actinides	Cf, Cm
(f)	Gases	Cl

29. Classify the elements as metals, metalloids, or nonmetals:

	Metals	Metalloids	Nonmetals
N			X
Na	X		
Ni	X		
Ne			X
Np	X		

Molecular Formulas and Models

31. The formula for sulfuric acid is H_2SO_4. The molecule is **not flat**. The O atoms are arranged around the sulfur at the corners of a tetrahedron—that is the O-S-O angles would be about 109 degrees. The hydrogen atoms are connected to two of the oxygen atoms also with angles (H-O-S) of approximately 109 degrees.

33. The molecular formula is $C_4H_8N_2O_3$. The structural formula is:

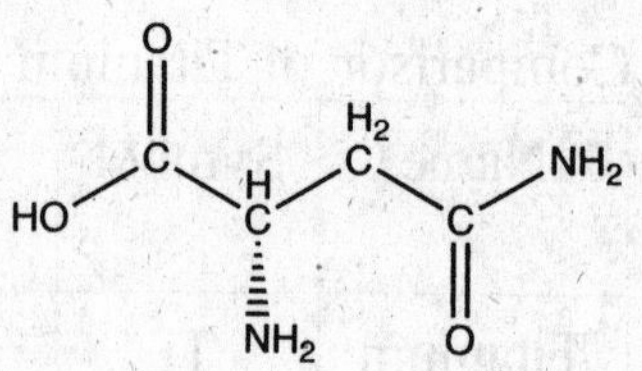

Ions and Ion Charges

35. Most commonly observed ion for:

    (a) Magnesium: 2+ —like all the alkaline earth metals

    (b) Zinc: 2+

    (c) Nickel: 2+

    (d) Gallium: 3+ (an analog of Aluminum)

37. The symbol and charge for the following ions:

    (a) barium ion Ba^{2+}

    (b) titanium(IV) ion Ti^{4+}

    (c) phosphate ion PO_4^{3-}

    (d) hydrogen carbonate ion HCO_3^{-}

    (e) sulfide ion S^{2-}

    (f) perchlorate ion ClO_4^{-}

    (g) cobalt(II) ion Co^{2+}

    (h) sulfate ion SO_4^{2-}

39. When potassium becomes a monatomic ion, potassium—like all alkali metals—**loses 1 electron.** The noble gas atom with the same number of electrons as the potassium ion is **argon**.

Ionic Compounds

41. Barium is in Group 2A, and is expected to form a 2+ ion while bromine is in group 7A and
    expected to form a 1- ion. Since the compound would have to have an **equal amount** of
    negative and positive charges, the formula would be $BaBr_2$.

43. Formula, Charge, and Number of ions in:

	cation	# of	anion	# of
(a) K_2S	K^+	2	S^{2-}	1
(b) $CoSO_4$	Co^{2+}	1	SO_4^{2-}	1
(c) $KMnO_4$	K^+	1	MnO_4^-	1
(d) $(NH_4)_3PO_4$	NH_4^+	3	PO_4^{3-}	1
(e) $Ca(ClO)_2$	Ca^{2+}	1	ClO^-	2
(f) $NaCH_3CO_2$	Na^+	1	$CH_3CO_2^-$	1

45. Regarding cobalt oxides: Cobalt(II) oxide $\quad$ CoO $\qquad$ cobalt ion : $\quad Co^{2+}$

$\qquad\qquad\qquad\qquad\qquad\quad$ Cobalt(III) oxide $\quad Co_2O_3 \qquad\qquad\qquad Co^{3+}$

47. Provide correct formulas for compounds:

   (a) $AlCl_3$ The tripositive aluminum ion requires three chloride ions.

   (b) KF $\quad$ Potassium is a monopositive cation. Fluoride is a mononegative anion.

   (c) Ga_2O_3 is correct; Ga is a 3+ ion and O forms a 2- ion

   (d) MgS $\quad$ is correct; Mg forms a 2+ ion and S forms a 2- ion

Naming Ionic Compounds

49. Names for the ionic compounds

   (a) K_2S $\qquad\qquad\qquad$ potassium sulfide

   (b) $CoSO_4$ $\qquad\qquad\quad$ cobalt(II) sulfate

   (c) $(NH_4)_3PO_4$ $\qquad\quad$ ammonium phosphate

   (d) $Ca(ClO)_2$ $\qquad\qquad$ calcium hypochlorite

51. Formulas for the ionic compounds

   (a) ammonium carbonate $(NH_4)_2CO_3$

   (b) calcium iodide $\qquad\qquad CaI_2$

   (c) copper(II) bromide $\qquad CuBr_2$

   (d) aluminum phosphate $\quad AlPO_4$

   (e) silver(I) acetate $\qquad\quad AgCH_3CO_2$

53. Names and formulas for ionic compounds:

cation	anion CO_3^{2-}	anion I^-
Na^+	Na_2CO_3 sodium carbonate	NaI sodium iodide
Ba^{2+}	$BaCO_3$ barium carbonate	BaI_2 barium iodide

Coulomb's Law

55. The fluoride ion has a smaller radius than the iodide ion. Hence the distance between the sodium and fluoride ions will be less than the comparable distance between sodium and iodide. Coulomb's Law indicates that the attractive force becomes greater as the distance between the charges grows smaller—hence NaF will have stronger forces of attraction.

Naming Binary, Nonmetal Compounds

57. Names of binary nonionic compounds
     (a) NF_3 nitrogen trifluoride

     (b) HI   hydrogen iodide

     (c) BI_3 boron triiodide

     (d) PF_5 phosphorus pentafluoride

59. Formulas for:
     (a) sulfur dichloride SCl_2
     (b) dinitrogen pentaoxide N_2O_5
     (c) silicon tetrachloride $SiCl_4$
     (d) diboron trioxide B_2O_3

Atoms and the Mole

61. The mass, in grams of:

(a) 2.5 mol Al: $\dfrac{2.5 \text{ mol Al}}{1} \cdot \dfrac{26.98 \text{ g Al}}{1 \text{ mol Al}} = 67 \text{ g Al (2 sf)}$

(b) 1.25×10^{-3} mol Fe: $\dfrac{1.25 \times 10^{-3} \text{mol Fe}}{1} \cdot \dfrac{55.85 \text{ g Fe}}{1 \text{ mol Fe}} = 0.0698 \text{ g Fe (3 sf)}$

(c) 0.015 mol Ca: $\dfrac{0.015 \text{ mol Ca}}{1} \cdot \dfrac{40.1 \text{ g Ca}}{1 \text{ mol Ca}} = 0.60 \text{ g Ca (2 sf)}$

(d) 653 mol Ne: $\dfrac{653 \text{ mol Ne}}{1} \cdot \dfrac{20.18 \text{ g Ne}}{1 \text{ mol Ne}} = 1.32 \times 10^4 \text{ g Ne (3 sf)}$

Note that, whenever possible, one should use a molar mass of the substance that contains **one more** significant figure than the data, to reduce round-off error.

63. The amount (moles) of substance represented by:

(a) 127.08 g Cu: $\dfrac{127.08 \text{ g Cu}}{1} \cdot \dfrac{1 \text{ mol Cu}}{63.546 \text{ g Cu}} = 1.9998 \text{ mol Cu (5 sf)}$

(b) 0.012 g Li: $\dfrac{0.012 \text{ g Li}}{1} \cdot \dfrac{1 \text{ mol Li}}{6.94 \text{ g Li}} = 1.7 \times 10^{-3} \text{ mol Li (2 sf)}$

(c) 5.0 mg Am: $\dfrac{5.0 \text{ mg Am}}{1} \cdot \dfrac{1 \text{ g Am}}{10^{3} \text{ mg Am}} \cdot \dfrac{1 \text{ mol Am}}{243 \text{ g Am}} = 2.1 \times 10^{-5} \text{ mol Am (2 sf)}$

(d) 6.75 g Al $\dfrac{6.75 \text{ g Al}}{1} \cdot \dfrac{1 \text{ mol Al}}{26.98 \text{ g Al}} = 0.250 \text{ mol Al (3 sf)}$

65. 1-gram samples of He, Fe, Li, Si, C:

Which sample contains the **largest number** of atoms? ...the **smallest number** of atoms?

If we calculate the number of atoms of any one of these elements, say He, the process is:

$$\dfrac{1.0 \text{ g He}}{1} \cdot \dfrac{1 \text{ mol He}}{4.0026 \text{ g He}} \cdot \dfrac{6.0221 \times 10^{23} \text{ atoms He}}{1 \text{ mol He}} = 1.5 \times 10^{23} \text{ atoms He}$$

All the calculations proceed analogously, with the ONLY numerical difference attributable to the molar mass of the element. Therefore the element with the **smallest** molar mass (He) will have the **largest number** of atoms, while the element with the **largest** molar mass (Fe) will have the **smallest number** of atoms. This is a great question to answer by **thinking** rather than calculating.

67. Analysis of a 10.0-g sample of apatite contained 3.99g Ca, 1.85g P, 41.4g O, and 0.02g H.

Calculating moles of each element gives:

$3.99 \text{ g Ca} \cdot \dfrac{1 \text{ mol Ca}}{40.08 \text{ g Ca}} = 0.0996 \text{ mol Ca}$ and $1.85 \text{ g P} \cdot \dfrac{1 \text{ mol P}}{30.97 \text{ g P}} = 0.0597 \text{ mol P}$

$41.4 \text{ g O} \cdot \dfrac{1 \text{ mol O}}{16.00 \text{ g O}} = 2.59 \text{ mol O}$ and $0.02 \text{ g H} \cdot \dfrac{1 \text{ mol H}}{1.0 \text{ g H}} = 0.02 \text{ mol H}$

From smallest to largest # moles: 0.02 mol H < 0.0597 mol P < 0.0996 mol Ca < 2.59 mol O

Molecules, Compounds, and the Mole

69. Molar mass of the following: (with atomic weights expressed to 4 significant figures)

(a) Fe_2O_3 (2)(55.85) + (3)(16.00) = 159.7

(b) BCl_3 (1)(10.81) + (3)(35.45) = 117.2

(c) $C_6H_8O_6$ (6)(12.01) + (8)(1.008) + (6)(16.00) = 176.1

71. Molar mass of the following: (with atomic weights expressed to 4 significant figures)

(a) $Ni(NO_3)_2 \cdot 6H_2O$ (1)(58.69) + (2)(14.01) + 6(16.00) + (12)(1.008) + (6)(16.00)

= 290.8

(b) $CuSO_4 \cdot 5H_2O$ $(1)(63.55) + (1)(32.07) + 4(16.00) + (10)(1.008) + (5)(16.00)$

$$= 249.7$$

73. Mass represented by 0.0255 moles of the following compounds:

Molar masses are calculated as before. To determine the mass represented by 0.0255 moles we recall that 1 mol of a substance has a mass equal to the molar mass expressed in units of grams. We calculate for (a) the mass represented by 0.0255 moles:

$$\frac{0.0255 \text{ mol } C_3H_7OH}{1} \cdot \frac{60.10 \text{ g } C_3H_7OH}{1 \text{ mol } C_3H_7OH} = 1.53 \text{ g } C_3H_7OH$$

Compound	Molar mass	Mass of 0.0255 moles
(a) C_3H_7OH	60.10	1.53
(b) $C_{11}H_{16}O_2$	180.2	4.60
(c) $C_9H_8O_4$	180.2	4.60
(d) C_3H_6O	58.08	1.48

Masses are expressed to 3 sf, since the # of moles has 3.

75. Regarding sulfur trioxide:

1. Amount of SO_3 in 1.00 kg: $\dfrac{1.00 \times 10^3 \text{g } SO_3}{1} \cdot \dfrac{1 \text{ mol } SO_3}{80.07 \text{ g } SO_3} = 12.5 \text{ mol } SO_3$

2. Number of SO_3 molecules: $\dfrac{12.5 \text{ mol } SO_3}{1} \cdot \dfrac{6.022 \times 10^{23} \text{ molecules}}{1 \text{ mol } SO_3} = 7.52 \times 10^{24}$

3. Number of S atoms: With 1 S atom per SO_3 molecule $- 7.52 \times 10^{24}$ S atoms

4. Number of O atoms: With 3 O atoms per SO_3 molecule $- 3 \times 7.52 \times 10^{24}$ O atoms

$$\text{or } 2.26 \times 10^{25} \text{ O atoms}$$

77. $C_8H_9NO_2$, the molecular formula for acetominophen has a molecular mass of 151.17 g/mol.

Two 500 mg tablets [or $2 \times (500 \times 10^{-3})$g = 1.00g] would contain:

1.00g$\cdot \dfrac{1 \text{ mol acetominophen}}{151.17\text{g acetominophen}} \cdot \dfrac{6.02 \times 10^{23} \text{ molecules acetominophen}}{1 \text{ mol acetominophen}} = 3.98 \times 10^{21}$ molecules

$$= 4 \times 10^{21} \text{ molecules (1 sf)}$$

Percent Composition

79. Mass percent for: [4 significant figures]

(a) PbS: $(1)(207.2) + (1)(32.06) = 239.3$ g/mol

$\%Pb = \dfrac{207.2 \text{ g Pb}}{239.3 \text{ g PbS}} \times 100 = 86.60 \%$

%S = 100.00 - 86.60 = 13.40 %

(b) C_3H_8: (3)(12.01) + (8)(1.008) = 44.09 g/mol

$$\%C = \frac{36.03 \text{ g C}}{44.09 \text{ g } C_3H_8} \times 100 = 81.71 \%$$

%H = 100.00 - 81.71 = 18.29 %

(c) $C_{10}H_{14}O$: (10)(12.01) + (14)(1.008) + (1)(16.00) = 150.21 g/mol

$$\%C = \frac{120.1 \text{ g C}}{150.21 \text{ g } C_{10}H_{14}O} \times 100 = 79.96 \%$$

$$\%H = \frac{14.112 \text{ g H}}{150.21 \text{ g } C_{10}H_{14}O} \times 100 = 9.394 \%$$

%O = 100.00 - (79.96 + 9.394) = 10.65 %

81. Mass of CuS to provide 10.0 g of Cu:

To calculate the weight percent of Cu in CuS, we need the respective atomic weights:
Cu = 63.546   S = 32.066   adding CuS = 95.612

The % of Cu in CuS is then: $\frac{63.546 \text{ g Cu}}{95.612 \text{ g CuS}} \times 100 = 66.46 \% \text{ Cu}$

Now with this fraction (inverted) calculate the mass of CuS that will provide 10.0 g of Cu:

$$\frac{10.0 \text{ g Cu}}{1} \cdot \frac{95.612 \text{ g CuS}}{63.546 \text{ g Cu}} = 15.0 \text{ g CuS}$$

Empirical and Molecular Formulas

83. The empirical formula ($C_2H_3O_2$) would have a mass of 59.04 g.

Since the molar mass is 118.1 g/mol we can write

$$\frac{1 \text{ empirical formula}}{59.04 \text{ g succinic acid}} \cdot \frac{118.1 \text{ g succinic acid}}{1 \text{ mol succinic acid}} = \frac{2.0 \text{ empirical formulas}}{1 \text{ mol succinic acid}}$$

So the molecular formula contains 2 empirical formulas (2 x $C_2H_3O_2$) or $C_4H_6O_4$.

85. Provide the empirical or molecular formula for the following, as requested:

	Empirical Formula	Molar Mass (g/mol)	Molecular Formula
(a)	CH	26.0	C_2H_2
(b)	CHO	116.1	$C_4H_4O_4$
(c)	CH_2	112.2	C_8H_{16}

Note that we can calculate the mass of an empirical formula by adding the respective atomic weights (13 for CH, for example). The molar mass (26.0 for part (a)) is obviously twice that for an empirical formula, so the molecular formula would be 2 x empirical formula (or C_2H_2 in part (a)).

87. Calculate the empirical formula of acetylene by calculating the atomic ratios of carbon and hydrogen in 100 g of the compound.

$$92.26 \text{ g C} \cdot \frac{1 \text{ mol C}}{12.011 \text{g C}} = 7.681 \text{ mol C} \quad \text{and } 7.74 \text{ g H} \cdot \frac{1 \text{ mol H}}{1.008 \text{ g H}} = 7.678 \text{ mol H}$$

Calculate the atomic ratio: $\quad \dfrac{7.68 \text{ mol C}}{7.68 \text{ mol H}} = \dfrac{1 \text{ mol C}}{1 \text{ mol H}}$

The atomic ratio indicates that there is 1 C atom for 1 H atom (1:1). **The empirical formula is then CH**. The formula mass is 13.01. Given that the molar mass of the compound is 26.02 g/mol, there are two formula units per molecular unit, hence the **molecular formula for acetylene is C_2H_2** .

89. Determine the empirical and molecular formulas of cumene:

The percentage composition of cumene is 89.94% C and (100.00-89.94) or 10.06%H.

We can calculate the ratio of mol C: mol H as done in SQ87.

$$89.94 \text{ g C} \cdot \frac{1 \text{ mol C}}{12.011 \text{g C}} = 7.489 \text{ mol C}$$

$$10.06 \text{ g H} \cdot \frac{1 \text{ mol H}}{1.008 \text{ g H}} = 9.981 \text{ mol H}$$

Calculating the atomic ratio:

$$\frac{9.981 \text{ mol H}}{7.489 \text{ mol C}} = \frac{1.33 \text{ mol H}}{1.00 \text{ mol C}} \text{ or a ratio of 3C : 4H}$$

So the empirical formula for cumene is C_3H_4 , with a formula mass of 40.06.

If the molar mass is 120.2 g/mol, then dividing the "empirical formula mass" into the molar mass gives: 120.2/40.06 or 3 empirical formulas **per** molar mass. The **molecular formula** is then 3 x C_3H_4 or C_9H_{12}.

91. Empirical and Molecular formula for Mandelic Acid:

$$63.15 \text{ g C} \cdot \frac{1 \text{ mol C}}{12.0115 \text{ g C}} = 5.258 \text{ mol C}$$

$$5.30 \text{ g H} \cdot \frac{1 \text{ mol H}}{1.0079 \text{ g H}} = 5.26 \text{ mol H}$$

$$31.55 \text{ g O} \cdot \frac{1 \text{ mol O}}{15.9994 \text{ g O}} = 1.972 \text{ mol O}$$

Using the smallest number of atoms, we calculate the ratio of atoms:

$$\frac{5.258 \text{ mol C}}{1.972 \text{ mol O}} = \frac{2.666 \text{ mol C}}{1 \text{ mol O}} \text{ or } \frac{22/3 \text{ mol C}}{1 \text{mol O}} \text{ or } \frac{8/3 \text{ mol C}}{1 \text{ mol O}}$$

So 3 mol O combine with 8 mol C and 8 mol H so the empirical formula is $C_8H_8O_3$.

The formula mass of $C_8H_8O_3$ is 152.15. Given the data that the molar mass is 152.15 g/mL, the molecular formula for mandelic acid is $C_8H_8O_3$.

Determining Formulas from Mass Data

93. Given the masses of xenon involved, we can calculate the number of moles of the element:

$$0.526 \text{ g Xe} \cdot \frac{1 \text{mol Xe}}{131.29 \text{ g Xe}} = 0.00401 \text{ mol Xe}$$

The mass of fluorine present is: 0.678 g compound – 0.526 g Xe = 0.152 g F

$$0.152 \text{ g F} \cdot \frac{1 \text{ mol F}}{19.00 \text{ g F}} = 0.00800 \text{ mol F}$$

Calculating atomic ratios:

$$\frac{0.00800 \text{ mol F}}{0.00401 \text{ mol Xe}} = \frac{2 \text{ mol F}}{1 \text{ mol Xe}} \text{ indicating that the empirical formula is } XeF_2$$

95. Formula of compound formed between zinc and iodine:

Calculate the amount of zinc and iodine present:

$$\frac{2.50 \text{ g Zn}}{1} \cdot \frac{1 \text{ mol Zn}}{65.39 \text{ g Zn}} = 3.82 \times 10^{-02} \text{mol Zn} \text{ and}$$

$$\frac{9.70 \text{ g I}_2}{1} \cdot \frac{1 \text{ mol I}_2}{253.8 \text{ g I}_2} = 3.82 \times 10^{-02} \text{ mol I}_2 \text{ (recall that the iodine is a diatomic specie, and}$$

would be the form of iodine reacting). Note that the amount of Zinc and I_2 combined are identical, making the formula for the compound ZnI_2 . An **alternative** way of solving the problem would be to use the **atomic mass** of iodine (126.9g/ mol I) to represent 7.64×10^{-02} mol of I. The ratio of Zn:I would then be 1:2.

GENERAL QUESTIONS

97. Symbol	^{58}Ni	^{33}S	^{20}Ne	^{55}Mn
Number of protons	28	16	10	25
Number of neutrons	30	17	10	30
Number of electrons in the neutral atom	28	16	10	25
Name of element	nickel	sulfur	neon	manganese

99. Crossword puzzle: Clues:

Horizontal

1-2 A metal used in ancient times: tin (Sn)

3-4 A metal that burns in air and is found in Group 5A: bismuth (Bi)

Vertical

1-3 A metalloid: antimony (Sb)

2-4 A metal used in U.S. coins: nickel (Ni)

Single squares:

1. A colorful nonmetal: sulfur (S)

2. A colorless gaseous nonmetal: nitrogen (N)

3. An element that makes fireworks green: boron (B)

4. An element that has medicinal uses: iodine (I)

Diagonal:

1-4 An element used in electronics: silicon (Si)

2-3 A metal used with Zr to make wires for superconducting magnets: niobium (Nb)

Using these solutions, the following letters fit in the boxes:

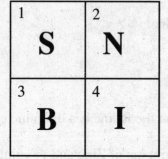

101. (a) The average mass of one copper atom:

One mole of copper (with a mass of 63.546 g) contains 6.0221×10^{23} atoms. So the

average mass of **one** copper atom is: $\dfrac{63.546 \text{ g Cu}}{6.0221 \times 10^{23} \text{atoms Cu}} = 1.0552 \times 10^{-22}$ g/Cu atom

(b) Given the cost data: \$41.70 for 7.0 g and the mass of a Cu atom (from part (a)), the cost

of one Cu atom is:

$\dfrac{\$41.70}{7.0 \text{ g Cu}} \cdot \dfrac{1.0552 \times 10^{-22} \text{ g Cu}}{1 \text{ Cu atom}} = 6.286 \times 10^{-22}$ dollars/Cu atom

or 6.3×10^{-22} dollars/Cu atom (to 2 sf)

103. Identify the element that:

(a) Is in Group 2A and the 5th period: Strontium

(b) Is in the 5th period and Group 4B: Zirconium

(c) Is in the second period in Group 4A: Carbon

(d) Is an element in the 4th period of Group 5A: Arsensic

(e) Is a halogen (Group 7A) in the 5th period: Iodine

(f) Is an alkaline earth element (Group 2A) in the 3rd period: Magnesium

(g) Is a noble gas (Group 8A) in the 4th period: Krypton

(h) Is a nonmetal in Group 6A and the 3rd period: Sulfur

(i) Is a metalloid in the 4th period: Germanium or Arsenic

105. Which of the following has the greater mass:

(a) 0.5 mol Na, 0.5 mol Si, 0.25 mol U

Easily done by observation and a "mental" calculation. Examine the atomic masses of each element. 0.5 mol of any element has a mass that is one-half the atomic mass. One quarter mol of U (atomic mass approximately 238 g) will have the greatest mass of these three.

(b) 9.0 g of Na, 0.5 mol Na, 1.2 x 10^{22} atoms Na: 0.5 mol Na will have a mass of approximately 12.5 g Na; One mole of Na will have 6.0 x 10^{23} atoms, so 1.2 x 10^{22} atoms Na will be $\dfrac{1.2 \times 10^{22}}{6.0 \times 10^{23}}$ = 0.02 mol Na and a mass of (0.02mol)(23 g/mol) = 0.46g Na; 0.5 mol Na will have the greatest mass of these three choices.

(c) 10 atoms of Fe or 10 atoms of K

As in (a) this is easily done by a visual inspection of atomic masses. Fe has a greater atomic mass, so 10 atoms of Fe would have a greater mass than 10 atoms of K.

107. Arrange the elements from least massive to most massive:

Calculate a common unit by which to compare the substances (say grams?)

(a) $\dfrac{3.79 \times 10^{24} \text{atoms Fe}}{1} \cdot \dfrac{1 \text{ mol Fe}}{6.0221 \times 10^{23} \text{atom Fe}} \cdot \dfrac{55.845 \text{ g Fe}}{1 \text{ mol Fe}}$ = 351 g Fe

(b) $\dfrac{19.921 \text{ mol H}_2}{1} \cdot \dfrac{2.0158 \text{ g H}_2}{1 \text{ mol H}_2}$ = 40.157 g H$_2$

(c) $\dfrac{8.576 \text{ mol C}}{1} \cdot \dfrac{12.011 \text{ g C}}{1 \text{ mol C}}$ = 103.0 g C

(d) $\dfrac{7.4 \text{ mol Si}}{1} \cdot \dfrac{28.0855 \text{ g Si}}{1 \text{ mol Si}}$ = 210 g Si

(e) $\dfrac{9.221 \text{ mol Na}}{1} \cdot \dfrac{22.9898 \text{ g Na}}{1 \text{ mol Na}}$ = 212.0 g Na

(f) $\dfrac{4.07 \times 10^{24}\text{atoms Al}}{1} \cdot \dfrac{1 \text{ mol Al}}{6.0221 \times 10^{23}\text{atom Al}} \cdot \dfrac{26.9815 \text{ g Al}}{1 \text{ mol Al}} = 182 \text{ g Al}$

(g) $\dfrac{9.2 \text{ mol Cl}_2}{1} \cdot \dfrac{70.9054 \text{ g Cl}_2}{1 \text{ mol Cl}_2} = 650 \text{ g Cl}_2$

In ascending order of mass: H_2, C, Al, Si, Na, Fe, Cl_2

109. (a) Using our present atomic weights (based on carbon-12) the relative masses of O:H are:

$\dfrac{\text{at. mass O}}{\text{at. mass H}} = \dfrac{15.9994}{1.00794} = 15.873$

If $H \equiv 1.0000$ u, the atomic mass of O would be $15.8729 \cdot 1.0000 = 15.873$ u

Similarly for carbon:

$\dfrac{\text{at. mass C}}{\text{at. mass H}} = \dfrac{12.011}{1.00794} = 11.916$

If H is 1.0000 u, the atomic mass of C would be $11.916 \cdot 1.0000 = 11.916$ u

The number of particles associated with one mole is:

$\dfrac{11.916}{12.0000} = \dfrac{X}{6.02214199 x 10^{23}}$ and $X = 5.9802 \times 10^{23}$ particles

(b) Using the ratio from part a

$\dfrac{\text{at. mass H}}{\text{at. mass O}} = \dfrac{1.00794}{15.9994} = 0.0629986$

If $O \equiv 16.0000$ u, the atomic mass of H would be

$0.0629986 \cdot 16.0000 = 1.00798$ u

Similarly for carbon, the ratios of the atomic masses of C to O is:

$\dfrac{\text{at. mass C}}{\text{at. mass O}} = \dfrac{12.011}{15.9994} = 0.75071$, and the atomic mass of C is

$0.75071 \cdot 16.0000 = 12.011$ u

The number of particles associated with one mole is:

$\dfrac{12.011}{12.0000} = \dfrac{X}{6.02214199 x 10^{23}}$ and $X = 6.0279 \times 10^{23}$ particles

111. Possible compounds from ions:

	CO_3^{2-}	SO_4^{2-}
NH_4^+	$(NH_4)_2CO_3$	$(NH_4)_2SO_4$
Ni^{2+}	$NiCO_3$	$NiSO_4$

Compounds are electrically neutral—hence the total positive charge contributed by the cation (+ion) has to be equal to the total negative charge contributed by the anion (- ion). Since both carbonate and sulfate are di-negative anions, two ammonium ions are required, while only one nickel(II) ion is needed.

113. Compound from the list with the highest weight percent of Cl: One way to answer this question is to calculate the %Cl in each of the five compounds. An observation that each compound has the same number of Cl atoms provides a "non-calculator" approach to answering the question.

Since 3 Cl atoms will contribute the same TOTAL mass of Cl to the formula weights, the compound with the highest weight percent of Cl will also have the **lowest** weight percent of the other atom. Examining the atomic weights of the "other" atoms:

B	As	Ga	Al	P
10.81	74.92	69.72	26.98	30.97

B contributes the smallest mass of these five atoms, hence the smallest contribution to the molar masses of the five compounds—so BCl_3 has the highest weight percent of Cl.

115. To determine the greater mass, let's first ask the question about the molar mass of Adenine. The formula for adenine is: $C_5H_5N_5$ with a molar mass of 135.13 g. The number of molecules requested is exactly 1/2 mole of adenine molecules. So 1/2 mol of adenine molecules would have a mass of 1/2(135.13g) or 67.57 g. So 1/2 mol of adenine has a greater mass than 40.0 g of adenine.

117. A drop of water has a volume of 0.050 mL. Assuming the density of water is 1.00 g/cm^3, the number of molecules of water may be calculated by first determining the mass of water present.

$$\frac{0.050 \text{ mL}}{1} \cdot \frac{1 \text{ cm}^3}{1 \text{ mL}} \cdot \frac{1.00 \text{ g}}{1 \text{ cm}^3} = 0.050 \text{ g water.}$$ The molar mass of water is 18.02 g. The

number of moles of water is then: $\frac{0.050 \text{ g water}}{1} \cdot \frac{1 \text{ mol water}}{18.02 \text{ g water}} = 2.77 \times 10^{-3}$ mol.

The number of molecules is then: 2.77×10^{-3} mol $\times 6.02 \times 10^{23}$ molecules/mol =

1.7×10^{21} molecules (2 sf).

119. Molar mass and mass percent of the elements in $Cu(NH_3)_4SO_4 \cdot H_2O$:

Molar Mass: $(1)(Cu) + (4)(N) + 12(H) + (1)(S) + 4(O) + (2)(H) + (1)(O)$.

Combining the hydrogens and oxygen from water with the compound:

$\quad$ (1)(Cu) + $\quad$ (4)(N) + $\quad\quad$ 14(H) + $\quad\quad$ (1)(S) + $\quad\quad$ (5)(O) =

(1)(63.546) + $\quad$ (4)(14.0067) + 14(1.0079) + (1)(32.066) + (5)(15.9994) = 245.75 g/mol

The mass percents are:

$\quad\quad\quad$ Cu: (63.546/245.75) x 100 = 25.86% Cu

$\quad\quad\quad$ N: (56.027/245.75) x 100 = 22.80% N

$\quad\quad\quad$ H: (14.111/245.75) x 100 = 5.742% H

$\quad\quad\quad$ S: (32.066/245.75) x 100 = 13.05% S

$\quad\quad\quad$ O: (79.997/245.75) x 100 = 32.55% O

The mass of copper and of water in 10.5 g of the compound:

For Copper: $\dfrac{10.5 \text{ g compound}}{1} \cdot \dfrac{25.86 \text{ g Cu}}{100.00 \text{ g compound}} = 2.72 \text{ g Cu}$

For Water: $\dfrac{10.5 \text{ g compound}}{1} \cdot \dfrac{18.02 \text{ g H}_2\text{O}}{245.72 \text{ g compound}} = 0.770 \text{ g H}_2\text{O}$

121. The empirical formula of malic acid, if the ratio is: $C_1H_{1.50}O_{1.25}$

Since we prefer all subscripts to be integers, we ask what "multiplier" we can use to convert each of these subscripts to integers **while** retaining the ratio of C:H:O that we're given. Multiplying each subscript by 4 (we need to convert the 0.25 to an integer) gives a ratio of $C_4H_6O_5$.

123. A compound $Fe_x(CO)_y$ is 30.70 % Fe: This implies that the balance of the mass (69.30 %) is attributable to the CO molecules. One approach is to envision CO as **one** atom, with an atomic weight of (12 + 16) 28g. Assuming we have 100 grams, the ratios of masses are then:

$\dfrac{30.70 \text{ g Fe}}{1} \cdot \dfrac{1 \text{ mol Fe}}{55.845 \text{ g Fe}} = 0.5497$ mol Fe and for the "element" CO,

$\dfrac{69.30 \text{ g CO}}{1} \cdot \dfrac{1 \text{ mol CO}}{28.010 \text{ g CO}} = 2.474$ mol CO and the ratio of the particles is:

(dividing by 0.5497) 1 Fe: 4.5 CO, so an empirical formula that is $Fe(CO)_{4.5}$.

Knowing that we don't typically like fractional atoms, we can express the atomic ratio by multiplying both subscripts by 2: $Fe_2(CO)_9$.

125. For the molecule saccharin:

(a) The formula is $C_7H_5NO_3S$

(b) Mol of saccharin associated with 124 mg:

$$\frac{125 \text{ mg saccharin}}{1} \cdot \frac{1 \text{ g saccharin}}{1000 \text{ mg saccharin}} \cdot \frac{1 \text{ mol saccharin}}{183.19 \text{ g saccharin}} = 6.82 \times 10^{-4} \text{ mol saccharin}$$

(c) Mass of S in 125 mg saccharin:

$$\frac{125 \times 10^{-3} \text{ g saccharin}}{1} \cdot \frac{32.07 \text{ g S}}{183.19 \text{ g saccharin}} = 0.02188 \text{ g S or } 21.9 \text{ mg S}$$

127. Formulas for compounds; identify the ionic compounds

(a)	sodium hypochlorite	NaClO	ionic
(b)	boron triiodide	BI_3	
(c)	aluminum perchlorate	$Al(ClO_4)_3$	ionic
(d)	calcium acetate	$Ca(CH_3CO_2)_2$	ionic
(e)	potassium permanganate	$KMnO_4$	ionic
(f)	ammonium sulfite	$(NH_4)_2SO_3$	ionic
(g)	potassium dihydrogen phosphate	KH_2PO_4	ionic
(h)	disulfur dichloride	S_2Cl_2	
(i)	chlorine trifluoride	ClF_3	
(j)	phosphorus trifluoride	PF_3	

The ionic compounds are identified by noting the presence of a **metal**.

129. Empirical and molecular formulas:

(a) For Fluorocarbonyl hypofluorite:

In a 100.00 g sample there are:

$$\frac{14.6 \text{ g C}}{1} \cdot \frac{1 \text{ mol C}}{12.0115 \text{ g C}} = 1.215 \text{ mol C}$$

$$\frac{39.0 \text{ g O}}{1} \cdot \frac{1 \text{ mol O}}{15.9994 \text{ g O}} = 2.438 \text{ mol O}$$

$$\frac{46.3 \text{ g F}}{1} \cdot \frac{1 \text{ mol F}}{18.9984 \text{ g F}} = 2.437 \text{ mol F}$$

Dividing all three terms by 1.215 gives a ratio of O:C of 2:1. Likewise F:C is 2:1

The empirical formula would be $C_1O_2F_2$ with an "empirical mass" of 82.0 g/mol Since the molar mass is also 82.0 g/mol, the molecular formula is also CO_2F_2.

(b) For Azulene:

Given the information that azulene is a hydrocarbon, if it is 93.71 % C, it is also (100.00 - 93.71) or 6.29 % H.

In a 100.00 g sample of azulene there are

$$93.71 \text{ g C} \cdot \frac{1 \text{ mol C}}{12.0115 \text{ g C}} = 7.802 \text{ mol C and}$$

$$6.29 \text{ g H} \cdot \frac{1 \text{ mol H}}{1.0079 \text{ g H}} = 6.241 \text{ mol H}$$

The ratio of C to H atoms is: 1.25 mol C : 1 mol H or a ratio of 5 mol C:4 mol H (C_5H_4).

The mass of such an empirical formula is ≈ 64. Given that the molar mass is ~128 g/mol, the molecular formula for azulene is $C_{10}H_8$.

131. Molecular formula of cadaverine:

Calculate the amount of each element in the compound (assuming that you have 100 g)

$$58.77 \text{ g C} \cdot \frac{1 \text{ mol C}}{12.0115 \text{ g C}} = 4.893 \text{ mol C}$$

$$13.81 \text{ g H} \cdot \frac{1 \text{ mol H}}{1.0079 \text{ g H}} = 13.70 \text{ mol H}$$

$$27.40 \text{ g N} \cdot \frac{1 \text{ mol N}}{14.0067 \text{ g N}} = 1.956 \text{ mol N}$$

The ratio of C:H:N can be found by dividing each by the smallest amount (1.956): to give $C_{2.50}H_7N_1$ and converting each subscript to an integer (multiplying by 2) $C_5H_{14}N_2$. The weight of this "empirical formula" would be approximately 102, hence the molecular formula is also $C_5H_{14}N_2$.

133. The empirical formula for MMT:

Moles of each atom present in 100. g of MMT:

$$49.5 \text{ g C} \cdot \frac{1 \text{ mol C}}{12.0115 \text{ g C}} = 4.12 \text{ mol C}$$

$$3.2 \text{ g H} \cdot \frac{1 \text{ mol H}}{1.0079 \text{ g H}} = 3.2 \text{ mol H}$$

$$22.0 \text{ g O} \cdot \frac{1 \text{ mol O}}{15.9994 \text{ g O}} = 1.38 \text{ mol O}$$

$$25.2 \text{ g Mn} \cdot \frac{1 \text{ mol Mn}}{54.938 \text{ g Mn}} = 0.459 \text{ mol Mn}$$

The ratio of C:H:O:Mn can be found by dividing each by the smallest amount (0.459): to give $MnC_9H_7O_3$.

135. Chromium oxide has the formula Cr_2O_3.

The weight percent of Cr in Cr_2O_3 is: $[(2 \times 52.00)/((2 \times 52.00) + (3 \times 16.00))] \times 100$ or

$(104.00/152.00) \times 100$ or 68.42% Cr.

[The numerator is the sum of the mass of 2 atoms of Cr, while the denominator is the sum

of the mass of 2 atoms of Cr and 3 atoms of O.]

The weight of chromium oxide necessary to produce 850 kg Cr:

$$\frac{850 \text{ kg Cr}}{1} \cdot \frac{100 \text{ kg Cr}_2\text{O}_3}{68.42 \text{ kg Cr}} = 1,200 \text{ kg Cr}_2\text{O}_3 \text{ to 2 sf}$$

The second fraction represents the %Cr in the oxide. Dividing the desired mass of Cr by the

percent Cr (or multiplying by the reciprocal of that percentage) gives the mass of oxide

needed.

137. I_2 + Cl_2 → I_xCl_y
        0.678 g   (1.246 - 0.678)   1.246 g

Calculate the ratio of I : Cl atoms

$0.678 \text{ g I} \cdot \dfrac{1 \text{ mol I}}{126.9 \text{ g I}} = 5.34 \times 10^{-3} \text{ mol I atoms}$

$0.568 \text{ g Cl} \cdot \dfrac{1 \text{ mol Cl}}{35.45 \text{ g Cl}} = 1.6 \times 10^{-2} \text{ mol Cl atoms}$

The ratio of Cl:I is: $\dfrac{1.6 \times 10^{-2} \text{mol Cl atoms}}{5.34 \times 10^{-3} \text{mol I atoms}} = 3.00 \dfrac{\text{Cl atoms}}{\text{I atoms}}$

The empirical formula is ICl_3 (FW = 233.3) Given that the molar mass of I_xCl_y was

467 g/mol, we can calculate the number of empirical formulas per mole:

$$\frac{467 \text{g/mol}}{233.3 \text{g/empiricalformula}} = 2 \frac{\text{empirical formulas}}{\text{mol}} \text{ for a molecular formula of } I_2Cl_6.$$

139. Mass of Fe in 15.8 kg of FeS_2:

$\% \text{ Fe in FeS}_2 = \dfrac{55.85 \text{gFe}}{119.97 \text{gFeS}_2} \times 100 = 46.55 \text{ \% Fe}$

and in 15.8 kg FeS_2: $15.8 \text{ kg FeS}_2 \cdot \dfrac{46.55 \text{ kg Fe}}{100.00 \text{ kg FeS}_2} = 7.35 \text{ kg Fe}$

141. The formula of barium molybdate is $BaMoO_4$. What is the formula for sodium molybdate?

This question is easily answered by observing that the compound indicates ONE barium ion. Since the barium ion has a 2+ charge, 2 Na+ cations would be needed, making the formula for sodium molybdate Na_2MoO_4 or choice (d).

143. Mass of Bi in two tablets of Pepto-Bismol™ ($C_{21}H_{15}Bi_3O_{12}$):

Moles of the active ingredient:

$$\frac{2 \text{ tablets}}{1} \cdot \frac{300. \times 10^{-3} \text{g } C_{21}H_{15}Bi_3O_{12}}{1 \text{ tablet}} \cdot \frac{1 \text{ mol } C_{21}H_{15}Bi_3O_{12}}{1086 \text{ g } C_{21}H_{15}Bi_3O_{12}} = 5.52 \times 10^{-4} \text{ mol } C_{21}H_{15}Bi_3O_{12}$$

Mass of Bi:

$$\frac{2 \text{ tablets}}{1} \cdot \frac{300. \times 10^{-3} \text{g } C_{21}H_{15}Bi_3O_{12}}{1 \text{ tablet}} \cdot \frac{1 \text{ mol } C_{21}H_{15}Bi_3O_{12}}{1086 \text{ g } C_{21}H_{15}Bi_3O_{12}} \cdot$$

$$\frac{3 \text{ mol Bi}}{1 \text{ mol } C_{21}H_{15}Bi_3O_{12}} \cdot \frac{208.98 \text{ g Bi}}{1 \text{ mol Bi}} = 0.346 \text{ g Bi}$$

145. What is the molar mass of ECl_4 and the identity of E?

2.50 mol of ECl_4 has a mass of 385 grams. The molar mass of ECl_4 would be:

$$\frac{385 \text{ g } ECl_4}{2.50 \text{ mol } ECl_4} = 154 \text{ g/mol } ECl_4.$$

Since the molar mass is 154, and we know that there are 4 chlorine atoms per mole of the compound, we can subtract the mass of 4 chlorine atoms to determine the mass of E. 154 - 4(35.5) = 12. The element with an atomic mass of 12 g/mol is **carbon**.

147. For what value of n, will Br compose 0.105% of the mass of the polymer,

$Br_3C_6H_3(C_8H_8)_n$?

Knowing that the 3Br atoms comprises 0.105% of the formula weight of the polymer, we can write the fraction:

$$\frac{3 \cdot Br}{\text{formula weight}} = 0.00105 \text{ (where 3*Br) is 3 times the atomic weight of Br. Substituting}$$

we get $\dfrac{239.7}{\text{formula weight}} = 0.00105$ or $\dfrac{239.7}{0.00105} = $ formula weight $= 2.283 \times 10^5$ g.

Noting that the "fixed" part of the formula contains 3 Br atoms, 6 C atoms, and 3 H atoms, we can calculate the mass associated with this part of the molecule (239.7 + 75.09 = 314.79). Subtracting this mass from the total (2.283×10^5) gives 2.280×10^5 u as the mass corresponding to the "C_8H_8" units. Since each such unit has a mass of 104.15 (8C + 8H), we can divide the mass of *one* C_8H_8 unit into the 2.280×10^5 mass remaining:

$$\frac{2.280 \times 10^5 \text{ u}}{104.15 \text{ u/C}_8\text{H}_8} = 2189 \text{ C}_8\text{H}_8 \text{ units to give a value for } n \text{ of } 2.19 \times 10^3 \text{ (to 3 sf).}$$

149. For the Zn-64 atom:

(a) Calculate the density of the nucleus: We can do so provided we make two assumptions:

1. The mass of the Zn atom is identical to the mass of the nucleus of the Zn atom. Given the very small masses of the electrons, this isn't a bad assumption.

2. Assume the nucleus of the Zn atom is a sphere (whose volume would be 4/3 Πr^3.) Given the desired units, convert the radius to units of

$$\text{cm:} \frac{4.8 \times 10^{-6} \text{nm}}{1} \bullet \frac{100 \text{ cm}}{1 \times 10^9 \text{ nm}} = 4.8 \times 10^{-13} \text{cm}$$

$$V = \frac{4 \bullet 3.1416 \bullet \left(4.8 \times 10^{-13}\right)^3}{3} = 4.6 \times 10^{-37} \text{ cm}^3$$

$$D = \frac{1.06 \times 10^{-22} \text{g}}{4.6 \times 10^{-37} \text{ cm}^3} = 2.3 \times 10^{14} \text{ g/cm}^3$$

(b) The density of the space occupied by the electrons: First express the radius of the atom in cm, as in part (a). 0.125 nm = 1.25×10^{-8} cm.

Calculate the volume of that sphere: $V = \dfrac{4 \bullet 3.1416 \bullet \left(1.25 \times 10^{-8}\right)^3}{3} = 8.18 \times 10^{-24} \text{ cm}^3$

Mass of 30 electrons (It is zinc, yes?) : 30 electrons x 9.11×10^{-28} g

$$D = \frac{2.733 \times 10^{-26} \text{g}}{8.18 \times 10^{-24} \text{ cm}^3} = 3.34 \times 10^{-3} \text{ g/cm}^3$$

(c) As we've learned, the mass of the atom is concentrated in the nucleus—borne out by these densities.

151. Calculate:

(a) moles of nickel—found by density **once** the volume of foil is calculated.

$$V = 1.25 \text{ cm} \times 1.25 \text{ cm} \times 0.0550 \text{ cm} = 8.59 \times 10^{-2} \text{ cm}^3$$

$$\text{Mass} = \frac{8.908 \text{ g}}{1 \text{ cm}^3} \bullet 8.59 \times 10^{-2} \text{ cm}^3 = 0.766 \text{ g Ni}$$

$$0.766 \text{ g Ni} \bullet \frac{1 \text{ mol Ni}}{58.69 \text{ g Ni}} = 1.30 \times 10^{-2} \text{ mol Ni}$$

(b) Formula for the fluoride salt:

Mass F = (1.261 g salt - 0.766 g Ni) = 0.495 g F

$$\text{Moles F} = 0.495 \text{ g F} \cdot \frac{1 \text{ mol F}}{19.00 \text{ g F}} = 2.60 \times 10^{-2} \text{ mol F},$$

so 1.30×10^{-2} mol Ni combines with 2.60×10^{-2} mol F , indicating a formula of NiF_2

(c) Name: Nickel(II) fluoride

IN THE LABORATORY

153. Molecules of water per formula unit of $MgSO_4$:

From 1.687 g of the hydrate, only 0.824 g of the magnesium sulfate remain.

The mass of water contained in the solid is: (1.687-0.824) or 0.863 grams

Use the molar masses of the solid and water to calculate the number of moles of each substance present:

$$0.824 \text{g} \cdot \frac{1 \text{ mol } MgSO_4}{120.36 \text{ g } MgSO_4} = 6.85 \times 10^{-3} \text{ mol of magnesium sulfate}$$

$$0.863 \text{ g } H_2O \cdot \frac{1 \text{ mol } H_2O}{18.02 \text{ g } H_2O} = 4.79 \times 10^{-2} \text{ mol of water ;the ratio of water to } MgSO_4 \text{ is:}$$

$$\frac{4.79 \times 10^{-2} \text{ mol water}}{6.85 \times 10^{-3} \text{ mol magnesium sulfate}} = 6.99 \text{ So we write the formula as } MgSO_4 \cdot 7 \, H_2O.$$

155. The volume of a cube of Na containing 0.125 mol Na:

First we need to know the **mass** of 0.125 mol Na:

$$0.125 \text{mol Na} \cdot \frac{22.99 \text{ g Na}}{1 \text{ mol Na}} = 2.87 \text{ g Na (3 sf)}$$

Now we can calculate the volume that contains 2.87 g Na: (using the density given)

$$2.87 \text{g Na} \cdot \frac{1 \text{ cm}^3}{0.97 \text{ g Na}} = 3.0 \text{ m}^3$$

If the cube is a perfect cube (that is each side is equivalent in length to any other side), what is the length of one edge?

$$3.0 \text{ cm}^3 = l \times l \times l \quad \text{so} \quad 1.4 \text{ cm} = \text{length of one edge}$$

157. Using the student data, let's calculate the number of moles of $CaCl_2$ and mole of H_2O:

$$0.739 \text{ g } CaCl_2 \cdot \frac{1 \text{ mol } CaCl_2}{111.0 \text{ g } CaCl_2} = 0.00666 \text{ mol } CaCl_2$$

$$(0.832 \text{ g - } 0.739 \text{ g}) \text{ or } 0.093 \text{ g } H_2O \cdot \frac{1 \text{ mol } H_2O}{18.02 \text{ g } H_2O} = 0.0052 \text{ mol } H_2O$$

38

The number of moles of water/mol of calcium chloride is then

$$\frac{0.0052 \text{ mol H}_2\text{O}}{0.00666 \text{ mol CaCl}_2} = 0.78$$

This is a sure sign that they should **(c) heat the crucible again, and then reweigh it**.

159. (a) Species responsible for lines at 50 and 52:

$^{12}\text{CH}_3{}^{35}\text{Cl}$ gives rise to line at 50 (12 + 3 + 35)

$^{12}\text{CH}_3{}^{37}\text{Cl}$ gives rise to line at 52 (12 + 3 + 37)

The differential in sizes of the lines is due to the relative distribution of isotopes of Cl, with Cl-37 being about 1/3 of the abundance of Cl-35.

(b) The line at 51 is due to C-13 in place of C-12, with the accompanying smaller size of the peak.

Summary and Conceptual Questions

161. Necessary information to calculate the number of atoms in one cm^3 of iron:

A sample calculation to arrive at an exact value is shown below:

$$\frac{1.00 \text{ cm}^3}{1} \cdot \frac{7.86 \text{ g Fe}}{1 \text{ cm}^3} \cdot \frac{1 \text{ mol Fe}}{55.845 \text{ g Fe}} \cdot \frac{6.0221 \times 10^{23} \text{ atoms Fe}}{1 \text{ mol Fe}}$$

Note that we needed: (b) atomic weight of Fe, (c) Avogadro's number, and (d) density.

163. Given the greater reactivity of Ca over Mg with water, one would anticipate that Ba would be even more reactive than Ca or Mg—with a more vigorous release of hydrogen gas. Reactivity of these metals increases down the group. Mg is in period 3, Ca in period 4, and Ba is in period 6. This trend is noted for Group IA as well.

165. The hydrated salt loses water upon heating. The anhydrous cobalt salt is a deep blue, and therefore visible.

Applying Chemical Principles

1. A flask of nitrogen has a mass of 0.20389 g. Nitrogen has a density of 1.25718 g/L under standard conditions of temperature and pressure. What is the volume of the flask (in cm^3)?

$$\frac{1\ L}{1.25718\ g} \cdot \frac{0.20389\ g}{1} \cdot \frac{1000\ cm^3}{1\ L} = 162.18\ cm^3$$

3. The isotopic data is as follows:

Isotope	Atomic Mass (u)	Abundance(%)
^{36}Ar	35.967545	0.337
^{38}Ar	37.96732	0.063
^{40}Ar	?	?

Since there are 3 stable isotopes, the SUM of their abundances must be 100%.
So the % abundance of ^{40}Ar has to be 100.000-(0.337+0.063) = 99.600%.

Since the atomic weight is a **weighted average** of all isotopes, we can solve for the atomic mass of ^{40}Ar as follows:

$(1.00 \cdot 39.948u) = (0.99600 \cdot x) + (0.00337 \cdot 35.967545) + (0.00063 \cdot 37.96732)$
$39.948u - (0.00337 \cdot 35.967545) - (0.00063 \cdot 37.96732) = (0.99600 \cdot x)$
$39.803 = (0.99600 \cdot x)$ and dividing by 0.99600 = 39.963u

5. Given that the density of argon is 1.78 g/L under standard conditions of temperature and pressure, how many argon atoms are present in a room with dimensions 4.0 m × 5.0 m × 2.4 m?
The volume of the room is (4.0 x 5.0 x 2.4) or $48m^3$
Converting m^3 to cm^3 will allow us to use the density (recall that 1L = 1000 cm^3)

$$48\ m^3 \cdot \frac{1 \times 10^6 cm^3}{1\ m^3} \cdot \frac{1\ L}{1 \times 10^3 cm^3} \cdot \frac{1.78\ g\ Ar}{1\ L} = 85440\ g\ Ar$$

We know that 1 mol of Ar atoms has a mass of 39.948 g so
85440g Ar x (1 mol Ar/39.948 g Ar) x (6.02 x 10^{23} atoms Ar/1mol Ar)
$= 1.3 \times 10^{27}$ atoms Ar

Chapter 3
Chemical Reactions

PRACTICING SKILLS

Balancing Equations

Balancing equations can be a matter of "running in circles" if a reasonable methodology is not employed. While there isn't one "right place" to begin, generally you will suffer fewer complications if you begin the balancing process using a substance that contains the **greatest number** of elements **or** the **largest subscript** values. Noting that you must have at least that many atoms of each element involved, coefficients can be used to increase the "atomic inventory". In the next few questions, you will see one **emboldened** substance in each equation. This emboldened substance is the one that I judge to be a "good" starting place. One last hint--modify the coefficients of uncombined elements, i.e. those not in compounds, <u>after</u> you modify the coefficients for compounds containing those elements -- <u>not before</u>!

1. (a) $2Al \ (s) + \textbf{Fe}_2\textbf{O}_3 \ (s) \rightarrow 2 \ Fe \ (s) + Al_2O_3 \ (s)$

    1. Note the need for <u>at least</u> 2 Fe and 3 O atoms as products.

    2. A coefficient of 2 for Fe (in products) balances the Fe inventory.

    3. Al_2O_3 (as product) takes care of the O inventory, and mandates 2 Al atoms .

    4. 2 Al would give 2 Al atoms on both sides of the equation.

    (b) $H_2O \ (l) + C \ (s) \rightarrow CO \ (g) + H_2 \ (g)$

    Note this equation has 2 H atoms both left and right; 1 O atom on both sides and 1 C atom on both sides. Hooray—it's already balanced.

    (c) $\textbf{SiCl}_4 \ (l) + 2 \ Mg \ (s) \rightarrow Si \ (s) + 2 \ MgCl_2 \ (s)$

    1. Each $SiCl_4$ (reactant) will require 4 Cl atoms (as product). Since $MgCl_2$ has 2 Cl atoms per unit, we'll need 2 $MgCl_2$ units.

    2. A coefficient of 2 for $MgCl_2$ mandates 2 Mg atoms (as reactants).

    3. A coefficient of 2 for elemental Mg (in reactants) balances the Mg inventory.

3. (a) $4 \ Cr \ (s) + 3 \ O_2 \ (g) \rightarrow 2 \ \textbf{Cr}_2\textbf{O}_3 \ (s)$

    1. Note the need for <u>at least</u> 2 Cr and 3 O atoms.

    2. Oxygen is diatomic -- we'll need an <u>even</u> number of oxygen atoms, so try : 2 Cr_2O_3.

    3. 3 O_2 would give 6 O atoms on both sides of the equation.

    4. 4 Cr would give 4 Cr atoms on both sides of the equation.

(b) Cu_2S (s) + O_2 (g) → 2 Cu(s) + SO_2 (g)

1. A minimum of 2 O in SO_2 is required, and is provided with one molecule of elemental oxygen.

2. 2 Cu atoms (on the right) indicates 2 Cu (on the left).

(c) $C_6H_5CH_3$ (l) + 9 O_2 (g) → 4 H_2O (l) + 7 CO_2 (g)

1. A minimum of 7 C and 8 H is required.

2. 7 CO_2 furnishes 7 C and 4 H_2O furnishes 8 H atoms.

3. 4 H_2O and 7 CO_2 furnish a total of 18 O atoms, making the coefficient of O_2 = 9.

5. Balance and name the reactants and products:

(a) Fe_2O_3(s) + 3 Mg(s) → 3 MgO(s) + 2 Fe(s)

1. Note the need for <u>at least</u> 2 Fe and 3 O atoms.

2. 2 Fe atoms would provide the proper iron atom inventory.

3. 3 MgO would give 3 O atoms on both sides of the equation.

4. 3 Mg would give 3 Mg atoms on both sides of the equation.

Reactants: iron(III) oxide and magnesium
Products: magnesium oxide and iron

(b) $AlCl_3$(s) + 3 NaOH(aq) → $Al(OH)_3$(s) + 3 NaCl(aq)

1. Note the need for <u>at least</u> 1 Al and 3 Cl atoms.

2. 3 NaCl molecules would provide the proper Cl atom inventory.

3. 3 NaCl would require 3 Na atoms on the left side—a coefficient of 3 for NaOH is needed.

4. 3 OH groups (from $Al(OH)_3$) would give 3 OH groups needed on both sides of the equation—so a coefficient of 3 for NaOH is needed to provide that balance.

Reactants: aluminum chloride and sodium hydroxide
Products: aluminum hydroxide and sodium chloride.

(c) 2 $NaNO_3$(s) + H_2SO_4(aq) → Na_2SO_4(s) + 2 HNO_3(aq)

1. Note the need for <u>at least</u> 2 Na and 1 S and 4 O atoms.

2. 2 $NaNO_3$ will provide the proper Na atom inventory.

3. The coefficient of 2 in front of $NaNO_3$ requires a coefficient of 2 for HNO_3—providing a balance for N atoms.

4. The implied coefficient of 1 for Na_2SO_4 suggests a similar coefficient for H_2SO_4—to balance the S atom inventory.

5. O atom inventory is done "automatically" when we balanced N and S inventories.

Reactants: sodium nitrate and sulfuric acid
Products: sodium sulfate and nitric acid

(d) **NiCO3**(s) + 2 HNO3(aq) → Ni(NO3)2(aq) + CO2(g) + H2O(l)

1. Note the need for <u>at least</u> 1 Ni atom on both sides. This inventory will mandate 2 NO3 groups on the right—and also on the left. Since these come from HNO3 molecules, we'll need 2 HNO3 on the left.

2. The 2 H from the acid and the CO3 from nickel carbonate, provide 2H, 1 C and 3 O atoms. 1 H2O takes care of the 2H, and **one** of the O atoms, 1 CO2 consumes the 1 C and the remaining 2 O atoms.

   Reactants: nickel(II) carbonate and nitric acid

   Products: nickel(II) nitrate, carbon dioxide, and water

Chemical Equilibrium

7. The greater electrical conductivity of the HCl solution at equilibrium indicates a greater concentration of ions (H_3O^+ and Cl^-), indicating that the HCl solution is more product-favored at equilibrium than the HCO_2H solution.

Ions and Molecules in Aqueous Solution

9. What is an electrolyte? What are experimental means for discriminating between weak and strong electrolytes?

   An electrolyte is a substance whose aqueous solution conducts an electric current.

   As to experimental means for discriminating between weak and strong electrolytes, refer to the apparatus in the Active Figure 5.2. NaCl is a strong electrolyte and would cause the bulb to glow brightly—reflecting a large number of ions in solution while aqueous ammonia or vinegar (an aqueous solution of acetic acid) would cause the bulb to glow only dimly—indicating a smaller number of ions in solution.

11. Predict water solubility:

    (a) $CuCl_2$ is expected to be soluble, while CuO and $FeCO_3$ are not. Chlorides are generally water soluble, while oxides and carbonates are not.

    (b) $AgNO_3$ is soluble. AgI and Ag_3PO_4 are not soluble. Nitrate salts are soluble. Phosphate salts are generally insoluble. While halides are generally soluble, those of Ag^+ are not.

    (c) K_2CO_3, KI and $KMnO_4$ are soluble. In general, salts of the alkali metals are soluble.

13. Ions produced when the compounds dissolve in water.

Compound	Cation	Anion
(a) KOH	K^+	OH^-
(b) K_2SO_4	$2 K^+$	SO_4^{2-}
(c) $LiNO_3$	Li^+	NO_3^-

	Compound	Cation	Anion
(d)	$(NH_4)_2SO_4$	$2\,NH_4^+$	SO_4^{2-}

15.

	Compound	Water Soluble	Cation	Anion
(a)	Na_2CO_3	yes	$2\,Na^+$	CO_3^{2-}
(b)	$CuSO_4$	yes	Cu^{2+}	SO_4^{2-}
(c)	NiS	no		
(d)	$BaBr_2$	yes	Ba^{2+}	$2\,Br^-$

Precipitation Reactions and Net Ionic Equations

17. $CdCl_2(aq) + 2\,NaOH(aq) \rightarrow Cd(OH)_2(s) + 2\,NaCl(aq)$

Net ionic equation: $Cd^{2+}(aq) + 2\,OH^-(aq) \rightarrow Cd(OH)_2(s)$

19. Balanced equations for precipitation reactions:

(a) $NiCl_2(aq) + (NH_4)_2S(aq) \rightarrow NiS(s) + 2\,NH_4Cl(aq)$

Net ionic equation: $Ni^{2+}(aq) + S^{2-}(aq) \rightarrow NiS(s)$

(b) $3\,Mn(NO_3)_2(aq) + 2\,Na_3PO_4(aq) \rightarrow Mn_3(PO_4)_2(s) + 6\,NaNO_3(aq)$

Net ionic equation: $3\,Mn^{2+}(aq) + 2\,PO_4^{3-}(aq) \rightarrow Mn_3(PO_4)_2(s)$

Acids and Bases and Their Reactions

21. $HNO_3(aq) + H_2O(l) \rightarrow H_3O^+(aq) + NO_3^-(aq)$

alternatively: $HNO_3(aq) \rightarrow H^+(aq) + NO_3^-(aq)$

23. $H_2C_2O_4(aq) \rightarrow H^+(aq) + HC_2O_4^-(aq)$

$HC_2O_4^-(aq) \rightarrow H^+(aq) + C_2O_4^{2-}(aq)$

25. $MgO(s) + H_2O(l) \rightarrow Mg(OH)_2(s)$ (metal oxide reacts with water to form a base)

27. Complete and Balance

(a) $2\,CH_3CO_2H(aq) + Mg(OH)_2(s) \rightarrow Mg(CH_3CO_2)_2(aq) + 2\,H_2O(l)$

    acetic   magnesium   magnesium   water
    acid   hydroxide   acetate

(b) $HClO_4(aq) + NH_3(aq) \rightarrow NH_4ClO_4(aq)$

    perchloric   ammonia   ammonium
    acid   perchlorate

29. Write and balance the equation for barium hydroxide reacting with nitric acid:

$Ba(OH)_2(s) + 2\ HNO_3(aq) \rightarrow Ba(NO_3)_2(aq) + 2\ H_2O(l)$

barium	nitric	barium	water
hydroxide	acid	nitrate	

31. $HNO_3(aq) + H_2O(l) \rightleftharpoons H_3O^+ (aq) + NO_3^- (aq)$

    BA   BB   BA   BB   [BA = Brönsted Acid; BB = Brönsted base]

Since nitric acid is a STRONG acid, the equilibrium favors the PRODUCTS.

33. Show H_2O reacting (with HBr) as a Brönsted base:

$HBr(aq) + H_2O(l) \rightarrow H_3O^+ (aq) + Br^- (aq)$ (water accepts a proton from HBr)

  (BA)   (BB)

Show H_2O reacting (with NH_3) as a Brönsted acid: (water donates a proton to NH_3)

$NH_3(aq) + H_2O(l) \rightarrow NH_4^+ (aq) + OH^- (aq)$

  (BB)   (BA)

Writing Net Ionic Equations

35. (a) $(NH_4)_2CO_3(aq) + Cu(NO_3)_2(aq) \rightarrow CuCO_3(s) + 2\ NH_4NO_3(aq)$

    (net) $CO_3^{2-}(aq) + Cu^{2+}(aq) \rightarrow CuCO_3(s)$

(b) $Pb(OH)_2(s) + 2\ HCl(aq) \rightarrow PbCl_2(s) + 2\ H_2O(l)$

    (net) $Pb(OH)_2(s) + 2\ H_3O^+(aq) + 2\ Cl^-(aq) \rightarrow PbCl_2(s) + 4\ H_2O(l)$

(c) $BaCO_3(s) + 2\ HCl(aq) \rightarrow BaCl_2(aq) + H_2O(l) + CO_2(g)$

    (net) $BaCO_3(s) + 2\ H^+(aq) \rightarrow Ba^{2+}(aq) + H_2O(l) + CO_2(g)$

    alternatively: $BaCO_3(s) + 2\ H_3O^+(aq) \rightarrow Ba^{2+}(aq) + 3\ H_2O(l) + CO_2(g)$

(d) $2\ CH_3CO_2H(aq) + Ni(OH)_2(s) \rightarrow Ni(CH_3CO_2)_2(aq) + H_2O(l)$

    net: $2\ CH_3CO_2H(aq) + Ni(OH)_2(s) \rightarrow Ni^{2+}(aq) + 2\ CH_3CO_2^-(aq) + 2\ H_2O(l)$

37. (a) $AgNO_3(aq) + KI(aq) \rightarrow AgI(s) + KNO_3(aq)$

    (net) $Ag^+(aq) + I^-(aq) \rightarrow AgI(s)$

(b) $Ba(OH)_2(aq) + 2\ HNO_3(aq) \rightarrow 2\ H_2O(l) + Ba(NO_3)_2(aq)$

    (net) $OH^- (aq) + H_3O^+ (aq) \rightarrow 2\ H_2O(l)$

(c) $2\ Na_3PO_4(aq) + 3\ Ni(NO_3)_2(aq) \rightarrow Ni_3(PO_4)_2(s) + 6\ NaNO_3(aq)$

    (net) $2\ PO_4^{3-}(aq) + 3\ Ni^{2+}(aq) \rightarrow Ni_3(PO_4)_2(s)$

39. (a) $HNO_2(aq) + NaOH(aq) \rightarrow H_2O(l) + NaNO_2(aq)$

    (net) $HNO_2(aq) + OH^-(aq) \rightarrow H_2O(l) + NO_2^-(aq)$

(b) $Ca(OH)_2(s) + 2\ HCl(aq) \rightarrow 2\ H_2O(l) + CaCl_2(aq)$ (note: aqueous medium isn't stated)

(net) $Ca(OH)_2(s) + 2\ H_3O^+(aq) \rightarrow Ca^{2+}(aq) + 4\ H_2O(l)$

Gas-Forming Reactions

41. Write and balance the equation for iron(II) carbonate reacting with nitric acid:

$FeCO_3(s) + 2\ HNO_3(aq) \rightarrow Fe(NO_3)_2(aq) + H_2O(l) + CO_2(g)$

| iron(II) carbonate | nitric acid | iron(II) nitrate | water | carbon dioxide |

43. Overall, balanced equation for reaction of $(NH_4)_2S$ with HBr:

$(NH_4)_2S(s) + 2\ HBr(aq) \rightarrow H_2S(g) + 2\ NH_4Br(aq)$

| Ammonium sulfide | Hydrobromic acid | Hydrogen sulfide | Ammonium bromide |

Oxidation Numbers

45. For questions on oxidation number, read the symbol (x) as "the oxidation number of x."

$\quad$ (a) BrO_3^- $\qquad$ $(Br) + 3(O) = -1$

Since oxygen almost always has an oxidation number of -2, we can substitute this value and solve for the oxidation number of Br.

$$(Br) + 3(-2) = -1$$
$$(Br) = +5$$

$\quad$ (b) $C_2O_4^{2-}$ $\qquad$
$$2\ (C) + 4\ (O) = -2$$
$$2\ (C) + 4\ (-2) = -2$$
$$2\ (C) + -8 = -2$$
$$2\ (C) = +6$$
$$(C) = +3$$

$\quad$ (c) F^- $\qquad$ The oxidation number for any monatomic ion is the charge on the ion. So $(F) = -1$

$\quad$ (d) CaH_2 $\qquad$
$$(Ca) + 2\ (H) = 0$$
$$(Ca) + 2\ (-1) = 0$$
$$(Ca) = +2$$

$\quad$ (e) H_4SiO_4 $\qquad$
$$4(H) + (Si) + 4(O) = 0$$
$$4(+1) + (Si) + 4(-2) = 0$$
$$(Si) = +4$$

(f) HSO_4^- $(H) + (S) + 4(O) = -1$

$(+1) + (S) + 4(-2) = -1$

(S) $= +6$

Oxidation-Reduction Reactions

47. (a) Oxidation-Reduction: Zn(s) has an oxidation number of 0, while Zn^{2+}(aq) has an oxidation number of +2—hence Zn is being oxidized. N in NO_3^- has an oxidation number of +5, while N in NO_2 has an oxidation number of +4—hence N is being reduced.

(b) Acid-Base reaction: There is no change in oxidation number for any of the elements in this reaction—hence it is NOT an oxidation-reduction reaction. H_2SO_4 is an acid, and $Zn(OH)_2$ acts as a base.

(c) Oxidation-Reduction: Ca(s) has an oxidation number of 0, while Ca^{2+}(aq) has an oxidation number of +2—hence Ca is being oxidized. H in H_2O has an oxidation number of +1, while H in H_2 has an oxidation number of 0—hence H is being reduced.

49. Determine which reactant is oxidized and which is reduced:

(a) $C_2H_4(g) + 3 O_2(g) \rightarrow 2 CO_2(g) + 2 H_2O(l)$

specie	ox. number before	after	has experienced	functions as the
C	-2	+4	oxidation	(C_2H_4) reducing agent
H	+1	+1	no change	
O	0	-2	reduction	(O_2) oxidizing agent

(b) $Si(s) + 2 Cl_2(g) \rightarrow SiCl_4(l)$

specie	ox. number before	after	has experienced	functions as the
Si	0	+4	oxidation	(Si) reducing agent
Cl	0	-1	reduction	(Cl_2) oxidizing agent

Types of Reactions in Aqueous Solution

51. Precipitation (PR), Acid-Base (AB), or Gas-Forming (GF)

(a) $Ba(OH)_2(aq) + 2 HCl(aq) \rightarrow BaCl_2(aq) + 2 H_2O(l)$ AB

(b) $2 HNO_3(aq) + CoCO_3(s) \rightarrow Co(NO_3)_2(aq) + H_2O(l) + CO_2(g)$ GF

(c) $2 Na_3PO_4(aq) + 3 Cu(NO_3)_2(aq) \rightarrow Cu_3(PO_4)_2(s) + 6 NaNO_3(aq)$ PR

53. Precipitation (PR), Acid-Base (AB), or Gas-Forming (GF)

    (a) $MnCl_2(aq) + Na_2S(aq) \rightarrow MnS(s) + 2\,NaCl(aq)$ PR

    (net) $Mn^{2+}(aq) + S^{2-}(aq) \rightarrow MnS(s)$

    (b) $K_2CO_3(aq) + ZnCl_2(aq) \rightarrow ZnCO_3(s) + 2\,KCl(aq)$ PR

    (net) $CO_3^{2-}(aq) + Zn^{2+}(aq) \rightarrow ZnCO_3(s)$

55. Balance the following and classify them as PR, AB, GF, or OR:

    (a) $CuCl_2(aq) + H_2S(aq) \rightarrow CuS(s) + 2\,HCl(aq)$ PR

    (b) $H_3PO_4(aq) + 3\,KOH(aq) \rightarrow 3\,H_2O(l) + K_3PO_4(aq)$ AB

    (c) $Ca(s) + 2\,HBr(aq) \rightarrow H_2(g) + CaBr_2(aq)$ OR

    (d) $MgCl_2(aq) + 2\,NaOH(aq) \rightarrow 3\,Mg(OH)_2(s) + 2\,NaCl(aq)$ PR

57. Identify the reactants (x and y) and write the complete balanced equation for each.

    (a) $x + y \rightarrow H_2O(l) + CaBr_2(aq)$

    (complete) $Ca(OH)_2(s) + 2\,HBr(aq) \rightarrow 2\,H_2O(l) + CaBr_2(aq)$

    (b) $x + y \rightarrow Mg(NO_3)_2(aq) + CO_2(g) + H_2O(l)$

    (complete) $MgCO_3(aq) + 2\,HNO_3(aq) \rightarrow Mg(NO_3)_2(aq) + CO_2(g) + H_2O(l)$

    (c) $x + y \rightarrow BaSO_4(s) + NaCl(aq)$

    (complete) $BaCl_2(aq) + Na_2SO_4(aq) \rightarrow BaSO_4(s) + 2\,NaCl(aq)$

    (d) $x + y \rightarrow NH_4^+(aq) + OH^-(aq)$

    (complete) $NH_3(aq) + H_2O(l) \rightarrow NH_4^+(aq) + OH^-(aq)$

GENERAL QUESTIONS

59. Balance:

    (a) Synthesis of urea:

    $CO_2(g) + 2\,NH_3(g) \rightarrow CO(NH_2)_2(s) + H_2O(l)$

      1. Note the need for two NH_3 in each molecule of urea, so multiply NH_3 by 2.

      2. $2\,NH_3$ provides the two H atoms for a molecule of H_2O.

      3. Each CO_2 provides the O atom for a molecule of H_2O.

    (b) synthesis of uranium(VI) fluoride

    $UO_2(s) + 4\,HF(aq) \rightarrow UF_4(s) + 2\,H_2O(l)$

    $UF_4(s) + F_2(g) \rightarrow UF_6(s)$

      1. The 4 F atoms in UF_4 requires 4 F atoms from HF.   (equation 1)

      2. The H atoms in HF produce 2 molecules of H_2O.   (equation 1)

      3. The 1:1 stoichiometry of $UF_6 : UF_4$ provides a simple balance. (equation 2)

(c) synthesis of titanium metal from TiO_2:

$TiO_2(s) + 2 Cl_2(g) + 2 C(s) \rightarrow TiCl_4(l) + 2 CO(g)$

$TiCl_4(l) + 2 Mg(s) \rightarrow Ti(s) + 2 MgCl_2(s)$

1. The O balance mandates 2 CO for each TiO_2.   (equation 1)

2. A coefficient of 2 for C provides C balance.   (equation 1)

3. The Ti balance (TiO_2:$TiCl_4$) requires 4 Cl atoms, hence 2 Cl_2 (equation 1)

4. The Cl balance requires 2 $MgCl_2$, hence 2 Mg.   (equation 2)

61. Formula for the following compounds:

(a) soluble compound with Br^- ion: almost any bromide compound with the exception of Ag^+, Hg_2^{2+} and Pb^{2+}

(b) insoluble hydroxide: almost any hydroxide except salts of NH_4^+ and the alkali metal ions

(c) insoluble carbonate: almost any carbonate except salts of NH_4^+ and the alkali metal ions

(d) soluble nitrate-containing compound: all nitrate-containing compounds are soluble

The listing of soluble and insoluble compounds in your text will provide general guidelines for predicting the solubility of compounds.

(e) a weak Bronsted acid: the carboxylic acids are weak acids: CH_3CO_2H (acetic)

63. For the following copper salts:

Water soluble: $Cu(NO_3)_2$, $CuCl_2$ — nitrates and chlorides are soluble

Water insoluble: $CuCO_3$, $Cu_3(PO_4)_2$ — carbonates and phosphates are insoluble

65. **Spectator ions** in the following equation and the net ionic equation:

$2 H_3O^+(aq) + 2 \mathbf{NO_3^-(aq)} + Mg(OH)_2(s) \rightarrow 4 H_2O(l) + Mg^{2+}(aq) + 2 \mathbf{NO_3^-(aq)}$

The emboldened nitrate ions are the spectator ions. The net ionic equation would be the first equation shown above without the spectator ions:

$2 H_3O^+(aq) + Mg(OH)_2(s) \rightarrow 4 H_2O(l) + Mg^{2+}(aq)$ [An acid-base exchange]

67. For the reaction of chlorine with NaBr: $Cl_2(g) + 2 NaBr(aq) \rightarrow 2 NaCl(aq) + Br_2(l)$

(a) Oxidized: **bromine's** oxidation number is changed from -1 to 0

Reduced: **chlorine's** oxidation number is changed from 0 to −1

(b) Oxidizing agent: $\mathbf{Cl_2}$ removes the electrons from NaBr

Reducing agent: **NaBr** provides the electrons to the chlorine.

69. Reaction: $MgCO_3(s) + 2\ HCl(aq) \rightarrow CO_2(g) + MgCl_2(aq) + H_2O(l)$

(a) The net ionic equation: $MgCO_3(s) + 2\ H_3O^+(aq) \rightarrow CO_2(g) + Mg^{2+}(aq) + 3\ H_2O(l)$

The spectator ion is the chloride ion (Cl^-).

(b) The production of $CO_2(g)$ characterizes this as a gas-forming reaction.

71. Species present in aqueous solutions of:

compound	types of species	species present
(a) NH_3	molecules (weak base)	NH_3, NH_4^+, OH^-
(b) CH_3CO_2H	molecules (weak acid)	CH_3CO_2H, $CH_3CO_2^-$, H^+
(c) $NaOH$	ions (strong base)	Na^+ and OH^-
(d) HBr	ions (strong acid)	H_3O^+ and Br^-

In every case, H_2O will be present (but omitted in this list)

73. Balance and classify each as PR, AB, GF

(a) $K_2CO_3(aq) + 2\ HClO_4(aq) \rightarrow 2\ KClO_4(aq) + CO_2(g) + 2\ H_2O(l)$ GF

Products: potassium perchlorate, carbon dioxide, water (respectively)

Net ionic equation: $CO_3^{2-}(aq) + 2\ H^+(aq) \rightarrow CO_2(g) + H_2O(l)$

(b) $FeCl_2(aq) + (NH_4)_2S(aq) \rightarrow FeS(s) + 2\ NH_4Cl(aq)$ PR

Products: iron(II) sulfide, ammonium chloride (respectively)

Net ionic equation: $Fe^{2+}(aq) + S^{2-}(aq) \rightarrow FeS(s)$

(c) $Fe(NO_3)_2(aq) + Na_2CO_3(aq) \rightarrow FeCO_3(s) + 2\ NaNO_3(aq)$ PR

Products: iron(II) carbonate, sodium nitrate (respectively)

Net ionic equation: $Fe^{2+}(aq) + CO_3^{2-}(aq) \rightarrow Fe\ CO_3(s)$

(d) $3\ NaOH(aq) + FeCl_3(aq) \rightarrow 3\ NaCl(aq) + Fe(OH)_3(s)$ PR

Products: sodium chloride, iron(III) hydroxide (respectively)

Net ionic equation: $Fe^{3+}(aq) + 3\ OH^-(aq) \rightarrow Fe\ (OH)_3(s)$

75. Which of the following compounds in each pair could be separated by stirring with water?

(a) NaOH dissolves in water, with $Ca(OH)_2$ remaining as undissolved.

(b) $MgCl_2$ will dissolve in water, while MgF_2 will remain undissolved.

(c) KI dissolves—as do all the salts of Group IA cations, while AgI will remain undissolved.

(d) NH_4Cl dissolves readily—as a salt containing BOTH ammonium ions AND chloride ions, while $PbCl_2$ is an insoluble chloride.

77. Compound(s) in each set that creates a solution that is **only** a weak conductor:

    (a) NH_3 reacts only slightly with water to produce a weakly conducting solution. The sodium and barium hydroxides are strong bases, providing a strong conductor. $Fe(OH)_3$ does not dissolve to an appreciable extent, so it will not produce a weakly conducting solution.

    (b) Both CH_3CO_2H and HF are weak acids, and will produce only a weakly conducting solution when dissolved in water. Na_3PO_4 will dissolve to a great extent—providing a strongly conducting solution, and HNO_3 is a strong acid—producing a strongly conducting solution.

79. Na_2S was treated with acid, producing a gas:

    $$S^{2-} (aq) + 2 H^+ (aq) \rightarrow H_2S (g)$$

    A black precipitate forms when the gas is bubbled into lead nitrate solution:

    $$Pb^{2+} (aq) + H_2S (g) \rightarrow PbS (s) + 2 H^+(aq)$$

IN THE LABORATORY

81. For the reaction:

    $$2 NaI(s) + 2 H_2SO_4(aq) + MnO_2(s) \rightarrow Na_2SO_4(aq) + MnSO_4(aq) + I_2 (g) + 2 H_2O(l)$$

    (a) Oxidation number of each atom in the equation: (ox.numbers shown in order)

    Reactants: NaI (+1,-1) H_2SO_4 (+1, +6, -2) MnO_2 (+4,-2)

    Products: Na_2SO_4 (+1,+6,-2) $MnSO_4$ (+2,+6,-2) I_2(0) H_2O(+1,-2)

    (b) Oxidizing agent: MnO_2 Oxidized: I in NaI

    Reducing agent: NaI   Reduced: Mn (in MnO_2)

    (c) The formation of gaseous iodine "drives" the process—product-favored

    (d) Names of reactants and products:

NaI	H_2SO_4	MnO_2	Na_2SO_4	$MnSO_4$	I_2	H_2O
sodium iodide	sulfuric acid	manganese(IV) oxide	sodium sulfate	manganese(II) sulfate	iodine	water

83. Another way to prepare $MgCl_2$: Given the reactivity of both elemental magnesium and chlorine, one can bring the two elements into contact (carefully!)

    $$Mg(s) + Cl_2(g) \rightarrow MgCl_2(s)$$

85. In the reaction:

$C_6H_{12}O_6$ (aq) + 2 Ag$^+$ (aq) + 2 OH$^-$ (aq) → $C_6H_{12}O_7$ (aq) + 2 Ag(s) + H_2O(l)

Oxidized: $C_6H_{12}O_6$ is oxidized to $C_6H_{12}O_7$ (simple observation—note that O is added)

Reduced: Ag$^+$(aq) is reduced to Ag(s) (oxidation number changes from +1 to 0)

Oxidizing agent: Ag$^+$ (aq) oxidizes the sugar

Reducing agent: $C_6H_{12}O_6$ reduces Ag$^+$

SUMMARY AND CONCEPTUAL QUESTIONS

87. A simple experiment to prove that lactic acid is a weak acid (ionizing to a small extent) is to test the conductivity of the solution. A conductivity apparatus (e.g. a light bulb) will indicate only a small current flow (a light bulb will glow only dimly).

To prove that the establishment of equilibrium is reversible, add strong acid (H_3O^+). The shift of equilibrium to the left should result in the molecular acid precipitating from solution.

89. Using the reagents: $BaCl_2$, $BaCO_3$, $Ba(OH)_2$, H_2SO_4, Na_2SO_4,

Prepare barium sulfate by:

a precipitation reaction

The reaction of $BaCl_2$ with Na_2SO_4 will perform this task:

$BaCl_2$ (aq) + Na_2SO_4(aq) → $BaSO_4$(s) + 2 NaCl (aq)

a gas-forming reaction

$BaCO_3$(s) + H_2SO_4(l) → $BaSO_4$(s) + H_2O(l) + CO_2(g)

One might think about using $Ba(OH)_2$ as one reactant for part (a), but the substance isn't very water-soluble. $BaCl_2$ is much more water-soluble.

91. Begin by treating the percentages as mass (out of 100g of the compound), and calculate the number of moles of each atom:

$$\frac{20.315 \text{ g Ni}}{1} \cdot \frac{1 \text{ mol Ni}}{58.6934 \text{ g Ni}} = 0.34612 \text{ mol Ni}$$

$$\frac{33.258 \text{ g C}}{1} \cdot \frac{1 \text{ mol C}}{12.0107 \text{ g C}} = 2.7690 \text{ mol C}$$

$$\frac{4.884 \text{ g H}}{1} \cdot \frac{1 \text{ mol H}}{1.00794 \text{ g H}} = 4.846 \text{ mol H}$$

$$\frac{22.151 \text{ g O}}{1} \cdot \frac{1 \text{ mol O}}{15.9994 \text{ g O}} = 1.3845 \text{ mol O}$$

$$\frac{19.395 \text{ g N}}{1} \cdot \frac{1 \text{ mol N}}{14.0067 \text{ g N}} = 1.3845 \text{ mol N}$$

Using the # of moles of nickel, calculate ratios for mole element X/mol Ni

$$\frac{2.7690 \text{ mol C}}{0.34612 \text{ mol Ni}} = 8.000 \text{ mol C/mol Ni}$$

$$\frac{4.846 \text{ mol H}}{0.34612 \text{ mol Ni}} = 14.001 \text{ mol H/mol Ni}$$

$$\frac{1.3845 \text{ mol O}}{0.34612 \text{ mol Ni}} = 4.000 \text{ mol O/mol Ni and}$$

$$\frac{1.3845 \text{ mol N}}{0.34612 \text{ mol Ni}} = 4.000 \text{ mol N/mol Ni so the empirical formula is: } Ni_1C_8H_{14}O_4N_4$$

93. (a) Oxidation states for As,S,N in the reaction:

Reactants: $As_2S_3(s)$: As = +3, S = -2; HNO_3: N = +5

Products: H_3AsO_4: As = +5 ; NO: N = +2 ; S = 0

(b) Formula for AgxAsOy:

Using the composition, As, 16.199% and Ag, 69.964%, we know that the remaining percent belongs to O, so 100.000 – (16.199 + 69.964) = 13.837 % O. As in 91 above, express the mass of each element (out of 100g), and calculate the # moles of As, Ag, and O.

$$\frac{16.199 \text{ g As}}{1} \cdot \frac{1 \text{ mol As}}{74.9216 \text{ g As}} = 0.21621 \text{ mol As}$$

$$\frac{69.964 \text{ g Ag}}{1} \cdot \frac{1 \text{ mol Ag}}{107.868 \text{ g Ag}} = 0.64861 \text{ mol Ag}$$

$$\frac{13.837 \text{ g O}}{1} \cdot \frac{1 \text{ mol O}}{15.9994 \text{ g O}} = 0.86484 \text{ mol O}$$

Express these as mol of element X/mol As:

$$\frac{0.64861 \text{ mol Ag}}{0.21621 \text{ mol As}} = 2.9999 \text{ mol Ag/mol As and } \frac{0.86484 \text{ mol O}}{0.21621 \text{ mol As}} = 4.0000 \text{ mol O/mol As}$$

The compound has the formula: Ag_3AsO_4.

Applying Chemical Principles

1. What is the value of x in a sample of $La_{2-x}Ba_xCuO_4$?

    Given: %La = 63.43, %Ba = 5.085, %Cu = 15.69, and %O = 15.80.

    Noting that the percentages supplied total to 100%, we can calculate the # moles of each element present, assuming we have 100.g of the substance:

    $$\frac{63.43 \text{ g La}}{1} \cdot \frac{1\text{mol La}}{138.91\text{g La}} = 0.4566 \text{ mol La}; \quad \frac{5.085 \text{ g Ba}}{1} \cdot \frac{1\text{mol Ba}}{137.33 \text{ g Ba}} = 0.03703 \text{ mol Ba}$$

    $$\frac{15.69 \text{ g Cu}}{1} \cdot \frac{1\text{mol Cu}}{63.546 \text{ g Cu}} = 0.2469 \text{ mol Cu}; \quad \frac{15.80 \text{ g O}}{1} \cdot \frac{1\text{mol O}}{15.9994\text{g O}} = 0.9875 \text{ mol O}$$

    Now unlike our usual procedure of dividing each number of moles by the **smallest**, we can exercise a bit of reasoning here. Note that the formula of the superconductor shows the subscript of Cu as 1! So let's force our calculations to have 1 has the subscript by dividing each of the amounts of La, Ba, and O by the # moles of Cu:

    $$\frac{0.4566 \text{ mol La}}{0.2469 \text{ mol Cu}} = 1.849 \text{ mol La/mol Cu} \qquad \frac{0.03703 \text{ mol Ba}}{0.2469 \text{ mol Cu}} = 0.1500 \text{ mol Ba/mol Cu}$$

    $$\frac{0.9875 \text{ mol O}}{0.2469 \text{ mol Cu}} = 4.000 \text{ mol O/mol Cu}.$$ Note that this ratio simply **confirms** our formula for the superconductor! Since Ba is x, and we found x = 0.1500, then the subscript for La is (2-x) or 1.85! The formula for the semiconductor is then $La_{1.85}Ba_{0.15}CuO_4$!

3. Charges present on the copper ions in $YBa_2Cu_3O_7$?

    Since all the charges must add to 0, and since we're confident that each O will have a charge of 2-, we can add the charges for Y, Ba, and O to obtain (3+) + 2(2+) + 7(2-) = 7-.

    The three Cu ions must have charges that add to 7+. So if we TWO Cu ions at 2+ and ONE Cu ion at 3+, we accomplish that requirement.

5. What mass of oxygen is required to convert 1.00 g $YBa_2Cu_3O_{6.50}$ to $YBa_2Cu_3O_{6.93}$?

    Moles of YBCO: 1(Y) + 2(Ba) + 3(Cu) + 6.50(O) = 1(88.91)+2(137.327) + 3(63.546) + 6.50(15.9994) = 658.20 (Note that the atomic mass of Y limits the sum.)

    $$\frac{1.00 \text{ g YBCO}}{1} \cdot \frac{1 \text{ mol YBCO}}{658.20 \text{ g YBCO}} = 0.001519 \text{ mol YBCO}.$$ Adding (6.93-6.50) or 0.43 mol

    $$O_2 \text{ requires } \boxed{\frac{32.00 \text{ g O}_2}{1 \text{ mol O}_2} \cdot \frac{1 \text{ mol O}_2}{2 \text{ mol O}} \cdot \frac{0.43 \text{ mol O}}{1 \text{ mol YBCO}} \cdot \frac{0.001519 \text{ mol YBCO}}{1} = 0.011 \text{ g O}_2}$$

Chapter 4
Stoichiometry: Quantitative Information about Chemical Reactions

PRACTICING SKILLS
Mass Relationships in Chemical Reactions: Basic Stoichiometry

1. Moles of oxygen needed to react with 6.0 mol of Al:

$$4 \text{ Al (s)} + 3 \text{ O}_2 \text{ (g)} \rightarrow 2 \text{ Al}_2\text{O}_3 \text{ (s)}$$

$$6.0 \text{ mol Al} \cdot \frac{3 \text{ mol O}_2}{4 \text{ mol Al}} = 4.5 \text{ mol O}_2 \text{ ; What mass of Al}_2\text{O}_3 \text{ should be produced?}$$

$$6.0 \text{ mol Al} \cdot \frac{2 \text{ mol Al}_2\text{O}_3}{4 \text{ mol Al}} \cdot \frac{102 \text{ g Al}_2\text{O}_3}{1 \text{mol Al}_2\text{O}_3} = 310 \text{ g Al}_2\text{O}_3 \text{ (to 2 sf)}$$

3. Quantity of Br_2 to react with 2.56 g of Al:

   According to the balanced equation, 2 mol of Al react with 3 mol of Br_2 .

   Calculate the # of moles of Al, then multiply by 3/2 to obtain # mol of Br_2 required.

$$2.56 \text{ g Al} \cdot \frac{1 \text{mol Al}}{26.98 \text{gAl}} \cdot \frac{3 \text{ mol Br}_2}{2 \text{ mol Al}} \cdot \frac{159.8 \text{ g Br}_2}{1 \text{ mol Br}_2} = 22.7 \text{ g Br}_2$$

   Mass of Al_2Br_6 expected:

   This could be solved in several ways. The simplest is to recognize that—according to the Law of Conservation of Matter--mass is conserved in a reaction. If 22.7 g of bromine react with exactly 2.56 g of aluminum, the total products would also have a mass of (22.7 g + 2.56 g) or 25.3 g Al_2Br_6.

5. For the reaction of methane burning in oxygen:
   (a) The products of the reaction are the oxides of carbon and of hydrogen.
      $$CH_4 + O_2 \rightarrow CO_2 + H_2O$$

   (b) The balanced equation requires 4 hydrogen atoms (so 2 water molecules), and 2 water molecules and 1 carbon dioxide mandate 2 oxygen molecules:
      $$CH_4 + 2 O_2 \rightarrow CO_2 + 2 H_2O$$

   (c) The mass of O_2 required for complete combustion of 25.5 g of methane:
      $$\frac{25.5 \text{g CH}_4}{1} \cdot \frac{1 \text{mol CH}_4}{16.04 \text{g CH}_4} \cdot \frac{2 \text{mol O}_2}{1 \text{mol CH}_4} \cdot \frac{32.00 \text{g O}_2}{1 \text{ mol O}_2} = 102 \text{g O}_2$$

   (d) As in SQ 4.3, the simplest method to determine the total mass of products expected is to recognize that mass is conserved in a reaction. If 25.5g of methane reacts with exactly 102g of oxygen, the total products would have a mass of (25.5g + 102g) or 127.5g (or 128g to 3 sf).

Amounts Tables and Chemical Stoichiometry

7. Emboldened quantities indicate given data

Equation	2 PbS(s) +	3 O_2(g) →	2 PbO(s) +	2SO_2(g)
Initial amount (mol)	**2.50**			
Change in amount upon reaction (mol)	-2.50	-3/2(2.50) = 3.75	+2.50	+2.50
Amount after complete reaction (mol)	0	0	2.50	2.50

The amount of O_2 required is 1.5 times the amount of PbS (note the 3:2 ratio). With the required amount of oxygen present, the amount of PbO and SO_2 formed is equal to the amount of PbS consumed (note the ratios of 2:2 for both PbS:PbO and PbS:SO_2)

9. For the reaction of elemental Cr with oxygen to produce chromium(III) oxide:

Equation	4 Cr(s)	+ 3 O_2(g) →	2 Cr_2O_3(s)
Initial amount (g)	**0.175**		
Initial amount (mol)	3.37×10^{-3}		
Change in amount upon reaction (mol)	-3.37×10^{-3}	$-3/4(3.37 \times 10^{-3})$ = 2.52×10^{-4}	$+1/2(3.37 \times 10^{-3})$ = 1.68×10^{-3}
Amount after complete reaction (mol)	0	0	$+2.40 \times 10^{-2}$
Amount after complete reaction (g)	0	0	0.256

$$\frac{0.175 \text{ g Cr}}{1} \cdot \frac{1 \text{ mol Cr}}{52.00 \text{ g Cr}} = 3.37 \times 10^{-3} \text{ mol Cr}$$

The balanced equation indicates that for each mol of elemental Cr, 1/2 mol of Cr_2O_3 is produced (the 4:2 ratio), so 3.37×10^{-3} mol Cr produces 1.68×10^{-3} mol Cr_2O_3

$$\frac{1.68 \times 10^{-3} \text{ mol } Cr_2O_3}{1} \cdot \frac{152.0 \text{ g } Cr_2O_3}{1 \text{ mol } Cr_2O_3} = 0.256 \text{ g } Cr_2O_3$$

The amount of oxygen required is 3/4 of the amount of elemental Cr (4:3 ratio), with each mole of oxygen having a mass of 32.00 g.

$$\frac{3.37 \times 10^{-3} \text{ mol Cr}}{1} \cdot \frac{3 \text{ mol } O_2}{4 \text{ mol Cr}} \cdot \frac{32.00 \text{ g } O_2}{1 \text{ mol } O_2} = 0.0808 \text{ g } O_2$$

Limiting Reactants

11. The reaction to produce sodium sulfide: $Na_2SO_4(aq) + 4\ C(s) \rightarrow Na_2S(aq) + 4\ CO(g)$

    With 15 g of Na_2SO_4 and 7.5 g of C, what is the limiting reactant?

    The amounts table:

Equation	$Na_2SO_4(aq)$	$+ 4\ C(s) \rightarrow$	$Na_2S(aq) +$	$4\ CO(g)$
Initial amount (g)	**15**	**7.5**		
Initial amount (mol)	0.11	0.62		
Change in amount upon reaction (mol)	-0.11	-4(0.11) = 0.44	+0.11	+0.44
Amount after complete reaction (mol)	0	0.18	+0.11	+0.44

$$\frac{15\ \text{g Na}_2\text{SO}_4}{1} \bullet \frac{1\ \text{mol Na}_2\text{SO}_4}{142\ \text{g Na}_2\text{SO}_4} = 0.11\ \text{mol Na}_2\text{SO}_4 \qquad \frac{7.5\ \text{g C}}{1} \bullet \frac{1\ \text{mol C}}{12.00\ \text{g C}} = 0.63\ \text{mol C}$$

The coefficients of the balanced equation indicate the need for 4 mol of C for 1 mol of Na_2SO_4. The actual ratio of C: Na_2SO_4 is 0.63:0.11 (or 5.6:1), indicating that Na_2SO_4 is the limiting reagent. With Na_2SO_4 as the limiting reagent, the amount of Na_2S produced is identical (on a mol:mol basis). The mass of Na_2S produced is:

$$\frac{15\ \text{g Na}_2\text{SO}_4}{1} \bullet \frac{1\ \text{mol Na}_2\text{SO}_4}{142\ \text{g Na}_2\text{SO}_4} \bullet \frac{78.05\ \text{g Na}_2\text{S}}{1\ \text{mol Na}_2\text{S}} = 8.24\ \text{g Na}_2\text{S} \text{ or } 8.2\ \text{g (2 sf)}$$

The limiting reactant can be alternatively identified using the mol-rxn concept.

$$\frac{15\ \text{g Na}_2\text{SO}_4}{1} \bullet \frac{1\ \text{mol Na}_2\text{SO}_4}{142\ \text{g Na}_2\text{SO}_4} \bullet \frac{1\ \text{mol-rxn}}{1\ \text{mol Na}_2\text{SO}_4} = 0.11\ \text{mol-rxn}$$

$$\frac{7.5\ \text{g C}}{1} \bullet \frac{1\ \text{mol C}}{12.00\ \text{g C}} \bullet \frac{1\ \text{mol-rxn}}{4\ \text{mol C}} = 0.156\ \text{mol-rxn}$$

Note that with Na_2SO_4, the mol-rxn is smaller, making Na_2SO_4 the limiting reactant.

13. Identify the limiting reactant when 1.6 mol of S_8 and 35 mol of F_2 react:

    The balanced equation is: $S_8 + 24\ F_2 \rightarrow 8\ SF_6$

    Determine the required ratio (from the balanced equation): $\dfrac{24\ \text{mol F}_2}{1\ \text{mol S}_8}$

    Determine the ratio of actual amounts: $\dfrac{35\ \text{mol F}_2}{1.6\ \text{mol S}_8} = \dfrac{21.9\ \text{mol F}_2}{1\ \text{mol S}_8}$

    Since the ratio of actual fluorine available is **less than** the required ratio, **fluorine is the limiting reactant.**

15. For the reaction of methane with water: The amounts table:

Equation	$CH_4(g)$ +	$H_2O(g)$ →	$CO_2(g)$ +	$3 H_2(g)$
Initial amount (g)	**995**	**2510**		
Initial amount (mol)	62.0	139.3		
Change in amount upon reaction (mol)	-62.0	-62.0	+62.0	+3(62.0)
Amount after complete reaction (mol)	0	139.3 - 62.0 = 77.3		+186

(a) Limiting reagent:

$$995 \text{ g CH}_4 \cdot \frac{1 \text{ mol CH}_4}{16.04 \text{ g CH}_4} = 62.0 \text{ mol CH}_4$$

$$2510 \text{ g H}_2\text{O} \cdot \frac{1 \text{ mol H}_2\text{O}}{18.02 \text{ g H}_2\text{O}} = 139.3 \text{ mol H}_2\text{O}$$

The required ratio of methane to water: $\dfrac{1 \text{ mol H}_2\text{O}}{1 \text{ mol CH}_4}$ and

the actual ratio: $\dfrac{139.3 \text{ mol H}_2\text{O}}{62.0 \text{ mol CH}_4} = 2.25$

The actual ratio indicates that **methane is the limiting reactant.**

(b) Maximum mass of H_2 possible:

$$62.0 \text{ mol CH}_4 \cdot \frac{3 \text{ mol H}_2}{1 \text{ mol CH}_4} \cdot \frac{2.016 \text{ g H}_2}{1 \text{ mol H}_2} = 375 \text{ g H}_2$$

(c) Mass of water remaining:

Since 1 mol of methane reacts with 1 mol of water, we know that 62.0 mol of CH_4 reacts

with 62.0 mol of water. The amount of water remaining is: (139.3 – 62.0) = 77.3 mol H_2O. The mass of this amount of water is:

$$77.3 \text{ mol H}_2\text{O} \cdot \frac{18.02 \text{ g H}_2\text{O}}{1 \text{ mol H}_2\text{O}} = 1{,}393 \text{ g H}_2\text{O or } 1390 \text{g (to 3 sf)}$$

17. Hexane gas burns in air (O_2) to give CO_2 and H_2O.

The amounts table:

Equation	$2 C_6H_{14}(g)$ +	$19 O_2(g)$ →	$12 CO_2(g)$	+ $14 H_2O(g)$
Initial amount (g)	**215**	**215**		
Initial amount (mol)	2.49	6.72		
Change in amount upon reaction (mol)	-2/19(6.72) = -0.707	-6.72	+12/19(6.72) = 4.24	+14/19(6.72) = 4.95
Amount after complete reaction (mol)	2.49-0.707 = 1.78	0	4.24	4.95

(a) The balanced equation for the reaction.

$2\ C_6H_{14}(g) + 19\ O_2(g) \rightarrow 12\ CO_2(g) + 14\ H_2O(g)$. A coefficient of 6 for CO_2 seems a reasonable start, with a coefficient of 7 for H_2O providing the balance for H. However, a quick examination of the count of O atoms on the right gives an **odd** number. This indicates that doubling the coefficient for hexane will provide 12 C (on the right), and 14 as a coefficient for H_2O. With a total of 38 O atoms (on the right), 19 O_2 molecules provide the balance.

(b) If 215 g of C_6H_{14} is mixed with 215 g of O_2, what masses of CO_2 and H_2O are produced in the reaction?

Moles of each reactant: $\dfrac{215\ \text{g}\ C_6H_{14}}{1} \bullet \dfrac{1\ \text{mol}\ C_6H_{14}}{86.18\ \text{g}\ C_6H_{14}} = 2.49\ \text{mol}\ C_6H_{14}$ and

$\dfrac{215\ \text{g}\ O_2}{1} \bullet \dfrac{1\ \text{mol}\ O_2}{32.0\ \text{g}\ O_2} = 6.72\ \text{mol}\ O_2$

Determining the required and actual ratios:

$\dfrac{19\ \text{mol}\ O_2}{2\ \text{mol}\ C_6H_{14}} = \dfrac{9.5\ \text{mol}\ O_2}{1\ \text{mol}\ C_6H_{14}}$ and the actual ratio: $\dfrac{6.72\ \text{mol}\ O_2}{2.49\ \text{mol}\ C_6H_{14}} = \dfrac{2.69\ \text{mol}\ O_2}{1\ \text{mol}\ C_6H_{14}}$

indicating that **oxygen is the limiting reagent.**

The stoichiometric ratios provide:

$6.72\ \text{mol}\ O_2 \bullet \dfrac{12\ \text{mol}\ CO_2}{19\ \text{mol}\ O_2} \bullet \dfrac{44.01\ \text{g}\ CO_2}{1\ \text{mol}\ CO_2} = 187\ \text{g}\ CO_2$, and for the 4.95 mol of

water:

$\dfrac{4.95\ \text{mol}\ H_2O}{1} \bullet \dfrac{18.02\ \text{g}\ H_2O}{1\ \text{mol}\ H_2O} = 89.2\ \text{g}\ H_2O$

(c) What mass of the excess reactant remains after the hexane has been burned?

With 1.78 mol of C_6H_{14} remaining, the mass is:

$\dfrac{1.78\ \text{mol}\ C_6H_{14}}{1} \bullet \dfrac{86.18\ \text{g}\ C_6H_{14}}{1\ \text{mol}\ C_6H_{14}} = 154\ \text{g}\ C_6H_{14}$

(d) The amounts table is written at the beginning of the solution to this study question.

Percent Yield

19. Percent yield of CH3OH: $\dfrac{\text{actual}}{\text{theoretical}} = \dfrac{332\ \text{g}\ CH_3OH}{407\ \text{g}\ CH_3OH} = 81.6\%$ yield

21. In the formation of $Cu(NH_3)_4SO_4$:

(a) The theoretical yield of $Cu(NH_3)_4SO_4$ from 10.0 g of $CuSO_4$:

$$10.0 \text{ g } CuSO_4 \; \bullet \; \frac{1 \text{ mol } CuSO_4}{159.6 \text{ g } CuSO_4} \; \bullet \; \frac{1 \text{ mol } Cu(NH_3)_4SO_4}{1 \text{ mol } CuSO_4}$$

$$\bullet \; \frac{227.7 \text{ g } Cu(NH_3)_4SO_4}{1 \text{ mol } Cu(NH_3)_4SO_4} = 14.3 \text{ g } Cu(NH_3)_4SO_4$$

(b) Percentage yield of the compound:

$$\frac{12.6 \text{ g compound}}{14.3 \text{ g compound}} \bullet 100 = 88.3\%$$

Analysis of Mixtures

23. Mass percent of $CuSO_4 \bullet 5 H_2O$ in the mixture:

Mass of H_2O = 1.245 g - 0.832 g = 0.413 g H_2O

Since this water was a part of the hydrated salt, let's calculate the mass of that salt present: In 1 mol of $CuSO_4 \bullet 5 H_2O$ there are 90.10 g H_2O and 159.61 g $CuSO_4$ or 249.71 g $CuSO_4 \bullet 5 H_2O$. These masses correspond to the molar masses of anhydrous $CuSO_4$ and 5• mol H_2O. So:

$$0.413 \text{ g } H_2O \; \bullet \; \frac{249.71 \text{ g } CuSO_4 \bullet 5H_2O}{90.10 \text{ g} H_2O} = 1.14 \text{ g hydrated salt}$$

$$\% \text{ hydrated salt} = \frac{1.14 \text{ g hydrated salt}}{1.245 \text{ g mixture}} \bullet 100 = 91.9\%$$

25. Mass percent of $CaCO_3$ in a limestone sample:

Note that the ratio of carbon dioxide to calcium carbonate is 1:1 (from balanced equation). Calculate the amount of carbon dioxide represented by 0.558 g CO_2.

$$0.558 \text{ g } CO_2 \; \bullet \; \frac{1 \text{ mol } CO_2}{44.01 \text{ g } CO_2} = 0.01268 \text{ mol } CO_2$$

Since there would have been 0.01268 mol of $CaCO_3$, the mass is:

$$0.01268 \text{ mol } CaCO_3 \bullet \frac{100.1 \text{ g } CaCO_3}{1 \text{ mol } CaCO_3} = 1.269 \text{ g } CaCO_3.$$

The percent of $CaCO_3$ in the sample is:

$$\frac{1.269 \text{ g } CaCO_3}{1.506 \text{ g sample}} \bullet 100 = 84.3\%$$

27. Mass percent of Tl_2SO_4 in 10.20 g sample:

Begin by balancing the equation:

$$Tl_2SO_4 + 2NaI(aq) \rightarrow 2TlI(s) + Na_2SO_4(aq)$$

This equation tells us that for each mol of Tl_2SO_4, we expect **two** moles of TlI.

Determine the amount of TlI: $0.1964 \text{ g TlI} \cdot \dfrac{1 \text{ mol TlI}}{331.29 \text{ g TlI}} = 5.928 \times 10^{-4} \text{ mol TlI}$

Use the ratio described by the balanced equation:

$5.928 \times 10^{-4} \text{ mol TlI} \cdot \dfrac{1 \text{ mol } Tl_2SO_4}{2 \text{ mol TlI}} = 2.964 \times 10^{-4} \text{ mol } Tl_2SO_4$

Using the molar mass of thallium(I) sulfate gives the mass of the sulfate present in the sample: $2.964 \times 10^{-4} \text{ mol } Tl_2SO_4 \cdot \dfrac{504.83 \text{ g } Tl_2SO_4}{1 \text{ mol } Tl_2SO_4} = 0.1496 \text{ g } Tl_2SO_4$

The mass percent of Tl_2SO_4 in the sample is $\dfrac{0.1496 \text{ g } Tl_2SO_4}{10.20 \text{ g sample}} \cdot 100 = 1.467\%$

Using Stoichiometry to determine Empirical and Molecular Formulas

29. The basic equation is:

$$C_xH_y + O_2 \rightarrow x\, CO_2 + \frac{y}{2} H_2O$$

Without balancing the equation, one can see that all the C in CO_2 comes from the styrene as does all the H in H_2O. Let's use the percentage of C in CO_2 to determine the mass of C in styrene, and the percentage of H in H_2O to provide the mass of H in styrene.

$1.481 \text{ g } CO_2 \cdot \dfrac{12.01 \text{ g C}}{44.01 \text{ g } CO_2} = 0.404 \text{ g C}$

Similarly:

$0.303 \text{ g } H_2O \cdot \dfrac{2.02 \text{ g H}}{18.02 \text{ g } H_2O} = 0.0340 \text{ g H}$

Alternatively, the mass of H could be determined by subtracting the mass of C from the 0.438 g styrene

Mass H = 0.438 g styrene - 0.404 g C

Establish the ratio of C atoms to H atoms

$0.0340 \text{ g H} \cdot \dfrac{1 \text{ mol H}}{1.008 \text{ g H}} = 0.0337 \text{ mol H}$

$0.404 \text{ g C} \cdot \dfrac{1 \text{ mol C}}{12.011 \text{ g C}} = 0.0336 \text{ mol C}$

This number of H and C atoms indicates an empirical formula for styrene of 1:1 or C_1H_1.

31. Combustion of Cyclopentane:

    The approach is similar to that for study question 37.

    The general equation is $C_xH_y + O_2 \rightarrow x\,CO_2 + \frac{y}{2}\,H_2O$

    All the carbon from cyclopentane eventually resides in the CO_2 produced.

    Calculate the amount of C in cyclopentane:

    $0.300 \text{ g } CO_2 \cdot \dfrac{1 \text{ mol } CO_2}{44.01 \text{ g } CO_2} \cdot \dfrac{1 \text{ mol C}}{1 \text{ mol } CO_2} = 0.00682 \text{ mol C}$

    Similarly the amount of H in cyclopentane:

    $0.123 \text{ g } H_2O \cdot \dfrac{1 \text{ mol } H_2O}{18.01 \text{ g } H_2O} \cdot \dfrac{2 \text{ mol H}}{1 \text{ mol } H_2O} = 0.0137 \text{ mol H}$

    (a) The empirical formula is then: $\dfrac{0.0137 \text{ mol H}}{0.00682 \text{ mol C}} = \dfrac{2.00 \text{ mol H}}{1.00 \text{ mol C}}$

    C_1H_2

    (b) If the molar mass is 70.1 g/mol, the molecular formula is:

    Since the molecular formula represents some **multiple** of the empirical formula,

    calculate the "empirical formula mass".

    For C_1H_2 that mass is 14.03 g/empirical formula [(1 x 12.0115) + (2 x 1.0079)]

    Calculate the # of empirical formulas in a molecular formula:

    $\dfrac{70.1 \text{ g/molecular formula}}{14.03 \text{ g/empirical formula}} = 5$, giving a molecular formula of C_5H_{10}.

33. An unknown compound has the formula CxHyOz.

    When 0.0956 g of the compound burns in air, 0.1356 g of CO_2 and 0.0833 g of H_2O result.

    What is the empirical formula of the compound?

    All the C that originated in the compound resides in the CO_2 produced, and all the H in the

    water produced originated in the compound. The same can not be said for the O. While some

    of the O in both carbon dioxide and water originated in the compound, some was

    incorporated from the atmosphere during the combustion process. For that reason, we can

    calculate the mass of C and H that resides in the two products, and obtain **by difference**

    [total mass of compound – (mass of C + mass of H)]the mass of O originally contained in the

    compound.

    $\dfrac{0.1356 \text{ g } CO_2}{1} \cdot \dfrac{12.01 \text{ g } CO_2}{44.01 \text{ g } CO_2} = 0.03700 \text{ g C}$ and similarly for H

    $\dfrac{0.0833 \text{ g } H_2O}{1} \cdot \dfrac{2.02 \text{ g H}}{18.02 \text{ g } H_2O} = 0.00934 \text{ g H};$

Now we can determine the mass of O originating in the compound: 0.0956 g C,H, and O (in the compound)- 0.03700 g C – 0.00934 g H = 0.04926 g O

Determine the number of moles of each of the 3 elements:

$$0.03700 \text{ g C} \cdot \frac{1 \text{ mol C}}{12.011 \text{ g C}} = 3.081 \times 10^{-3} \text{ mol C}$$

$$0.00934 \text{ g H} \cdot \frac{1 \text{ mol H}}{1.008 \text{ g H}} = 9.26 \times 10^{-3} \text{ mol H}$$

$$0.04926 \text{ g O} \cdot \frac{1 \text{ mol O}}{16.00 \text{ g O}} = 3.079 \times 10^{-3} \text{ mol O}$$

Establish a small whole-number ratio of the amounts of C,H,O:

Note that C and O are present in equi-molar amounts (3.08×10^{-3}). Dividing the number of moles of H by the moles of C (or O), gives a ratio of 3 H: 1 C or 3 H:1 O—for an empirical formula of $C_1H_3O_1$, more commonly written CH_3O,

If the molar mass if 62.1 g/mol, what is the molecular formula?

Adding the atomic weights of 1C, 3H, and 1 O gives approximately (12+3+16)=31 for an "empirical formula" weight.

The molecular formula is then : $\dfrac{62.1 \text{ g}}{1 \text{ mol}} \cdot \dfrac{1 \text{ empirical formula}}{31 \text{ g}} = \dfrac{2 \text{ empirical formulas}}{1 \text{ mol}}$ so the molecular formula is $C_2H_6O_2$.

35. Formula of the carbonyl compound formed with nickel:

$$Ni_x(CO)_y(s) + O_2 \rightarrow x\ NiO\ (s) + y\ CO_2(g)$$

Given the mass of NiO formed, we can calculate the mass of Ni in NiO (and also in the nickel carbonyl compound).

Quantity of Ni:

$$0.0426 \text{ g NiO} \cdot \frac{1 \text{ mol NiO}}{74.6924 \text{ g NiO}} = 5.70 \times 10^{-4} \text{ mol NiO}$$ (and an equal number of moles of nickel since the oxide has 1mol Ni:1mol NiO). That number of moles of Ni would have a mass of: $5.70 \times 10^{-4} \text{ mol Ni} \cdot \dfrac{58.693 \text{g Ni}}{1 \text{ mol Ni}} = 0.03347 \text{ g Ni}$.

Quantity of CO contained in the nickel carbonyl compound:

0.0973 g compound – 0.03347 g Ni = 0.06383 g CO

Amount of CO contained in the compound:

$$0.06383 \text{ g CO} \cdot \frac{1 \text{ mol CO}}{28.010 \text{ g CO}} = 2.279 \times 10^{-3} \text{ mol CO}$$

The ratio of Ni:CO is: $\dfrac{2.279 \times 10^{-3} \text{molCO}}{5.70 \times 10^{-4} \text{ mol Ni}} = 4$ and the empirical formula is $Ni(CO)_4$.

Solution Concentration

37. Molarity of Na_2CO_3 solution:

$$6.73 \text{ g } Na_2CO_3 \bullet \frac{1 \text{ mol } Na_2CO_3}{106.0 \text{ g } Na_2CO_3} = 0.0635 \text{ mol } Na_2CO_3$$

$$\text{Molarity} \equiv \frac{\# \text{ mol}}{\text{Liter}} = \frac{0.0635 \text{ mol } Na_2CO_3}{0.250 \text{ L}} = 0.254 \text{ M } Na_2CO_3$$

Concentration of Na^+ and CO_3^{2-} ions:

$$\frac{0.254 \text{ mol } Na_2CO_3}{L} \bullet \frac{2 \text{ mol } Na^+}{1 \text{ mol } Na_2CO_3} = 0.508 \text{ M } Na^+$$

$$\frac{0.254 \text{ mol } Na_2CO_3}{L} \bullet \frac{1 \text{ mol } CO_3^{2-}}{1 \text{ mol } Na_2CO_3} = 0.254 \text{ M } CO_3^{2-}$$

39. Mass of $KMnO_4$:

$$\frac{0.0125 \text{ mol } KMnO_4}{L} \bullet \frac{0.250 \text{ L}}{1} \bullet \frac{158.0 \text{ g } KMnO_4}{1 \text{ mol } KMnO_4} = 0.494 \text{ g } KMnO_4$$

41. Volume of 0.123 M NaOH to contain 25.0 g NaOH:

Calculate moles of NaOH in 25.0 g:

$$\frac{25.0 \text{ g NaOH}}{1} \bullet \frac{1 \text{ mol NaOH}}{40.00 \text{ g NaOH}} = 0.625 \text{ mol NaOH}$$

The volume of 0.123 M NaOH that contains 0.625 mol NaOH:

$$0.625 \text{ mol NaOH} \bullet \frac{1 \text{ L}}{0.123 \text{ mol NaOH}} \bullet \frac{1 \times 10^3 \text{ mL}}{1 L} = 5.08 \times 10^3 \text{ mL}$$

43. Identity and concentration of ions in each of the following solutions:

(a) 0.25 M $(NH_4)_2SO_4$ gives rise to (2 x 0.25) 0.50 M NH_4^+ ions & 0.25 M SO_4^{2-} ions

(b) 0.123 M Na_2CO_3 gives rise to (2 x 0.123) 0.246 M Na^+ ions & 0.123 M CO_3^{2-}.

(c) 0.056 M HNO_3 gives rise to 0.056 M H^+ ions & 0.056 M NO_3^- ions.

Preparing Solutions

45. Prepare 500. mL of 0.0200 M solution of Na_2CO_3:

Decide what amount (#moles) of sodium carbonate are needed:

$$\frac{0.0200 \text{ M } Na_2CO_3}{1 \text{ L}} \bullet \frac{0.500 \text{ L}}{1} = 0.0100 \text{ mol } Na_2CO_3$$

What mass does 0.0100 mol sodium carbonate have?

MW = 106.0 g Na_2CO_3/1 mol Na_2CO_3 and using 1/100 mol would require:

0.0100 mol Na_2CO_3 • 106.06 g/mol Na_2CO_3 or *1.06 g Na_2CO_3*.

To prepare the solution, take 1.06 g of Na_2CO_3, and transfer it to the volumetric flaks. Rinse all the solid into the flask. Add a bit of distilled water and stir carefully until all the solid Na_2CO_3 has dissolved. Once the solid has dissolved, add distilled water to the calibrated mark on the neck of the volumetric flask. Stopper and stir to assure complete mixing!

47. Molarity of HCl in the diluted solution: We can calculate the molarity if we know the number of moles of HCl in the 25.0 mL solution.

   1. Moles of HCl in 25.0 mL of 1.50 M HCl:

   $$M \times V = \frac{1.50 \text{ mol HCl}}{L} \bullet \frac{25.0 \times 10^{-3}L}{1} = 0.0375 \text{ mol HCl}$$

   2. When that number of moles is distributed in 500. mL: $\frac{0.0375 \text{ mol HCl}}{0.500 \text{ L}} = 0.0750$ M HCl

   Perhaps a shorter way to solve this problem is to note the number of moles (found by multiplying the original molarity times the volume) is distributed in a given volume, resulting in the diluted molarity. Mathematically: $M_1 \times V_1 = M_2 \times V_2$

49. Using the definition of molarity we can calculate the amount of H_2SO_4 in the following:

   $$\frac{0.125 \text{mol} H_2SO_4}{L} \bullet 1.00 \text{ L} = 0.125 \text{ moles of } H_2SO_4$$

   If we dilute 20.8 mL of 6.00 M H_2SO_4 to one liter we obtain:

   $$\frac{6.00 \text{ M } H_2S O_4}{1 \text{ L}} \bullet 0.0208L = 0.1248 \text{ mol } H_2SO_4 \text{ or a } 0.125 \text{ M solution of } H_2SO_4$$

   Method (a) is correct!

Serial Dilutions

51. The serial dilutions are all based upon the concept that one extracts a certain amount (moles of a substance), and moves it into another container with an added substance (typically solvent). One need only monitor the # of moles of substance extracted (concentration x volume). The first dilution dilutes a volume of 25.00 mL to a total volume of 100.00mL—a factor of 4. Hence the concentration of the original 0.136M HCl is reduced by a factor of 4—to 0.0340M HCl.

Mathematically this is expressed:

$$M_c \times V_c = (0.136 \text{mol/L} \cdot 25.00 \text{mL}) = M \times 100.00 \text{mL} \text{ or } \frac{(0.136 \text{mol/L} \cdot 25.00 \text{mL})}{100.00 \text{mL}} = 0.0340 M$$

Note from the expression above that the **ratios** of volumes provide us with the luxury of expressing the two volumes in any units (i.e. both L or both mL) as the units will cancel . The second dilution is a 10-fold dilution (10.00mL is diluted to 100.00mL), so the concentration will be reduced by a factor of ten.

The final concentration will be 1/10 • 0.0340M or 0.00340M (3.40 x 10^{-3}M)

Calculating and using pH

53. The hydrogen ion concentration of a wine whose pH = 3.40:

Since pH is defined as $-\log[H^+]$; $[H^+] = 10^{-3.40} = 4.0 \times 10^{-4}$ M H$^+$

The solution has a pH < 7.00, so it is **acidic**.

55. The $[H^+]$ and pH of a solution of 0.0013 M HNO$_3$:

Since nitric acid is a strong acid (and strong electrolyte), we can state that:

$[H^+]$ = 0.0013 M; the pH = - log (0.0013) or 2.89

57. Make the interconversions and decide if the solution is acidic or basic:

	pH	$[H^+]$	Acidic/Basic
(a)	**1.00**	0.10	Acidic
(b)	**10.50**	3.2 x 10^{-11} M	Basic
(c)	4.89	**1.3 x 10^{-5} M**	Acidic
(d)	7.64	**2.3 x 10^{-8} M**	Basic

pH values less than 7 indicate an acidic solution while those greater than 7 indicate a basic solution. Similarly solutions for which the $[H^+]$ is greater than 1.0 x 10^{-7} are acidic while those with hydrogen ion concentrations LESS THAN 1.0 x 10^{-7} are basic.

Stoichiometry of Reactions in Solution

59. Volume of 0.109 M HNO$_3$ to react with 2.50 g of Ba(OH)$_2$:

Need several steps:

1. Calculate mol of barium hydroxide in 2.50 g: $2.50 \text{ g Ba(OH)}_2 \cdot \dfrac{1 \text{ mol Ba(OH)}_2}{171.3 \text{ g Ba(OH)}_2}$

2. Calculate mole of HNO$_3$ needed to react with that # of mol of barium hydroxide

$$2.50 \text{ g Ba(OH)}_2 \cdot \frac{1 \text{ mol Ba(OH)}_2}{171.3 \text{ g Ba(OH)}_2} \cdot \frac{2 \text{ mol HNO}_3}{1 \text{ mol Ba(OH)}_2}$$

3. Calculate volume of 0.109 M HNO_3 that contains that # of mol of nitric acid.

$$2.50 \text{ g Ba(OH)}_2 \cdot \frac{1 \text{ mol Ba(OH)}_2}{171.3 \text{ g Ba(OH)}_2} \cdot \frac{2 \text{ mol HNO}_3}{1 \text{ mol Ba(OH)}_2} \cdot \frac{1 \text{ L}}{0.109 \text{ M HNO}_3} = 0.268 \text{ L}$$

or 268 mL.

61. Mass of NaOH formed from 15.0 L of 0.35 M NaCl:

$$\frac{0.35 \text{ mol NaCl}}{1 \text{ L}} \cdot \frac{15.0 \text{ L}}{1} \cdot \frac{2 \text{ mol NaOH}}{2 \text{ mol NaCl}} \cdot \frac{40.0 \text{ g NaOH}}{1 \text{ mol NaOH}} = 210 \text{ g NaOH}$$

Mass of Cl_2 obtainable:

$$\frac{0.35 \text{ mol NaCl}}{1 \text{ L}} \cdot \frac{15.0 \text{ L}}{1} \cdot \frac{1 \text{ mol Cl}_2}{2 \text{ mol NaCl}} \cdot \frac{70.9 \text{ g Cl}_2}{1 \text{ mol Cl}_2} = 186 \text{ g Cl}_2 \text{ or } 190 \text{ g Cl}_2 \text{ (2 sf)}$$

63. Volume of 0.0138M $Na_2S_2O_3$ to dissolve 0.225 g of AgBr:

Calculate:　1. mol of AgBr

　　　　　　2. mol of $Na_2S_2O_3$ needed to react (balanced equation)

　　　　　　3. volume of 0.0138 M $Na_2S_2O_3$ containing that number of moles.

$$0.225 \text{ g AgBr} \cdot \frac{1 \text{ mol AgBr}}{187.8 \text{ g AgBr}} \cdot \frac{2 \text{ mol Na}_2\text{S}_2\text{O}_3}{1 \text{ mol AgBr}} \cdot \frac{1 \text{ L}}{0.0138 \text{ mol Na}_2\text{S}_2\text{O}_3}$$
$$\cdot \frac{1000 \text{ mL}}{1 \text{ L}} = 174 \text{ mL Na}_2\text{S}_2\text{O}_3$$

65. The balanced equation:

$$Pb(NO_3)_2(aq) + 2 \text{ NaCl(aq)} \rightarrow PbCl_2(s) + 2 \text{ NaNO}_3(aq)$$

Volume of 0.750 M $Pb(NO_3)_2$ needed:

$$\frac{2.25 \text{ mol NaCl}}{1 \text{ L}} \cdot \frac{1.00 \text{ L}}{1} \cdot \frac{1 \text{ mol Pb(NO}_3)_2}{2 \text{ mol NaCl}} \cdot \frac{1 \text{ L}}{0.750 \text{ mol Pb(NO}_3)_2} \cdot \frac{1000 \text{ mL}}{1 \text{ L}}$$

$$= 1500 \text{ mL or } 1.50 \times 10^3 \text{ mL}$$

Titrations

67. To calculate the volume of 0.812 M HCl needed, we calculate the moles of NaOH in 1.45 g, then use the stoichiometry of the balanced equation:

$$HCl(aq) + NaOH(aq) \rightarrow NaCl(aq) + H_2O(l)$$

$$1.45 \text{ g NaOH} \cdot \frac{1 \text{ mol NaOH}}{40.00 \text{ g NaOH}} \cdot \frac{1 \text{ mol HCl}}{1 \text{ mol NaOH}} \cdot \frac{1 \text{ L}}{0.812 \text{ mol HCl}} \cdot \frac{1000 \text{ mL}}{1 \text{ L}} = 44.6 \text{ mL HCl}$$

69. Calculate:

    1. moles of Na_2CO_3 corresponding to 2.150 g Na_2CO_3

    2. moles of HCl that react with that number of moles (using the balanced equation)

    3. the molarity of HCl containing that number of moles of HCl in 38.55 mL.

The balanced equation is:

$$Na_2CO_3(aq) + 2\,HCl(aq) \rightarrow 2\,NaCl(aq) + H_2O(l) + CO_2(g)$$

$$2.150\ g\ Na_2CO_3 \cdot \frac{1\ mol\ Na_2CO_3}{106.0\ g\ Na_2CO_3} \cdot \frac{2\ mol\ HCl}{1\ mol\ Na_2CO_3} \cdot \frac{1}{0.03855\ L\ sol} = 1.052\ M\ HCl$$

71. Molar Mass of an acid if 36.04 mL of 0.509 M NaOH will titrate 0.954 g of an acid H_2A:

Note that the acid is a diprotic acid, and will require 2 mol of NaOH for each mol of the acid.

Calculate the # moles of NaOH: $\dfrac{0.509\ mol\ NaOH}{1\ L} \cdot 0.03604\ L = 0.0183\ mol\ NaOH$

The number of moles of acid will be **half** of the number of moles of NaOH:

$$0.0183\ mol\ NaOH \cdot \frac{1\ mol\ H_2A}{2\ mol\ NaOH} = 0.00917\ mol\ H_2A$$

Since we know the mass corresponding to this number of moles of acid, we can calculate the

molar mass (# g/mol): $\dfrac{0.954\ g\ H_2A}{0.00917\ mol\ H_2A} = 104\ g/mol$

73. Mass percent of iron in a 0.598 gram sample that requires 22.25 mL of 0.0123 M $KMnO_4$:

Note from the balanced, net ionic equation that 1 mol MnO_4^- requires 5 mol of Fe^{2+}.

The # mol of MnO_4^- is: $\dfrac{0.0123\ M\ MnO_4^-}{1\ L} \cdot 0.02225\ L = 2.74 \times 10^{-4}\ mol\ MnO_4^-$

Using the ratio of permanganate ion to iron(II) ion we obtain:

$$2.74 \times 10^{-4}\ mol\ MnO_4^- \cdot \frac{5\ mol\ Fe^{2+}}{1\ mol\ MnO_4^-} = 1.37 \times 10^{-3}\ mol\ Fe^{2+}$$

Using the mass of the iron(II) ion, we can calculate the mass of iron present:

$$1.37 \times 10^{-3}\ mol\ Fe^{2+} \cdot \frac{55.847\ g\ Fe^{2+}}{1\ mol\ Fe^{2+}} = 0.0764\ g\ Fe^{2+}$$

The mass percent of iron is then: $\dfrac{0.0764\ g\ Fe^{2+}}{0.598\ g\ sample} \cdot 100 = 12.8\ \%\ Fe^{2+}$

Spectrophotometry

75. The following data were collected for a dye:

Dye Concentration (x 10^6)M	Absorbance (at 475 nm)
0.50	0.24
1.5	0.36
2.5	0.44
3.5	0.59
4.5	0.70

(a) The calibration plot, slope and intercept:

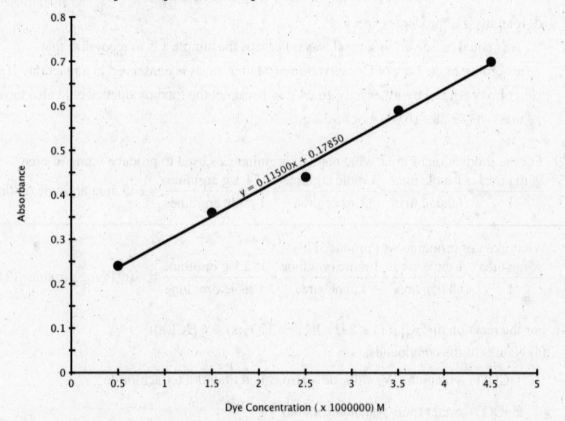

The equation for the best-fit straight line is: $y = 0.115x + 0.1785$. Note that this graph has the dye concentrations multiplied by 10^6. Multiplying the slope (above) by this factor gives a slope of 1.2×10^5 with an intercept of 0.18.

(b) The dye concentration in a solution with A = 0.52:

One can calculate the value by substituting into the equation and rearranging:

$0.52 = 0.115x + 0.179$, so $(0.52 - 0.179)/0.115 = 3.0 \times 10^{-6}$ (to 2 sf).

[Recall that all concentrations had been multiplied by the factor (10^6)

GENERAL QUESTIONS ON STOICHIOMETRY

77. For the reaction of benzene with oxygen:

(a) the products of the reaction:

Combination of C_6H_6 with O_2 gives the oxide of C and the oxygen of H:

$CO_2(g) + H_2O(g)$

(b) the balanced equation for the reaction is:

$2\ C_6H_6\ (l) + 15\ O_2(g) \rightarrow 12\ CO_2(g) + 6\ H_2O(g)$

(c) mass of oxygen, in grams, needed to completely consume the 16.04 g C_6H_6.

$$16.04\ \text{g}\ C_6H_6 \cdot \frac{1\ \text{mol}\ C_6H_6}{78.11\ \text{g}\ C_6H_6} \cdot \frac{15\ \text{mol}\ O_2}{2\ \text{mol}\ C_6H_6} \cdot \frac{31.999\ \text{g}\ O_2}{1\ \text{mol}\ O_2} = 49.28\ \text{g}\ O_2\ .$$

(d) total mass of products expected:

This could be solved in several ways. Perhaps the simplest is to recognize that—according to the Law of Conservation of Matter, mass is conserved in a reaction. If 49.28 g of oxygen react with exactly 16.04 g of benzene, the total products would also have a mass of (49.28 +16.04) g or 65.32 g.

79. For the production of urea, what mass of arginine was used to produce 95mg of urea?

$$\frac{95\text{mg urea}}{1} \cdot \frac{1\ \text{mole urea}}{60.05\text{g urea}} \cdot \frac{1\ \text{mole arginine}}{1\ \text{mole urea}} \cdot \frac{174.1\ \text{g arginine}}{1\ \text{mole arginine}} = 275.5\text{mg arginine (280mg to 2 sf)}$$

What mass of ornithine was produced?

$$\frac{95\text{mg urea}}{1} \cdot \frac{1\ \text{mole urea}}{60.05\text{g urea}} \cdot \frac{1\ \text{mole ornithine}}{1\ \text{mole urea}} \cdot \frac{132.1\ \text{g ornithine}}{1\ \text{mole ornithine}} = 209.06\text{mg ornithine (210mg to 2 sf)}$$

81. For the reaction of $TiCl_4(l) + 2\ H_2O(l) \rightarrow TiO_2(s) + 4\ HCl(g)$:

(a) Names of the compounds:

$TiCl_4(l)$ – titanium(IV) chloride—also called titanium tetrachloride

$H_2O(l)$ – water (non-systematic name)

$TiO_2(s)$ – titanium(IV) oxide—also known as titanium dioxide

$HCl(g)$ – hydrogen chloride (the aqueous form is known as hydrochloric acid)

(b) Mass of water to react with 14.0 mL of (d = 1.73 g/mL)

$$\frac{14.0\ \text{mL}\ TiCl_4}{1} \cdot \frac{1.73\ \text{g}\ TiCl_4}{1\ \text{mL}\ TiCl_4} \cdot \frac{1\ \text{mol}\ TiCl_4}{189.7\text{g}\ TiCl_4} \cdot \frac{2\ \text{mol}\ H_2O}{1\ \text{mol}\ TiCl_4} \cdot \frac{18.02\ \text{g}\ H_2O}{1\ \text{mol}\ H_2O} = 4.60\ \text{g}\ H_2O$$

(c) Mass of products expected:

$$\frac{14.0 \text{ mL TiCl}_4}{1} \cdot \frac{1.73 \text{ g TiCl}_4}{1 \text{ mL TiCl}_4} \cdot \frac{1 \text{ mol TiCl}_4}{189.7 \text{ g TiCl}_4} \cdot \frac{1 \text{ mol TiO}_2}{1 \text{ mol TiCl}_4} \cdot \frac{79.866 \text{ g TiO}_2}{1 \text{ mol TiO}_2} = 10.2 \text{ g TiO}_2$$

$$\frac{14.0 \text{ mL TiCl}_4}{1} \cdot \frac{1.73 \text{ g TiCl}_4}{1 \text{ mL TiCl}_4} \cdot \frac{1 \text{ mol TiCl}_4}{189.7 \text{ g TiCl}_4} \cdot \frac{4 \text{ mol HCl}}{1 \text{ mol TiCl}_4} \cdot \frac{36.46 \text{ g HCl}}{1 \text{ mol HCl}} = 18.6 \text{ g HCl}$$

83. Production of Sodium azide: $NaNO_3 + 3 \ NaNH_2 \rightarrow NaN_3 + 3 \ NaOH + NH_3$

Mass of NaN_3 produced when 15.0 g of $NaNO_3$ (85.0 g/mol) reacts with 15.0 g of $NaNH_2$:

$$\frac{15.0 \text{ g NaNO}_3}{1} \cdot \frac{1 \text{ mol NaNO}_3}{85.0 \text{ g NaNO}_3} = 0.176 \text{ mol NaNO}_3$$

$$\frac{15.0 \text{ g NaNH}_2}{1} \cdot \frac{1 \text{ mol NaNH}_2}{39.0 \text{ g NaNH}_2} = 0.385 \text{ mol NaNH}_2$$

Determine the limiting reagent:

Required ratio: $\dfrac{3 \text{ mol NaNH}_2}{1 \text{ mol NaNO}_3} = 3$ Available ratio: $\dfrac{0.385 \text{ mol NaNH}_2}{0.176 \text{ mol NaNO}_3} = 2.19$

The ratios indicate that $NaNH_2$ is the limiting reagent.

$$\frac{0.385 \text{ mol NaNH}_2}{1} \cdot \frac{1 \text{ mol NaN}_3}{3 \text{ mol NaNH}_2} \cdot \frac{65.0 \text{ g NaN}_3}{1 \text{ mol NaN}_3} = 8.33 \text{ g NaN}_3$$

85. Determine the mass percent of saccharin in the sample of sweetener:

1. Determine the S in $BaSO_4$:
$$\frac{0.2070 \text{ g BaSO}_4}{1} \cdot \frac{32.066 \text{ g S}}{223.393 \text{ g BaSO}_4} = 0.02844 \text{ g S} \ (4 \text{ sf})$$

2. Determine the mass of saccharin that contains this mass of S:
$$\frac{0.02844 \text{ g S}}{1} \cdot \frac{183.17 \text{ g saccharin}}{32.066 \text{ g S}} = 0.1625 \text{ g saccharin}$$

3. The mass percent of saccharin is then:
$$\frac{0.1625 \text{ g saccharin}}{0.2140 \text{ g sweetener}} \cdot 100 = 75.92 \ \% \text{ saccharin}$$

87. To determine the empirical formula, we need the ratios of silicon and hydrogen atoms in the compound. Knowing that 6.22 g of the compound produced 11.64 g SiO_2 let's calculate the moles of silicon in that 11.64 g.

$$\frac{11.64 \text{ g SiO}_2}{1} \cdot \frac{1 \text{ mol SiO}_2}{60.0843 \text{ g SiO}_2} \cdot \frac{1 \text{ mol Si}}{1 \text{ mol SiO}_2} = 0.1937 \text{ mol Si}$$

The mass of water formed can be used to calculate the amount of H in the compound.

$$\frac{6.980 \text{ g H}_2\text{O}}{1} \cdot \frac{1 \text{ mol H}_2\text{O}}{18.01 \text{ g H}_2\text{O}} \cdot \frac{2 \text{ mol H}}{1 \text{ mol H}_2\text{O}} = 0.7749 \text{ mol H}$$

Now we can determine the ratio of the moles of atoms of the elements present:

$$\frac{0.7749 \text{ mol H}}{0.1937 \text{ mol Si}} = 4.00$$

The ratio of Si:H is 1 mol Si:4 mol H, giving an empirical formula of SiH_4.

89. Empirical formula of quinone:

Similarly to the SQ above (87) we calculate the mass of C and H in the original compound. Knowing the total mass, and the masses of C and H allows one to calculate (by difference) the amount of O in the compound.

$$\frac{0.257 \text{ g CO}_2}{1} \cdot \frac{12.01 \text{ g C}}{44.01 \text{ g CO}_2} = 0.0701 \text{ g C}$$

$$\frac{0.0350 \text{ g H}_2\text{O}}{1} \cdot \frac{2.016 \text{ g H}}{18.01 \text{ g H}_2\text{O}} = 0.00392 \text{ g H}$$

The amount of O present is: 0.105g compound – (0.0701g C + 0.00392g H)= 0.03095 g O

Converting each of these masses to moles yields:

$$\frac{0.03095 \text{g O}}{1} \cdot \frac{1 \text{ mol O}}{15.9994 \text{ g O}} = 0.00193 \text{ mol O}$$

$$\frac{0.0701 \text{g C}}{1} \cdot \frac{1 \text{ mol C}}{12.011 \text{ g C}} = 0.00584 \text{ mol C} \text{ and for H}$$

$$\frac{0.00392 \text{ g H}}{1} \cdot \frac{1 \text{ mol H}}{1.00797 \text{ g H}} = 0.003889 \text{ mol H}$$

The ratio of C:H:O is 0.00584: 0.003889: 0.00193, and dividing by the smallest number (to obtain integral values): 3:2:1 for an empirical formula of $C_3H_2O_1$

91. Theoretical yield of sulfuric acid from 3.00 kg of Cu_2S:

First determine the # mol of cuprite, then the # mol of H_2SO_4 (since there 1 H_2SO_4 molecule per molecule of cuprite). The mass of H_2SO_4 is then determined from the # mol of the acid.

$$\frac{3.00 \times 10^3 g \text{ Cu}_2\text{S}}{1} \cdot \frac{1 \text{ mol Cu}_2\text{S}}{159.2 \text{ g Cu}_2\text{S}} \cdot \frac{1 \text{ mol H}_2\text{SO}_4}{1 \text{ mol Cu}_2\text{S}} \cdot \frac{98.08 \text{ g H}_2\text{SO}_4}{1 \text{ mol H}_2\text{SO}_4} = 1848 \text{ g H}_2\text{SO}_4$$

or 1.85 kg H_2SO_4 (to 3 s.f.)

93. A metal forms an oxide, MO_2.

The mass of oxide (0.452 g) – mass of metal (0.356 g) = 0.096 g of oxygen.

$$\frac{0.096 \text{ g O}}{1} \cdot \frac{1 \text{ mol O}}{15.9994 \text{ g O}} \cdot \frac{1 \text{ mol M}}{2 \text{ mol O}} = 0.00300 \text{ mol M}$$

The third factor is arrived at by the formula (given in the problem).

We know that this number of moles (0.00300 mol) corresponds to 0.356 g, so we can calculate the approximately atomic weight of the metal: $\dfrac{0.356 \text{ g M}}{0.00300 \text{ mol M}} = 119$ g/mol .

The metal with atomic weight of approximately 119 is tin, Sn.

95. The problem seems complex owing to the series of reactions that must occur to produce the desired product: $KClO_4$. Ask the question how do I produce $KClO_4$? The answer is with the third equation given:

$4 \text{ KClO}_3 \rightarrow 3 \text{ KClO}_4 + \text{KCl}$. To produce $KClO_4$ requires $KClO_3$, which can be produced with the 2nd equation given. To simplify the OVERALL equation, multiply that equation by 4:

$12 \text{ KClO} \rightarrow 8 \text{ KCl} + 4 \text{ KClO}_3$. To produce KClO however, we need to carry out the first equation. Since we need 12 KClO in the OVERALL equation, let's multiply that equation by 12:

$12 \text{ Cl}_2 + 24 \text{ KOH} \rightarrow 12 \text{ KCl} + 12 \text{ KClO} + 12 \text{ H}_2\text{O}$. Add these 3 equations:

$4 \cancel{\text{KClO}_3} \rightarrow 3 \text{ KClO}_4 + \text{KCl}$

$12 \cancel{\text{KClO}} \rightarrow 8 \text{ KCl} + 4 \cancel{\text{KClO}_3}$

$\underline{12 \text{ Cl}_2 + 24 \text{ KOH} \rightarrow 12 \text{ KCl} + 12 \cancel{\text{KClO}} + 12 \text{ H}_2\text{O}}$ producing an overall equation:

$12 \text{ Cl}_2 + 24 \text{ KOH} \rightarrow 3 \text{ KClO}_4 + 21 \text{ KCl} + 12 \text{ H}_2\text{O}$ and noting that all coefficients are divisible by 3, the equation reduces to:

$4 \text{ Cl}_2 + 8 \text{ KOH} \rightarrow 1 \text{ KClO}_4 + 7 \text{ KCl} + 4 \text{ H}_2\text{O}$. This equation tells us that each mol of $KClO_4$ requires 4 mol of Cl_2. With this factor, we can write the mathematical expression:

$$\dfrac{234 \times 10^3 \text{g KClO}_4}{1} \bullet \dfrac{1 \text{ mol KClO}_4}{138.55 \text{ g KClO}_4} \bullet \dfrac{4 \text{ mol Cl}_2}{1 \text{ mol KClO}_4} \bullet \dfrac{70.905 \text{ g Cl}_2}{1 \text{ mol Cl}_2} = 479 \times 10^3 \text{g Cl}_2$$

or 479 kg of Cl_2.

97. Mass of lime obtainable from 125 kg of limestone:

The concentration of CaO in the limestone is 95.0% (95.0 g of CaO per 100.0 g of limestone).

We can find the amount of $CaCO_3$ in 125 kg of limestone, and then the amount of CaO in $CaCO_3$

$$\dfrac{125 \times 10^3 \text{ g limestone}}{1} \bullet \dfrac{95.0 \text{ g CaCO}_3}{100.0 \text{ g limestone}} \bullet \dfrac{56.07 \text{ g CaO}}{100.1 \text{ g CaCO}_3} = 66.5 \times 10^3 \text{ g CaO}$$

or 66.5 kg lime

99. A mixture of butene, C_4H_8, and butane, C_4H_{10}, has a mass if 2.86 g.

8.80 g of CO_2 and 4.14 g of H_2O result upon combustion.

What is the weight percent of butene and butane in the mixture? The balanced equations for

the combustions: Butene: $C_4H_8(g) + 6\ O_2(g) \rightarrow 4\ CO_2(g) + 4\ H_2O(g)$

Butane: $2\ C_4H_{10}(g) + 13\ O_2(g) \rightarrow 8\ CO_2(g) + 10\ H_2O(g)$

Establish 2 equations with 2 unknowns: Let x = g C_4H_8 and y = g C_4H_{10}

Then x + y = 2.86 g and

$$\frac{x}{56.1072\ g\ C_4H_8} \bullet \frac{4\ mol\ CO_2}{1\ mol\ C_4H_8} + \frac{y}{58.123\ g\ C_4H_{10}} \bullet \frac{8\ mol\ CO_2}{2\ mol\ C_4H_{10}} = 8.8\ g\ CO_2 \bullet \frac{1\ mol\ CO_2}{44.01\ g\ CO_2}$$

Note that the fractions $\dfrac{4\ mol\ CO_2}{1\ mol\ C_4H_8}$ and $\dfrac{8\ mol\ CO_2}{2\ mol\ C_4H_{10}}$ result from the stoichiometry of the

two combustion equations. Rearrange the first equation: y = (2.86 – x) and substitute into the

2nd equation:

$$\frac{x}{56.1072\ g\ C_4H_8} \bullet \frac{4\ mol\ CO_2}{1\ mol\ C_4H_8} + \frac{(2.86 - x)}{58.123\ g\ C_4H_{10}} \bullet \frac{8\ mol\ CO_2}{2\ mol\ C_4H_{10}} = 8.8\ g\ CO_2 \bullet \frac{1\ mol\ CO_2}{44.01\ g\ CO_2}$$

Grouping terms we get:

$$\frac{4\ x}{56.1072\ g\ C_4H_8} + \frac{8 \bullet (2.86 - x)}{2 \bullet 58.123\ g\ C_4H_{10}} = \frac{8.8\ g\ CO_2}{44.01\ g\ CO_2}\ \text{solving gives}$$

x = 1.27 g C_4H_8 and (2.86 – x) or 1.59 g C_4H_{10}

101. The weight percent of CuS and Cu_2S in the ore:

We need to determine the mass of Cu originally contained in CuS (and similarly in Cu_2S).

Knowing that the ore is 11.0% impure let's us write: 100.0g ore – 11.0g impurity = 89.0 g of

Cu_xS. We are also told that the mass of **pure** copper (when reduced) is 89.5% of 75.4 g Cu.

(or 67.48 g). Since we have two unknowns (the mass of CuS and the mass of Cu_2S) we need

two equations:

(1) Cu (from CuS) + Cu (from Cu_2S) = 67.48 g.

(2) If we let x = mass of CuS, then 89.0 – x = mass of Cu_2S.

Further we know that the % of Cu in each of the salts can be calculated from the

formulas, and represented by (mass Cu)/(mass Cu_xS salt).

Rewriting equation (1):

$$\frac{63.546\ g\ Cu}{95.612\ g\ CuS} \bullet x + \frac{127.10\ g\ Cu}{159.158\ g\ Cu_2S} \bullet (89.0 - x) = 67.48 .$$

[For those of us who like to keep track of units, note that x and the term (89.0-x) will have units of g Cu salt, so those units will "cancel" with the denominator of the respective fractions.] For simplicity's sake let's reduce the two fractions to a decimal:

$0.6646 x + 0.7985(89.0 - x) = 67.48$ and $0.6646 x + 71.067 - 0.7985 x = 67.48$

Combining and simplifying: $-0.1339 x = -3.587$ and $x = 3.587/0.1339$ or 26.79 g CuS. Knowing the mass of CuS, the mass of $Cu_2S = 89.0 - 26.79 = 62.2$ g Cu_2S

The weight percent of CuS in the ORE: $= 26.8\%$ and of $Cu_2S = 62.2\%$

103. Reaction: $MgCO_3(s) + 2 HCl(aq) \rightarrow CO_2(g) + MgCl_2(aq) + H_2O(l)$

(a) The net ionic equation: $MgCO_3(s) + 2 H^+(aq) \rightarrow CO_2(g) + Mg^{2+}(aq) + H_2O(l)$

The spectator ion is the chloride ion (Cl^-).

(b) The production of $CO_2(g)$ characterizes this as a gas-forming reaction.

(c) Mass of $MgCO_3(s)$ to react with 125 mL of HCl whose pH = 1.56:

The solution with pH = 1.56 has a $[H^+]$ of $10^{-1.56}$ or 2.75×10^{-2} M.

$$\frac{0.0275 \text{ mol } H^+}{1 \text{ L}} \cdot \frac{0.125 \text{ L}}{1} \cdot \frac{1 \text{ mol } MgCO_3}{2 \text{ mol } H^+} \cdot \frac{84.31 \text{g } MgCO_3}{1 \text{ mol } MgCO_3} = 0.15 \text{ g } MgCO_3$$

Since the pH given only has 2 sf, our answer has 2.

105. Limiting reagent between 125 mL of 0.15 M CH_3CO_2H and 15.0 g $NaHCO_3$:

moles CH_3CO_2H: $\frac{0.15 \text{ mol } CH_3CO_2H}{1L} \cdot 0.125 \text{ L} = 0.01875 \text{ mol } CH_3CO_2H$

moles $NaHCO_3$: $15.0 \text{ g } NaHCO_3 \cdot \frac{1 \text{ mol } NaHCO_3}{84.01 \text{ g } NaHCO_3} = 0.179 \text{ mol } NaHCO_3$

Since the reaction proceeds with a 1:1 stoichiometry, *CH_3CO_2H is the limiting reagent*. The reaction will also produce 1 mol of $NaCH_3CO_2$ for each mol of CH_3CO_2H.

$$\frac{0.01875 \text{ mol } CH_3CO_2H}{1} \cdot \frac{1 \text{ mol } NaCH_3CO_2}{1 \text{ mol } CH_3CO_2H} \cdot \frac{82.03 \text{ g } NaCH_3CO_2}{1 \text{ mol } NaCH_3CO_2} = 1.5 \text{ g } NaCH_3CO_2$$

107. Weight percent of $Na_2S_2O_3$ in 3.232 g sample of material:

$$\frac{0.246 \text{ mol } I_2}{1 \text{ L}} \cdot \frac{0.04021 \text{ L}}{1} \cdot \frac{2 \text{ mol } Na_2S_2O_3}{1 \text{ mol } I_2} \cdot \frac{158.11 \text{ g } Na_2S_2O_3}{1 \text{ mol } Na_2S_2O_3} = 3.13 \text{ g } Na_2S_2O_3$$

Now that we know the amount of sodium thiosulfate in the sample, we can calculate the percent of the compound in the impure mixture: $\frac{3.13 \text{ g } Na_2S_2O_3}{3.232 \text{ g mixture}} \cdot 100 = 96.8 \%$

109. For a solution of HCl:

   (a) the pH of a 0.105 M HCl solution: Since HCl is considered a strong acid, the concentration of the hydronium ion will also be 0.105 M, and the pH= -log[0.105] or 0.979

   (b) What is the hydronium ion concentration of a solution with pH = 2.56? Since the pH= 2.56, $[H_3O^+]= 10^{-2.56}$ or 2.8×10^{-3}, an acidic solution

   (c) Solution has a pH of 9.67:

   Hydronium ion concentration = $10^{-9.67}$ or 2.1×10^{-10} M

   Is solution acidic or basic? With a pH greater than 7, this solution is considered **basic**

   (d) pH of solution formed by diluting 10.0-mL of 2.56M HCl to 250. mL:

   As we are simply diluting the solution, the HCl solution will have an amount of HCl that is: M x V = 2.56 mol/L x 0.010 L or 0.0256 mol HCl

   This amount will be contained in 250. mL , so a concentration of 0.0256 mol/0.250 L or 0.102 M, and a pH of -log[0.102] or 0.990 (close to that of the solution in part (a))

111. Much like SQ 109, you need to keep track of the hydronium ion being contributed by the two solutions. Calculate the amount of H_3O^+ in each of the solutions, add those amounts, and determine the concentration in 750. mL:

$$\frac{(0.500L) \bullet \left(2.50\ ^{mol}\!/_L\right) + (0.250L) \bullet \left(3.75\,^{mol}\!/_L\right)}{0.750\ L} = \frac{1.25\ mol + 0.925\ mol}{0.750\ L} = 2.90\ M$$

   The pH is then: -log[2.90] = -0.462

113. 2.56 g $CaCO_3$ in a beaker containing 250. mL 0.125 M HCl.

   After reaction, does any $CaCO_3$ remain?

   The amount of HCl present is: $\dfrac{0.125\ mol\ HCl}{1\ L} \bullet \dfrac{0.250\ L}{1} = 0.0313\ mol\ HCl$

   The amount of HCl that reacts with the $CaCO_3$ corresponding to 2.56 g is:

   $\dfrac{2.56\ g\ CaCO_3}{1} \bullet \dfrac{1\ mol\ CaCO_3}{100.09\ g\ CaCO_3} \bullet \dfrac{2\ mol\ HCl}{1\ mol\ CaCO_3} = 0.0512\ mol\ HCl$

   This calculation shows that the amount of HCl present is NOT SUFFICIENT to consume the $CaCO_3$ corresponding to 2.56 g of the solid.

   Mass of $CaCl_2$ produced:

   Since HCl is the limiting reagent, the amount of $CaCl_2$ that can be produced is:

   $\dfrac{0.0313\ mol\ HCl}{1} \bullet \dfrac{1\ mol\ CaCl_2}{2\ mol\ HCl} \bullet \dfrac{110.98\ g\ CaCl_2}{1\ mol\ CaCl_2} = 1.73\ g\ CaCl_2$

115. Determine the # mol of methylene blue:

$$\frac{1.0 \text{ g C}_{16}\text{H}_{18}\text{ClN}_3\text{S}}{1} \cdot \frac{1 \text{ mol C}_{16}\text{H}_{18}\text{ClN}_3\text{S}}{319.86 \text{ g C}_{16}\text{H}_{18}\text{ClN}_3\text{S}} = 0.0031 \text{ mol C}_{16}\text{H}_{18}\text{ClN}_3\text{S}$$

The measurement indicates that the molar concentration of methylene blue is 4.1×10^{-8} M. The concentration of methylene blue in the measured sample will be equal to that of methylene blue in the pool--assuming that the methylene blue has been uniformly distributed throughout the pool water before the sample was taken. Recalling that M = # mol/V, and knowing that we know BOTH the molarity and the # mol, we can calculate the volume.

$$\frac{0.0031 \text{ mol methylene blue}}{1} \cdot \frac{1 \text{ L}}{4.1 \times 10^{-8} \text{ mol methylene blue}} = 76,000 \text{ L (2 sf)}$$

Note that we ignored the 50 mL of solution that originally contained the 1.0 g of methylene blue.

117. For the reaction in which Au is dissolved by treatment with sodium cyanide:
    (a) Oxidizing agent: elemental O_2 Reducing agent: elemental Au

     Substance oxidized: elemental Au (oxidation state changes from 0 to +1)
     Substance reduced: elemental O_2 (oxidation state changes from 0 to -2)

    (b) Volume of 0.075 M NaCN to extract gold from 1000 kg of rock:

     Mass of gold in the rock: $\dfrac{0.019 \text{ g gold}}{100 \text{ g rock}} \cdot \dfrac{10^3 \text{ g rock}}{1 \text{ kg rock}} \cdot 10^3 \text{ kg rock} = 190 \text{ g gold}$

     Amount of NaCN needed (from the balanced equation):

     $190 \text{ g Au} \cdot \dfrac{1 \text{ mol Au}}{196.9 \text{ g Au}} \cdot \dfrac{8 \text{ mol NaCN}}{4 \text{ mol Au}} = 1.9 \text{ mol NaCN}$

     Volume of 0.075 M NaCN that contains that amount of NaCN:

     $1.9 \text{ mol NaCN} \cdot \dfrac{1 \text{ L}}{0.075 \text{ mol NaCN}} = 26 \text{ L}$

119. The % atom economy for the desired product, $CH_3CH_2CH_2CH_2Br$:

     We need the molar masses for all atoms used in the reaction:

     Alcohol [(4 x C) + (10 x H) + (1 x O)] = 74

     NaBr [(1 x Na) + (1 x Br)] = 103

     H_2SO_4 [(2 x H) + (1 x S) + (4 x O)] = 98   Adding these gives 275

     Calculating the molar mass of the desired product: [(4 x C) + (9 x H) + (1 x Br)] = 137

     So the % atom economy = $\dfrac{137}{275}$ = 0.498, and converting to percent gives 49.8%

IN THE LABORATORY

121. The concentration of the resulting solution can be determined by two identical processes.

First dilution: $\dfrac{0.110 \text{ mol Na}_2\text{CO}_3}{1 \text{ L}} \bullet \dfrac{25.0 \text{ mL}}{100.0 \text{ mL}} = 0.0275 \dfrac{\text{mol Na}_2\text{CO}_3}{\text{L}}$

Note that the ratio of volumes can both be expressed in units of milliliters, without a conversion to units of Liters.

Second dilution: $\dfrac{0.0275 \text{ mol Na}_2\text{CO}_3}{1 \text{ L}} \bullet \dfrac{10.0 \text{ mL}}{250. \text{ mL}} = \dfrac{1.10 \times 10^{-3} \text{mol Na}_2\text{CO}_3}{\text{L}}$

123. Weight percent of Cu in 0.251 g of a copper containing alloy:
    (a) Oxidizing and reducing agents in the two equations:

$$2 \text{ Cu}^{2+}(aq) + 5 \text{ I}^- (aq) \rightarrow 2 \text{ CuI}(s) + \text{I}_3^-(aq)$$

I^- reduces the Cu(II) ion to the Cu(I) ion—so I^- acts as the reducing agent.

Cu^{2+} oxidizes I^- to I_3^- (I_2 and I^-)–and acts as the oxidizing agent.

$$\text{I}_3^-(aq) + 2 \text{ S}_2\text{O}_3^{2-}(aq) \rightarrow \text{S}_4\text{O}_6^{2-}(aq) + 3\text{I}^-(aq)$$

$\text{S}_2\text{O}_3^{2-}$ reduces the I_3^- ion(I_2 and I^-)–to the I^- ion—so $\text{S}_2\text{O}_3^{2-}$ acts as the reducing agent.

I_3^- oxidizes $\text{S}_2\text{O}_3^{2-}$ to $\text{S}_4\text{O}_6^{2-}$ –and acts as the oxidizing agent.

In $\text{S}_4\text{O}_6^{2-}$, the oxidation state of S can be thought of as (+2.5), and in $\text{S}_2\text{O}_3^{2-}$ as (+2)

(b) Several steps in this problem:
 (1) Note the relationship between Cu (Cu^{2+}) and I_3^- formed (2 mol Cu^{2+}:1 mol I_3^-)
 (2) Excess I_3^- reacts with $\text{S}_2\text{O}_3^{2-}$ in a 1 mol I_3^- to 2 mol $\text{S}_2\text{O}_3^{2-}$ ratio
 (3) Calculate the amount of thiosulfate in 26.32 mL of 0.101 M solution:

$\dfrac{0.101 \text{ mol S}_2\text{O}_3^{2-}}{1 \text{ L}} \bullet 0.02632 \text{ L} = 0.00266 \text{ mol S}_2\text{O}_3^{2-}$ (to 3 sf)

(4) This amount can now be used to calculate I_3^- produced (step 2), and the amount of I_3^- related to the amount of Cu present (step 1):

$0.00266 \text{ mol S}_2\text{O}_3^{2-} \bullet \dfrac{1 \text{ mol I}_3^-}{2 \text{ mol S}_2\text{O}_3^{2-}} \bullet \dfrac{2 \text{ mol Cu}^{2+}}{1 \text{ mol I}_3^-} \bullet \dfrac{1 \text{ mol Cu}}{1 \text{ mol Cu}^{2+}} \bullet \dfrac{63.546 \text{ g Cu}}{1 \text{ mol Cu}} = 0.169 \text{ g Cu}$

The weight percent is then: $\dfrac{0.169 \text{ g Cu}}{0.251 \text{ g alloy}} \bullet 100 = 67.3 \text{ \% Cu}$

125. The equation: $Cr(NH_3)_xCl_3(aq) + x\ HCl\ (aq) \rightarrow x\ NH_4^+(aq) + Cr^{3+}(aq) + (x+3)Cl^-(aq)$

The amount of HCl: 24.26 mL of 1.500 M HCl (1.500 mol/L x 0.02426L) = 0.03639 mol

We don't know the ratio between moles of salt and moles of HCl. What we do know is

that for each mol of HCl that reacts, 1 mol of NH_3 is present in the salt (and 1 mol of NH_4^+

is produced in the reaction). So knowing the amount of NH_3 we can calculate its mass.

$$\frac{0.03639\ mol\ NH_3}{1} \bullet \frac{17.030\ g\ NH_3}{1\ mol\ NH_3} = 0.6197\ g\ NH_3$$

Given the mass of the salt (1.580 g), we can determine the mass of the remaining

components (Cr and Cl). 1.580 g $CrCl_3$ – 0.6197 g NH_3 = 0.9603 g $CrCl_3$.

Now calculate the amount of the $CrCl_3$ salt: MM = 51.9961 + 3(35.4527) = 158.3542 g/mol

$$\frac{0.9603\ g\ CrCl_3}{1} \bullet \frac{1\ mol\ CrCl_3}{158.3542\ g\ CrCl_3} = 6.064\ x\ 10^{-3}\ mol\ CrCl_3$$

Now we can determine the ratio of ammonia to chromium(III) chloride:

$$\frac{3.639x10^{-2}mol\ NH_3}{6.064x10^{-3}mol\ CrCl_3} = 6.000,\ \text{so the formula for the salt is: } Cr(NH_3)_6Cl_3.$$

127. A 1.236-g sample of the herbicide liberated the chlorine as Cl ion, and the ion precipitated
as AgCl, with a mass of 0.1840 g. The mass percent of 2,4-D in the sample:
First, determine the moles of chlorine present. Noting that each molecule of 2,4D has 2 atoms
of Cl, allows you to calculate the number of moles of 2,4D (each with a molar mass of
221.04 g/mol)

$$\frac{0.1840\ g\ AgCl}{1} \bullet \frac{1\ mol\ AgCl}{143.32\ g\ AgCl} \bullet \frac{1\ mol\ Cl}{1\ mol\ AgCl} \bullet \frac{1\ mol\ 2,4D}{2\ mol\ Cl} \bullet \frac{221.04\ g\ 2,4D}{1\ mol\ 2,4D} = 0.1419\ g\ 2,4D$$

The mass percent of 2,4D in the sample is: $\frac{0.1419\ g\ 2,4D}{1.236\ g\ sample} \bullet 100 = 11.48\ \%\ 2,4D$

129. The amount of water per mol of $CaCl_2$ can be determined by realizing that in the 150 g
sample of the partially hydrated material there is **both** $CaCl_2$ and H_2O. Identify the amount
of $CaCl_2$ remaining in the water:
Determine the solubility/80g (the data provides solubility/100 g).

$$\frac{74.5\ g\ CaCl_2}{100\ g\ water} = \frac{X}{80\ g\ water}\qquad X = 59.6\ g\ CaCl_2$$

From the amount precipitated: $\dfrac{74.9 \text{ g CaCl}_2 \cdot 6\text{H}_2\text{O}}{1} \cdot \dfrac{110.98 \text{ g CaCl}_2}{219.07\text{g CaCl}_2 \cdot 6\text{H}_2\text{O}} = 37.9 \text{ g CaCl}_2$

Now the total mass of $CaCl_2$ is 59.6 + 37.9 or 97.5 g $CaCl_2$

The partially hydrated material had a mass of 150g, so by difference we can calculate the amount of water present in that sample: 150g – 97.5 = 52.5 g of water.

Calculate the moles of water and of $CaCl_2$

$\dfrac{52.5 \text{ g H}_2\text{O}}{1} \cdot \dfrac{1 \text{ mol H}_2\text{O}}{18.02 \text{ g H}_2\text{O}} = 2.91 \text{ mol H}_2\text{O}$ and

$\dfrac{97.5 \text{ g CaCl}_2}{1} \cdot \dfrac{1 \text{ mol CaCl}_2}{110.98 \text{ g CaCl}_2} = 0.879 \text{ mol CaCl}_2$

with a mol ratio of 2.91 mol H_2O: 0.879 mol $CaCl_2$ or 3.3 mol water/mol $CaCl_2$

131. The data given are plotted below:

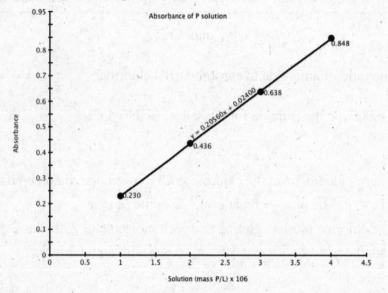

(a) The equation for the line is shown: y = 0.20560x + 0.02400 in which y = Absorbance note that the data on the x axis is multiplied by 10^6, so the slope is 205600 (2.06×10^5) and the intercept is 0.02400.

(b) The mass of P/L of urine: The absorbance for the urine sample is 0.518. Using the equation for the line, we can calculate the concentration of phosphorus/L.

$\dfrac{(0.518 - 0.0240)}{0.2056} = 2.40 \ (\text{x } 10^{-6})\text{M P/L}$

Knowing that this concentration was arrived at by diluting 1.00-mL of the 1122 sample to 50 mL, we can calculate the concentration in the original (i.e. 1122-mL) sample:

$\dfrac{2.40 \times 10^{-6} g \ P \ / \ L}{1} \cdot \dfrac{50.00 \text{ mL}}{1.00 \text{ mL}} = 1.20 \times 10^{-4} \text{ g P/L}$

(c) Mass of Phosphate excreted per day:

The concentration of P calculated above x volume of urine in 24 hr (1122-mL) gives an

overall mass of : 1.20×10^{-4} g P/L x 1.122 L = 1.35×10^{-4} g P,

Since we want mass of *phosphate*, we need to calculate the mass of PO_4^{3-} that

corresponds to this mass of phosphorus:

$$\frac{1.35 \times 10^{-4} \text{g P}}{1} \cdot \frac{94.97 \text{ g } PO_4^{3-}}{30.97 \text{ g P}} = 4.13 \times 10^{-4} \text{g } PO_4^{3-}$$

SUMMARY AND CONCEPTUAL QUESTIONS

133. The total mass of the beakers and solutions after reaction will be equal to the mass of the

beakers and solutions before the reaction: 161.170 g.

135. Each of the 3 flasks contains 0.100 mol of HCl. The mass of Zn differs. Calculate the # of

mol Zn in each flask.

Flask 1: 7.00 g Zn $\bullet \dfrac{1 \text{ mol Zn}}{65.39 \text{ g Zn}}$ = 0.107 mol Zn

Flask 2: 3.27 g Zn $\bullet \dfrac{1 \text{ mol Zn}}{65.39 \text{ g Zn}}$ = 0.0500 mol Zn

Flask 3: 1.31 g Zn $\bullet \dfrac{1 \text{ mol Zn}}{65.39 \text{ g Zn}}$ = 0.0200 mol Zn

The balanced equation tells us that for each mol of Zn, we need 2 mol of HCl. In flask 1,

0.107 mol Zn exceeds the amount of HCl available. The reaction consumes all the HCl, and

leaves unreacted Zn metal. In flask 2, the 0.0500 mol Zn reacts **exactly** with the 0.100 mol of

HCl, leaving **no** unreacted Zn or HCl. In flask 3, the 0.0200 mol of Zn react with 0.0400 mol

of HCl—completely consuming **all the Zn**, and leaving unreacted HCl. The smaller amount

of Zn present (0.0200 mol) produces a smaller amount of H_2 (0.0200 mol)—thus not totally

inflating the balloon.

137. To decide the relative concentrations, calculate the dilutions. Let's assume that the HCl has

a concentration of 0.100 M. Then calculate the diluted concentrations in each case.

Student 1 :   20.0 mL of 0.100 M HCl is diluted to 40.0 mL total

The diluted molarity is: $\dfrac{0.100 \text{ mol HCl}}{1 \text{ L}} \bullet 0.0200 \text{ L} = 0.400 \text{ L} \bullet M = 0.050 \dfrac{\text{mol HCl}}{\text{L}}$

Student 2 :   20.0 mL of 0.100 M HCl is diluted to 80.0 mL total

The diluted molarity is : $\dfrac{0.100 \text{ mol HCl}}{1 \text{ L}} \bullet 0.0200 \text{ L} = 0.0800 \text{ L} \bullet M = 0.025 \dfrac{\text{mol HCl}}{\text{L}}$

So the second student's prepared HCl solution is **(c) half the concentration of the first student**'s. **However**, since the total number of moles of HCl in both solutions *is identical*, they will calculate **the same concentration for the original HCl solution.**

139. (a) What is the % atom economy for the synthesis of maleic anhydride from benzene by this reaction?

We need the molar masses for all atoms used in the reaction:

Benzene$[(6 \times C) + (6 \times H)] = 78$ (atomic masses expressed to integral values)

Oxygen $[(\, ^9/_2 \times O_2) = 9 \times 16] = 144$

Adding these gives 222

Calculating the molar mass of the desired product: $[(4 \times C) + (2 \times H) + (3 \times O)] = 98$

So the % atom economy $= \dfrac{98}{222} = 0.441$, and converting to percent gives 44.1%

(b) Percent yield of the anhydride:

Theoretical yield:

$$1000 \text{ g C}_6\text{H}_6 \cdot \frac{1 \text{ mol C}_6\text{H}_6}{78 \text{ g C}_6\text{H}_6} \cdot \frac{1 \text{ mol C}_4\text{H}_2\text{O}_3}{1 \text{ mol C}_6\text{H}_6} \cdot \frac{98\text{g C}_4\text{H}_2\text{O}_3}{1 \text{ mol C}_4\text{H}_2\text{O}_3} = 1256 \text{ g C}_4\text{H}_2\text{O}_3$$

Percent yield: $\dfrac{972\text{g C}_4\text{H}_2\text{O}_3}{1256\text{g C}_4\text{H}_2\text{O}_3} = 0.774$ or 77.4% (and to 1 sf—limited by 1kg—80%)

What mass of the byproduct CO_2 is also produced?

Theoretical yield:

$$1000 \text{ g C}_6\text{H}_6 \cdot \frac{1 \text{ mol C}_6\text{H}_6}{78 \text{ g C}_6\text{H}_6} \cdot \frac{2 \text{ mol CO}_2}{1 \text{ mol C}_6\text{H}_6} \cdot \frac{44\text{g CO}_2}{1 \text{ mol CO}_2} = 1128 \text{ g CO}_2$$

and with a percent yield of 80%: $(0.80 \times 1128\text{g}) = 900 \text{ g CO}_2$ (to 1 sf)

Applying Chemical Principles

1. Which of the compounds listed produce a gas when reacting with HCl?

   The hydrogen carbonate salts ($NaHCO_3$ and $KHCO_3$) will react with HCl by the general

   equation: $NaHCO_3 (s) + HCl(aq) \rightarrow NaCl (aq) + H_2O(g) + CO_2 (g)$

   $CaCO_3(s)$ will react with HCl in a similar manner:

   $CaCO_3(s) + 2 HCl(aq) \rightarrow CaCl_2 (aq) + H_2O(g) + CO_2 (g)$

   The hydroxide compounds will react to form chloride salts, but produce no gas.

3.(a) A balanced chemical equation for the reaction of $Mg(OH)_2$ and HCl:

   A simple acid-base reaction: $Mg(OH)_2 (s) + 2 HCl (aq) \rightarrow MgCl_2(aq) + 2 H_2O (l)$

   (b) Mass of $Mg(OH)_2$ present in one tablet:

   Amount of HCl used: 29.52×10^{-3} L • 0.500 mol HCl/L = 0.01476 mol HCl

   We need to subtract the amount of HCl that is consumed by the calcium carbonate present.
   That reaction can be written: $CaCO_3(s) + 2 HCl(aq) \rightarrow CaCl_2 (aq) + H_2O(g) + CO_2 (g)$

   $$\frac{0.550 \text{ g } CaCO_3}{1} \cdot \frac{1 \text{ mol } CaCO_3}{100.071 \text{ g } CaCO_3} = 0.00550 \text{ mol } CaCO_3$$

   This will require 2 • 0.0550 mol of HCl (stoichiometry of reaction) = 0.0110 mol HCl

   The difference in HCl (initially added) and that used for reaction with calcium carbonate:
   0.01476 mol HCl - 0.0110 mol HCl = 0.00376 mol HCl.
   The equation (in 3(a)) shows that we will consume 2 mol HCl/mol $Mg(OH)_2$ present.

   Mol $Mg(OH)_2$ = 0.00376 mol HCl /2 = 0.00188 mol $Mg(OH)_2$
   The mass of $Mg(OH)_2$ corresponding to this amount is:
   $$\frac{58.3088 \text{ g } Mg(OH)_2}{1 \text{ mol } Mg(OH)_2} \cdot \frac{0.00188 \text{ mol } Mg(OH)_2}{1} = 0.1096 \text{ g } Mg(OH)_2 \text{ or } 0.110 \text{ mg } (3 \text{ sf})$$

5. Neutralizes the greatest amount of acid?:
   One tablet of Tums contains 500. mg $CaCO_3$
   $$\frac{500 \times 10^{-3} \text{ g } CaCO_3}{1} \cdot \frac{1 \text{ mol } CaCO_3}{100.071 \text{ g } CaCO_3} = 5.00 \times 10^{-3} \text{ mol } CaCO_3. \text{ From the equation}$$
   in 3(b) above, you see that this will neutralize 2 x this number of moles of HCl:
   1.00×10^{-2} mol HCl

   One table of Rolaids consumes 29.52 mL of 0.500 M HCl (3b above). This corresponds
   to 29.52×10^{-3} L • 0.500 mol HCl/L = 0.01476 mol HCl

One teaspoon of Maalox contains a mixture of 200. mg of $Al(OH)_3$ and 200. mg of $Mg(OH)_2$.

$$\frac{0.200 \text{ g } Mg(OH)_2}{1} \cdot \frac{1 \text{ mol } Mg(OH)_2}{58.3088 \text{ g } Mg(OH)_2} \cdot \frac{2 \text{ mol HCl}}{1 \text{ mol } Mg(OH)_2} = 0.00686 \text{ mol HCl}$$

$$\frac{0.200 \text{ g } Al(OH)_3}{1} \cdot \frac{1 \text{ mol } Al(OH)_3}{78.00 \text{ g } Al(OH)_3} \cdot \frac{3 \text{ mol HCl}}{1 \text{ mol } Al(OH)_3} = 0.00769 \text{ mol HCl}$$

The total amount of HCl consumed would be: $0.00769 + 0.00686 = 0.01455 \text{ mol HCl}$

The antacids rank as follows—in consumption of HCl:
Tums: 0.0100 mol HCl
Rolaids: 0.01476 mol HCl
Maalox: 0.01455 mol HCl

Chapter 5
Principles of Chemical Reactivity:
Energy and Chemical Reactions

PRACTICING SKILLS
Energy:Some Basic Principles

1. Define the terms system and surroundings.

   SYSTEM: Is nothing more than the "thing" we are studying in a particular situation. If we are concerned with the energy change associated with a stick of wood burning, then the SYSTEM is the stick of wood.

   SURROUNDINGS: The surroundings are defined as all that matter which is in thermal contact with the system. So if we were burning our stick of wood in a fireplace, the surroundings would include the air, and the fireplace materials in thermal contact with the stick of wood.

   So when we say that a system and its surroundings are in thermal equilibrium, we mean that the energy (heat) flow between the system and the surroundings is in a steady state—that the amount of energy "leaving" the system and energy "leaving" the surroundings is equal.

3. Identify whether the following processes are exothermic or endothermic.

   (a) combustion of methane: Since we frequently burn hydrocarbons as a source of energy (think propane grill!), the combustion of methane is **exothermic.**

   (b) melting of ice: Ice melts (solid becomes liquid) as it absorbs thermal energy from the surroundings—a process we call **endothermic.**

   (c) raising the temperature of water from 25 °C to 100 °C:We do this by **increasing** the energy of the water molecules—and this occurs in an **endothermic** process.

   (d) heating $CaCO_3(s)$ to form $CaO(s)$ and $CO_2(g)$: Just as in heating water (c above) we add thermal energy to $CaCO_3(s)$ in an **endothermic** process.

Specific Heat Capacity

5. What is the specific heat capacity of mercury, if the molar heat capacity is 28.1 J/mol • K?

   Note that the difference in **units** of these two quantities is in the amount of substance. In one case, moles, while in the other grams. $28.1\dfrac{J}{mol \bullet K} \bullet \dfrac{1\ mol}{200.59\ g} = 0.140\ \dfrac{J}{g \bullet K}$

7. Heat energy to warm 168 g copper from -12.2 °C to 25.6 °C:

Heat = mass x heat capacity x ΔT

For copper = $(168\ g)(\dfrac{0.385\ J}{g\bullet K})[25.6°C - (-12.2)\ °C]\bullet\dfrac{1\ K}{1\ °C}$ = 2.44 x 10^3 J or 2.44 kJ

9. The final temperature of a 344 g sample of iron when 2.25 kJ of heat are added to a sample originally at 18.2 °C. The energy added is:

q_{Fe} = (mass)(heat capacity)(ΔT)

2.25 x 10^3 J = $(344\ g)(0.449\ \dfrac{J}{g\bullet K})(x)$ and solving for x we get:

14.57 K = x and since 1K = 1°C, ΔT = 14.57 °C.

The final temperature is (14.57 + 18.2)°C or 32.8°C.

11. Final T of copper-water mixture:

We must **assume** that **no energy** will be transferred to or from the beaker containing the water. Then the **magnitude** of energy lost by the hot copper and the energy gained by the cold water will be equal (but opposite in sign).

q_{copper} = -q_{water}

Using the heat capacities of H_2O and copper, and expressing the temperatures in Kelvin (K = °C + 273.15) we can write:

$(45.5\ g)(0.385\ \dfrac{J}{g\bullet K})(T_{final} - 372.95\ K)$ = $-(152.\ g)(4.184\ \dfrac{J}{g\bullet K})(T_{final} - 291.65K)$

Simplifying each side gives:

$17.52\ \dfrac{J}{K}\bullet T_{final} - 6533\ J$ = $-636.0\ \dfrac{J}{K}\bullet T_{final} + 185,480\ J$

$653.52\ \dfrac{J}{K}\bullet T_{final}$ = 192013 J

T_{final} = 293.81 K or (293.81 - 273.15) or 20.7 °C

Don't forget: **Round numbers only at the end**.

13. Final temperature of water mixture:

This problem is solved almost exactly like question 11. The difference is that both samples are samples of water. From a mechanical standpoint, the heat capacity of both samples will be identical—and can be omitted from both sides of the equation:

q_{water} (at 95 °C) = -q_{water} (at 22 °C)

$(85.2\ g)(4.184\ \dfrac{J}{g\bullet K})(T_{final} - 368.15\ K)$ = $-(156\ g)(4.184\ \dfrac{J}{g\bullet K})(T_{final} - 295.15\ K)$ or

$(85.2\ g)(T_{final} - 368.15\ K)$ = $-(156\ g)(T_{final} - 295.15\ K)$ or

$$85.2 \text{ J/K } T_{final} - 31366.38 \text{ J} = -156 \text{ J/K } T_{final} + 46043.4 \text{ J}$$

rearranging: $241.2 \text{ J/K} \cdot T_{final} = 77409.78 \text{ J}$ so $321.0 \text{ K} = T_{final}$ or $47.8\,^\circ C$

15. Here the warmer Zn is losing heat to the water: $q_{metal} = -q_{water}$

Remembering that $\Delta T = T_{final} - T_{initial}$, we can calculate the change in temperature for the water and the metal. Further, since we know the final and initial for both the metal and the water, we can calculate the temperature difference in units of Celsius degrees, since the **change** in temperature on the **Kelvin** scale would be numerically identical.

For the metal: $\Delta T = T_{final} - T_{initial} = (27.1 - 98.8)$ or $-71.7\,^\circ C$ or -71.7 K.

For the water: $\Delta T = T_{final} - T_{initial} = (27.1 - 25.0)$ or $2.1\,^\circ C$ or 2.1 K (recalling that a Celsius degrees and a Kelvin are the same "size").

$$(13.8 \text{ g})(C_{metal})(-71.7 \text{ K}) = -(45.0 \text{ g})(4.184 \frac{J}{g \cdot K})(2.1 \text{ K})$$

$$- 989.46 \text{ g} \cdot K(C_{metal}) = - 395 \text{ J}$$

$$C_{metal} = 0.40 \frac{J}{g \cdot K} \quad \text{(to 2 significant figures)}$$

Changes of State

17. Quantity of energy evolved when 1.0 L of water at $0\,^\circ C$ solidifies to ice:

The mass of water involved: If we assume a density of liquid water of 1.000 g/mL, 1.0 L of water (1000 mL) would have a mass of 1000 g.

To freeze 1000 g water: $1000 \text{ g ice} \cdot \frac{333 \text{ J}}{1.000 \text{ g ice}} = 333. \times 10^3 \text{ J}$ or 330 kJ (to 2sf)

19. Heat required to vaporize (convert liquid to gas) 125 g C_6H_6:

The heat of vaporization of benzene is 30.8 kJ/mol.

Convert mass of benzene to moles of benzene: $125 \text{ g} \cdot 1 \text{ mol}/78.11 \text{ g} = 1.60 \text{ mol}$

Heat required: $1.60 \text{ mol } C_6H_6 \cdot 30.8 \text{ kJ/mol} = 49.3 \text{ kJ}$

NOTE: No sign has been attached to the amount of heat, since we wanted to know the **amount**. If we want to assign a **direction** of heat flow in this question, then we would add a (+) to 49.3 kJ to indicate that heat is being **added** to the liquid benzene.

21. To calculate the quantity of heat for the process described, think of the problem in two steps:

    1) cool liquid from $23.0\,^\circ C$ to liquid at $-38.8\,^\circ C$

    2) freeze the liquid at its freezing point ($-38.8\,^\circ C$)

Note that the specific heat capacity is expressed in units of mass, so convert the volume of liquid mercury to **mass**. $1.00 \text{ mL} \cdot 13.6 \text{ g/mL} = 13.6 \text{ g Hg}$ (Recall:$1 \text{ cm}^3 = 1 \text{ mL}$)

1) The energy to cool 13.6 g of Hg from 23.0 °C to liquid at – 38.8 °C is:

ΔT = (234.35 K - 296.15 K) or - 61.8 K

13.6 g Hg $\bullet$ 0.140 $\dfrac{J}{g \bullet K}$ $\bullet$ - 61.8 K = - 118 J

2) To convert liquid mercury to solid Hg at this temperature:

- 11.4 J/g $\bullet$ 13.6 g = - 155 J (The (-) sign indicates that heat is being removed from the Hg.

The total energy released by the Hg is: [- 118 J + - 155 J] = - 273 J and since $q_{mercury}$ = - $q_{surroundings}$ the amount released to the surroundings is 273 J.

23. To accomplish the process, one must:

1) heat the ethanol from 20.0 °C to 78.29 °C (ΔT = 58.29 K)

2) boil the ethanol (convert from liquid to gas) at 78.29 °C

Using the specific heat for ethanol, the energy for the first step is:

(2.44 $\dfrac{J}{g \bullet K}$)(1000 g)(58.29 K) = 142,227.6 J (142,000 J to 3 sf)

To boil the ethanol at 78.29 °C, we need:

855 $\dfrac{J}{g}$ $\bullet$ 1000 g = 855,000 J

The total heat energy needed (in J) is (142,000 + 855,000) = 997,000 or 9.97 x 10^5 J

Enthalpy Changes

Note that in this chapter, I have left negative signs with the value for heat released

(heat released = - ; heat absorbed = +)

25. For a process in which the $\Delta H°$ is negative, that process is **exothermic**.

To calculate heat released when 1.25 g NO react, note that the energy shown (-114.1 kJ) is released when **2** moles of NO react, so we'll need to account for that:

1.25 g NO $\bullet$ $\dfrac{1 \text{ mol NO}}{30.01 \text{ g NO}}$ $\bullet$ $\dfrac{-114.1 \text{ kJ}}{2 \text{ mol NO}}$ = -2.38 kJ

27. The combustion of isooctane (IO) is **exothermic**. The molar mass of IO is: 114.2 g/mol. The heat evolved is:

1.00 L of IO $\bullet$ $\dfrac{0.69 \text{ g IO}}{1 mL}$ $\bullet$ $\dfrac{1 \times 10^3 \text{ mL}}{1 \text{ L}}$ $\bullet$ $\dfrac{1 \text{ mol IO}}{114.2 \text{ g IO}}$ $\bullet$ $\dfrac{-10922 \text{ kJ}}{2 \text{ mol IO}}$ = -3.3 x 10^4 kJ

Calorimetry

29. 100.0 mL of 0.200 M CsOH and 50.0 mL of 0.400 M HCl each supply 0.0200 moles of base and acid respectively. If we assume the specific heat capacities of the solutions are 4.2 J/g • K, the **heat evolved** for 0.200 moles of CsOH is:

$q = (4.2 \text{ J/g} \bullet \text{K})(150. \text{ g})(24.28\ ^\circ\text{C} - 22.50^\circ\text{C})$ [and since $1.78^\circ\text{C} = 1.78$ K]

$q = (4.2 \text{ J/g} \bullet \text{K})(150. \text{ g})(1.78 \text{ K})$

$q = 1120$ J

The molar enthalpy of neutralization is: $\dfrac{-1120 \text{ J}}{0.0200 \text{ mol CsOH}} = -56000$ J/mol (to 2 sf)

or -56 kJ/mol

31. For the problem, we'll assume that the coffee-cup calorimeter absorbs **no** heat.
    Since $q_{metal} = -q_{water}$
    Remembering that $\Delta T = T_{final} - T_{initial}$, we can calculate the change in temperature for the water and the metal. Further, since we know the final and initial for both the metal and the water, we can calculate the temperature difference in units of Celsius degrees, since the **change** in temperature on the **Kelvin** scale would be numerically identical.
    For the metal : $\Delta T = T_{final} - T_{initial} = (24.3 - 99.5)$ or -75.2˚C or -75.2K.

For the water: $\Delta T = T_{final} - T_{initial} = (24.3 - 21.7)$ or 2.6˚C or 2.6 K (recalling that a Celsius degrees and a Kelvin are the same "size".

$(20.8 \text{ g})(C_{metal})(-75.2 \text{ K}) = -(75.0 \text{ g})(4.184 \dfrac{\text{J}}{\text{g} \bullet \text{K}})(2.6 \text{ K})$

$-1564.16 \text{ g} \bullet \text{K}(C_{metal}) = -816$ J

$C_{metal} = 0.52 \dfrac{\text{J}}{\text{g} \bullet \text{K}}$ (to 2 significant figures)

33. Enthalpy change when 5.44 g of NH_4NO_3 is dissolved in 150.0 g water at 18.6 ˚C.
    Calculate the heat released by the solution: $\Delta T = (16.2 - 18.6)$ or -2.4 ˚C or -2.4 K
    $(155.4 \text{ g})(4.2 \dfrac{\text{J}}{\text{g} \bullet \text{K}})(-2.4 \text{ K}) = -1566$ J or -1600 J(to 2 sf)
    Calculate the amount of NH_4NO_3: $5.44 \text{ g } NH_4NO_3 \bullet \dfrac{1 \text{ mol } NH_4NO_3}{80.04 \text{ g } NH_4NO_3} = 0.0680$ mol
    Recall that the energy that was released by the solution is **absorbed** by the ammonium nitrate, so we change the sign from (-) to (+).

The enthalpy change has been requested in units of kJ, so divide the energy (in J) by 1000:

Enthalpy of dissolving $= \dfrac{1.566 \text{ kJ}}{0.0680 \text{ mol}} = 23.0$ kJ/mol or 23 kJ/mol (to 2 sf)

35. Calculate the heat evolved (per mol SO_2) for the reaction of sulfur with oxygen to form SO_2
There are several steps:

1) Calculate the heat transferred to the water:

$$815 \text{ g} \bullet 4.184 \dfrac{J}{g \bullet K} \bullet (26.72 - 21.25)°C \bullet 1K/1°C = 18{,}700 \text{ J}$$

2) Calculate the heat transferred to the bomb calorimeter

$$923 \text{ J/K} \bullet (26.72 - 21.25)°C \bullet 1K/1°C = 5{,}050 \text{ J}$$

3) Amount of sulfur present: $2.56 \text{ g} \bullet \dfrac{1 \text{ mol S}_8}{256.536 \text{ g S}_8} = 0.010$ mol S_8

Note from the equation that 8 mol of SO_2 form from each mole of S_8

4) Calculating the quantity of heat related per mol of SO_2 yields:

$$\dfrac{(18{,}700 \text{ J} + 5{,}050 \text{ J})}{0.08 \text{ mol SO}_2} = 297{,}000 \text{ J/mol SO}_2 \text{ or } 297 \text{ kJ/mol SO}_2$$

37. Quantity of heat evolved in the combustion of benzoic acid:

Let's approach this in several steps:

1) Calculate the heat transferred to the water:

$$775 \text{ g} \bullet 4.184 \dfrac{J}{g \bullet K} \bullet (31.69 - 22.50)°C \bullet 1K/1°C = 29{,}800 \text{ J}$$

2) Calculate the heat transferred to the bomb calorimeter

$$893 \text{ J/K} \bullet (31.69 - 22.50)°C \bullet 1K/1°C = 8{,}210 \text{ J}$$

3) Amount of benzoic acid:

$$1.500 \text{ g benzoic acid} \bullet \dfrac{1 \text{ mol benzoic acid}}{122.1 \text{ g benzoic acid}} = 1.229 \times 10^{-2} \text{ mol benzoic acid}$$

4) Heat evolved per mol of benzoic acid is:

$$\dfrac{(29{,}800 \text{ J} + 8{,}210 \text{ J})}{1.229 \times 10^{-2} \text{ mol}} = 3.09 \times 10^6 \text{ J/mol or } 3.09 \times 10^3 \text{ kJ/mol}$$

39. Heat absorbed by the ice : $\dfrac{333 \text{ J}}{1.00 \text{ g ice}} \bullet 3.54 \text{ g ice} = 1{,}180 \text{ J}$ (to 3 sf)

Since this energy (1180 J) is released by the metal, we can calculate the heat capacity of the metal: heat   = heat capacity x mass x ΔT

-1180 J $= C \times 50.0 \text{ g} \times (273.2 \text{ K} - 373 \text{ K})$ [Note that ΔT is negative!]

$0.236 \dfrac{J}{g \bullet K}$ $= C$

Note that the heat released (left side of equation) has a negative sign to indicate the **directional flow** of the energy.

Hess's Law

41. (a) Hess's Law allows us to calculate the overall enthalpy change by the appropriate combination of several equations. In this case we add the two equations, reversing the second one (with the concomitant reversal of sign).

$$CH_4\ (g) + 2\ O_2\ (g) \rightarrow CO_2\ (g) + 2\ H_2O\ (g) \qquad \Delta H° = -\ 802.4\ kJ$$

$$CO_2\ (g) + 2\ H_2O\ (g) \rightarrow CH_3OH(g) + 3/2\ O_2\ (g) \qquad \Delta H° = +\ 676\ kJ$$

$$CH_4\ (g) + 1/2\ O_2\ (g) \rightarrow CH_3OH(g) \qquad \Delta H° = -\ 126\ kJ$$

(b) A graphic description of the energy change:

$$CH_4\ (g) + 1/2\ O_2\ (g)$$

$$+\ 3/2\ O_2\ (g)$$

$$\Delta H° = -\ 126.4\ kJ$$

$$CH_3\ OH(g)$$

$$\Delta H° = -\ 802.4\ kJ$$

$$\Delta H° = +\ 676\ kJ$$

$$CO_2(g) + 2\ H_2O\ (g)$$

43. The overall enthalpy change for $1/2\ N_2\ (g) + 1/2\ O_2\ (g) \rightarrow NO\ (g)$

For the overall equation, note that elemental nitrogen and oxygen are on the "left" side of the equation, and NO on the "right" side of the equation. Noting that equation 2 has 4 ammonia molecules consumed, let's multiply equation 1 by 2:

$$2\ N_2\ (g) + 6\ H_2\ (g) \rightarrow 4\ NH_3\ (g) \qquad \Delta H = (2)(-\ 91.8\ kJ)$$

The second equation has NO on the right side :

$$4\ NH_3\ (g) + 5\ O_2\ (g) \rightarrow 4\ NO\ (g) + 6\ H_2O\ (g) \qquad \Delta H = -\ 906.2 kJ$$

The third equation has water as a product, and we need to "consume" the water formed in equation two, so let's reverse equation 3—changing the sign—AND multiply it by 6

$6 H_2O (g) \rightarrow 6 H_2 (g) + 3 O_2 (g)$ $\Delta H = (+241.8)(6)$ kJ

Adding these 3 equations gives

$2 N_2 (g) + 2 O_2 (g) \rightarrow 4 NO (g)$ $\Delta H = 361$ kJ

So, dividing all the coefficients by 4 provides the desired equation with

a $\Delta H = +361$ kJ • 0.25 or 90.3 kJ

Standard Enthalpies of Formation

45. The equation requested requires that we form **one** mol of product liquid CH_3OH from its elements—each in their standard state.

Begin by writing a balanced equation:

$2 C (s,graphite) + O_2(g) + 4 H_2(g) \rightarrow 2 CH_3OH(l)$

Now express the reaction so that you form one mole of CH_3OH—divide coefficients by 2.

$C (s) + 1/2 O_2(g) + 2 H_2(g) \rightarrow CH_3OH(l)$

And from Appendix L, the $\Delta_fH°$ is reported as −238.4 kJ/mol

47. (a) The equation of the formation of Cr_2O_3 (s) from the elements:

$2 Cr (s) + 3/2 O_2 (g) \rightarrow Cr_2O_3 (s)$

from Appendix L $\Delta_fH°$ is reported as: -1134.7 kJ/mol for the oxide.

(b) The enthalpy change if 2.4 g of Cr is oxidized to Cr_2O_3 (g) is:

$$2.4 \text{ g Cr} \cdot \frac{1 \text{ mol Cr}}{52.0 \text{ g Cr}} \cdot \frac{-1134.7 \text{ kJ}}{2 \text{ mol Cr}} = -26 \text{ kJ} \quad (\text{to 2 sf})$$

49. Calculate $\Delta_rH°$ for the following processes:

(a) 1.0 g of white phosphorus burns:

$P_4 (s) + 5 O_2 (g) \rightarrow P_4O_{10} (s)$ from Appendix L: $\Delta_fH°$ -2984.0 kJ/mol

$$1.0 \text{ g } P_4 \cdot \frac{1.0 \text{ mol } P_4}{123.89 \text{ g } P_4} \cdot \frac{-2984.0 \text{ kJ}}{1 \text{ mol } P_4} = -24 \text{ kJ}$$

(b) 0.20 mol NO (g) decomposes to N_2 (g) and O_2 (g):

From Appendix L: $\Delta_fH°$ for NO = 90.29 kJ/mol

Since the reaction requested is the **reverse** of $\Delta_fH°$, we change the sign to − 90.29 kJ/mol.

The enthalpy change is then $\dfrac{-90.29 \text{ kJ}}{1 \text{ mol}}$ • 0.20 mol = -18 kJ

(c) 2.40 g NaCl is formed from elemental Na and elemental Cl_2:

From Appendix L: $\Delta_fH°$ for NaCl (s) = - 411.12 kJ/mol

The amount of NaCl is: $2.40 \text{ g NaCl} \cdot \dfrac{1 \text{ mol NaCl}}{58.44 \text{ g NaCl}} = 0.0411 \text{ mol}$

The overall energy change is: $-411.12 \text{ kJ/mol} \cdot 0.0411 \text{ mol} = -16.9 \text{ kJ}$

(d) 250 g of Fe oxidized to Fe_2O_3 (s):

From Appendix L: $\Delta_fH°$ for Fe_2O_3 (s) = -825.5 kJ/mol

The overall energy change is:

$$\dfrac{250 \text{ g Fe}}{1} \cdot \dfrac{1 \text{ mol Fe}}{55.845 \text{ g Fe}} \cdot \dfrac{1 \text{ mol Fe}_2\text{O}_3}{2 \text{ mol Fe}} \cdot \dfrac{-825.5 \text{ kJ}}{1 \text{ mol Fe}_2\text{O}_3} = -1.8 \times 10^3 \text{kJ}$$

51. (a) The enthalpy change for the reaction:

$$4\,NH_3(g) + 5\,O_2(g) \quad \rightarrow \quad 4\,NO(g) + 6\,H_2O(g)$$

$\Delta_fH°$ (kJ/mol)   -45.90   0   +90.29   -241.83

$\Delta_rH° = [(4 \text{ mol})(+90.29 \dfrac{\text{kJ}}{\text{mol}}) + (6 \text{ mol})(-241.83 \dfrac{\text{kJ}}{\text{mol}})] -$

$$[(4 \text{ mol})(-45.90 \dfrac{\text{kJ}}{\text{mol}}) + (5 \text{ mol})(0)]$$

= (-1089.82 kJ) - (-183.6 kJ)

= -906.2 kJ. **The reaction is exothermic.**

(b) Heat **evolved** when 10.0 g NH_3 react:

The balanced equation shows that 4 mol NH_3 result in the release of 906.2 kJ.

$$10.0 \text{ g NH}_3 \cdot \dfrac{1 \text{ mol NH}_3}{17.03 \text{ g NH}_3} \cdot \dfrac{-906.2 \text{ kJ}}{4 \text{ mol NH}_3} = -133 \text{ kJ}$$

53. (a) The enthalpy change for the reaction:

$2\,BaO_2$ (s) $\rightarrow$ $2\,BaO$ (s) + O_2 (g)

Given $\Delta_fH°$ for BaO is: -553.5 kJ/mol and $\Delta_fH°$ for BaO_2 is: -634.3 kJ/mol

This equation can be seen as the summation of the two equations:

(1) $2\,Ba$ (s) + O_2 (g) $\rightarrow$ $2\,BaO$ (s)

(2) $\underline{2\,BaO_2 \text{ (s)} \rightarrow 2\,Ba \text{ (s)} + O_2 \text{ (g)}}$

Equation (1) corresponds to the formation of BaO x 2 while equation(2) corresponds to (2x) the **reverse** of the formation of BaO_2

$\Delta_rH° = (2\Delta_fH°$ for BaO$) + -2(\Delta_fH°$ for $BaO_2) =$

$\Delta_rH° = 161.6$ kJ and the reaction is **endothermic.**

(b) Energy level diagram for the equations in question:

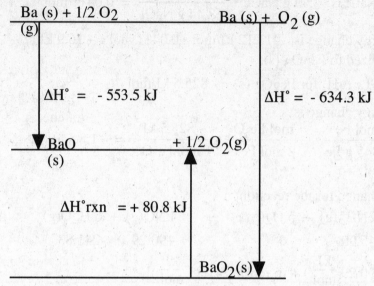

55. The molar enthalpy of formation of naphthalene can be calculated since we're given the enthalpic change for the reaction:

$$C_{10}H_8 (s) + 12 O_2(g) \rightarrow 10 CO_2(g) + 4 H_2O(l)$$

$\Delta H°_f$(kJ/mol)   ?   0   -393.509   -285.83

$\Delta H°_{rxn}$ = $\sum \Delta H°_f$ products $-$ $\sum \Delta H°_f$ reactants

-5156.1 kJ = [(10 mol)(-393.509 $\frac{kJ}{mol}$) + (4 mol)(-285.83 $\frac{kJ}{mol}$)] $-$ [$\Delta H°_f$ C$_{10}$H$_8$]

-5156.1 kJ = (-5078.41 kJ) $-$ $\Delta H°_f$ C$_{10}$H$_8$

- 77.7 kJ = $-$ $\Delta H°_f$ C$_{10}$H$_8$

77.7 kJ = $\Delta H°_f$ C$_{10}$H$_8$

GENERAL QUESTIONS

57. Define and give an example of:

(a) Exothermic and Endothermic—the suffix "thermic" talks about heat, and the prefixes "exo" and "endo" tell us whether heat is ADDED to the surroundings from the system (exo) or REMOVED from the surroundings to the system (endo). Combustion reactions (e.g. gasoline burning in your automobile) are EXOthermic reactions, while ice melting is an ENDOthermic reaction.

(b) System and Surroundings—The "system" is the reactant(s) and product(s) of a reaction, while the "surroundings" is EVERYTHING else. Suppose we burn gasoline in an internal combustion engine. The gasoline (and air) in the cylinder(s) composes the "system",

94

while the engine, and the air contacting the engine are the "surroundings". Together the "system" and "surroundings" compose the "universe"—at least in thermodynamics.

(c) Specific heat capacity—is the quantity of heat required to change the temperature of 1g of a substance by 1 degree Celsius. Water has a specific heat capacity of about 4.2 J/g•K, meaning that 1 gram of water at 15 degrees C, to which 4.2 J of energy is added, will have a temperature of 16 degrees C. (or 14 degrees C—if 4.2 J of energy is removed).

(d) State function—Any parameter that is dependent ONLY on the initial and final states. Chemists typically use CAPITAL letters to indicate state functions (e.g. H, S,) while non-state functions are indicated with LOWER CASE letters (e.g. q, w). Your checking account balance is a state function!

(e) Standard state—Defined as the MOST STABLE (PHYSICAL) STATE for a substance at at a pressure of 1 bar and at a specified temperature. (Typically 298K) The standard state for elemental nitrogen at 25 °C (298K) is **gas**.

(f) Enthalpy change—the heat transferred in a process that is carried out under constant pressure conditions is the enthalpy change, ΔH.
The enthalpy change upon the formation of 1 mol of water($\grave{U}$) is –285.8 kJ, meaning that 285.8 kJ is released upon the formation of 1 mol of liquid water from 1 mol of hydrogen (g) and 1/2 mol oxygen (g).

(g) Standard Enthalpy of Formation—the enthalpy change for the formation of 1 mol of a compound directly from its component elements, each in their standard states. The standard enthalpy of formation of nitrogen gas (N_2) = 0 kJ/mol .

59. Define system and surroundings for each of the following, and give direction of heat transfer:

(a) Methane is burning in a gas furnace in your home:
   (System) methane + oxygen (Surroundings) components of furnace and the air in your home. The heat flows from the methane + oxygen to the furnace and air.

(b) Water drops on your skin evaporate:
   (System) water droplets (Surroundings) your skin and the surrounding air. The heat flows from your skin and the air to the water droplet.

(c) Liquid water at 25 °C is placed in freezer:
   (System) water (Surroundings) freezer. The heat flows from the water to the freezer.

(d) Aluminum and Fe_2O_3 react in a flask on a lab bench:
   (System) Al and Fe_2O_3 (Surroundings) flask, lab bench, and air around flask. The heat flows from the reaction of Al and Fe_2O_3 into the surroundings.

95

61. Standard Enthalpies of Formation for O(g), O_2 (g), O_3 (g).

Substance	$\Delta_f H$ (at 298K) kJ/mol
O(g)	249.170
O_2 (g)	0
O_3 (g)	142.67

What is the standard state of O_2? The standard state of oxygen (O_2) is as a gas.

Is the formation of O from O_2 exothermic?

$\Delta_r H = \Sigma \Delta_f H_{products} - \Sigma \Delta_f H_{reactants}$ ($O_2 \rightarrow 2O$)

$\Delta_r H = (2 \text{ mol})(249.170) - (1 \text{ mol})(0) = 498.340$ kJ (endothermic)

What is the ΔH for 3/2 O_2(g) $\rightarrow O_3$ (g)

$\Delta_r H = \Sigma \Delta_f H_{products} - \Sigma \Delta_f H_{reactants}$

$\Delta_r H = (1 \text{ mol})(142.67 \text{ kJ/mol}) - (3/2 \text{ mol})(0) = 142.67$ kJ

63. Determine whether heat is evolved or required, and whether work was done on the system or whether the system does work on the surroundings, in the following processes at constant pressure:

(a) Liquid water at 100 °C is converted to steam at 100 °C. **Heat will be required** to convert liquid water to gaseous water. The gaseous water will occupy more volume than the liquid water, hence **work is done on the surroundings by the system**.

(b) Dry ice, CO_2(s), sublimes to give CO_2(g). **Heat will be required** to convert solid carbon dioxide into gaseous carbon dioxide. In the same way that the gasification of water (in (a.) above) results in a volume expansion, the gaseous carbon dioxide will occupy a greater volume than the solid, hence **work is done on the surroundings by the system.**

65. Enthalpy change that occurs when 1.00 g of $SnCl_4$(l) reacts with excess H_2O(l):

$\Delta_r H = \Sigma \Delta_f H_{products} - \Sigma \Delta_f H_{reactants}$

$\Delta_r H = [(1 \text{ mol})(-577.63 \frac{\text{kJ}}{\text{mol}}) + (4 \text{ mol})(-167.159 \frac{\text{kJ}}{\text{mol}})] -$

$[(1 \text{ mol})(-511.3 \frac{\text{kJ}}{\text{mol}}) + (2 \text{ mol})(-285.83 \frac{\text{kJ}}{\text{mol}})]$

$\Delta_r H = -1246.266 \text{ kJ} - (-1082.96 \text{ kJ}) = -163.306 \text{ kJ}$ for 1 mol of $SnCl_4(l)$.

Convert this amount into an amount **per gram** of $SnCl_4(l)$, by dividing by the mass of $SnCl_4$.

$$\frac{-163.306 \text{ kJ}}{1 \text{ mol } SnCl_4} \cdot \frac{1 \text{ mol } SnCl_4}{260.5208 \text{ g}} = -0.6268 \frac{\text{kJ}}{\text{g}} \text{ or } -0.627 \text{ kJ/g} \text{ (3 sf)}$$

67. If 187 J raises the temperature of 93.45 g of Ag from 18.5°C to 27.0°C, what is the specific heat capacity of silver?

Recall that $q = m \cdot c \cdot \Delta t$; so 187 J = 93.45 g $\cdot$ c $\cdot$ (27.0 – 18.5)°C. and

$$C_{Ag} = \frac{187 \text{ J}}{93.45 \text{ g} \cdot (27.0 - 18.5)°C} = 0.24 \text{ J/g} \cdot \text{K}$$

69. Addition of 100.0 g of water at 60 °C to 100.0 g of ice at 0.00 °C. The water cools to 0.00 °C. How much ice has melted?

As the ice absorbs heat from the water, two processes occur: (1) the ice melts and (2) the water cools. We can express this with the equation $q_{water} = -q_{ice}$

The melting of ice can be expressed with the heat of fusion of ice, 333 J/g, as $q = m \cdot 333 \text{ J/g}$.

The cooling of the water may be expressed: $q = m \cdot c \cdot \Delta T$. Setting these quantities equal gives:

$m \cdot c \cdot \Delta T = m \cdot 333 \text{ J/g}$ or $100.0 \text{ g} \cdot (4.184 \text{ J/g} \cdot K) \cdot (0 - 60)K = -x \text{ g} \cdot 333 \text{ J/g}$

[x = quantity of ice that melts. Note that since Celsius degrees and kelvin are the same "size", Δt is –60°C or –60 K]

$100.0 \text{ g} \cdot (4.184 \text{ J/g} \cdot K) \cdot (0 - 60)K = -x \cdot 333 \text{ J/g}$

$-25104 \text{ J} = -x \cdot 333 \text{ J/g}$ or $\dfrac{-25104 \text{ J}}{-333 \text{ J/g}} = x$ or 75.4 g of ice.

71. 90 g (two 45 g cubes) of ice cubes (at 0 °C) are dropped into 500. mL tea at 20.0 °C (Assume a density of 1.00 g/mL for tea). What is the final temperature of the mixture if all the ice melts?

Since $q_{water} = -q_{ice}$ and we can set up the expression.

$m \cdot c \cdot \Delta T = -m \cdot 333 \text{ J/g}$ NOTE however, that not only does all the ice melt, but the melted ice warms to a temperature above 0 °C. We add a term to account for that:

$m_{tea} \cdot c \cdot \Delta T_{tea} = -[m_{ice} \cdot 333 \text{ J/g} + m_{ice} \cdot c \cdot \Delta T_{ice}]$

$500. \text{ g} \cdot (4.184 \text{ J/g} \cdot K) \cdot (F - 293.2 \text{ K}) = -[(90 \text{ g} \cdot 333 \text{ J/g}) +$

$(90\ g \bullet(4.184\ J/g\bullet K) \bullet (F - 273.2\ K))]$

where F is the final temperature of the tea and melted ice.

2092F J - 613,374 J = - [29970 J + 377F J - 102,876 J] Simplifying:

2092F J - 613,374 J + 29970 J + 377F J - 102,876 J = 0

(2092F J + 377F J) + (- 613,374 J + 29970 J - 102,876 J) = 0

2469 F J + -686,280 = 0 or 2469 F J = 686,280 and F = (686,280/2469) = 278 K

and noting that 45 g of ice cube has 2 sf, we report a final temperature of 280 K.

73. One can arrive at the desired answer if you recall the **definition** of $\Delta H°_f$. The definition is the enthalpy change associated with the formation of **one mole** of the substance (in this case B_2H_6) from its elements—each in their standard state (s for boron and g for hydrogen).

(a) Note that the 1st equation given uses **four** moles of B as a reactant — and we'll need only 2, so divide the first equation by 2 to give:

$2\ B\ (s) + 3/2\ O_2\ (g) \rightarrow B_2O_3\ (s)$ $\Delta H° = 1/2(-2543.8\ kJ) = -1271.9\ kJ$

The formation of 1 mole of B_2H_6 will require the use of 6 moles of H (or 3 moles of H_2), so multiply the second equation by 3 to give:

$3\ H_2\ (g) + 3/2\ O_2\ (g) \rightarrow 3\ H_2O\ (g)$ $\Delta H° = 3(-241.8\ kJ) = -725.4\ kJ$

Finally the third equation given has B_2H_6 as a **reactant and not a product**. So reverse the third equation to give:

$B_2O_3\ (s) + 3\ H_2O\ (g) \rightarrow B_2H_6\ (g) + 3\ O_2\ (g)$ $\Delta H° = -(-2032.9\ kJ)= +2032.9kJ$

(b) Adding the three equations gives the equation:

$2\ B\ (s) +3\ H_2\ (g) \rightarrow B_2H_6\ (g)$ with a $\Delta H° = (-1271.9 + -725.4 + 2032.9)kJ$

or a $\Delta H°$ for B_2H_6 (g) of + 35.6 kJ

(c) Energy level diagram for the reactions:

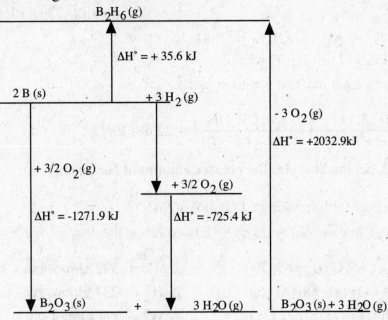

(d) Formation of B_2H_6 (g) is **endothermic**

75. (a) Enthalpy change for:

$$C(s) + \quad H_2O(g) \quad \rightarrow \quad CO(g) + \quad H_2(g)$$

$\Delta H°_f$ (kJ/mol)   0   -241.83   -110.525   0

$$\Delta H°_{rxn} = [(1\ mol)(-110.525\ \frac{kJ}{mol}) + 0] - [\ 0\ +\ (1\ mol)(-241.83\ \frac{kJ}{mol})]$$

$$= +131.31\ kJ$$

(b) The process is **endothermic**.

(c) Heat involved when 1.0 metric ton (1000.0 kg) of C is converted to coal gas:

$$1000.0\ kg\ C \cdot \frac{1000\ g\ C}{1\ kg\ C} \cdot \frac{1\ mol\ C}{12.011\ g\ C} \cdot \frac{+131.31\ kJ}{1\ mol\ C} = 1.0932 \times 10^7\ kJ$$

77. For the combustion of C_8H_{18}:

$$C_8H_{18}(l) + 25/2\ O_2(g) \rightarrow 8\ CO_2(g) + 9\ H_2O(l)$$

$$\Delta_r H° = [(8\ mol)(-393.509\ \frac{kJ}{mol}) + (9\ mol)(-285.83\ \frac{kJ}{mol})] - [(1\ mol)(-259.2\ \frac{kJ}{mol}) + 0]$$

$$\Delta_r H° = -5461.3\ kJ$$

Expressed on a gram basis: $-5461.3\ \frac{kJ}{mol} \cdot \frac{1\ mol\ C_8H_{18}}{114.2\ g\ C_8H_{18}} = -47.81\ kJ/g$

For the combustion of CH_3OH:

$$2\ CH_3OH(l) + 3\ O_2(g) \rightarrow 2\ CO_2(g) + 4\ H_2O(l)$$

$$\Delta_r H° = [(2mol)(-393.509 \text{ kJ/mol}) + (4 \text{ mol})(-285.83 \text{ kJ/mol})]$$
$$- [(2mol)(-238.4 \text{ kJ/mol}) + 0]$$

$$= [(-787.0) + (-967.2)] + 477.4 \text{ kJ}$$

$$= -1453.5 \text{ kJ or } -726.77 \text{ kJ/mol}$$

Express this on a per mol and per gram basis:

$$\frac{-1453.5 \text{ kJ}}{2 \text{ mol CH}_3\text{OH}} \cdot \frac{1 \text{ mol CH}_3\text{OH}}{32.04 \text{ g CH}_3\text{OH}} = -22.682 \text{ kJ/g}$$

On a per gram basis, **octane liberates the greater amount** of heat energy.

79. (a) Enthalpy change for formation of 1.00 mol of $SrCO_3$

Sr (s) + C (graphite) + 3/2 O_2 (g) → $SrCO_3$ (s) using the data:

Sr(s) + 1/2 O_2 (g) → SrO(s) $\Delta_f H° = -592$ kJ/mol-rxn
SrO(s) + CO_2 (g) → $SrCO_3$ (s) $\Delta_r H° = -234$ kJ/mol-rxn
C (graphite) + O_2 (g) → CO_2 (g) $\Delta_f H° = -394$ kJ/mol-rxn

Let's add the equations to give our desired overall equation.

Sr(s) + 1/2 O_2 (g) → S̶r̶O̶(s) $\Delta_f H° = -592$ kJ/mol-rxn
S̶r̶O̶(s) + C̶O̶₂(g) → $SrCO_3$ (s) $\Delta_r H° = -234$ kJ/mol-rxn
C (graphite) + O_2 (g) → C̶O̶₂ (g) $\Delta_f H° = -394$ kJ/mol-rxn
Sr (s) + C (graphite) + 3/2 O_2 (g) → $SrCO_3$ (s) $\Delta_r H° = -1220.$ kJ/mol-rxn

(b) Energy diagram relating the energy quantities:

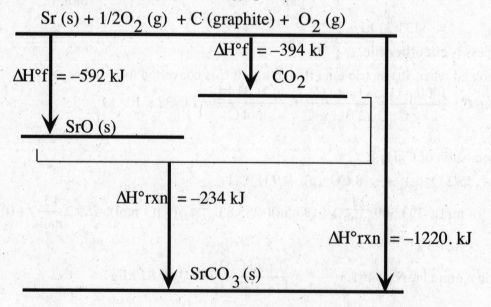

$$\text{Sr (s)} + 1/2 O_2 \text{ (g)} + \text{C (graphite)} + O_2 \text{ (g)}$$

ΔH°f = −394 kJ

ΔH°f = −592 kJ CO_2

SrO (s)

ΔH°rxn = −234 kJ

ΔH°rxn = −1220. kJ

$SrCO_3$ (s)

81. The desired equation is: $CH_4 (g) + 3 Cl_2 (g) \rightarrow 3 HCl (g) + CHCl_3 (g)$

Begin with equation 1 (the combustion of methane)

$CH_4 (g) + 2 O_2 (g) \rightarrow 2 H_2O (l) + CO_2 (g)$ $\Delta H= -890.4$ kJ $= -890.4$ kJ/mol-rxn

Noting that we form HCl as one of the products, using the second equation, we need to **reverse** it and (to adjust the coefficient of HCl to 3), multiply by 3/2 to give:

$3/2 H_2 (g) + 3/2 Cl_2 (g) \rightarrow 3 HCl$ $\Delta H = -3/2(+184.6)$ kJ/mol-rxn $= -276.9$ kJ

Note that CO_2 formed in equation 1 doesn't appear in the overall equation so let's use the equation for the formation of CO_2 (reversed) to "consume" the CO_2:

$CO_2 (g) \rightarrow C (graphite) + O_2 (g)$ $\Delta H = -1(-393.5)$ kJ $= + 393.5$ kJ

Noting also that equation 1 produces 2 water molecules, let's "consume" them by using the equation for the formation of water (reversed) multiplied by 2:

$2 H_2O (l) \rightarrow 2 H_2 (g) + O_2 (g)$ $\Delta H = -2(-285.8)$ kJ $= + 571.6$ kJ

and finally we need to produce $CHCl_3$ which we can do with the equation that represents the ΔH_f for $CHCl_3$:

$C(graphite) + 1/2 H_2 (g) + 3/2 Cl_2 (g) \rightarrow CHCl_3 (g)$ $\Delta H = -103.1$ kJ

The overall enthalpy change would then be:

$\Delta H = -890.4$ kJ $- 276.9$ kJ $+ 393.5$ kJ $+ 571.6$ kJ -103.1 kJ $= -305.3$ kJ

IN THE LABORATORY

83. q_{metal} = heat capacity x mass x ΔT

$q_{metal} = C_{metal} \cdot 27.3$ g $\cdot (299.47$ K $- 372.05$ K$)$

Note that ΔT is negative, since T_{final} of the metal is LESS THAN $T_{initial}$

and $q_{water} = 15.0$ g $\cdot \dfrac{4.184 \text{ J}}{\text{g} \cdot \text{K}} \cdot (299.47$ K $- 295.65$ K$) = 239.7$ J

Setting $q_{metal} = - q_{water}$

$C_{metal} \cdot 27.3$ g $\cdot (-72.58$ K$) = - 239.7$ J and solving for C gives:

$C_{metal} = 0.121 \dfrac{\text{J}}{\text{g} \cdot \text{K}}$

85. Calculate the enthalpy change for the precipitation of AgCl (in kJ/mol):

1) How much AgCl is being formed?

250. mL of 0.16 M $AgNO_3$ will contain (0.250L $\cdot$ 0.16 mol/L) 0.040 mol of $AgNO_3$

125 mL of 0.32 M NaCl will contain (0.125L $\cdot$ 0.32 mol/L) 0.040 mol of NaCl.

Given the stoichiometry, we anticipate the formation of 0.040 mol of AgCl.

2) How much energy is evolved?

[(250 +125ml) • 1g/mL = 375g of water

$$375 \text{ g} \bullet 4.2 \frac{J}{g \bullet K} \bullet (296.05 \text{ K} - 294.30) \text{ K} = 2,800 \text{ J (to 2 sf)}$$

The enthalpy change is then - 2800 J (since the reaction **releases** heat).

The change in kJ/mol is $\dfrac{2800 \text{ J}}{0.040 \text{ mol}} \bullet \dfrac{1 \text{ kJ}}{1000 \text{ J}} = -69 \text{ kJ/mol}$

87. Heat evolved when ammonium nitrate is decomposed:

$\Delta T = (20.72-18.90) = 1.82 \,°C$ (or 1.82 K).

Heat absorbed by the calorimeter: 155 J/K • 1.82K = 282 J

Heat absorbed by the water: $415 \text{ g} \bullet 4.18 \frac{J}{g \bullet K} \bullet 1.82 \text{ K} = 3160 \text{ J}$

Total heat **released** by the decomposition: 3160 J + 282 J = 3,440 J (to 3 sf)

With 7.647 g NH_4NO_3 = 0.09554 mol, heat released = $\dfrac{3440 \text{ J}}{0.09554 \text{ mol}} = 36.0 \text{ kJ/mol}$

89. The enthalpy change for the reaction:

$$Mg(s) + 2 H_2O(l) \rightarrow Mg(OH)_2 (s) + H_2(g)$$
$\Delta_fH° $ (kJ/mol) 0   -285.83   -924.54   0

$\Delta_rH° = (1 \text{ mol})(-924.54 \frac{kJ}{mol}) - (2 \text{ mol})(-285.83 \frac{kJ}{mol})$

$= -352.88 \text{ kJ or } -3.5288 \times 10^5 \text{ J}$

Each mole of magnesium releases 352.88 kJ of heat energy.

Calculate the heat required to warm 25 mL of water from 25 to 85 °C.

heat = heat capacity x mass x ΔT

$= (4.184 \frac{kJ}{mol})(25 \text{ mL})(\frac{1.00 \text{ g}}{1 \text{ mL}})(60 \text{ K})$

= 6276 or 6300 J or 6.3 kJ (to 2 sf)

Magnesium required:

$6.3 \text{ kJ} \bullet \dfrac{1 \text{ mol Mg}}{352.88 \text{ kJ}} \bullet \dfrac{24.3 \text{ g Mg}}{1 \text{mol Mg}} = 0.43 \text{ g Mg}$

SUMMARY AND CONCEPTUAL QUESTIONS

91. Without doing calculations, decide whether each is exo- or endothermic:

    (a) combustion of natural gas—oxidation reactions of carbon and hydrogen typically
        release heat--this process is exothermic.

    (b) Decomposition of sugar to form carbon and water- When you burn sugar to form
        carbon and water, heat is evolved--the reaction is exothermic.

93. Determine the value of ΔH for the reaction:
$$Ca(s) + 1/8\ S_8\ (s) + 2\ O_2(g) \rightarrow CaSO_4(s)$$
    Imagine this as the sum of several processes:

    1) $Ca(s) + 1/2\ O_2(g) \rightarrow CaO(s)$
    2) $1/8\ S_8\ (s) + 3/2\ O_2(g) \rightarrow SO_3(g)$
    3) $CaO(s) + SO_3(g) \rightarrow CaSO_4(s)$ $\Delta H = -402.7$ kJ

Note that the SUM of the three processes is the DESIRED equation (the formation of

$CaSO_4(s)$). The OVERALL ΔH is then the SUM of the ΔH for process (1) and ΔH for

process (2). We know that the $\Delta_r H°$ for (3) = -402.7 kJ or

$\Delta_r H° = \Delta_f H°\ CaSO_4(s) - [\Delta_f H°\ CaO(s) + \Delta_f H°\ SO_3(g)]$. Since we know the $\Delta_r H°$ and

BOTH the $\Delta_f H°$ for CaO(s) and $SO_3(g)$, we can calculate the $\Delta_f H°\ CaSO_4(s)$.

From Appendix L we find,

$\Delta_f H°$ for CaO(s) = -635.09 kJ/mol and $\Delta_f H°$ for $SO_3(g)$ = - 395.77 kJ/mol

$\Delta_r H° = \Delta_f H°\ CaSO_4(s) - [\Delta_f H°\ CaO(s) + \Delta_f H°\ SO_3(g)]$

- 402.7 kJ = $\Delta_f H°\ CaSO_4(s)$ - [-635.09 kJ/mol + - 395.77 kJ/mol]

- 1,433.6 kJ = $\Delta_f H°\ CaSO_4(s)$

95. The molar heat capacities for Al, Fe, Cu, and Au are:

$$0.897\ \frac{J}{g \cdot K} \cdot \frac{26.98\ g\ Al}{1\ mol\ Al} = 24.2\ \frac{J}{mol \cdot K}$$

$$0.449\ \frac{J}{g \cdot K} \cdot \frac{55.85\ gFe}{1\ mol\ Fe} = 25.1\ \frac{J}{mol \cdot K}$$

$$0.385\ \frac{J}{g \cdot K} \cdot \frac{63.55\ g\ Cu}{1\ mol\ Cu} = 24.5\ \frac{J}{mol \cdot K}$$

$$0.129\ \frac{J}{g \cdot K} \cdot \frac{197.0\ g\ Au}{1\ mol\ Au} = 25.4\ \frac{J}{mol \cdot K}$$

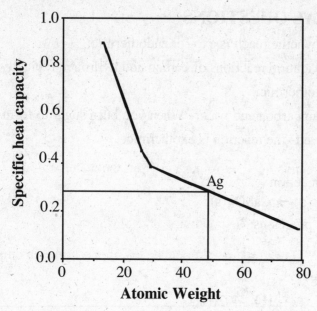

The graph shown is a plot of specific heat capacity versus atomic weight. As you can see, no simple linear relationship exists for these metals. The plot of the specific heat of Cu (atomic weight 63.55) and Au (atomic weight 197) does show a **decreasing** value of specific heat capacity as the atomic weight of the element increases. If you estimate the atomic weight to be about 100 (exact value is about 108), one could **estimate** a value of approximately 0.28 as the specific heat (compared to the experimental value of 0.236).

Alternatively, a quick examination of the values for the four metals above indicates that they are **quite similar**, with an average of 24.8 J/mol • K. This translates into:

$$24.8 \frac{J}{mol \cdot K} \cdot \frac{1 \text{ mol Au}}{107.9 \text{ g Au}} = 0.230 \text{ J/g} \cdot K$$

97. Mass of methane needed to heat the air from 15.0 to 22.0 °C:
    Calculate the volume of air, then with the density and average molar mass, the moles of air present:

$$275 \text{ m}^2 \cdot 2.50 \text{ m} \cdot \frac{1000 \text{ L}}{1 \text{ m}^3} \cdot \frac{1.22 \text{ g air}}{1 \text{ L air}} \cdot \frac{1 \text{ mol air}}{28.9 \text{ g air}} = 2.90 \times 10^4 \text{ mol air}$$

The energy needed to change the temperature of that amount of air by $(22.0 - 15.0)°C$:

$$2.90 \times 10^4 \text{ mol air} \cdot 29.1 \frac{J}{mol \cdot K} \cdot 7.0 \text{ K} = 5.9 \times 10^6 \text{ J}$$

What quantity of energy does the combustion of methane provide?
The reaction may be written: $CH_4 (g) + 2 O_2 (g) \rightarrow 2 H_2O (g) + CO_2 (g)$

Using data from Appendix L:

$\Delta_r H = [(2\ mol)(-241.83\ kJ/mol) + (1mol)(-393.509\ kJ/mol)]$

$- [(1\ mol)(-74.87\ kJ/mol) + (2\ mol)(0)] = -802.3\ kJ$

The amount of methane necessary is:

$5.9 \times 10^6\ J \cdot \dfrac{1\ kJ}{1000\ J} \cdot \dfrac{1\ mol\ CH_4}{802.3\ kJ} \cdot \dfrac{16.0\ g\ CH_4}{1\ mol\ CH_4} = 120\ g\ CH_4$ (2 sf)

99. Calculate the quantity of heat transferred to the surroundings from the water vapor

condensation as rain falls.

Calculate the volume of water that falls, and then the mass of that water:

From the conversion factors listed in your textbook, calculate the area of $1mi^2$ in cm^2

[1 km = 0.62137 mi and 1 km = 10^5 cm.]

$\dfrac{1\ mi^2}{1} \cdot \dfrac{(1\ km)^2}{(0.62137\ mi)^2} \cdot \dfrac{(10^5\ cm)^2}{(1\ km)^2} = \dfrac{10^{10} cm^2}{0.38610} = 2.59 \times 10^{10}\ cm^2$

1 in = 2.54 cm so the VOLUME of water is $2.59 \times 10^{10}\ cm^2 \times 2.54\ cm = 6.6 \times 10^{10}\ cm^3$.

The mass of water is: $6.6 \times 10^{10}\ cm^3 \times 1.0\ g/cm^3$ or 6.6×10^{10} g of water.

The amount of heat: $\dfrac{6.6 \times 10^{10}g\ water}{1} \cdot \dfrac{1\ mol\ water}{18.02\ g\ water} \cdot \dfrac{44.0\ kJ}{1\ mol\ water} = 1.6 \times 10^{11}\ kJ$

Note the much larger energy for this process than for the detonation of a ton of dynamite.

101. (a) The diagram is:

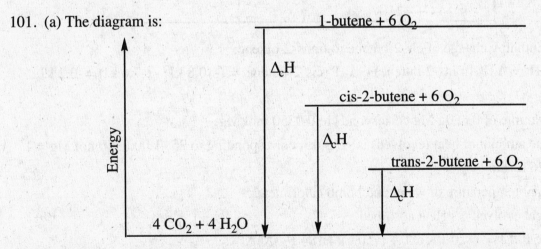

(b)For the combustion reaction: $C_4H_8(g) + 6\ O_2(g) \rightarrow 4\ CO_2(g) + 4\ H_2O(l)$

$\Delta_c H° = 4 \cdot \Delta_f H° (CO_2) + 4 \cdot \Delta_f H° (H_2O) - \Delta_f H° (C_4H_8) + 6 \cdot \Delta_f H°(O_2)$ Note that the last term

will be 0 in all cases. Substitute the thermodynamic data for each of the three isomers:

cis-2-butene:

$\Delta_c H° = 4 \cdot \Delta_f H° (CO_2) + 4 \cdot \Delta_f H°(H_2O) - \Delta_f H°(C_4H_8)$

1 mol • -2709.8kJ/mol =

$\qquad$ 4 mol • -393.509kJ/mol+ 4 mol • -285.83 kJmol – 1 mol • $\Delta_f H°(C_4H_8)$

and solving for $\Delta_f H°(C_4H_8)$ yields -7.6 kJ

trans-2-butene:

1 mol • -2706.6kJ/mol =

$\qquad$ 4 mol • -393.509kJ/mol+ 4 mol • -285.83 kJmol – 1 mol • $\Delta_f H°(C_4H_8)$

and solving for $\Delta_f H°(C_4H_8)$ yields -10.8 kJ

1-butene:

1 mol • -2716.8kJ/mol =

$\qquad$ 4 mol • -393.509kJ/mol+ 4 mol • -285.83 kJmol – 1 mol • $\Delta_f H°(C_4H_8)$

and solving for $\Delta_f H°(C_4H_8)$ yields -0.6 kJ

(c)Relation of Enthalpies of Isomers to the elements:

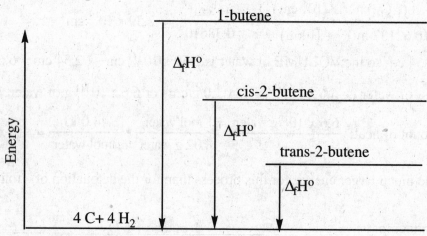

(d)Enthalpy change of cis-2-butene to trans-2-butene:

$\Delta_r H° = \Delta_f H°$ (trans-2-butene) - $\Delta_f H°$(cis-2-butene) = (-10.8 kJ) - (-7.6 kJ) = -3.2 kJ

103. (a)A sample of 0.850g Mg corresponds to 0.0350 mol Mg.

The amount of heat (evolved) is -25.4 kJ, corresponding to 25.4kJ/0.0350mol Mg = -726 kJ/mol

(b) Final temperature of water and bomb calorimeter:

Heat evolved = - Heat absorbed

-25400 J = -(820 J/K)ΔT + (750.g)(4.184 J/g•K)ΔT

-25400 J = -(820 J/K)ΔT + (3138 J/K)ΔT and –25400 J = -3958 J/K ΔT

-25400 J/-3958 J/K = 6.41 K   (or 6.41 ˚C—since a K and a ˚C are the same "size")

The new temperature of water will be 18.6 ˚C + 6.41˚C = 25.0 ˚C

105. (a) The energy diagram shown here
indicates that methane liberates
955.1 kJ/mol while methanol
liberates only 676.1 kJ/mol.

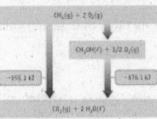

(b) Energy per gram:

For methane: $\dfrac{-955.1 \text{ kJ}}{1 \text{ mol}} \cdot \dfrac{1 \text{ mol}}{16.04 \text{ g}} = -59.54 \text{ kJ/g}$

For methanol: $\dfrac{-676.1 \text{ kJ}}{1 \text{ mol}} \cdot \dfrac{1 \text{ mol}}{32.04 \text{ g}} = -21.10 \text{kJ/g}$

(c) Enthalpy conversion from methane to methanol: The diagram indicates that the
difference in enthalpy for these two substances is the difference between the two "top
boxes". Hence $\Delta H = -955.1 \text{ kJ} - (-676.1 \text{ kJ}) = -279 \text{ kJ/mol}$

(d) The equation for conversion of methane to methanol: $CH_4(g) + 1/2\ O_2(g) \rightarrow CH_3OH(l)$

107. (a) Piece of metal to heat? To cool to achieve a maximum T? Final temperature of water?
To convey maximum heat per gram, one needs a metal with the greatest specific heat—
so of these 3 metals, Al, is the best candidate, and the larger piece of metal (1) would
convey MORE heat than the smaller piece(2) of Al. To minimize the heat absorbed by
the "cooler" metal, one needs a metal with the lesser specific heat—and the smaller the
better—so the smaller piece of Au (4)is a prime candidate. As to final T:

Heat loss (by warm metal) = Heat gain (by cool metal and water)

Note that the SIGNS of the two will be opposite, so let's (arbitrarily) place a (-) sign in
front of the "heat loss" side.

$-(100.0 \text{ g})(0.9002 \text{ J/g}\bullet\text{K})(T_f - 373) =$

$(50.0 \text{ g})(0.1289 \text{ J/g}\bullet\text{K})(T_f-263) + (300.\text{g})(4.184 \text{ J/g}\bullet\text{K})(T_f-294)$

$-90.02T_f + 33577 = 6.445T_f - 1695.035 + 1255.2T_f - 369028.8$

Collecting T_f terms:

$-90.02T_f + -6.445T_f + -1255.2T_f = -369028.8 + -33577 + -1695.035$ or

$-1351.665\ T_f = -404300.835$ and a $T_f = 299.1\text{K}$ or $(299-273) = 26°C$

(b) Process is similar to that in (a) but we want *minimal* T change:

Consider the following table of data and calculations:

Specific heat	Mass	Heat Capacity	Metal	ΔT for metal heated to 100°C	ΔT for metal cooled to - 10°C	Heat lost upon cooling to 21 °C	Heat gained upon warming to 21°C
0.9002	50.0	45.0	Al	79	31	3555.79	**1395.31**
0.3860	50.0	19.3	Zn	79	31	**1524.7**	598.3

Note that the amount of heat lost by cooling 50.0g Zn and the heat gained by warming 50.0g of Al is approximately equal. Obviously you could do these calculations for all the combinations of metals (both type and mass). Let's see how these two compute!

- Heat lost = Heat gained

- $(50.0 \text{ g})(0.3860 \text{ J/g}\bullet\text{K})(T_f - 373) =$

$$(50.0 \text{ g})(0.9002 \text{ J/g}\bullet\text{K})(T_f - 263) + (300.\text{g})(4.184 \text{ J/g}\bullet\text{K})(T_f - 294)$$

$-19.3T_f + 71989 = 45.01T_f - 11837.6 + 1255.2T_f - 369028.8$

Collecting T_f terms:

$-19.3T_f + -45.01T_f + -1255.2T_f = -369028.8 + -11837.6 - 71989$ or

$-1319.54 \ T_f = -388065.3$ and a $T_f = 294$K or (294 - 273) = 21°C

109. The work done on the surroundings as carbon dioxide sublimes:

When the solid is converted to gas, the **change** in volume is essentially 0.36L (since 1.0g of carbon dioxide(s) will occupy only a small volume. The work is then:

$w = -P \times \Delta V$.

$w = -1\text{atm} \times 0.36\text{L} \times \dfrac{101.3 \text{ J}}{1 \text{ L}\cdot\text{atm}} = -36 \text{ J} \ (2 \text{ sf})$

Applying Chemical Principles

1. (a) What is the Enthalpy change for: $2 KNO_3(s) + 3 C(s) + S(s) \rightarrow K_2S(s) + N_2(g) + 3 CO_2(g)$

$\Delta_r H = [(1mol)(-376.6 \text{ kJ/mol}) + (1mol)(0 \text{ kJ/mol}) + (3 \text{ mol})(-393.509 \text{ kJ/mol})]$

$- [(2 \text{ mol})(-494.6 \text{ kJ/mol}) + (3 \text{ mol})(0 \text{ kJ/mol}) + (1 \text{ mol})(0 \text{ kJ/mol})] = -802.3 \text{ kJ}$

$\Delta_r H = [(-1557.127 \text{ kJ}) - (-989.2 \text{ kJ})] = -567.9 \text{ kJ}$

(b) Using the assumption that 1 mol of black powder is: 2 mol KNO_3, 3 mol of C, and 1 mol of S, black powder would have a MM of $(2 \times 101.09) + (3 \times 12.011) + (1 \times 32.066) = 270.3$ g/mol.

$$\frac{-567.9 \text{ kJ}}{1 \text{ mol powder}} \cdot \frac{1 \text{ mol powder}}{270.3 \text{ g}} = -2.10 \text{ kJ/g}$$

3. (a) Balanced equation for the decomposition of nitroglycerin:

$4 C_3H_5N_3O_9(s) \rightarrow 12 CO_2(g) + 10 H_2O(g) + 6 N_2(g) + O_2(g)$

If you begin with a coefficient of 1 for "nitro", 3C in "nitro" gives 3 CO_2, 2.5 H_2O, 1.5 N_2, and 0.25 O_2. So the question is, "What coefficient converts ALL these coefficients into integers"? Multiplying all coefficients by 4 solves things nicely!

(b) The decomposition releases 6.23 kJ/g nitro. Sum the atomic weights of the involved atoms in nitroglycerine to get 227.0872 g/mol. The energy released PER MOLE of nitro is: 6.23 kJ/g • 227.0872 g/mol nitro = -1414.754 kJ/mol nitro. -1410 kJ/mol (3 sf)

Solve the equation for the Enthalpy change for the process:

$4 C_3H_5N_3O_9(s) \rightarrow 12 CO_2(g) + 10 H_2O(g) + 6 N_2(g) + O_2(g)$

$\Delta_r H = [(12 \text{ mol})(-393.509 \text{ kJ/mol})]) + (10 \text{ mol})(-241.83 \text{ kJ/mol}) + (6 \text{ mol})(0 \text{ kJ/mol}) + (1 \text{ mol})(0 \text{ kJ/mol})] - [(4 \text{ mol})(\Delta_r H \text{ nitro})]$

$-5640 \text{ kJ} = [(-7140.408 \text{ kJ}) - (4 \text{ mol})(\Delta_r H \text{ nitro})] =$

$-1500.408 \text{ kJ} = 4 \cdot \Delta_r H \text{ nitro}$, and dividing by 4 gives: -375 kJ/mol nitro $= \Delta_r H$ nitro

Chapter 6
The Structure of Atoms

PRACTICING SKILLS
Electromagnetic Radiation

1. Using Figure 6.2:

    (a) Microwave radiation is less energetic than X-ray radiation.

    (b) Red light uses higher frequency light than radar.

    (c) Infrared radiation is of longer wavelength than ultraviolet.

3. (a) The higher frequency light is the green 500 nm light.

    Recall that frequency and wavelength are inversely related.

    (b) The frequency of amber light (595 nm) is:

    $$\text{frequency} = \frac{\text{speed of light}}{\text{wavelength}} = \frac{2.9979 \times 10^8 \text{m/s}}{595 \text{ nm}} \cdot \frac{1.00 \times 10^9 \text{nm}}{1.00 \text{ m}}$$

    $$= 5.04 \times 10^{14} \text{ s}^{-1}$$

Electromagnetic Radiation and Planck's Equation

5. To calculate the energy of one photon of light with 500 nm wavelength, we need to first calculate the frequency of the radiation:

    $$\text{frequency} = \frac{\text{speed of light}}{\text{wavelength}} = \frac{2.9979 \times 10^8 \text{ m/s}}{5.0 \times 10^2 \text{ nm}} \cdot \frac{1.00 \times 10^9 \text{ nm}}{1.00 \text{ m}}$$

    $$= 6.0 \times 10^{14} \text{ s}^{-1}$$

    And the energy is then $E = h\upsilon$ or $(6.626 \times 10^{-34} \text{ J} \cdot \text{s} \cdot \text{photons}^{-1})(6.0 \times 10^{14} \text{ s}^{-1})$

    $$= 4.0 \times 10^{-19} \text{ J} \cdot \text{photons}^{-1}$$

    Energy of 1.00 mol of photons $= 4.0 \times 10^{-19} \text{ J} \cdot \text{photon}^{-1} \cdot \dfrac{6.0221 \times 10^{23} \text{ photons}}{1.00 \text{ mol photons}}$

    $$= 2.4 \times 10^5 \text{ J/mol photon}$$

7. The frequency of the line at 396.15 nm:

    $$\text{frequency} = \frac{\text{speed of light}}{\text{wavelength}} = \frac{2.9979 \times 10^8 \text{ m/s}}{3.9615 \times 10^2 \text{ nm}} \cdot \frac{1.00 \times 10^9 \text{nm}}{1.00 \text{ m}}$$

    $$= 7.5676 \times 10^{14} \text{ s}^{-1}$$

    The energy of a photon of this light may be determined: $E = h\upsilon$

    Planck's constant, h, has a value of $6.626 \times 10^{-34} \text{ J} \cdot \text{s} \cdot \text{photons}^{-1}$

    $E = (6.626 \times 10^{-34} \text{ J} \cdot \text{s} \cdot \text{photon}^{-1})(7.5676 \times 10^{14} \text{ s}^{-1}) = 5.0144 \times 10^{-19} \text{ J} \cdot \text{photon}^{-1}$

Energy of 1.00 mol of photons $= 5.0144 \times 10^{-19}$ J $\bullet$ photon$^{-1} \bullet \dfrac{6.0221 \times 10^{23} \text{photons}}{1.00 \text{ mol photons}}$

$= 3.02 \times 10^{5}$ J/mol photon or 302 kJ/mol photon.

9. Since energy is proportional to frequency ($E = h\upsilon$), we can arrange the radiation in order of increasing energy per photon by listing the types of radiation in increasing frequency (or decreasing wavelength).

$\rightarrow$ Energy increasing $\rightarrow$

FM music	microwave	yellow light	x-rays

$\rightarrow$ Frequency (υ) increasing $\rightarrow$

$\leftarrow$ Wavelength (λ) increasing $\leftarrow$

Photoelectric Effect

11. Energy $= 2.0 \times 10^{2}$ kJ/mol $\bullet \dfrac{1 \text{ mol}}{6.0221 \times 10^{23} \text{ photons}} \bullet \dfrac{1.00 \times 10^{3} \text{ J}}{1.00 \text{ kJ}} = 3.3 \times 10^{-19}$ J $\bullet$ photons^{-1}

What wavelength of light would provide this energy ?

$E = h\upsilon = \dfrac{hc}{\lambda}$ or $\lambda = \dfrac{hc}{E} = \dfrac{(6.626 \times 10^{-34} \text{ J} \bullet \text{s} \bullet \text{photons}^{-1})(2.9979 \times 10^{8} \text{ m} \bullet \text{s}^{-1})}{3.3 \times 10^{-19} \text{ J} \bullet \text{photons}^{-1}}$

$= 6.0 \times 10^{-7}$ m or 6.0×10^{2} nm

Radiation of this λ , (**visible** region) of the electromagnetic spectrum--would appear **orange**.

Atomic Spectra and the Bohr Atom

13. (a) The **most energetic light** would be represented by the light of **shortest wavelength** (253.652 nm).

(b) The frequency of this light is : $\dfrac{2.9979 \times 10^{8} \text{ m/s}}{253.652 \text{ nm}} \bullet \dfrac{1.00 \times 10^{9} \text{ nm}}{1.00 \text{ m}} = 1.18190 \times 10^{15}$ s^{-1}

The energy of 1 photon with this wavelength is:

$E = h\upsilon = (6.62608 \times 10^{-34} \dfrac{\text{J} \bullet \text{s}}{\text{photon}})(1.18190 \times 10^{15} \text{ s}^{-1}) = 7.83139 \times 10^{-19} \dfrac{\text{J}}{\text{photon}}$

(c) The line emission spectrum of mercury shows the visible region between $\approx$ 400 and 750 nm. The lines at 404 and 436 nm are present while the lines at 253 nm, 365 nm and 1013 nm lie outside the visible region. The 404 nm line is violet, while the 436 nm line is blue.

15. The Balmer series of lines terminates with $n_f = 2$. According to Figure 6.10, the transition originates at $n_i = 6$. Light of wavelength 410.2 nm would be violet.

17. (a) <u>Transitions from</u>   <u>to</u>

$$n = 5 \qquad\qquad n = 4, 3, 2, \text{ or } 1 \qquad (4 \text{ transitions})$$
$$n = 4 \qquad\qquad n = 3, 2, \text{ or } 1 \qquad\quad (3 \text{ transitions})$$
$$n = 3 \qquad\qquad n = 2 \text{ or } 1 \qquad\qquad\quad (2 \text{ transition})$$
$$n = 2 \qquad\qquad n = 1 \qquad\qquad\qquad\quad (1 \text{ transition})$$

Ten transitions are possible from these five quantum levels, providing 10 emission lines.

(b) Photons of the highest **frequency** are emitted in a transition from level of n = _5_ to a level with n = _1_ . Recalling that energy is directly proportional to frequency (E = hυ), the highest frequency will correspond to the highest energy—a transition from the two levels that differ most in energy.

(c) Emission line having the longest wavelength corresponds to a transition from level of n = _5_ to a level with n = _4_ . Levels 4 and 5 are closer in energy than other possibilities given here, so a transition between the two would be of longest wavelength (lowest E).

19. (a) Photons of the lowest energy will be emitted in a transition from the level with **n = 3** to the level **n = 2**. This is easily seen with the aid of the equation

$$\Delta E = Rhc\left(\frac{1}{n_f^{\,2}} - \frac{1}{n_i^{\,2}}\right).$$

Since R, h, and c are constant for any transition, inspection shows that the smaller change in energy results if $n_f = 3$ and $n_i = 2$. (The fractions in the equation above would correspond to $\left(\frac{1}{4} - \frac{1}{16}\right)$ for the 4→2 transition or $\left(\frac{1}{4} - \frac{1}{9}\right)$ for the 3→2 transition.

(b) Once again, using the equation above, the fraction in parenthesis (and hence ΔE) will be greater for the transition from $n_i = 4$ to $n_f = 1$ than for the transition from 5 →2.

21. The wavelength of emitted light for the transition n = 3 to n = 1.

$$\Delta E = -Rhc\left(\frac{1}{1^2} - \frac{1}{3^2}\right) \text{ and the value of Rhc = 1312 kJ/mol, so}$$

$$\Delta E = -1312 \text{ kJ/mol} \left(\frac{1}{1^2} - \frac{1}{3^2}\right) \text{ or } = -1312 \text{ kJ/mol } (8/9) = -1166 \text{ kJ/mol}$$

To calculate the frequency and wavelength, we use E = hυ. Recall that we must first express the energy **per photon** (as opposed to a mole of photons).

$$\Delta E = \frac{-1166 \text{ kJ/mol photons}}{6.022 x 10^{23} \text{ photons/1 mol photons}} \cdot \frac{10^3 J}{1 \text{ kJ}} = 1.936 \times 10^{-18} \text{ J/photon}$$

and solving for frequency, $\upsilon = \dfrac{1.936 \times 10^{-18} \text{ J/photon}}{6.626 \times 10^{-34} \text{ J} \cdot \text{ s /photon}} = 2.923 \times 10^{15} \text{ s}^{-1}$

Substituting into $\lambda\upsilon = c$, we get $\dfrac{2.998 \times 10^8 \text{ m}}{2.923 \times 10^{15} \text{ s}^{-1}} = 1.0257 \times 10^{-7} \text{m}$ (102.6 nm—far UV)

DeBroglie and Matter Waves

23. Mass of an electron: 9.11×10^{-31} kg

    Planck's constant: 6.626×10^{-34} J • s • photon^{-1}

    Velocity of the electron: 2.5×10^{8} cm • s^{-1} or 2.5×10^{6} m • s^{-1}

    $$\lambda = \frac{h}{mv} = \frac{6.626 \times 10^{-34} \text{ J} \cdot \text{s}}{(9.11 \times 10^{-31} \text{ kg} \cdot 2.5 \times 10^{6} \text{ m} \cdot \text{s}^{-1})}$$

    $$= 2.9 \times 10^{-10} \text{ m} = 2.9 \text{ Angstroms} = 0.29 \text{ nm}$$

25. The wavelength can be determined exactly as in Question 23:

    $$\lambda = \frac{h}{mv} = \frac{6.626 \times 10^{-34} \text{ J} \cdot \text{s}}{(1.0 \times 10^{-1} \text{ kg} \cdot 30. \text{ m} \cdot \text{s}^{-1})}$$

    $$= 2.2 \times 10^{-34} \text{ m} \quad \text{or} \quad 2.2 \times 10^{-25} \text{ nm}$$

    Velocity to have a wavelength of 5.6×10^{-3} nm

    (First convert the wavelength to units of meters)

    $$5.6 \times 10^{-3} \text{ nm} \cdot \frac{1 \text{ m}}{1 \times 10^{9} \text{ nm}} = 5.6 \times 10^{-12} \text{ m}$$

    Then rewriting the above equation:

    $$v = \frac{h}{m \cdot \lambda} = \frac{6.626 \times 10^{-34} \text{ J} \cdot \text{s}}{(1.0 \times 10^{-1} \text{ kg} \cdot 5.6 \times 10^{-12} \text{ m})} = 1.2 \times 10^{-21} \frac{\text{m}}{\text{s}}$$

Quantum Mechanics

27. (a) $n = 4$ possible l values = 0,1,2,3   (l = 0,1,... (n - 1))

    (b) $l = 2$ possible m_l values = -2,-1,0,+1,+2 (-l ...,0,....+ l)

    (c) orbital = 4s   possible values for n = 4; l = 0; m_l = 0

    (d) orbital = 4f   possible values for n = 4; l = 3; m_l = -3,-2,-1,0,+1,+2,+3

29. An electron in a 4p orbital must have n = 4 and l = 1. The possible m_l values give rise to the

    following sets of n, l, and m_l

n	l	m_l	
4	1	-1	Note that the **three values** of m describe
4	1	0	**three orbital orientations**.
4	1	+1	

31. Subshells in the electron shell with n = 4:

    There are 4: s, p, d, and f sublevels corresponding to l= 0, 1, 2, and 3 respectively.

    Recall that values of l from 0 to a maximum of (n-1) are possible.

33. Explain why each of the following is not a possible set of quantum numbers for an electron in an atom.
    (a) $n = 2$, $l = 2$, $m_l = 0$ For $n = 2$, maximum value of l is one (1).
    (b) $n = 3$, $l = 0$, $m_l = -2$ For $l = 0$, possible value of m_l is 0.
    (c) $n = 6$, $l = 0$, $m_l = 1$ For $l = 0$, possible value of m_l is 0.

35.

quantum number designation	maximum number of orbitals
(a) $n = 3$; $l = 0$; $m_l = +1$	none; for $l = 0$, the only possible value of $m_l = 0$
(b) $n = 5$; $l = 1$	3 ("**p**" orbitals)
(c) $n = 7$; $l = 5$	eleven; the # of orbitals is "$2 l +1$"
(d) $n = 4$; $l = 2$; $m_l = -2$	1 (one of the three 4 "**p**" orbitals)

37. Explain why the following sets of quantum numbers are not valid:
    (a) $n = 4$, $l = 2$, $m_l = 0$, $m_s = 0$:

    The possible values of m_s can only be +1/2 or -1/2. Change m_s to either +1/2 or -1/2.
    (b) $n = 3$, $l = 1$, $m_l = -3$, $m_s = -1/2$:

    The possible values for m_l is $-1......0.....+1$. Changing m_l to -1,0,+1 would give a valid set of quantum numbers.
    (c) $n = 3$, $l = 3$, $m_l = -1$, $m_s = +1/2$

    The maximum value of l is (n-1). Changing l to 2 would provide a valid set of quantum numbers.

39. Which of the following orbitals cannot exist and why:

2s exists	$n = 2$	permits l values as large as 1 ($l = 0$ is an s sublevel)
2d cannot exist		$l = 2$ is not permitted for $n < 3$ ($l = 2$ is a d sublevel)
3p exists	$n = 3$	permits l values as large as 2 ($l = 1$ is a p sublevel)
3f cannot exist		$l = 3$ is not permitted for $n < 4$ ($l = 3$ is an f sublevel)
4f exists	$l = 4$	permits l values as large as 3 ($l = 3$ is an f sublevel)
5s exists	$n = 5$	permits l values as large as 4 ($l = 0$ is an s sublevel)

41. The complete set of quantum numbers for :

	$\underline{n}$	$\underline{l}$	$\underline{m_l}$	
(a) 2p	2	1	$-1, 0, +1$	(3 orbitals)
(b) 3d	3	2	$-2, -1, 0, +1, +2$	(5 orbitals)
(c) 4f	4	3	$-3, -2, -1, 0, +1, +2, +3$	(7 orbitals)

43. With an $n = 4$, and $l = 2$, this orbital belongs in the 4th level ($n = 4$) and with an $l = 2$, this must be a "d" type orbital—so (d) 4d.

45. The number of nodal surfaces possessed by an orbital is equal to the value of the "l" quantum number, so for each of the following:

orbital		number of planar nodes
(a) 2s	(l = 0)	0
(b) 5d	(l = 2)	2
(c) 5f	(l = 3)	3

GENERAL QUESTIONS ON ATOMIC STRUCTURE

47. Concerning the photoelectric effect:

(a) Light is electromagnetic radiation—correct

(b) Intensity of light beam related to frequency—incorrect. The intensity is related to the number of photons of light.

(c) Light can be thought of as mass-less particles—correct.

49. Number of nodal surfaces for the following orbital types:

Orbital type	Nodal surfaces
s	0 (because l = 0)
p	1 (because l = 1)
d	2 (because l = 2)
f	3 (because l = 3)

51. Orbital types associated with values of "l"

Orbital type	Values of "l"
f	l = 3
s	l = 0
p	l = 1
d	l = 2

53.

Orbital Type	Number of orbitals in a given Subshell	Number of Nodal Surfaces
s	1	0
p	3	1
d	5	2
f	7	3

55. Regarding red and green light emitted by a sign:

   (a) Green light has the shorter wavelength and therefore higher energy photons.

   (b) Green light has higher energy photons than red light, the shorter wavelength (500 nm) is green.

   (c) Since frequency and wavelength are inversely related, the shorter wavelength (green) must have the higher frequency.

57. For radiation of 850 MHz:

   (a) The wavelength: $\dfrac{3.00 \times 10^8 \text{ m/s}}{850 \times 10^6 \text{ Hz}} \cdot \dfrac{1 \text{ Hz}}{1 \text{ s}^{-1}} = 0.35 \text{ m}$

   (b) The energy of 1.0 mol of photons with $\upsilon = 850$ MHz:

   $$E = \frac{hc}{\lambda} = \frac{(6.626 \times 10^{-34} \text{ J} \cdot \text{s} \cdot \text{photons}^{-1})(2.9979 \times 10^8 \text{ m} \cdot \text{s}^{-1})}{0.35 \text{ m}} \cdot \frac{6.0221 \times 10^{23} \text{photons}}{1 \text{ mol photons}}$$

   $= 0.34$ J/mol

   (c) The energy of a mole of photons of 420 nm light:

   $$E = \frac{hc}{\lambda} = \frac{(6.626 \times 10^{-34} \text{ J} \cdot \text{s} \cdot \text{photons}^{-1})(2.9979 \times 10^8 \text{ m} \cdot \text{s}^{-1})}{420 \times 10^{-9} \text{ m}} \cdot \frac{6.0221 \times 10^{23} \text{photons}}{1 \text{ mol photons}}$$

   $= 2.8 \times 10^5$ J/mol or 280 kJ/mol

   (d) The energy of a mole of photons of blue light is much greater than that of the corresponding photons from a cell phone.

59. Example 6.3 in your text illustrates the calculation of the ionization energy for H's electron

   $$E = \frac{-Z^2Rhc}{n^2} = -2.179 \times 10^{-18} \text{ J/atom} \Rightarrow -1312 \text{ kJ/mol}$$

   For He$^+$ the calculation yields

   $$E = \frac{-(2)^2(1.097 \times 10^7 \text{ m}^{-1})(6.626 \times 10^{-34} \text{ J} \cdot \text{s})(2.998 \times 10^8 \text{ m} \cdot \text{s}^{-1})}{(1)^2}$$

   $= -8.717 \times 10^{-18}$ J/ion and expressing this energy for a mol of ions

   $$= \frac{-8.717 \times 10^{-18} \text{J}}{\text{ion}} \cdot \frac{6.0221 \times 10^{23} \text{atoms}}{\text{mol}} \cdot \frac{1 \text{ kJ}}{1000 \text{ J}} = -5248 \frac{\text{kJ}}{\text{mol}}$$

   The energy to remove the electron is 5248 kJ/mol of ions. Note that this energy is four times that for H.

61. Orbitals in a H atom in order of increasing energy: 1s < 2s = 2p < 3s = 3p = 3d < 4s

   For the hydrogen atom, energy levels increase with increasing values of "n".

63. Wavelength and frequency for a photon with E = 1.173MeV:

Using the energy relationship: $E = h\upsilon$, we solve for frequency:

$$\upsilon = \frac{1.172 \times 10^6 \text{ ev}}{6.626 \times 10^{-34} \text{ J} \bullet \text{s}} \bullet \frac{9.6485 \times 10^4 \text{ J/mol}}{1 \text{ ev}} \bullet \frac{1 \text{ mol}}{6.022 \times 10^{23} \text{ photons}} = 2.836 \times 10^{20} \text{ s}^{-1}$$

Substituting into the wavelength-frequency relationship, we have:

$$\lambda = c/\upsilon = \frac{2.998 \times 10^8 \text{ m/s}}{2.836 \times 10^{20} \text{ s}^{-1}} = 1.057 \times 10^{-12} \text{ m}$$

65. Time for Sojourner's signal to travel 7.8×10^7 km:

If light travels at 2.9979×10^8 m $\bullet$ s^{-1}, we can calculate the time:

$$\frac{7.8 \times 10^7 \text{ km}}{1} \bullet \frac{1 \times 10^3 \text{ m}}{1 \text{ km}} \bullet \frac{1 \text{ s}}{2.9979 \times 10^8 \text{ m}} = 260 \text{ s}$$

Given that there are 60 s in 1 minute: $260 \text{ s} \bullet \frac{1 \text{ min}}{60 \text{ s}} = 4.3$ minutes

67. (a) The quantum number n describes the **size (and energy)** of an atomic orbital.

(b) The shape of an atomic orbital is given by the quantum number l.

(c) A photon of green light has **more energy** than a photon of orange light.

(d) The maximum number of orbitals that may be associated with the quantum numbers n= **4**, l = 3 is **seven.** (corresponding to m_l values of $\pm 3, \pm 2, \pm 1$, and 0)

(e) The maximum number of orbitals that may be associated with the quantum numbers n= **3**, l = **2**, and m_l = -2 is **one**.

(f) The orbital on the left is a **d orbital** and the one on the right is a **p orbital**, while the orbital in the middle is an **s orbital**.

(g) When n = 5, the possible values of l **are 0,1,2,3, and 4**. (Range is 0 ..... (n-1))

(h) The maximum number of orbitals that can be assigned to the n = 4 shell is **16**.

n = 4	l = 0	m_l = 0	1 orbital
n = 4	l = 1	m_l = -1,0,+1	3 orbitals
n = 4	l = 2	m_l = -2,-1,0,+1,+2	5 orbitals
n = 4	l = 3	m_l = -3,-2,-1,0,+1,+2,+3	7 orbitals
			16 orbitals

(i) The Co^{2+} ion has unpaired electrons. A sample of a Co(II) salt will be **paramagnetic**.

69. The diagram below, indicates that in (b) all the electrons are paired—giving rise to a diamagnetic substance—which is NOT strongly attracted to a magnetic field. In (a), the unpaired electron spins are aligned within the region (also known as **ferromagnetic**);

in (c) the unpaired spins are not aligned—a paramagnetic material. The ferromagnetic material (a), will be **strongly attracted** to the magnetic field.

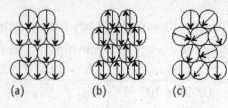

(a)   (b)   (c)

IN THE LABORATORY

71. The pickle glows since the materials in the pickle are being "excited" by the addition of the energy (electric current). Since the pickle has been soaked in brine (NaCl), the electrons in the sodium atom are excited and release energy as they "return" to lower energy states, providing "yellow" light. The same kind of light is visible in many street lamps.

73. (a) The wavelength of light at 2000 cm^{-1}? $\lambda = 1/2000$ cm^{-1}, or 0.0005 cm.

(b) Which is the low energy and high energy end of the spectrum?
Since E is proportion to the reciprocal wavelength, the larger the value of wavenumber, the greater the Energy. The low energy is the **right side** while the higher energy end is the **left side** of the spectrum.

(c) Which interaction (3300-3400 cm^{-1}) or (2800=3000 cm^{-1}) requires more energy?
Since E is proportional to wavenumber, the 3300-3400cm^{-1} requires more energy.

SUMMARY AND CONCEPTUAL QUESTIONS

75. The light visible from a sodium or mercury street light arises owing to (c) electrons are moving from a given energy level to one of lower n (and hence of lower energy).

77. The "wave-particle duality" describes the behavior of electrons—in which the particulate nature of the electron (e.g. as observed in the photoelectric effect) and the undulatory nature of the electron (e.g. the observation of diffraction patterns in the Davisson-Germer experiment) are both important to our explanation of natural phenomena. The implications of this duality arise in our understanding of electrons in atoms as waves existing in regions of high probability.

79. The Heisenberg Uncertainty Principle answers this question for us. This principle says that it is impossible to **simultaneously determine** both the position and energy of an electron. Reducing the uncertainty about one of these parameters increases the uncertainty of the measurement of the other parameter. So we can determine (a) OR (b), but not (c).

81. A photon with a wavelength of 93.8 nm can excite the electron up to level n = 6 (Figure 6.10). Transitions from n = 6 to lower values of n are possible. Transitions from level 6-> 5,4,3,2,1 — **5** transitions; from level 5-> 4,3,2,1 — **4** transitions; from level 4->3,2,1 — **3** transitions; from level 3->2,1 — **2** transitions; from level 2->1 — **1** transition. Total transitions: 5+4+3+2+1= 15 transitions.

83. Regarding Technetium:
(a) Technetium is in group 7B of the fifth period.
(b) Quantum numbers for an electron in the 5s subshell: n = 5, l = 0, m_l = 0
(c) Wavelength and frequency of a photon with energy of 0.141 MeV:
Using E = hυ, we solve for frequency:

$$\upsilon = \frac{0.141 \times 10^6 \text{ ev}}{6.626 \times 10^{-34} \text{J} \bullet \text{s}} \bullet \frac{9.6485 \times 10^4 \text{ J/mol}}{1 \text{ ev}} \bullet \frac{1 \text{ mol}}{6.022 \times 10^{23} \text{ photons}} = 3.41 \times 10^{19} \text{ s}^{-1}$$

Substituting into the wavelength-frequency relationship, we have:

$$\lambda = c/\upsilon = \frac{2.998 \times 10^8 \text{ m/s}}{3.41 \times 10^{19} \text{ s}^{-1}} = 8.79 \times 10^{-12} \text{m}$$

(d) In the preparation of $NaTcO_4$:
(i) The balanced equation: $HTcO_4$(aq) + NaOH(aq) → $NaTcO_4$(aq) + H_2O(l)
(ii) Mass of $NaTcO_4$ from 4.5 mg of Tc:

$$4.5 \times 10^{-3} \text{g Tc} \bullet \frac{1 \text{ mol Tc}}{98 \text{ g Tc}} \bullet \frac{1 \text{ mol NaTcO}_4}{1 \text{ mol Tc}} \bullet \frac{185 \text{ g NaTcO}_4}{1 \text{ mol NaTcO}_4} = 8.5 \times 10^{-3} \text{ g NaTcO}_4$$

Mass of NaOH required:

$$4.5 \times 10^{-3} \text{g Tc} \bullet \frac{1 \text{ mol Tc}}{98 \text{ g Tc}} \bullet \frac{1 \text{ mol HTcO}_4}{1 \text{ mol Tc}} \bullet \frac{1 \text{ mol NaOH}}{1 \text{ mol HTcO}_4} \bullet \frac{40.0 \text{ g NaOH}}{1 \text{ mol NaOH}} = 1.8 \times 10^{-3} \text{ g NaOH}$$

(e) What mass of $NaTcO_4$ corresponds to 1.5 micromol?

$$1.5 \times 10^{-6} \text{ mol NaTcO}_4 \bullet \frac{184.8944 \text{ g NaTcO}_4}{1 \text{ mol NaTcO}_4} = 0.00028 \text{ g or } 0.28 \text{ mg}$$

The concentration would be: 1.5×10^{-6} mol $NaTcO_4$/0.010 L or 1.5×10^{-4} M.

85. A photon of wavelength 97.3 nm strikes a H atom. Several emission lines are observed. Figure 6.10 indicates that, when a photon of this wavelength (energy corresponding to a transition of n=4 to n = 1) strikes a hydrogen atom, lines will be observed at three λ: 121.6nm, 486.1nm, and 656.3 nm—corresponding to electronic transitions between the higher energy levels. Transitions from n=4 give 3 lines (to n=3,2,1); from n=3 gives 2 lines (to n=2,1) and n=2 gives 1 line(to n=1) = 6 total.

Applying Chemical Principles

1. What is the frequency of light with wavelength = 587.6 nm?

   We know that wavelength and frequency of light are related by the equation: $\lambda \cdot \nu = c$.

   So $\nu = \dfrac{c}{\lambda} = \dfrac{3.000 \times 10^8 \, m/s}{587.6 \times 10^{-9} \, m} = 5.106 \times 10^{14} \, /s$

3. What is the energy(in Joules) of photons at 589.00 nm and 589.59 nm?

   $$E = h\nu = h \cdot \frac{c}{\lambda} = \frac{6.626 \times 10^{-34} \, J \cdot s/photon}{589.00 \times 10^{-9} m} \cdot 2.998 \times 10^8 m/s = 3.373 \times 10^{-19} \frac{J}{photon}$$

   and $\dfrac{6.626 \times 10^{-34} \, J \cdot s/photon}{589.59 \times 10^{-9} m} \cdot 2.998 \times 10^8 m/s = 3.369 \times 10^{-19} \dfrac{J}{photon}$

   What is the difference in energy between the two photons?

   So $\Delta E = (3.373 - 3.369) \times 10^{-19}$ J/photon or 4×10^{-22} J/photon

5. What are the final and initial electronic states (n) for the hydrogen line (in the Balmer series) labeled F in the figure? The line labeled F is at approximately 485 nm (4861 Å). What is certain is that the FINAL n value for this line is n=2 (which is true for all lines in the Balmer series). Refer to Figure 6.10 in your text to see that the line at 4861 Å results from the transition of an electron from n = 4 to n =2.

Chapter 7
The Structure of Atoms and Periodic Trends

PRACTICING SKILLS

Writing Electron Configurations of Atoms

1. The orbital box and spdf notation for P and Cl:

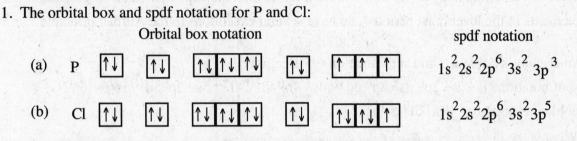

Orbital box notation   spdf notation

(a) P $1s^2 2s^2 2p^6 3s^2 3p^3$

(b) Cl $1s^2 2s^2 2p^6 3s^2 3p^5$

Note that Cl is in group 7A (17) indicating that there are SEVEN electrons in the outer shell, while P is in group 5A (15) indicating that there are FIVE electrons in the outer shell. Both Cl and P are on the "right side" of the periodic table—where elements have their "outermost" electrons in p subshells.

3. Electron configuration of chromium and iron:

(a) Cr: $1s^2 2s^2 2p^6 3s^2 3p^6 3d^5 4s^1$

(b) Fe: $1s^2 2s^2 2p^6 3s^2 3p^6 3d^6 4s^2$

The "surprising" configuration of elemental chromium—compared to the electron configuration of the preceding element, vanadium, arises from the stability of the "half-filled 3d sublevel" (which can be visualized as having a 4s electron occupy a 3d orbital—with the resultant $4s^1$ configuration).

5. (a) Arsenic's electron configuration: (33 electrons)

   spdf notation: $1s^2 2s^2 2p^6 3s^2 3p^6 3d^{10} 4s^2 4p^3$

   noble gas notation: $[Ar] 3d^{10} 4s^2 4p^3$

   (b) Krypton's electron configuration: (36 electrons)

   spdf notation: $1s^2 2s^2 2p^6 3s^2 3p^6 3d^{10} 4s^2 4p^6$

   noble gas notation: $[Kr]$

7. (a) Tantalum's noble gas and spdf notation (73 electrons)

   spdf notation: $1s^2 2s^2 2p^6 3s^2 3p^6 3d^{10} 4s^2 4p^6 4d^{10} 4f^{14} 5s^2 5p^6 5d^3 6s^2$

   noble gas notation: $[Xe] 4f^{14} 5d^3 6s^2$

(b) Platinum's noble gas and spdf notation (78 electrons)

  spdf notation: $1s^2\ 2s^2\ 2p^6\ 3s^2\ 3p^6\ 3d^{10}\ 4s^2\ 4p^6 4d^{10}\ 4f^{14}\ 5s^2\ 5p^6\ 5d^9 6s^1$

  noble gas notation: $[Xe]\ 4f^{14}\ 5d^9\ 6s^1$

Tantalum's configuration is expected, with Ta in period 5. and in group 5B. Platinum's configuration is a bit unexpected, but like other transition metals, it attempts to fill that d sublevel, resulting in a $5d^9 6s^1$ configuration rather than the expected $5d^8 6s^2$. Transition elements in the levels past period 4, do have several exceptions to the Aufbau principle.

9. Americium's noble gas and spdf notation (95 electrons)

  spdf notation: $1s^2\ 2s^2\ 2p^6\ 3s^2\ 3p^6\ 3d^{10}\ 4s^2\ 4p^6 4d^{10}\ 4f^{14}\ 5s^2\ 5p^6 5d^{10} 5f^7 6s^2 6p^6 7s^2$

  noble gas notation: $[Rn]\ 5f^7 7s^2$

Quantum Numbers and Electron Configurations

11. Maximum number of electrons associated with the following sets of quantum numbers:

	Characterized as	Maximum number of electrons
(a) $n = 4, l = 3, m_l = 1$	4f electrons	2 (an "f" orbital)
(b) $n = 6, l = 1, m_l = -1, m_s = -1/2$	6p electron	1
(c) $n = 3, l = 3, m_l = -3,$	NONE	With $n = 3$, the maximum value of l can be 2 (i.e. n-1)

13. The electron configuration for Mg using the orbital box method:

Mg: 1s   2s   2p   3s

Electron number   11 12

The noble gas notation: $[Ne]3s^2$

Electron number:	n	l	m_l	m_s
11	3	0	0	+ 1/2
12	3	0	0	- 1/2

15. The electron configuration for Gallium using the orbital box method:

Noble gas configuration

Ga [Ar]   3d   4s   4p $[Ar]\ 3d^{10}\ 4s^2\ 4p^1$

A possible set of quantum numbers for the highest energy electron:

n	l	m_l	m_s
4	1	-1	+ 1/2

Electron Configurations of Atoms and Ions and Magnetic Behavior

17. The orbital box representations for the following ions:

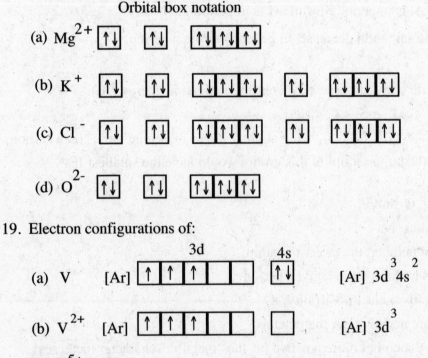

Orbital box notation

(a) Mg^{2+}

(b) K^+

(c) Cl^-

(d) O^{2-}

19. Electron configurations of:

(a) V   [Ar]   3d   4s   [Ar] $3d^3 4s^2$

(b) V^{2+} [Ar]   [Ar] $3d^3$

(c) V^{5+} [Ar]

Note that the V^{2+} ion contains unpaired electrons, and is therefore paramagnetic.

21. Manganese's orbital box and noble gas diagrams:

(a) Mn   [Ar]   [Ar] $3d^5 4s^2$

(b) Mn^{4+} [Ar]   [Ar] $3d^3$

(c) The ion is paramagnetic.

(d) As seen in part b, the ion contains 3 unpaired electrons.

Periodic Properties

23. Elements arranged in order of increasing size: C < B < Al < Na < K

Radii from the Figure 7.6 (in pm)   77 < 83 < 143 < 186 < 227

Since K is in period 4, we anticipate it being larger than Na, its analog in 1A (period 3).

Al is to the right of Na, so we expect it to be smaller than Na.

B and C are in period 2, with B to the right of C, and therefore larger than C.

25. The specie in each pair with the larger radius:

(a) Cl^- is larger than Cl -- The ion has more electrons/proton than the atom.

(b) Al is larger than O -- Al is in period 3, while O is in period 2.

(c) In is larger than I -- Atomic radii decrease, in general, across a period.

27. The group of elements with correctly ordered increasing ionization energy (IE):

(c) Li < Si < C < Ne.

Neon would have the greatest IE. Silicon, being slightly larger in atomic radius than carbon, has a lesser IE. Lithium, the largest atom of this group, would have the smallest IE.

29. For the elements Na, Mg, O, and P:

(a) The largest atomic radius: Na

The greater the period number, the larger the atom.

Radius also decreases to the right in a given period.

(b) The largest (most negative) electron affinity: O

Nonmetals have a more negative EA than metals.

Down a group, the EA becomes more positive (as the "metallic" character increases).

(c) Increasing ionization energy: Na < Mg < P < O

The ionization energy varies inversely with the atomic radius.

The smaller the atom, the greater the IE.

31. (a) Increasing ionization energy: S < O < F

Ionization energy is inversely proportional to atomic size.

(b) Largest ionization energy of O, S, or Se: O

Oxygen is the smallest of these Group 6A elements, and hence has the largest IE.

(c) Most negative electron affinity of Se, Cl, or Br: Cl

Chlorine is the smallest of these three elements. EA tends to increase on a diagonal from the lower left of the periodic table to the upper right..

(d) Largest radius of O^{2-}, F^-, F: O^{2-}

The oxide ion has the largest electron : proton ratio. If one considers the attraction of the nuclear species (protons) for the extranuclear species (electrons), the greater the number of protons/electron the smaller the specie—owing to an increased electron-proton attraction. So the oxide ion (with 10 electrons and 8 protons) has the fewest electrons (of these three species) per proton.

GENERAL QUESTIONS

33. The electron configuration for U and U^{4+} :

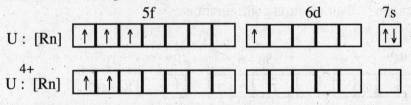

Both species have unpaired electrons and are therefore **paramagnetic**.

35. Using the spectroscopic notation the atom described would have an electron configuration:
$$1s^2 \, 2s^2 \, 2p^6 \, 3s^2 \, 3p^6 \, 4s^2$$

(a) Adding the electrons gives a sum of 20. Since this is a **neutral** atom, the # of protons and electrons wouldbe equal. Twenty protons gives this element an atomic number = 20.

(b) There are 2 s electrons in each of 4 shells : 8 s electrons total

(c) There are 6 p electrons in each of 2 shells:12 p electrons total

(d) There are 0 d electrons

(e) With its outer electrons in an "s" sublevel--specifically the 4s sublevel, the element is a **metal**-- specifically elemental **Calcium**.

37. Consider the sets of quantum numbers below:

	n	l	m_l	m_s	Possible elements (if allowable)
(a)	2	0	0	$-\frac{1}{2}$	Li or Be
(b)	1	1	0	$+\frac{1}{2}$	Not allowable, since max. "l" is n-1 (and n = 1)
(c)	2	1	-1	$-\frac{1}{2}$	B,C,N,O,F, or Ne
(d)	4	2	+2	$-\frac{1}{2}$	Y,Zr,Nb,Mo,Tc,Ru,Rh,Pd,Ag, or Cd

39. (a) Electron configuration for Nd, Fe, and B

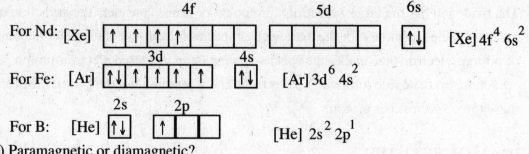

(b) Paramagnetic or diamagnetic?

All three species have unpaired electrons—therefore all are paramagnetic.

(c) For the ions Nd^{3+} and Fe^{3+} their electron configurations:

For the Nd^{3+} ion

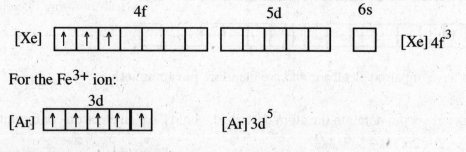

For the Fe^{3+} ion:

[Ar] [diagram: 3d with ↑ ↑ ↑ ↑ ↑]   [Ar] $3d^5$

Both ions have unpaired electrons, and are paramagnetic.

41. Arranged in order of increasing ionization energy for K, Ca, Si, and P.

As we move across Period 4, we find K, then Ca. In Period 3, we find Si, then P. With IE increasing across a period (which means that K < Ca and Si < P), and decreasing down a group (so elements in Period 4 generally have lower IE than analogues in Period 3), we expect the ionization energy to increase in the order: K < Ca < Si < P.

43. For element A = [Kr] $5s^1$ and B = [Ar] $3d^{10}4s^24p^4$

(a) Element A is a metal (1 electron in the s sublevel).

(b) Element B has the greater IE. B's atomic radius would be smaller than that of A.

(c) Element A has the less negative EA. In general EA for nonmetals is more positive than that of the metals in the same period.

(d) A would have the larger atomic radius—with its outermost electrons in the "5th" shell.

(e) As A has but one electron in the outer level, it would tend to form a 1+ ion. Similarly B has 6 electrons in the outer level, and would tend to form a 2- ion. The formula for the compound between these two ions would be A_2B. (Rb_2Se)

45. Ions not likely to be found include: In^{4+}, Fe^{6+}, and Sn^{5+}

Indium would form a cation by losing **three** electrons, forming the In^{3+} cation.

Iron can lose either 2 or 3 electrons, forming Fe^{2+} and Fe^{3+} ions. Loss of more electrons would necessitate loss of d electrons from a half-filled d subshell.

Tin can lose up to 4 electrons (2 s and 2 p) to form the 4+ cation. Loss of more electrons would require removal of electrons from a filled d subshell.

47. (a) Element with the largest atomic radius: Se has the largest radius—being in period 4, group 6A. S would be smaller. Cl would be smaller than S. Recall the fact that atomic radius decreases across a period.

(b) Br^- is larger than Br—having more electrons per proton.

(c) Na would have the largest **difference** between the 1st and 2nd IE. Removing one electron would provide the stable 1+ cation and provide a specie with a filled shell.

(d) Element with the largest ionization energy: N (IE is inversely proportional to atomic radius).

(e) Largest radius: N^{3-} would be the largest, since the number of electrons/proton is greatest for the N^{3-} ion and smallest for the F^- ion.

49. For the elements Na, B, Al, and C:

(a) Largest atomic radius: Na and Al have electrons in level 3, while B and C have valence electrons in level 2. So Na and Al are larger than B and C. With 3 electrons in the outer level, Al is going to be smaller than Na (with only 1 electron), so **Na** is the largest atom.

(b) Largest electron affinity: Since electron affinity increases across the periodic table, we anticipate that the electron affinity for C will be greater than that of B. Also EAs tend to decrease down a group, so we anticipate that Na and Al will be less than C or B, so **C** has the largest EA.

(c) Increasing order of ionization energy: With B and C being in period 2, and Na and Al in period 3, we would expect B and C to have greater IEs than that of Na and Al. With Na in group 1, and Al in group 3, we expect Al to have a greater IE than Na, so the order of increasing IE is Na < Al < B < C.

51. (a) The element (containing 27 electrons) is Cobalt

(b) The sample contains unpaired electrons—so it is paramagnetic

(c) The 3+ ion of cobalt would be formed by the loss of the two 4s electrons and 1 of the d electrons— leaving four unpaired electrons.

IN THE LABORATORY

53. (a) The theoretical yield of nickel(II) formate from 0.500g of nickel(II)acetate and excess formic acid:

$$\frac{0.500 \text{ g Ni(CH}_3\text{CO}_2)_2}{1} \cdot \frac{1 \text{ mol Ni(CH}_3\text{CO}_2)_2}{176.78 \text{ g Ni(CH}_3\text{CO}_2)_2} \cdot \frac{1 \text{ mol Ni(HCO}_2)_2}{1 \text{ mol Ni(CH}_3\text{CO}_2)_2} \cdot \frac{148.73\text{g Ni(HCO}_2)_2}{1 \text{ mol Ni(HCO}_2)_2} =$$

$$= 0.421 \text{ g Ni(HCO}_2)_2$$

(b) Nickel(II) formate is paramagnetic. The Ni(II) ion has 8 d electrons (2 of which are unpaired).

(c) Mass of nickel derived by heating 253 mg of nickel(II) formate:

$$\frac{0.253 \text{ g Ni(HCO}_2)_2}{1} \cdot \frac{1 \text{ mol Ni(HCO}_2)_2}{148.73 \text{ g Ni(HCO}_2)_2} \cdot \frac{58.69 \text{ g Ni}}{1 \text{ mol Ni(HCO}_2)_2} = 0.0998 \text{ g Ni}$$

or 99.8 mg Ni

Will the powder adhere to a magnet? Elemental nickel has two unpaired electrons, and the powder is expected to stick to the magnet.

SUMMARY AND CONCEPTUAL QUESTIONS

55. The radius of Li^+ is smaller than that of Li owing to the fact that the lithium ion has electrons in the 1st energy level, while atomic Li has electrons in the 1st and 2nd energy levels. Also note that the Li^+ has 3 protons and only 2 electrons, while atomic Li has 3 protons and 3 electrons. This proton:electron ratio would also "explain" the relatively larger size of the Li atom over its cation. Using similar logic, the radius of F^- is anticipated to be larger than that of F, since F^- has 9 protons and 10 electrons, while atomic F has 9 protons and 9 electrons. The proton:electron ratio would have us anticipate that the atomic F is smaller than the anion.

57. Using the data provided, deduce the group in the periodic table to which Element 1 and Element 2 belong:

IE (kJ/mol)	Element 1	Δ ({n}IP-{n-1}IP)	Element 2	Δ ({n}IP-{n-1}IP)
1st IP	1086.2		577.4	
		1266		1239
2nd IP	2352		1816.6	
		2268		928
3rd IP	4620		2744.6	
		1602		**8830**
4th IP	6222		11575	

The large difference between the 3rd IP and 2nd IP for Element 1 indicates that the 3rd electron is probably from an s subshell while the first 2 electrons are from a p subshell, so Element 1 is probably from **Group IVA** (e.g. C).

A quick look at Element 2's data shows a **very large** increase between the 3rd and 4th IP. This might be typical of an element in **Group III**, in which the 1st electron would be removed from a "p" sublevel while the 2nd and 3rd electrons would reside in an "s" sublevel.

59. Hund's Rule answers this question for us. The rule states that the most stable (lowest energy) configuration is obtained when electrons occupy orbitals singly with similar spins (d). The highest energy (least stable) would be (a) in which the electrons are paired in one orbital before all 3 orbitals have at least one electron.

61. Electron configurations for the two ionizations of the K atom:
Electron configuration for K: $1s^22s^22p^63s^23p^64s^1$. The loss of one electron (1st ionization) gives a configuration of: $1s^22s^22p^63s^23p^6$. The loss of the second electron (2nd ionization) gives a configuration of: $1s^22s^22p^63s^23p^5$. Note that the 1st ionization results in a specie that is isoelectronic with a noble gas (Ar), and is not anticipated to be very large. However, the loss of the 2nd electron disrupts the filled 3p sublevel—and we anticipate (and find to be experimentally true) that this 2nd IE will be much greater than the first IE.

63. Explain the trends in size:
    (a) The decrease in atomic size across a period is due to the increasing nuclear charge with an increasing number of protons. Given that the electrons are in the same outer energy level, the nuclear attraction for the electrons results in a diminishing atomic radius.
    (b) The slight decrease in atomic radius of the transition metals is a result of increased repulsions of (n-1)d electrons for (n)s electrons. This repulsion reduces the effects of the increasing nuclear charge across a period.

65. The existence of Mg^{2+} and O^{2-} rather than their monopositive and mononegative counterparts is easily argued from the experimental evidence that members of group 2A (Mg) typically lose 2 electrons while members of group 6A (O) typically gain 2 electrons—both with the aim of becoming isoelectronic with a noble gas. One form of experimental evidence is in measuring the relative melting points of MgO versus a +1 salt (e.g. NaCl). MgO melts about 2850 °C, while NaCl (a +1:-1) salt melts about 800 °C. The greater charges of the +2/-2 ions are one factor for the much greater melting point of MgO over that of the +1/-1 NaCl salt.

67. (a) Orbital energies decrease across period 2, owing to the increased effective nuclear charge. With the electrons in "p" sublevels—and hence about the same average distance from the nucleus, the increasing nuclear charge (with increasing numbers of protons in the nucleus) results in a greater attraction for the valence electrons—and a lower orbital energy.

(b) As the orbital energies decrease (become more negative), the amount of energy needed to remove an electron (the Ionization Energy) increases. With this decreasing orbital energy, an atom has a greater affinity to **gain** an electron.

(c) The data show:

Li	-520.0 kJ/mol	(2s)
Be	-899.2 kJ/mol	(2s)
B	-800.8 kJ/mol	(2p)
C	-1029 kJ/mol	(2p)

Li has the lowest ioinization energy, owing to the fact that the loss of the sole 2s electron results in a noble gas configuration for the ion. Be has a higher IE, since removal of an electron disrupts the pairing of electrons (in 2s), The slightly lower IE for B (compared to Be) isn't surprising, since removal of the 2p electron would give B an electron configuration equal to that of Be (with a pair of electrons). Finally the 2nd 2p electron (for C) is not spin paired with the first 2p electron, and has a higher effective nuclear charge than B, so we anticipate that C will have a greater IE than B.

69. A plot of the atomic radii of the elements K—V shows a decrease. With the mass of these elements increasing from K through V (as more protons, neutrons, and electrons are added), the density is expected to increase.

71. (a) Electron configurations for elements 113 and 115
Element 113 is located in period 7, group 3A while 115 is in group 5A.
The anticipated electron configrations are:
113   $[Rn]5f^{14}6d^{10}7s^27p^1$
115   $[Rn]5f^{14}6d^{10}7s^27p^3$

(b) Name an element in the same periodic group as 113 and 115.
Thallium is a group member of 3A –like 113, while Bismuth is a group member of 5A—like 115.

(c) What atom could be used as a projectile to bombard Am to produce the element 113?

With 95 protons in an Americium nucleus, we'd need to add 18 protons to form the 113 nucleus. Hence an element with 18 protons would be called for (Argon).

73. (a) Orbital box notation for sulfur:

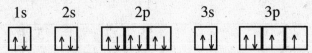

(b) Quantum numbers for the "last electron": $n = 3$, $l = 1$, $m_l = +1$ (or 0 or -1), $m_s = -1/2$

(c) Element with the smallest ionization energy: S

Element with the smallest radius: O

(d) S: Negative ions are always larger than the element from which they are derived.

(e) Grams of SCl_2 to make 675 g of $SOCl_2$:

$$\frac{675 \text{ g SOCl}_2}{1} \cdot \frac{1 \text{ mol SOCl}_2}{119.0 \text{ g SOCl}_2} \cdot \frac{1 \text{ mol SCl}_2}{1 \text{ mol SOCl}_2} \cdot \frac{102.97 \text{ g SCl}_2}{1 \text{ mol SCl}_2} = 584 \text{ g SCl}_2$$

(f) The theoretical yield of $SOCl_2$ if 10.0 g of SO_3 and 10.0 g of SCl_2 are used:

$$\text{Moles of SO}_3 = 10.0 \text{ g SO}_3 \cdot \frac{1 \text{ mol SO}_3}{80.06 \text{ g SO}_3} = 0.125 \text{ moles SO}_3$$

$$\text{Moles of SCl}_2 = 10.0 \text{ g SCl}_2 \cdot \frac{1 \text{ mol SCl}_2}{102.97 \text{ g SCl}_2} = 0.0971 \text{ moles SCl}_2$$

$$\text{Moles-available ratio: } \frac{0.125 \text{ mol SO}_3}{0.0971 \text{ mol SCl}_2} = \frac{1.29 \text{ mol SO}_3}{1 \text{ mol SCl}_2}$$

$$\text{Moles-required ratio: } \frac{1 \text{ mol SO}_3}{1 \text{ mol SCl}_2} \text{ so } \textbf{SCl}_2 \textbf{ is the limiting reagent.}$$

$$0.0971 \text{ moles SCl}_2 \cdot \frac{1 \text{ mol SOCl}_2}{1 \text{ moles SCl}_2} \cdot \frac{119.0 \text{ g SOCl}_2}{1 \text{ mol SOCl}_2} = 11.6 \text{ g SOCl}_2$$

(g) For the reaction:

$$SO_3(g) + SCl_2(g) \rightarrow SOCl_2(g) + SO_2(g) \qquad \Delta_r H^\circ = -96.0 \text{ kJ/mol SOCl}_2$$

$$\Delta_r H^\circ = [1 \cdot \Delta_f H^\circ \text{ SOCl}_2 \text{ (g)} + 1 \cdot \Delta_f H^\circ \text{ SO}_2(g)]$$

$$- [1 \cdot \Delta_f H^\circ \text{ SO}_3(g) + 1 \cdot \Delta_f H^\circ \text{ SCl}_2 \text{ (g)}]$$

$-96.0 \text{ kJ} = [-212.5 \text{ kJ} + -296.84 \text{ kJ}] - [-395.77 \text{ kJ} + 1 \cdot (\Delta_f H^\circ \text{ SCl}_2 \text{ (g)})]$

$-96.0 \text{ kJ} = [-509.34 \text{ kJ}] - [-395.77 \text{ kJ} + \Delta_f H^\circ \text{ SCl}_2(g)]$

$-96.0 \text{ kJ} = (395.77 \text{ kJ} - 509.34 \text{ kJ}) - \Delta_f H^\circ \text{ SCl}_2(g)$

$17.6 \text{ kJ} = -\Delta_f H^\circ \text{ SCl}_2(g)$ and $-17.6 \text{ kJ} = \Delta_f H^\circ \text{ SCl}_2(g)$

75. Examine the effective nuclear charge for:

(a) Calculate Z* for F, Ne: Relate the Z* to atomic radii and Ionization Energy

Z* (for F): 9 - [(2 • 0.85) + (6 • 0.35)]= 5.20

Z* (for Ne): 10 - [(2 • 0.85) + (7 • 0.35)]= 5.85

	O	F	Ne
Z*:	4.55	5.20	5.85
Atomic radius: (pm)	73	72	71
Ionization energy: (kJ/mol)	1314	1681	2081

Note the corresponding decrease in atomic radius as the Z* increases from O to F to Ne. Similarly the ionization energy also increases in that same order.

(b) Calculate Z* for Mn (3d) electron and for a (4s) electron.

For additional assistance, see J.C. Slater, *Phys.Rev*, **34**, 1293 (1929).

For a 3d electron in Mn, we calculate:

Mn: Rule 4 states that electrons in the same d or f group contribute 0.35, while all others (to the left) contribute 1.00. Those to the right (e.g. $(4s^2)$, do not shield. So we have 18 electrons which contribute 1.00 $[(1s^2)(2s^22p^6)(3s^23p^6)]$ and 4 electrons $[(3d^5)-1]$ which contribute 0.35 each:

$$Z* = 25- [(18 \times 1.00) +(4 • 0.35)] = 5.6$$

For a 4s electron:

$$Z* = 25 – [(10 • 1.00) + (13 • 0.85) + (1 • 0.35)] = 3.6$$

Examine the electron groupings for Mn: $[(1s^2)(2s^22p^6)(3s^23p^6)(3d^5)(4s^2)]$

There is 1 **other** 4s electron, 13(n-1) electrons $[(3s^23p^6)(3d^5)]$, 10(n-2) or lower electrons $[(1s^2)(2s^22p^6)]$.

In a manner parallel to O, F, and Ne in part (a), the lower Z* for the 4s electrons indicate that Mn loses those electrons more easily than the 3d electrons—owing to the greater effective nuclear charge on the 3d electrons.

Applying Chemical Principles

1. The most common oxidation state of a rare earth element is +3.

    (a) What is the ground state electron configuration of Sm^{3+}?

        The ground state configuration for elemental Sm is $[Xe]4f^6 6s^2$.

        [See Table 7.3] Losing three electrons would give an

        electron configuration of $[Xe]4f^5$, as the atom loses three higher energy electrons.

    (b) Write a balanced chemical equation for the reaction of $Sm(s)$ and $O_2(g)$.

        $4\ Sm(s) + 3\ O_2(g) \rightarrow 2\ Sm_2O_3\ (s)$

3. Gadolinium has eight unpaired electrons, the greatest number of any rare earth element.

    (a) Orbital box diagram of the ground state electron configuration of Gd

        The ground state configuration for elemental Gd is $[Xe]4f^7 5d^1 6s^2$. The orbital box

        diagram would look like:

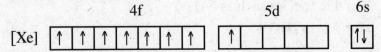

    (b) The most common oxidation state for Gd would be 3+, with the electron configuration

        for Gd^{3+} being $[Xe]4f^7$, (reflecting the stability of the half-filled f sublevel).

5. What is the frequency and energy (in J/photon) of light with wavelength 612 nm?

$$\text{frequency} = \frac{\text{speed of light}}{\text{wavelength}} = \frac{2.9979 \times 10^8 \text{ m/s}}{612 \text{ nm}} \cdot \frac{1.00 \times 10^9 \text{nm}}{1.00 \text{ m}}$$

$$= 4.90 \times 10^{14}\ s^{-1}$$

The energy of a photon of this light may be determined: $E = h\upsilon$

Planck's constant, h, has a value of 6.626×10^{-34} J $\bullet$ s $\bullet$ photons^{-1}

    $E = (6.626 \times 10^{-34}$ J $\bullet$ s $\bullet$ photon$^{-1})(4.90 \times 10^{14}\ s^{-1}) = 3.25 \times 10^{-19}$ J $\bullet$ photon^{-1}

Chapter 8
Bonding and Molecular Structure

PRACTICING SKILLS
Valence Electrons and the Octet Rule

1. | Element | Group Number | Number of Valence Electrons |
   |---------|--------------|------------------------------|
   | (a) O | 6A | 6 |
   | (b) B | 3A | 3 |
   | (c) Na | 1A | 1 |
   | (d) Mg | 2A | 2 |
   | (e) F | 7A | 7 |
   | (f) S | 6A | 6 |

3. | Group Number | Number of Bonds |
   |--------------|------------------|
   | 4A | 4 |
   | 5A | 3 (or 4 in species such as NH_4^+) |
   | 6A | 2 (or 3 as in H_3O^+) |
   | 7A | 1 |

Lewis Electron Dot Structures

5. (a) NF_3 : $[1(5) + 3(7)]$ = 26 valence electrons

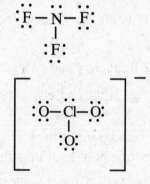

(b) ClO_3^- : $[1(7) + 3(6) + 1]$ = 26 valence electrons

         ↑

add electron due to ion charge

(c) HOBr: $[1(1) + 1(6) + 1(7)]$ = 14 valence electrons

$$H-\ddot{O}-\ddot{\underset{\cdot\cdot}{Br}}:$$

(d) SO_3^{2-} : $[1(6) + 3(6) + 2]$ = 26 valence electrons

         ↑

ion charge

7. (a) CHClF$_2$: [1(4) +1(1) + 1(7) + 2(7)] = 26 valence electrons

$$\overset{\displaystyle H}{\underset{\displaystyle :\overset{..}{\underset{..}{Cl}}:}{\overset{|}{\underset{|}{:\overset{..}{\underset{..}{F}}-C-\overset{..}{\underset{..}{F}}:}}}$$

(b) CH$_3$CO$_2$H: [3(1) + 2(4) + 2(6) + 1(1)] = 24 valence electrons

$$\overset{\displaystyle H \quad :\overset{..}{O}:}{\underset{\displaystyle H}{\overset{|\quad\;\; ||}{\underset{|}{H-C-C-\overset{..}{\underset{..}{O}}-H}}}}$$

(c) H$_3$CCN: [3(1) + 2(4) + 1(5)] = 16 valence electrons

$$\overset{\displaystyle H}{\underset{\displaystyle H}{\overset{|}{\underset{|}{H-C-C \equiv N:}}}}$$

(d) H$_2$CCCH$_2$: [4(1) + 3(4)] = 16 valence electrons

$$\overset{\displaystyle H \qquad\quad H}{\underset{}{\overset{|\qquad\quad\; |}{H-C=C=C-H}}}$$

9. Resonance structures for:

(a) SO$_2$:

$$:\overset{..}{\underset{..}{O}}-\overset{..}{S}=\overset{..}{\underset{..}{O} }\qquad\longleftrightarrow\qquad \overset{..}{\underset{..}{O}}=\overset{..}{S}-\overset{..}{\underset{..}{O}}:$$

(b) HNO$_2$:

$$H-\overset{..}{\underset{..}{O}}-\overset{}{N}=\overset{..}{\underset{..}{O}}\qquad\longleftrightarrow\qquad\overset{..}{\underset{..}{O}}=\overset{}{N}-\overset{..}{\underset{..}{O}}-H$$

(c) SCN$^-$:

$$\left[:\overset{..}{\underset{..}{N}}-C\equiv S:\right]^{-}\longleftrightarrow\left[\overset{..}{\underset{..}{N}}=C=\overset{..}{\underset{..}{S}}\right]^{-}\longleftrightarrow\left[:N\equiv C-\overset{..}{\underset{..}{S}}:\right]^{-}$$

11. (a) BrF$_3$: [1(7) + 3(7)] = 28 valence electrons

$$\overset{\displaystyle :\overset{..}{\underset{..}{F}}-\overset{..}{Br}-\overset{..}{\underset{..}{F}}:}{\underset{\displaystyle :\overset{..}{\underset{..}{F}}:}{|}}$$

(b) I$_3^-$: [3(7) + 1] = 22 valence electrons

$$\left[:\overset{..}{\underset{..}{I}}-\overset{..}{\underset{..}{I}}-\overset{..}{\underset{..}{I}}:\right]^{-}$$

135

(c) XeO_2F_2 : $[1(8) + 2(6) + 2(7)] = 34$ valence electrons

(d) XeF_3^+ : $[1(8) + 3(7) - 1] = 28$ valence electrons

Formal Charges

13. Formal charge on each atom in the following:

(a) N_2H_4

Atom	Formal Charge
H	$1 - 1/2(2) = 0$
N	$5 - 2 - 1/2(6) = 0$

(b) PO_4^{3-}

Atom	Formal Charge
P	$5 - 1/2(8) = +1$
O	$6 - 6 - 1/2(2) = -1$
	Sum = -3 (charge on ion)

(c) BH_4^-

Atom	Formal Charge
B	$3 - 1/2(8) = -1$
H	$1 - 1/2(2) = 0$
	Sum = -1 (charge on ion)

(d) NH_2OH

Atom	Formal Charge
N	$5 - 2 - 1/2(6) = 0$
H	$1 - 1/2(2) = 0$
O	$6 - 4 - 1/2(4) = 0$

15. Formal charge on each atom in the following:

(a) NO_2^+

Atom	Formal Charge
O	$6 - 4 - 1/2(4) = 0$
N	$5 - 0 - 1/2(8) = +1$

(b) NO_2^-

Atom	Formal Charge
O1	$6 - 4 - 1/2(4) = 0$
O2	$6 - 6 - 1/2(2) = -1$
N	$5 - 2 - 1/2(6) = 0$

(c) NF_3

Atom	Formal Charge
F	$7 - 6 - 1/2(2) = 0$
N	$5 - 2 - 1/2(6) = 0$

136

(d) HNO_3

Atom	Formal Charge
O1	6 - 4 -1/2(4) = 0
O2	6 - 6 - 1/2(2) = -1
O3	6 - 4- 1/2(4) = 0
N	5 - 0 - 1/2(8) = +1
H	1 - 1/2(2) = 0

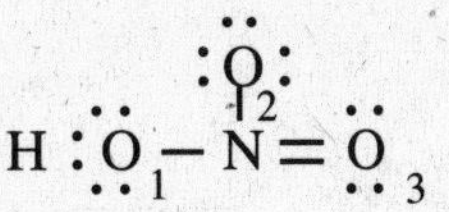

[Note: This is only 1 possible structure]

Molecular Geometry

17. Using the Lewis structure describe the Electron-pair and molecular geometry:

(a).

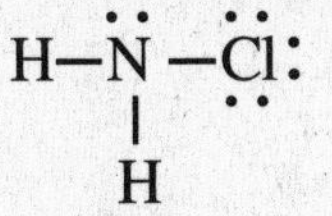

Electron-pair: tetrahedral
Molecular : trigonal pyramidal

(b)

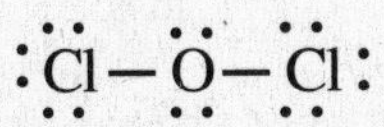

Electron-pair: tetrahedral
Molecular: bent or angular

(c)

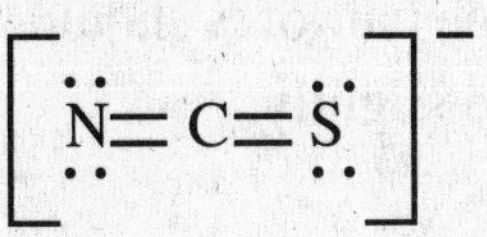

Electron-pair: linear
Molecular: linear

(d)

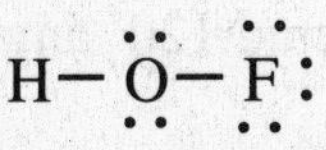

Electron-pair: tetrahedral
Molecular: bent or angular

19. Using the Lewis structure describe the Electron-pair and molecular geometry:

(a)

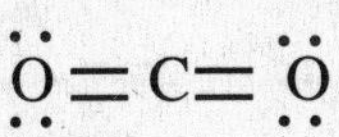

Electron-pair: linear
Molecular : linear

(b)

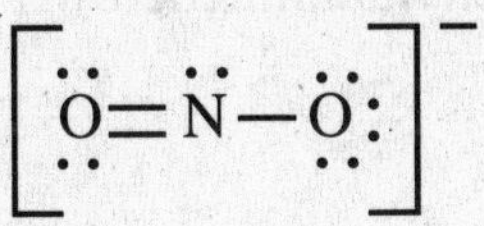

Electron-pair: trigonal planar
Molecular: bent or angular

(c)

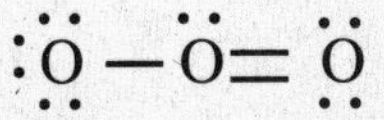

Electron-pair: trigonal planar
Molecular: bent or angular

(d)

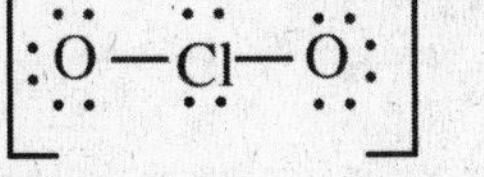

Electron-pair: tetrahedral
Molecular: bent or angular

For the species shown above, having at least one lone pair on the central atom **changes the molecular geometry from linear to bent.**

21. Using the Lewis structure describe the electron-pair and molecular geometry.

[Lone pairs on F have been omitted for clarity.]

(a)

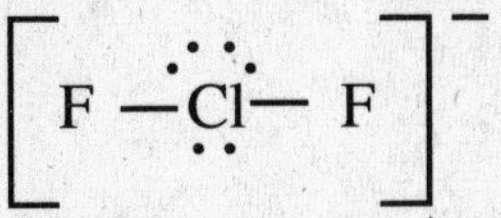

Electron-pair: trigonal bipyramidal
Molecular : linear

137

(b)

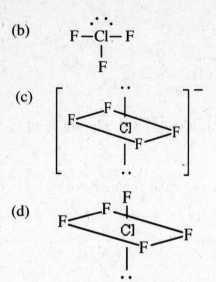

Electron-pair: trigonal bipyramidal
Molecular: T-shaped

(c)

Electron-pair: octahedral
Molecular: square planar

(d)

Electron-pair: octahedral
Molecular: square pyramidal

23. (a) O-S-O angle in SO$_2$:   Slightly less than 120°; The lone pair of S should
reduce the predicted 120° angle slightly.

(b) F-B-F angle in BF$_3$:   120°

(c) Cl-C-Cl in Cl$_2$CO   Slightly less than 120°; The two lone pairs of

electrons on O will reduce the predicted 120° angle

slightly.

(d) (1) H-C-H angle in CH$_3$CN: 109°

(2) C-C ≡N angle in CH$_3$CN: 180°

25. Estimate the values of the angles indicated in the model of phenylalanine below:

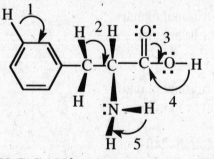

Angle 1:   H-C-C 120°   three groups around the C atom
Angle 2:   H-C-C 109°   four groups around the C atom
Angle 3:   O-C-O 120°   three groups around the C atom
Angle 4:   C-O-H 109°   four groups around the O atom
Angle 5:   H-N-H 109°   four groups around the N atom
The CH$_2$-CH(NH$_2$)-CO$_2$H can not be a straight line, since the first two carbons will have
bond angles of 109 degrees(with their connecting atoms) and the third C (the C of the CO$_2$H
group) has a 120 bond angle with the C and O on either side.

Bond Polarity, Electronegativity, and Formal Charge

27. Indicate the more polar bond (Arrow points toward the more negative atom in the dipole).

(a) C-O >   C-N
    →   →

(b) P-Cl >   P-Br
       →   →

(c) B-O >   B-S
    →   →

(d) B-F   >   B-I
       →   →

29.   For the bonds in acrolein the polarities are as follows:

$$\frac{\Delta \chi}{\Sigma \chi} \quad \begin{array}{ccc} \text{H-C} & \text{C-C} & \text{C=O} \\ 0.06 & 0 & 0.17 \end{array} \quad \text{(Note that } \chi \text{ represents electronegativity)}$$

(a) The C-C bonds are nonpolar, the C-H bonds are slightly polar, and the C=O bond is polar.

(b) The most polar bond in the molecule is the C=O bond, with the oxygen atom being the negative end of the dipole.

31. Atom(s) on which the negative charge resides in:

(a) OH^- Formal charges: O = -1 ; H = 0

Oxygen is much more electronegative than H so the negative charge resides on Oxygen

(b) BH_4^- See SQ 8-13(c) for a Lewis structure of this ion.

Formal charges: B = -1: H = 0   Hydrogen is only slightly more electronegative than B (2.2 compared to 2.0), so the negative charge would reside on the H atoms (although the B-H bonds are **not very polar**).

(c) $CH_3CO_2^-$

Formal charges: C = 0; O1 = 0, O2 = -1   Oxygen is more electronegative than C, so the charge would reside on the oxygens as opposed to the C. The picture shown here is a bit misleading, since **either oxygen** could have the double bond to the C, so **two resonance structures are available** with the negative charge distributed (delocalized) over both C-O bonds.

33. (a) Resonance structures of N_2O :

(b) Formal Charges:

N_1	$5 - 2 - 1/2(6) = 0$	$5 - 4 - 1/2(4) = -1$	$5 - 6 - 1/2(2) = -2$
N_2	$5 - 0 - 1/2(8) = +1$	$5 - 0 - 1/2(8) = +1$	$5 - 0 - 1/2(8) = +1$
O	$6 - 6 - 1/2(2) = -1$	$6 - 4 - 1/2(4) = 0$	$6 - 2 - 1/2(6) = +1$

(c) Of these three structures, the first is the most reasonable in that the most electronegative atom, O, bears a formal charge of -1.

35. Compare hydrogen carbonate ion and nitric acid electron dot structures:

(a) Structures isoelectronic?

HCO_3^{-1} has $1 + 4 + 3(6) + 1$ valence electrons = 24 valence electrons

HNO_3 has $1 + 5 + 3(6)$ valence electrons   = 24 valence electrons

so **the structures are isoelectronic.**

(b) Resonance structures for each substance:

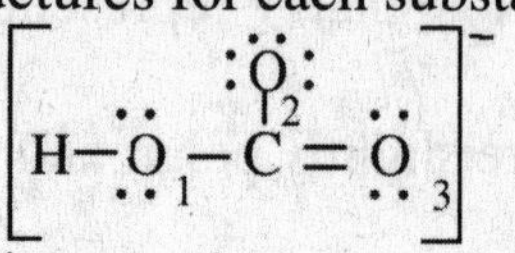

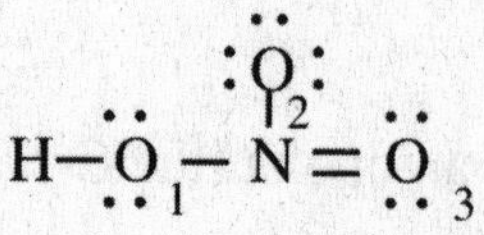

Note that the pictures above are **one** of **two** resonance structures that each substance possesses. The second structure for each would have two electron pairs between the C (or N) and the O labeled 2 in the diagrams.

(c) Formal charges of atoms in the structures: Let's calculate formal charges for the two structures shown above.

HNO_3

Atom	Formal Charge	[Calculated as Valence e – LP e – ½Bond e]
O1	$6 - 4 - 1/2(4) = 0$	
O2	$6 - 6 - 1/2(2) = -1$	
O3	$6 - 4 - 1/2(4) = 0$	
N	$5 - 0 - 1/2(8) = +1$	
H	$1 - 1/2(2) = 0$	

HCO_3^{-1}

Atom	Formal Charge	
O1	$6 - 4 - 1/2(4) = 0$	
O2	$6 - 6 - 1/2(2) = -1$	
O3	$6 - 4 - 1/2(4) = 0$	
C	$4 - 0 - 1/2(8) = 0$	
H	$1 - 1/2(2) = 0$	

(d) Nitric acid is a strong acid, and loses the hydrogen readily, forming a -1 ion.

Hydrogen carbonate bears a -1 charge, and the loss of the hydrogen causes the formation of a -2 ion (i.e. one of increased negative charge—and hence less "desirable"). So hydrogen carbonate is a weak acid!

37. For the nitrite ion and HNO_2:

(a)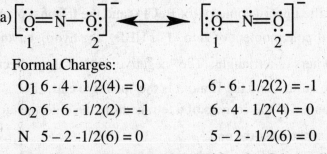

Formal Charges:

O_1 6 - 4 - 1/2(4) = 0   6 - 6 - 1/2(2) = -1

O_2 6 - 6 - 1/2(2) = -1   6 - 4 - 1/2(4) = 0

N   5 – 2 -1/2(6) = 0   5 – 2 - 1/2(6) = 0

(b) The preferred bonding of the H^+ ion for O is predictable owing to the negative formal charges on the O atoms as opposed to the 0 charge on the N atom.

(c) Resonance structures for HNO_2:

$$\left[\ddot{O}=\ddot{N}-\ddot{O}-H\right] \longleftrightarrow \left[:\ddot{O}-\ddot{N}=\ddot{O}-H\right]$$
$$\quad 1 \qquad 2 \qquad\qquad 1 \qquad 2$$

Formal Charges:

O_1 6 - 4 - 1/2(4) = 0   6 - 6 - 1/2(2) = -1

O_2 6 - 4 - 1/2(4) = 0   6 - 2 - 1/2(6) = 1

N   5 – 2 -1/2(6) = 0   5 – 2 - 1/2(6) = 0

H   1- 1/2(2) = 0   1- 1/2(2) = 0

The left structure is preferred, owing to the 0 formal charge on all atoms.

Molecular Polarity

39. For the molecules:

H_2O	NH_3	CO_2	ClF	CCl_4

(i) Using the electronegativities to determine bond polarity:

$\dfrac{\Delta\chi}{\sum\chi}$	$\dfrac{1.3}{5.7}$	$\dfrac{0.8}{5.2}$	$\dfrac{1.0}{6.0}$	$\dfrac{0.8}{7.2}$	$\dfrac{0.7}{5.7}$
$\dfrac{\Delta\chi}{\sum\chi}$	0.23	0.15	0.17	0.11	0.12

Reducing these fractions to a decimal form indicates that the H-O bonds in water are the most polar of these bonds. [Note: $\Delta\chi$ represents the **difference** in electronegativities between the elements while $\sum\chi$ is the **sum** of the electronegativities of the elements.]

(ii) The nonpolar compounds are:

CO_2 The O-C-O bond angle is 180°, thereby canceling the C-O dipoles.

CCl_4 The Cl-C-Cl bond angles are approximately 109°, with the Cl atoms directed at the corners of a tetrahedron. Such an arrangement results in a net dipole moment of zero.

(iii) The F atom in ClF is more negatively charged.(Electronegativity of F = 4.0, Cl = 3.2)

41. Molecular polarity of the following: (a) $BeCl_2$, (b) HBF_2, (c) CH_3Cl, (d) SO_3

    $BeCl_2$ and SO_3 are nonpolar—since the linear geometry (of $BeCl_2$) and the trigonal planar geometry (of SO_3) would give a net dipole moment of zero. For HBF_2, the hydrogen and fluorine atoms are arranged at the corners of a triangle. The "negative end" of the molecule lies on the plane between the fluorine atoms, and the H atom is the "positive end." For CH_3Cl, with the H and Cl atoms arranged at the corners of a tetrahedron, the chlorine atom is the negative end and the H atoms form the positive end.

Bond Order and Bond Length

43.
Specie	Number of bonds	Bond Order : Bonded Atoms
(a) H_2CO	3	1 : CH 2: C = O
(b) SO_3^{2-}	3	1 : SO
(c) NO_2^+	2	2 : NO
(d) NOCl	2	1: N-Cl 2: N=O

Bond order is calculated: (number of bonding electron pairs shared between two atoms).

45. In each case the shorter bond length should be between the atoms with smaller radii--if we assume that the bond orders are equal.

(a) B-Cl	B is smaller than Ga	(b) C-O	C is smaller than Sn
(c) P-O	O is smaller than S	(d) C=O	O is smaller than N

47. The bond order for NO_2^+ is 2, for NO_2^- is 3/2 while the bond order for NO_3^- is 4/3. The Lewis dot structure for the NO_2^+ ion indicates that both NO bonds are double, while in the nitrate ion, any resonance structure (there are three) shows one double bond and two single bonds. The nitrite ion has—in either resonance structure (there are two)—one double and one single bond. Hence the **NO bonds in the nitrate ion will be longest** while those in the **NO_2^+ ion will be shortest**.

Bond Strength and Bond Dissociation Enthalpy

49. The CO bond in carbon monoxide is shorter. The CO bond in carbon monoxide is a **triple bond**, thus it requires more energy to break than the CO double bond in H_2CO. (See bond order for HCHO in SQ8-43).

142

51. Estimate the enthalpy change for the reaction: $H_2C=CH_2(g) + H_2O(g) \rightarrow CH_3CH_2OH(g)$

    Begin by determining which bonds are broken, and which are formed. Those are tabulated below using dissociation enthalpies from Table 8.9

Bond	Enthalpy	# bonds	Total
C-H	413	4	1652
C=C	610	1	610
O-H	463	2	926
Energy of bonds broken		*total*	*3188*
C-H	413	5	2065
C-C	346	1	346
C-O	358	1	358
O-H	463	1	463
Energy of bonds formed		*total*	*3232*

So the overall enthalpy change is

$\Delta_r H = \Sigma\Delta H(\text{bonds broken}) - \Sigma\Delta H(\text{bonds formed}) = 3188 - 3232 \text{ kJ} = -44 \text{ kJ}$

How does this value compare to the value calculated from enthalpies of formation?
For the reaction: $H_2C=CH_2(g) + H_2O(g) \rightarrow CH_3CH_2OH(g)$

$\Delta_r H° = (\Delta_f H° \text{ for } CH_3CH_2OH(g)) - [(\Delta_f H° \text{ for } C_2H_4(g)) + (\Delta_f H° \text{ for } H_2O(g))]$

$\Delta_r H° = (-235.3 \text{ kJ/mol})(1\text{mol}) - [(52.47 \text{ kJ/mol})(1\text{mol}) + (-241.83 \text{ kJ/mol})(1 \text{ mol})]$

$\Delta_r H° = -45.9 \text{ kJ}$

53.

$$H_3C-CH_2-C=C-H \quad + \quad H_2 \quad \longrightarrow \quad H_3C-CH_2-\overset{\overset{H}{|}}{\underset{\underset{H}{|}}{C}}-\overset{\overset{H}{|}}{\underset{\underset{H}{|}}{C}}-H$$

Energy input:	1 mol C=C = 1 mol • 610 kJ/mol =	610 kJ
	1 mol H-H = 1 mol • 436 kJ/mol =	436 kJ
	Total input =	1046 kJ

Energy release:	1 mol C-C = 1 mol • 346 kJ/mol =	346 kJ
	2 mol C-H = 2 mol • 413 kJ/mol =	826 kJ
	Total released =	1172 kJ

| Energy change: | 1046 kJ - 1172 kJ = | - 126 kJ |

55. OF_2 (g) + H_2O (g) $\rightarrow$ O_2 (g) + 2 HF (g) ΔH = - 318 kJ

        Energy input :   2 mol O-F = 2 x (where x = O-F bond energy)

                               2 mol O-H = 2 mol • 463 kJ/mol = <u>926 kJ</u>

                                        Total input = (926 + 2x) kJ

        Energy release: 1 mol O=O = 1 mol • 498 kJ/mol = 498 kJ

                               2 mol H-F = 2 mol • 565 kJ/mol = <u>1130 kJ</u>

                                      Total release = 1628 kJ

              - 318 kJ   = 926 kJ + 2x - 1628 kJ

              384 kJ   = 2x

             192 kJ/mol = O-F bond energy

GENERAL QUESTIONS ON BONDING AND MOLECULAR STRUCTURE

57. Number of valence electrons for Li, Ti, Zn, Si, and Cl:

Specie	Li	Ti	Zn	Si	Cl
# valence electrons	1	4	2	4	7

The number of valence electrons for any main group (representative) element is easily determined by viewing the GROUP NUMBER in which that element resides. Ti—which is a transition element, has 2d electrons and 2s electrons, accounting for the 4—as the number of valence electrons. Zn also a transition element has a filled d sublevel and 2s electrons— giving rise to "2" valence electrons.

59. Which of the following do not have an octet surrounding the central atom:

    BF_4^-, SiF_4, SeF_4, BrF_4^-, XeF_4

A simple solution to this problem is gained by asking (a) how many valence electrons does the central atom have and (b) how many electrons are added by all the combined atoms.

    BF_4^-

        B has 3 electrons, 4 F each contribute 1 electron, -1 charge adds one electron: (3 + 4 + 1) = 8--an octet

    SiF_4,

        Si has 4 electrons, 4 F each contribute 1 electron (4 + 4) = 8 electrons--an octet.

SeF$_4$

Se has 6 electrons, 4 F each contribute 1 electron (6 + 4)= 10 electrons—NOT an octet

BrF$_4^-$

Br has 7 electrons, 4 F each contribute 1 elecron, -1 charge adds 1 electron (7 + 4 + 1) = 12 electrons—NOT an octet.

XeF$_4$

Xe has 8 electrons, 4 F each contribute 1 electron (8 + 4) = 12 electrons—NOT an octet.

61. For the formate ion, resonance structures and C-O bond order:

Bond order is calculated: (number of bonding electron pairs shared between two atoms).

C-O(1) bond order = 2 (a double bond)   C-O(2) bond order = 1 (a single bond)

The average bond order is then 3/2 order.

63. What bond dissociation enthalpies are needed for the reaction:

$$O_2 (g) + 2 H_2 (g) \rightarrow 2 H_2O (g)$$

One will need the H-H, the O-H and the O=O bond dissociation enthalpies.

Calculation for the enthalpy change for the reaction:

Energy input :   1 mol O=O bonds   498 kJ
                      2 mol H-H bonds   2 x 436 kJ
                                               1370 kJ

Energy release: 4 mol H-O bonds   4 x 463 = 1852 kJ

Energy change: 1370 kJ – 1852 kJ = -482 kJ

The $\Delta_r H°$ for the reaction is Σ (bonds broken) – Σ (bonds formed). Note that by subtracting the bond energies of those bonds formed, the **effect** will be to *change the sign* of the bond energies of those bonds formed (since bond formation is always exothermic).

65. Lewis structure(s) for the following: What are similarities and differences ?

(a) CO_2

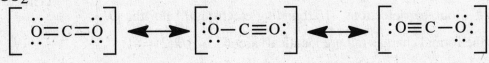

(b) N_3^-

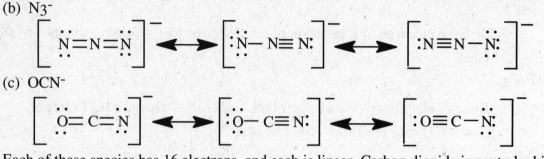

(c) OCN^-

Each of these species has 16 electrons, and each is linear. Carbon dioxide is neutral while the azide ion and isocyanate ion are charged. The isocyanate ion is also polar, while carbon dioxide and azide are not polar.

67. Bond orders in NO_2^- and NO_2^+:

The Lewis dot structure for the NO_2^+ ion (SQ8-15a) indicates that both NO bonds are double (bond order = 2).

The nitrite ion (NO_2^-) (SQ8-15b) has—in either resonance structure—one double and one single bond (bond order = 1.5), so 110 pm is the N-O bond distance in NO_2^+ and 124 pm is the N-O bond distance in NO_2^-.

69. Compare the F-Cl-F angles in ClF_2^+ and ClF_2^-

$$\left[F - \overset{..}{\underset{..}{Cl}} - F \right]^+ \quad \text{and} \quad \left[F - \overset{..}{\underset{..}{Cl}} - F \right]^-$$

Note that in the cation, there are (7 Cl electrons + 1 electron from each of 2 F + (-) an electron owing to the + charge) a total of 8 electrons—an octet around the Cl atom. In an anion, there are (7 Cl electrons + 1 electron from each of 2 F + 1 electron owing to the - charge) a total of 10 electrons. The cation will have 4 groups around the Cl atom—at a F-Cl-F bond angle of 109°. The anion will have 5 groups around the Cl atom, with a F-Cl-F bond angle of 180°.

71. Draw the electron dot for SO_3^{2-}. Does the H^+ ion attach to the S or the O atom?

The Lewis dot structure for sulfite ion shows single bonds between each O and S atom, and lone pair of electrons on the S atom. The formal charges show: for the S atom $(6 - 2 - 1/2(6) = +1)$.

[group # - lone pair electrons $-1/2$(bonding electrons)]. For the O atoms, the formal charges (all are identical) are $6 - 6 - 1/2(2) = -1$.

$$\left[\begin{array}{c} \overset{..}{\underset{..}{O}} - \overset{..}{\underset{|}{S}} - \overset{..}{\underset{..}{O}} \\ \overset{..}{\underset{..}{O}} \end{array} \right]^{2-}$$

So one would predict that the H^+ would attach itself to the more negative O atoms.

73. (a) Estimate the enthalpy change for the reaction and the $\Delta H_{combustion}$ for 1mol of

methanol.

$$2\ CH_3OH(g) + 3\ O_2\ (g) \rightarrow 2\ CO_2(g) + 4\ H_2O(g)$$

Energy input:	6 mol C-H	= 6 mol • 413 kJ/mol	= 2478 kJ
	2 mol C-O	= 2 mol • 358 kJ/mol	= 716 kJ
	2 mol O-H	= 2 mol • 463 kJ/mol	= 926 kJ
	3 mol O=O	= 3 mol • 498 kJ/mol	= 1494 kJ
		Total input	= 5614 kJ

Energy release:	4 mol C=O	= 4 mol • 745 kJ/mol	= 2980 kJ
	8 mol H-O	= 8 mol • 463 kJ/mol	= 3704 kJ
		Total released	= 6684 kJ

Energy change: = 5614 – 6684 = -1070 kJ/mol-rxn

Noting that the equation above indicates the combustion of 2 moles of methanol, the

$\Delta_c H$ for 1mole of methanol is –1070 kJ/mol-rxn/2mol = -535 kJ/mol CH_3OH

(b) The calculation using ΔH values:

$\Delta_r H = [2\ \Delta H_f\ (CO_2(g)) + 4\ \Delta H_f\ (H_2O(g))] – [2\ \Delta H_f\ (CH_3OH(g)) + 3\ \Delta H_f\ (O_2(g))]$

$\Delta_r H = [2\ mol • -393.509\ kJ/mol + 4mol • -241.83\ kJ/mol] – [2\ mol • -201.0\ kJ/mol + 0]$

$\Delta_r H = [-787.018\ kJ + -967.32\ kJ] – [\ -402.0\ kJ] = -1352.3\ kJ/mol-rxn$ and on a per mol

CH_3OH basis, the number is -1352.3 kJ/2 = -676.2 kJ/mol CH_3OH.

75. (a) Resonance structures for CNO^- with formal charges.

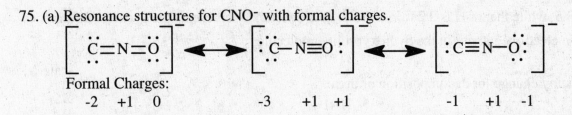

Formal Charges:
   -2   +1   0   -3   +1   +1   -1   +1   -1

(b) The most reasonable structure is the one at the far right because it is the one in which the

formal charges on the atoms are at a minimum, and oxygen has the negative formal

charge.

(c) The instability of the ion could be attributed to the fact that the least electronegative atom

in the ion bears a negative charge in all resonance structures.

77. (a) In XeF$_2$ the bonding pairs occupy the axial positions (of a trigonal bipyramid) with the
       lone pairs located in the equatorial plane. (See question 11(b) for an isoelectronic specie,
       (I$_3^-$).

   (b) In ClF$_3$ two of the three equatorial positions are occupied by the lone pairs of electrons
       on the Cl atom. (See question 21b)

79. Hydroxyproline has the structure:

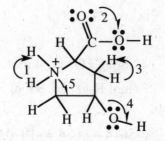

   (a) Values for the selected angles:
       Angle 1: 109° since the N has four groups of electrons around the atom.
       Angle 2: 120° since the C atom has three groups of electrons around it.
       Angle 3: 109° since the C atom has four groups of electrons around it.
       Angle 4: 109° since the O atom has four groups of electrons around it
           (2 bonding and 2 non-bonding pairs)
       Angle 5: 109° since the C atom has four groups of bonding pairs of electrons around it.

   (b) The most polar bonds in the molecule are the O-H bonds. The electronegativity of O is
       3.5, while that of H is 2.2 (a Δ of 1.3). This difference represents the **greatest** differences
       in electronegativity in the hydroxyproline molecule.

81. Enthalpy change for decomposition of urea:

Break N-C bonds (2)	$2 \cdot 305$ kJ/mol
C=O bond (1)	$1 \cdot 745$ kJ/mol bonds broken: $610 + 745 = 1355$ kJ
Make N-N bond (1)	$1 \cdot 163$ kJ/mol
C≡O bond (1)	$1 \cdot 1046$ kJ/mol bonds made: $163 + 1046 = 1209$ kJ

           Change in energy: 1355 kJ - 1209 kJ = 146 kJ

   Since there was no change in the number of N-H bonds (in urea) or hydrazine, those bond
   energies were omitted in the calculation.

83. (a) Bond energies for the conversion of acetone to dihydroxyacetone:

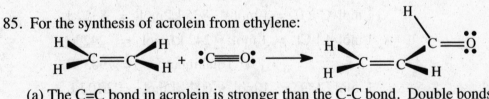

acetone   dihydroxyacetone

Energy input:

6 mol C-H	=	6 mol • 413 kJ/mol	=	2478 kJ
1 mol C=O	=	1 mol • 732 kJ/mol	=	732 kJ
2 mol C-C	=	2 mol • 346 kJ/mol	=	692 kJ
1 mol O=O	=	1 mol • 498 kJ/mol	=	498 kJ
		Total input	=	4400 kJ

Energy release:

1 mol C=O	=	1 mol • 732 kJ/mol	=	732 kJ
2 mol H-O	=	2 mol • 463 kJ/mol	=	926 kJ
2 mol C-C	=	2 mol • 346 kJ/mol	=	692 kJ
2 mol C-O	=	2 mol • 358 kJ/mol	=	716 kJ
4 mol C-H	=	4 mol • 413 kJ/mol	=	1652 kJ
		Total released	=	4718 kJ

Energy change = 4400 – 4718 = -318 kJ for the reaction, so the reaction is **exothermic**.

(b) The central C atom in both molecules contains a doubly-bonded O, with 2 lone pairs of electrons, giving the molecules an overall net polarity. Additionally dihydroxyacetone contains the two polar OH entities (with their 2 lone pairs of electrons), so **both** molecules are polar.

(c) Based on the electronegativity of O, the O-H bonds will be very polar, and the H attached to those O will be the most positive.

85. For the synthesis of acrolein from ethylene:

(a) The C=C bond in acrolein is stronger than the C-C bond. Double bonds are stronger than single bonds between the same two atoms.

(b) The C-C bond is longer than a C=C bond. The stronger the bond, the shorter the bond length.

(c) Polarity of ethylene and acrolein:

For the bonds in acrolein the polarities are as follows:

	H-C	C-C	C=O
$\frac{\Delta\chi}{\Sigma\chi}$	0.06	0	0.17

The C-C and C=C bonds are nonpolar, the C-H bonds are very slightly polar, and the C=O bond is polar. [$\Delta\chi$ represents the **difference** in electronegativities between the two bonded atoms while $\Sigma\chi$ represents the **sum** of the electronegativities of the two bonded atoms.]   The structure of ethylene (coupled with the relative lack of polarity of the C-H and C=C bonds) results in a molecule that is **nonpolar** (net dipole moment of 0). Acrolein however substitutes the polar C=O group (essentially inserted into a C-H bond, and gives a **polar** molecule.

(d) Is the reaction endothermic or exothermic? :

Careful examination of the structures reveals that to form acrolein from ethylene, we must break a C-H bond (which reforms in the product), and a C≡O bond (which forms C=O bond in the product). Using bond energies, the energy change is:

Break	C≡O bond (1)	1 • 1046 kJ/mol
	C-H bond (1)	1 • 413 kJ/mol
	Energy input:	1459 kJ
Make	C-C bond (1)	1 • 346 kJ/mol
	C-H bond (1)	1 • 413 kJ/mol
	C=O bond (1)	1 • 745 kJ/mol
	Energy released:	1504 kJ

Change in energy: 1459 kJ - 1504kJ = -45 kJ , so the reaction is **exothermic**.

87. For the reaction of acetylene with chlorine:

H—C≡C—H + Cl—Cl ⟶ (Cl)(H)C=C(Cl)(H)

Energy input: 2 mol C-H = 2 mol • 413 kJ/mol   = 826 kJ
                   1 mol C≡C = 1 mol • 835 kJ/mol   = 835 kJ
                   1 mol Cl-Cl = 1 mol • 242 kJ/mol = 242 kJ
                                    Total input   = 1903 kJ

Energy release: 1 mol C=C = 1 mol • 602 kJ/mol   = 610 kJ
                   2 mol C-H = 2 mol • 413 kJ/mol   = 826 kJ
                   2 mol C-Cl = 2 mol • 339 kJ/mol = 678 kJ
                                    Total released   = 2114 kJ

Energy change: = 1903 – 2114 = -211 kJ for the reaction, so the reaction is **exothermic**.

IN THE LABORATORY

89. Methanol is the more polar solvent, as it contains the polar O-H bond, and an angular (or bent) geometry around the C-O-H bonds. The O also contains 2 lone pairs of electrons.

    Toluene on the other hand contains only C-C and C-H bonds, neither of which are particularly polar. No lone pairs of electrons are present in the molecule.

91. (a) Lewis structure of DMS and bond angles:
    [The lone pairs of electrons on S–2 of them–
    have been omitted from this structure
    for clarity.

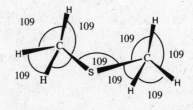

   (b) Location of positive and negative charges in the molecule.
       The molecule is not especially polar, owing to the similar electronegativities of C, H, and S. There are two pairs of electrons on the S atom, so there will be a *slightly* negative charge on the S. The C atoms would be only *slightly* positive.

   (c) Molecules of DMS present in 1.0 m³ of seawater with a concentration of 2.7nM DMS:

$$\frac{2.7 \times 10^{-9} \text{ mol DMS}}{1 \text{ L}} \cdot \frac{1 \text{ L}}{1.00 \times 10^{-3} \text{ m}^3} \cdot \frac{6.0 \times 10^{23} \text{ molecules DMS}}{1 \text{ mol DMS}} = 1.6 \times 10^{18} \text{ molecules DMS}$$

SUMMARY AND CONCEPTUAL QUESTIONS:

93. (a) To form PCl_3 from P_4 and Cl_2: P_4 (s) + 6 Cl_2 (g) → 4 PCl_3 (l)

   (b) Note differing physical states for this part of the problem
       $\Delta_r H° = (4 \cdot \Delta_f H°$ for $PCl_3(g)) - [(\Delta_f H°$ for P_4 (g)) + (6 $\cdot \Delta_f H°$ for Cl_2 (g))]
       $\Delta_r H° = (-287.3$ kJ/mol)(4mol) $- [(+58.9$ kJ/mol)(1mol) + (0 kJ/mol)(6 mol)]
       $\Delta_r H° = (-1148.0$ kJ) $- (+58.9$ kJ) = -1206.9 kJ

   (c) Estimate P-Cl bond energy from Enthalpy of reaction:
       $\Delta_r H° = \Sigma\Delta H$(bonds broken) $- \Sigma\Delta H$(bonds formed)
       $\Delta_r H° = [(6$ mol P-P bonds)(201 kJ/mol) + (6 mol Cl-Cl bonds)(242 kJ/mol)]
                                               - (12 mol P-Cl bonds)(X)

       Note that each mol of PCl_3 contains 3 P-Cl bonds!

       Also note that P_4 is a tetrahedron, with a **total** of 6 P-P bonds!

       Using the enthalpy change from part A and calculating the total known bond energies:
       -1206.9 kJ = (2658 kJ) – 12 mol P-Cl • X
       12 mol P-Cl • X = (2658 kJ) + (1206.9 kJ) = 3864.9 kJ
           and X = 3864.9 kJ/12 mol P-Cl bonds = 322.1 kJ/mol P-Cl bonds

   The P-Cl bond energy from Table 8.9 is 326 kJ/mol P-Cl bond.

95. (a) Odd electron molecules:

HBr	BrO	HOBr
$1 + 7 = 8$	$7 + 6 = 13$	$1 + 6 + 7 = 14$

(b) Estimate energy reactions of the three reactions:

$Br_2(g) \rightarrow 2\ Br(g)$ $1 \bullet$ Br-Br bond energy $= 1 \bullet 193$ kJ/mol $= 193$ kJ

$2\ Br(g) + O_2(g) \rightarrow 2\ BrO(g)$

Energy input:

O=O $=\ 1 \bullet 498$ kJ/mol   - 498 kJ

Energy released:

2 mol Br-O $= 2 \bullet 201$ kJ/mol $= 402$ kJ

Energy change: $498 - 402 = 96$ kJ

$BrO(g) + H_2O(g) \rightarrow HOBr(g) + OH(g)$

Energy input: 1 mol Br-O $= 1 \bullet 201$ kJ/mol $= 201$ kJ

2 mol H-O $= 2 \bullet 463$ kJ/mol $= \underline{926\ kJ}$

Total energy input: 1127 kJ

Energy released:

1 mol Br-O $= 1 \bullet 201$ kJ/mol $= 201$ kJ

2 mol H-O $= 2 \bullet 463$ kJ/mol $= \underline{926\ kJ}$

Total energy released: 1127 kJ

Energy change: $1127 - 1127 = 0$ kJ

(c) Molar heat of formation of HOBr:

$Br_2(g) + O_2(g) + H_2(g) \rightarrow 2\ HOBr(g)$

Energy input: 1 mol Br-Br $= 1 \bullet 193$ kJ/mol $=\ 193$ kJ

1 mol O=O $=\ 1 \bullet 498$ kJ/mol $=\ 498$ kJ

1 mol H - H $= 1 \bullet 436$ kJ/mol $= \underline{436\ kJ}$

Total energy input: 1127 kJ

Energy released:

2 mol Br-O $= 2 \bullet 201$ kJ/mol $= 402$ kJ

2 mol H-O $= 2 \bullet 463$ kJ/mol $=\ \underline{926\ kJ}$

Total energy released: 1328 kJ

Energy change: $1127 - 1328 = -201$ kJ

Noting that we form 2 moles of HOBr in this process, we divide to obtain –101 kJ/mol HOBr.

(d) The first two reactions in (b) are **endothermic**, while the 3rd is neither endo- nor exothermic. In (c), the process releases energy, therefore is **exothermic**.

Applying Chemical Principles

1. $(\chi_{Cl} - \chi_{H}) = ?$

   Using the formula: $(\chi_{Cl} - \chi_{H}) = 0.102\sqrt{\Delta H(Cl - H) - \dfrac{\Delta H(Cl - Cl) + \Delta H(H - H)}{2}}$

   $$= 0.102\sqrt{432 - \dfrac{242 + 436}{2}} = 0.984$$

   Compare this to the value obtained from Figure 8.11 : Cl – H = 3.2 - 2.2 or 1.0

3. Calculate electronegativity for S:

   $\chi_{S} = 1.97 \times 10^{-3}\,(IE - \Delta_{EA}H) + 0.19 = 1.97 \times 10^{-3}\,(1000 - (-200.41)) + 0.19$

   $\chi_{S} = 1.97 \times 10^{-3}(1200.41) + 0.19 = 2.55$. This compares favorably with the value of 2.6 listed in Figure 8.11.

Chapter 9
Bonding and Molecular Structure: Orbital Hybridization and Molecular Orbitals

PRACTICING SKILLS

Valence Bond Theory

1. The Lewis electron dot structure of $CHCl_3$:

    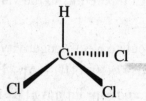

    Electron pair geometry = tetrahedral

    Molecular geometry = tetrahedral

    The H-C bonds are a result of the overlap of the

    hydrogen **s** orbital with **sp^3** hybrid orbitals on carbon.

    The Cl-C bonds are formed by the overlap of the **sp^3** hybrid orbitals on carbon with the

    p orbitals on chlorine.

    [Lone pairs on the chlorine atoms have been omitted for the sake of clarity.]

3. Lewis structure for hydroxylamine:

    Nitrogen and oxygen will BOTH use sp^3 hybridization in
    the molecule, as they both require 4 orbitals. N has bonds to
    3 atoms and 1 lp of electrons. O has bonds to 2 atoms and
    2 lp of electrons.
    One of the sp^3 hybrid orbitals on N will overlap with one of the sp^3 hybrid orbital's on O
    to form the N-O bond.

5. Lewis structure for carbonyl fluoride, COF_2.

    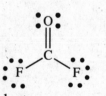

    Electron-pair geometry on C is trigonal planar.
    The molecular geometry on C is trigonal planar.
    The C utilizes sp^2 hybridization to form 3 equivalent
    orbitals. The O atom also utilizes sp^2 hybridization to form 3 equivalent
    orbitals. An sp^2 hybrid orbital for both C and O overlap to form the σ bond while the
    unhybridized p orbital forms the π bond.

7. Electron-Pair and Molecular Geometry; Orbital sets used by the underlined atoms:

	Atom Molecule	Electron-Pair Geometry	Molecular Geometry	Hybrid Orbitals	Groups to be attached to the atom
(a)	$\underline{B}Br_3$	Trigonal planar	Trigonal planar	sp^2	3
(b)	$\underline{C}O_2$	Linear	Linear	sp	2
(c)	$\underline{C}H_2Cl_2$	Tetrahedral	Tetrahedral	sp^3	4

(d)	$CO_3{}^{2-}$	Trigonal planar	Trigonal planar	sp^2	3

9. Hybrid orbital sets used by the underlined atoms:

(a) the C atoms and the O atom in dimethylether: <u>C</u>: sp^3 ; <u>O</u>: sp^3

In the case of either the carbon OR oxygen atoms in dimethylether, each atom is bound to *four other groups*. This would require **four orbitals**

(b) The carbon atoms in propene: <u>CH</u>3: sp^3 ; <u>C</u>H and <u>C</u>H2: sp^2

The methyl carbon is attached to four groups (three H and 1 C), so it needs four orbitals. The methylene and methine have bonds to four groups (2H and 2C) in the case of CH_2 and four groups (1H and 3 C) in the case of CH-.

(c) The C atoms and the N atom in glycine: <u>N</u>: sp^3; <u>C</u>H2: sp^3; <u>C</u>=O: sp^2

The N atom is attached to 4 groups (3 atoms and 1 lone pair), the CH_2 carbon has 4 groups attached (2 H atoms and 1N and 1 C atom). The carbonyl carbon is attached to only 3 groups (1C and 2 O atoms).

11. Hybrid orbital sets used by the underlined atoms:

(a) <u>Si</u>F6^{2-}: sp^3d^2 (b) <u>Se</u>F4: sp^3d (c) <u>I</u>Cl2$^-$: sp^3d (d) <u>Xe</u>F4: sp^3d^2

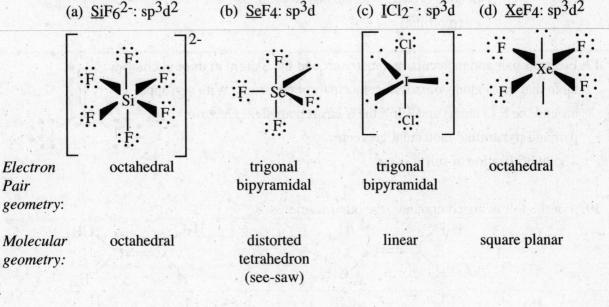

Electron Pair geometry: octahedral   trigonal bipyramidal   trigonal bipyramidal   octahedral

Molecular geometry: octahedral   distorted tetrahedron (see-saw)   linear   square planar

13. For the acid HPO_2F_2 and the anion $PO_2F_2^-$:

Structure

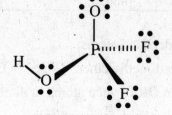 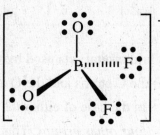

Valence electrons	1P (5) + 1H(1) + 2 O(6) + 2 F(7) = 32 electrons	1P (5) + 1charge (1) + 2 O(6) + 2 F(7) = 32 electrons
Molecular geometry	Tetrahedral around the P	Tetrahedral around the P
Hybridization of P	sp^3	sp^3

15. For the molecule $COCl_2$: Hybridization of C = sp^2

Bonding: 1 sigma bond between each chlorine and carbon
(**sp** hybrid orbitals)

1 sigma bond between carbon and oxygen
(**sp** hybrid orbitals)

1 pi bond between carbon and oxygen
(**p** orbital)

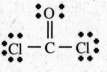

17. Electron-pair and molecular geometry around the S atom in thionyl chloride: Note that the S atoms possesses a lone pair of electrons. With 4 groups around the S (3 atoms and 1 lp), the S has **tetrahedral** electron-pair and **pyramidal** molecular geometry.

The hybridization of sulfur is **sp³**.

19. For the following compounds, the other isomer is:

(a)	H₃C—C=C—H with H and CH₃	H₃C—C=C—CH₃ with H and H
(b)	Cl—C=C—CH₃ with H and H	Cl—C=C—H with H and CH₃

Molecular Orbital Theory

21. Configuration for H_2^+: $(\sigma 1s)^1$

Bond order for H_2^+: $1/2$ (no. bonding e^- - no. antibonding e^-) = $1/2$

The bond order for molecular hydrogen is <u>one</u> (1), and the H-H bond is <u>stronger</u> in the H_2 molecule than in the H_2^+ ion.

23. The molecular orbital diagram for C_2^{2-}, the acetylide ion:

σ^*2p _____

π^*2p _____ _____

$\sigma 2p$ ↑↓

$\pi 2p$ ↑↓ ↑↓

σ^*2s ↑↓

$\sigma 2s$ ↑↓

There are 2 net pi bonds and 1 net sigma bond in the ion, giving a bond order of 3. On adding two electrons to C_2 (added to $\sigma 2p$) to obtain C_2^{2-}, the bond order increases by one. The ion is *not paramagnetic*.

25. Write the electron configuration for peroxide ion in molecular orbital terms, and then compare it with the electron configuration of the O_2 molecule with respect to the following criteria: For oxygen, the electron configuration O_2 is: $(\sigma 2s)^2(\sigma^*2s)^2(\pi 2p)^4(\sigma 2p)^2(\pi^*2p)^2$

The peroxide ion has 2 more electrons, hence: $(\sigma 2s)^2(\sigma^*2s)^2(\pi 2p)^4(\sigma 2p)^2(\pi^*2p)^4$

(a) magnetic character : molecular oxygen is paramagnetic (2 unpaired electrons) peroxide has no unpaired electrons, therefore diamagnetic.

(b) net number of σ and π bonds: 1 σ in peroxide versus 1 σ and 1π in oxygen

(c) bond order: for O_2: 2 [which is $1/2$ (bonding electrons-antibonding electrons)]

for O_2^{2-}: 1

(d) oxygen-oxygen bond length: O_2^{2-} bond will be longer than the O_2 bond—owing to the decreased bond order in the peroxide ion.

27. Which has the shortest and longest bond: Li_2, B_2, C_2, N_2, O_2?

Molecule	MO occupation	Bond order
Li_2	$(\sigma 2s)^2$	1
B_2	$(\sigma 2s)^2(\sigma^*2s)^2(\pi 2p)^2$	1
C_2	$(\sigma 2s)^2(\sigma^*2s)^2(\pi 2p)^4$	2
N_2	$(\sigma 2s)^2(\sigma^*2s)^2(\pi 2p)^4(\sigma 2p)^2$	3
O_2	$(\sigma 2s)^2(\sigma^*2s)^2(\pi 2p)^4(\sigma 2p)^2(\pi^*2p)^2$	2

In general bond length and bond order are inversely related. The greater the bond order, the shorter the bond. So N_2 (of these molecules) has the shortest bond. Table 9.1 confirms this logic. In general then, with MO as our only metric, the Li_2 and B_2 molecules would have equally long bonds.

29. (a) The electron configuration (showing only the outer level electrons) for CO is:
   (b) The HOMO is the $\sigma 2p$
   (c) There are no unpaired electrons, hence CO is *diamagnetic*.
   (d) There is one net sigma bond, and two net pi bonds for an overall bond order of 3.

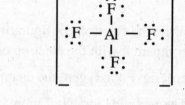

$\sigma^* 2p$ ____
$\pi^* 2p$ ____   ____
$\sigma 2p$ ↑↓
$\pi 2p$ ↑↓   ↑↓
$\sigma^* 2s$ ↑↓
$\sigma 2s$ ↑↓

GENERAL QUESTIONS

31. Lewis structure for AlF_4^-:

Electron-pair geometry: tetrahedral (4 groups)
Molecular geometry: tetrahedral
Orbitals on Al and F overlap: on Al: sp^3 and from F: p orbitals
Formal charges on the atoms?
   Formal charge on Al: $3 - 0 - 1/2(8) = -1$ No lone pairs (0) and four bonds (8 electrons)
   Formal charge on F (all are alike): $7 - 6 - 1/2(2) = 0$. Three lone pairs (6) and one bond (2 electrons)
Is this a reasonable charge distribution? No. Fluorine has a much greater electronegativity than does Al, so a reasonable charge distribution would have a *negative* formal charge on the F atoms, and a *positive* formal charge on the Al atom.

33. The O-S-O bond angle and the hybrid orbitals used by sulfur:
   (a) SO_2 120° angle   3 electron-pair groups (1 lp on S) sp^2
   (b) SO_3 120° angle   3 electron-pair groups sp^2
   (c) SO_3^{2-} 109° angle   4 electron-pair groups (1 lp on S) sp^3
   (d) SO_4^{2-} 109° angle   4 electron-pair groups sp^3

35. Resonance structures for the nitrite ion:

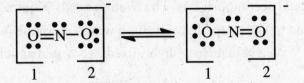

The electron-pair geometry of the ion is *trigonal planar*, and the molecular geometry is *bent* (or angular). With three electron-pairs around the N, the O-N-O bond angle is 120°. The average bond order is 3/2 (three bonds divided connecting the two O atoms). The hybridization associated with 3 electron-pair groups is *sp²*.

37. Resonance structures for N_2O:

Hybridization of N	N1:sp **N2:sp**		N1:sp² **N2:sp**	N1:sp³ **N2:sp**
Central N orbitals	The s & p orbitals		The s & p orbitals	The s & p orbitals

39. Regarding ethylene oxide, acetaldehyde, and vinyl alcohol:

(a) Formula:	C_2H_4O	C_2H_4O	C_2H_4O—these **are** isomers of one another
(b) Hybridization	C1 and C2 - sp³	C1 sp³ C2-sp²	C1 and C2 - sp²
(c) Bond angles	109 ° (anticipated) 60 °due to geometry	C1 109° C2-120°	120°
(d) Polarity	Polar	Polar	Polar
(e) Strongest bond		C-O bond (double bond)	C-C bond (double bond)

41. For the oxime shown:

(a) Hybridization of the C atoms and the N atom: The leftmost C in the diagram is attached to 3 H and another C, so 4 bonds are necessary—sp³ hybridization would accomplish this.

For the rightmost C atom, (attached to a C,H, and N), 3 bond orbitals are needed for the σ bond—sp^2 hybridization would accomplish this. The N atom needs 3 sigma bonding orbitals (1 lp, bond to C, bond to O), so sp^2 hybridization would accomplish this,

(b) Approximate C-N-O angle: With a N that is sp^2 hybridized, we would anticipate a bond angle of approximately 120°.

43. For phosphoserine:

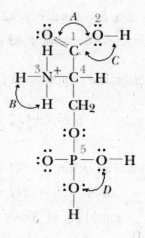

(a) Hybridizations of atoms 1-5:

Atom 1: 3 major groups attached(3 atoms): sp^2

Atom 2: 4 major groups attached (2 atoms; 2 lp): sp^3

Atom 3: 4 major groups attached (4 atoms): sp^3

Atom 4: 4 major groups attached (4 atoms): sp^3

Atom 5: 4 major groups attached (4 atoms): sp^3

(b) Approximate bond angles A-D:

Angle A: For an sp^2 hybridized C atom: 120°

Angle B: for an sp^3 hybridized N atom: 109°

Angle C: for an sp^3 hybridized O atom: 109°

Angle D: for an sp^3 hybridized P atom: 109°

(c) Most polar bonds in the molecule: Owing to differing electronegativities, we expect the P-O bonds and the O-H bonds to be the most polar.

45. For the molecule cinnamaldehyde:

(a) The C=O is the most polar bond (electronegativity differences).

(b) 2π bonds (outside the ring) and 3 π bonds (in the aromatic ring); There are 18 σ bonds.

(c) Cis-trans isomerism is possible. The two isomers are shown below.

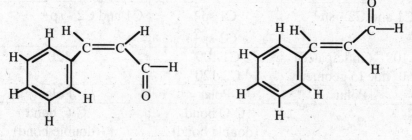

cis-isomer   *trans*-isomer

(d) Since each C is attached to 3 electron pairs, all C are sp^2 hybridized.

(e) Since the angles indicated have an sp^2 hybridized C at the center, all the angles are 120°.

47. (a) Hybridizations: In SbF_5, the Sb atom has five pairs of electrons around it, and the

hybridization would be sp^3d. In the SbF_6^- anion, with six

pairs of electrons around it, the hybridization would be sp^3d^2

(b) The Lewis structure for H_2F^+:

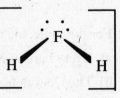

The electron-pair geometry is *tetrahedral* and the molecular

geometry is *bent (or angular)*. With 4 electron pairs around the F, the hybridization

would be sp^3.

49. (a) A Lewis dot structure of the peroxide ion indicates that the

bond order is 1.

(b) The molecular orbital electron configuration of the peroxide

ion is: (valence electrons) $(\sigma 2s)^2(\sigma^* 2s)^2(\sigma 2p)^2(\pi 2p)^4(\pi^* 2p)^4$.

Using the MO configuration, we have 8 bonding electrons and 6 antibonding electrons or

2 bonding electrons (Bond order =1).

(c) Both theories show *no unpaired electrons*, and a bond order of 1.

The magnetic character of both is the same: diamagnetic.

51. Consider the diatomic molecules of Li_2 through Ne_2:

molecule	electron configuration	magnetic property	bond order
Li_2	$(\sigma 2s)^2$	diamagnetic	1
Be_2	$(\sigma 2s)^2 (\sigma^* 2s)^2$	diamagnetic	0
B_2	$(\sigma 2s)^2(\sigma^* 2s)^2(\pi 2p)^2$	paramagnetic	1
C_2	$(\sigma 2s)^2(\sigma^* 2s)^2(\pi 2p)^4$	diamagnetic	2
N_2	$(\sigma 2s)^2(\sigma^* 2s)^2(\pi 2p)^4(\sigma 2p)^2$	diamagnetic	3
O_2	$(\sigma 2s)^2(\sigma^* 2s)^2(\pi 2p)^4(\sigma 2p)^2(\pi^* 2p)^2$	paramagnetic	2
F_2	$(\sigma 2s)^2(\sigma^* 2s)^2(\pi 2p)^4(\sigma 2p)^2(\pi^* 2p)^4$	diamagnetic	1
Ne_2	$(\sigma 2s)^2(\sigma^* 2s)^2(\pi 2p)^4(\sigma 2p)^2(\pi^* 2p)^4(\sigma 2p)^2$	diamagnetic	0

53. Using the orbital diagram we predict the electron configuration to be:

(with 4 electrons from C and 5 from N) $(\sigma 2s)^2(\sigma^* 2s)^2(\pi 2p)^4(\sigma 2p)^1$

(a) The highest energy MO to which an electron is assigned is the $\sigma 2p$.

(b) Bond order = 1/2 (# bonding electrons - # non-bonding electrons)

= 1/2 (7 - 2) or 5/2 or 2.5.

(c) There is a **net of 1/2** sigma bond $-(\sigma 2p)^1$, and 2 net π bonds $-(\pi 2p)^4$.

(d) The molecule has an unpaired electron so it is *paramagnetic*.

55. For the molecule menthol:

(a) Every carbon is surrounded by 4 electron pairs, so the hybridization is sp^3.

(b) The O atom has 4 electron pairs (two lone pairs and two bonded pairs), so the C-O-H bond angle is predicted to be the tetrahedral angle of 109°.

(c) With the pendant OH group, the molecule will be slightly polar.

(d) The ring is *not planar*. Each of the carbons is bonded to two other carbons with a 109° bond angle, and *unlike the benzene ring—which is planar*, this ring will be puckered.

IN THE LABORATORY

57. For BF_3 and NH_3BF_3 :

(a) With 3 groups attached to the B in BF_3, the geometry is trigonal planar. With 4 groups attached to the B in NH_3BF_3, the geometry is tetrahedral.

(b) In BF_3 the B is sp^2 hybridized, in NH_3BF_3 the B is sp^3 hybridized.

(c) The hybridization of B changes.

(d) The F atoms on BF_3 are very electronegative, giving the B atom a partially positive charge—and therefore attractive to the lone pair of electrons on the N atom.

(e) Given the partially positive charge on the B atom, the B atom will be attractive to a lone pair of electrons on the O atom of water:

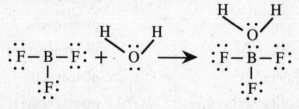

59. The sulfamate ion can be pictured as being formed by the reaction of the amide ion and sulfur trioxide:

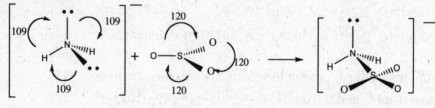

(a) Geometries of amide ion and of sulfur trioxide: With 4 groups (2 atoms and 2 lone pairs) around the nitrogen, the amide ion has a molecular geometry that is angular (or bent)— very much as water would be. The hybridization of the N would be sp^3. For sulfur

162

trioxide, there are three groups around the S, so the molecular geometry would be
trigonal planar and the hybridization of the S would be sp^2.

(b) The bond angles around the N and S in the sulfamate ion would be approximately 109°,
since both N and S would have 4 groups around them in the ion.

(c) While the N atom has sp^3 hybridization in both the amide and sulfamate ions, the
hybridization of the S changes from sp^2 in the molecule to sp^3 in the sulfamate ion.

(d) The SO_3 molecule is the acceptor of the electron pair during the reaction, as expected .
The negative area of the electron pair is attracted to the positive area of the sulfur trioxide
molecule.

61.

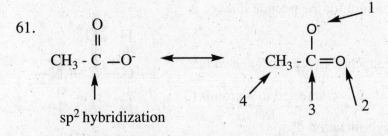

sp² hybridization

Using the right-most structure, I have arbitrarily labeled the three atoms attached to the

carboxylate C (labeled 3). O(1) has four orbitals occupied (1 for bonds to atoms and 3 for

the lone pairs of electrons) so it uses sp^3 hybridization, and overlaps one of those sp^3 hybrid

orbitals with the sp^2 hybrid orbitals on C(3) to form the σ bond. O(2) uses sp^2 hybridization

(1 for a bond to C(3) and 2 for orbitals for the lone pairs of electrons). C(3) uses an sp^2

hybrid to overlap with an sp^2 hybrid from O(2) to form the σ bond. The π bond between

C(3) and O(2) results from the overlap of the unhybridized p orbitals on the respective C

and O atoms. The σ bond between C(4) and C(3) results from overlapping an sp^2 hybrid

orbital from C(3) with an sp^3 hybrid orbital from C(4).

63. Resonance structures for SO_2:

See Figure 9.19a & b for an
analogous (isoelectronic) picture

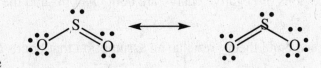

for O_3. In total the lone pairs and bonding pairs in the σ framework of SO_2 account for 7 of

the 9 valence electron pairs. The 3 unhybridized p orbitals (one from S, and two from O)

form 3 π MO. One of the remaining pairs of electrons occupies a bonding πp MO, the other

pair occupies a nonbonding πp MO. With one pair of electrons spread over two bonds, the bond order for those electrons is 0.5. The electrons in the σ framework account for a bond

order of 1, making the bond order for the bonds holding the O's to the S = 1.5.

SUMMARY AND CONCEPTUAL QUESTIONS

65. C has four valence orbitals (s and 3p). Hence hybridizing the **minimum** number would

provide 2 orbitals (sp), while the **maximum** number of hybrid orbitals would involve all of

the orbitals and result in **4 hybrid** orbitals (sp³).

67. The essential part of the Lewis structure for the peptide linkage is

shown here:

(a) The carbonyl carbon (C=O) is attached to 3 groups and so has

a hybridization of sp². The N atom is connected to 4 groups (3

atoms, 1 lone pair) and has a hybridization of sp³.

(b) Another structure is feasible:

This structure would have a formal charge on: carbonyl carbon

of 0, on the oxygen of -1, and on the N a formal charge +1.

Since the "preferred" structure has 0 formal charges on **all**

three atoms, the first structure is the more favorable.

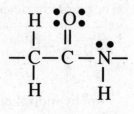

(c) If one views the "less preferred" resonance structure as a contributor, one can see that

both the carbonyl carbon and the N are sp² hybridized. This leaves one "p" orbital on

O,C, and N unhybridized, and capable of side-to-side overlap (also known as π type

overlap) between the C and the O and between the C and the N, forming a planar region

in the molecule. As illustrated by the electrostatic potential diagram for this dipeptide, the

regions of negative charge are both O atoms and the "free" –NH₂ group.

69. Valence bond theory rests on an assumption that bonds are formed by pairing of electrons in

orbitals on neighboring atoms. This theory has difficulties when one considers odd-electron

molecules. VB theory also fails in explaining the magnetic properties of some molecules, e.g.

O₂. MO theory, on the other hand, becomes more complex when molecules consist of more

than two atoms, and for this reason, we fall back on VB theory in such cases.

71. Place the three molecular orbitals for cyclobutadiene in order of increasing energy.
[Increasing number of orbital nodes]

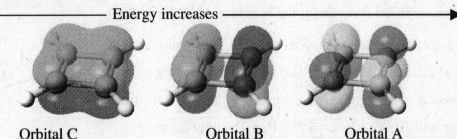

Orbital C   Orbital B   Orbital A

Note that Orbital C has no nodes, orbital B has two nodes, and orbital A has four nodes.

73. The anion from borax has the structure pictured here:

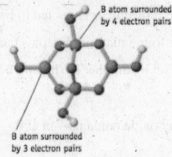

B atom surrounded by 4 electron pairs

B atom surrounded by 3 electron pairs

B atom with	3 electron pairs	4 electron pairs
Electronic geometry	Trigonal planar	Tetrahedral
Molecular geometry	Trigonal planar	Tetrahedral
Hybridization	sp^2	sp^3
Formal charge	$3 - \frac{1}{2}(6)$ $= 0$	$3 - \frac{1}{2}(8)$ $= -1$

Geometry: With 3 pairs of electrons around a B atom, VSEPR theory indicates trigonal planar geometry, utilizing 3 equivalent orbitals—hence sp^2 hybridization. For 4 pairs of electrons, we need tetrahedral geometry, and 4 equivalent orbitals—and sp^3 hybridization. Recall that formal charge is calculated by subtracting from the group number (3 for B) half the bonding electrons and all the lone pair electrons. There are no lone pairs on these B atoms.

75. Refer to "A closer look" on p427 for a pictorial representation of the HF_2^- ion. The bond order for the ion is 0.5, since two bonding electrons are spread over 2 bonds. The HF molecule has a bond order of 1. So the energy required to break an H-F bond in the ion should be **less than** the energy required to break the H-F bond in the molecule.

77. (a) The molecule has 45.69 % Br and 43.31 % F. In 100 g of the compound, we have 45.69 g Br and 43.31 g F. Calculate the # of moles of each type of element:

$$\frac{45.69 \text{ g Br}}{1} \cdot \frac{1 \text{ mol Br}}{79.904 \text{ g Br}} = 0.57181 \text{ mol Br} \; ; \; \frac{54.31 \text{ g F}}{1} \cdot \frac{1 \text{ mol F}}{18.9984 \text{ g F}} = 2.8587 \text{ mol F}$$

Expressing the ratio of mol F: mol Br: $\dfrac{2.8587 \text{ mol F}}{0.57181 \text{ mol Br}} = 4.999 \text{ mol F/mol Br}$, indicating

that the empirical formula for the compound is BrF_5.

(b) A suggested structure is shown here. The lone pair electrons
   on the F atoms are omitted. The electron-pair geometry is
   octahedral, with 6 groups of electrons around the Br (5 from
   atoms, and 1 from a lone-pair on Br). The hybridization
   would be d^2sp^3.

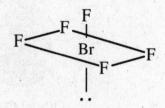

(c) The molecule is expected to have a small dipole moment. While the dipole moments of
   the four equatorial F atoms cancel each other, the dipole moment of the axial F, is not
   offset by another F (having a lone-pair of electrons in opposition to it).

79. (a) The oxide contains—in 100g—90.90 g Br, and 9.10 g O. The # mol of each atom would
   be:

$$\frac{90.90 \text{ g Br}}{1} \cdot \frac{1 \text{ mol Br}}{79.904 \text{ g Br}} = 1.1376 \text{ mol Br}; \frac{9.10 \text{ g O}}{1} \cdot \frac{1 \text{ mol O}}{15.9994 \text{ g O}} = 0.5688 \text{ mol O}$$

The ratio of Br: O is: $\dfrac{1.1376 \text{ mol Br}}{0.5688 \text{ mol O}} = 2.000 \text{ mol Br/mol O}$ with an empirical formula of

Br_2O. A proposed Lewis structure for this oxide would look like: $\cdot \ddot{B}r - \ddot{O} - \ddot{B}r \cdot$

The hybridization of the O atom would be sp^3 (with two atoms and two lone pairs of
electrons attached).

(b) Using Figure 9.18 as a template, we need to place 13 electrons (7 from Br, and 6 from
   O). Placing the electrons in the MO would give: $(\sigma_{4s})^2 (\sigma^*_{4s})^2 (\pi_{4p})^4 (\sigma_{4p})^2 (\pi^*_{4p})^3$. The
   HOMO would be the π^*_{4p} orbital.

Applying Chemical Principles

1. In the photoelectric effect, electrons are ejected when light strikes the surface of a <u>metal</u>.

3. Using the figure provided, the $\sigma(2p)$ has an ionization energy of 15.6 eV.

5. The ionization energy corresponding to 18.6 eV is from an antibonding orbital. The ionization energies of 15.6 eV and 16.7 eV correspond to electrons from bonding orbitals. Loss of these electrons results in a longer (and weaker) bond.

Chapter 10
Carbon: More Than Just Another Element

PRACTICING SKILLS

Alkanes and Cycloalkanes

1. The straight chain alkane with the formula C_7H_{16} is heptane. The prefix "hept" tells us that there are seven carbons in the chain.

3. Of the formulas given, which represents an alkane?

    Alkanes have saturated carbon atoms—with 4 bonds. Since bridging C atoms have two bonds to C, then the other 2 bonds are to H atoms. Terminal C atoms on the end of the chains have only 1 bond to the chain, leaving 3 bonds to H. The net result is that alkanes have the general formula: C_nH_{2n+2} Both (b) and (c) have this general formula.

5. Structure for 3-ethyl-2-methylhexane:

    If we number the "main chain" from left to right, placing a CH_3 group on C-2, and a CH_3CH_2 group on C-3 completes the structure. For a 5-carbon chain, one can envision the following: There are other isomeric structures. This structure would be named: 3,3-diethylpentane.

$$CH_3CH CHCH_2CH_2CH_3$$
$$\overset{\displaystyle CH_3}{|}$$
$$\underset{\displaystyle CH_2CH_3}{|}$$

$$CH_3CH_2 C CH_2CH_3$$
$$\overset{\displaystyle CH_2CH_3}{|}$$
$$\underset{\displaystyle CH_2CH_3}{|}$$

7. Systematic name for the alkane: Numbering the longest C chain from one end to the other indicates that the longest C chain has 4 carbons, the root name is therefore *butane*. On the 2nd and 3rd C there is a 1-C chain. Since we truncate the –ane ending and change it to –yl, these are *methyl groups*. Indicating that the methyl groups are on the 2nd and 3rd carbon atoms, we get 2,3-dimethylbutane.

9. Structures for the compounds:

   (a) 2,3-dimethylhexane

$$CH_3$$
$$|$$
$$CH_3CHCHCH_2CH_2CH_3$$
$$|$$
$$CH_3$$

   (b) 2,3-dimethyloctane

$$CH_3$$
$$|$$
$$CH_3CHCHCH_2CH_2CH_2CH_3$$
$$|$$
$$CH_3$$

   (c) 3-ethylheptane

$$CH_3CH_2CHCH_2CH_2CH_2CH_3$$
$$|$$
$$CH_2CH_3$$

   (d) 3-ethyl-2-methylhexane

$$CH_3$$
$$|$$
$$CH_3CHCHCH_2CH_2CH_3$$
$$|$$
$$CH_2CH_3$$

11. Structures for all compounds with a seven-carbon chain and one methyl substituent.

$$CH_3$$
$$|$$
$$CH_3CHCH_2CH_2CH_2CH_2CH_3$$

2-methylheptane

$$CH_3$$
$$|$$
$$CH_3CH_2CHCH_2CH_2CH_2CH_3$$

3-methylheptane

$$CH_3$$
$$|$$
$$CH_3CH_2CH_2CHCH_2CH_2CH_3$$

4-methylheptane

Assuming that we deal *only* with alkanes, there are 3 structures. *3-methylheptane has a chiral carbon*, since on C-3, there is a methyl group (CH_3) an ethyl group (CH_3CH_2), a H, and a butyl group ($CH_2CH_2CH_2CH_3$).

13. Chair form of cyclohexane, with axial and equatorial H's identified:

In this structure, I have labeled the axial hydrogen atoms as "a",

leaving the equatorial hydrogen atoms unlabeled.

15. Structures, and names of the two ethylheptanes:

$$CH_2CH_3$$
$$|$$
$$CH_3CH_2CHCH_2CH_2CH_2CH_3$$

3-ethylheptane

$$CH_2CH_3$$
$$|$$
$$CH_3CH_2CH_2CHCH_2CH_2CH_3$$

4-ethylheptane

Neither of the isomers have a chiral carbon.

17. Physical properties of C_4H_{10}: Assuming that we're talking about butane, the following properties are pertinent: mp = -138.4 °C, bp = −0.5°C (it's a colorless gas at room T). According to the CRC Handbook of Chemistry and Physics, it is slightly soluble in water, but more so in less polar ether, chloroform, and ethanol.

Predicted properties for $C_{12}H_{26}$: One would guess that the material is colorless (no features that would invoke color). Given the much longer chain, one could guess that it might be a liquid (It is! mp = -9.6°C and bp = 216°C). We would also guess that it would be less water soluble than butane--(and it is indeed reported *not* to be water soluble), but soluble in alcohol and ether.

Alkenes and Alkynes

19. Cis- and trans- isomers of 4-methyl-2-hexene:

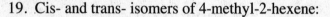

cis   trans

21. (a) Structures and names for alkenes with formula C_5H_{10}:

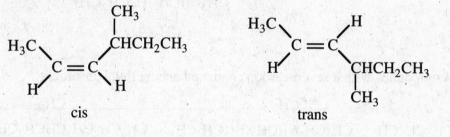

cis-2-pentene   trans-2-pentene

2-methyl-2-butene   1-pentene

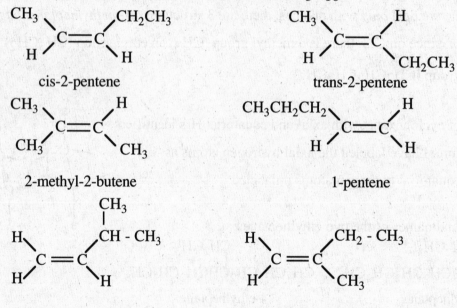

3-methyl-1-butene   2-methyl-1-butene

(b) The cycloalkane with the formula C_5H_{10}:

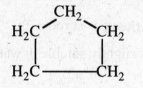

23. Structure and names for products of:
   (a) CH3CH=CH2 + Br2 →   CH3CHBrCH2Br
                                    1,2-dibromopropane

   (b) CH3CH2CH=CHCH3 + H2 → CH3CH2CH2CH2CH3
                                    pentane

25. The alkenes, which upon addition of HBr yields CH3CH2CHBrCH3 is:

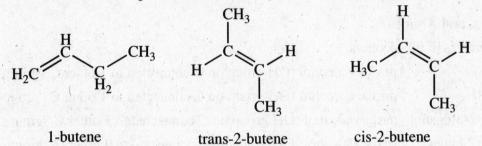

   1-butene   trans-2-butene   cis-2-butene

   The addition of HBr is accomplished by adding (H) on one "side" of the double bond and
   (Br) on the other. The equation for the reaction with 1-butene is:
                 CH3CH2CH=CH2 + HBr → CH3CH2CHBrCH3

27. Alkenes with the formula C3H5Cl:

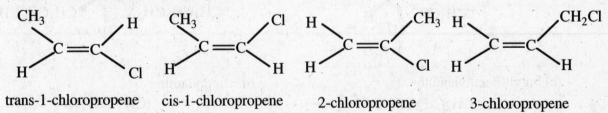

   trans-1-chloropropene   cis-1-chloropropene   2-chloropropene   3-chloropropene

29. The hydrogenation of 1-hexene proceeds, frequently catalyzed by Pd:
            CH3CH2CH2CH2CH=CH2 + H2 → CH3CH2CH2CH2CH2CH3

   The food industry finds this process useful in removing "unsaturated fats" from foods.

Aromatic Compounds

31. Structural formulas for:
   (a) m-dichlorobenzene   (b) p-bromotoluene

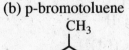

171

33. and 35. The alkylated product of p-xylene:

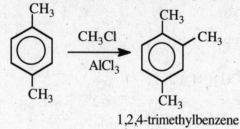

1,2,4-trimethylbenzene

Alcohols, Ethers, and Amines

37. Systematic names of the alcohols:

  (a) 1-propanol   primary alcohol (OH group on C connected to 1 other C "group"

  (b) 1-butanol   primary alcohol (OH group on C connected to 1 other C "group"

  (c) 2-methyl-2-propanol tertiary alcohol (OH group on C connected to 3 other C "groups"

  (d) 2-methyl-2-butanol   tertiary alcohol (OH group on C connected to 3 other C "groups"

39. Formulas and structures

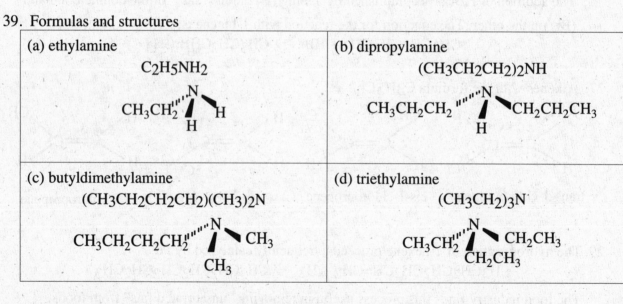

41. Structural formulas for alcohols with the formula $C_4H_{10}O$:

1-butanol $CH_3CH_2CH_2CH_2OH$

2-butanol $CH_3CH_2CHCH_3$ 2-methyl-2-propanol CH_3CCH_3
                            | | (OH above, CH_3 below)
                            OH

2-methyl-1-propanol CH_3CHCH_2OH
                              |
                              CH_3

43. Amines treated with acid:

   (a) $C_6H_5NH_2 + HCl \longrightarrow C_6H_5NH_3^+Cl^-$

   (b) $(CH_3)_3N + H_2SO_4 \longrightarrow (CH_3)_3NH^+ HSO_4^-$

45. Oxidation of the following alcohols—in the case of the primary alcohols (OH on the end) we form a carboxylic acid. For the secondary alcohol (b), we form a ketone. Note that the amine (d), is not oxidized:

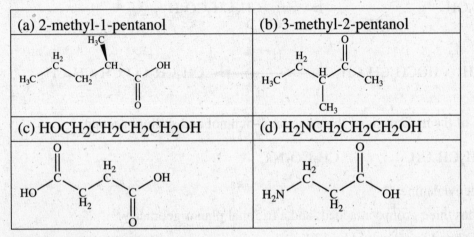

| (a) 2-methyl-1-pentanol | (b) 3-methyl-2-pentanol |
| (c) HOCH₂CH₂CH₂CH₂OH | (d) H₂NCH₂CH₂CH₂OH |

Compounds with a Carbonyl Group

47. Structural formulas for:

   (a) 2-pentanone   (b) hexanal

   $CH_3-CH_2-CH_2-\underset{O}{\overset{||}{C}}-CH_3$ $CH_3-CH_2-CH_2-CH_2-CH_2-\underset{O}{\overset{||}{C}}-H$

   (c) pentanoic acid

   $CH_3CH_2CH_2CH_2CO_2H$ or $CH_3(CH_2)_3CO_2H$

49. Name the following compounds:

   (a) $CH_3CH_2\underset{\underset{CH_3}{|}}{C}HCH_2CO_2H$ 3-methylpentanoic acid
                                                            a carboxylic acid

   (b) $CH_3CH_2\overset{\overset{O}{||}}{C}OCH_3$ methyl propanoate
                                                            an ester

   (c) $CH_3\overset{\overset{O}{||}}{C}OCH_2CH_2CH_2CH_3$ butyl ethanoate
                                                            an ester

   (d) 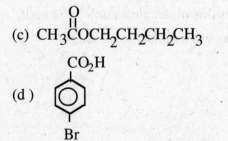   p-bromobenzoic acid
                                     an aromatic carboxylic acid

173

51. (a) The product of oxidation is pentanoic acid:

$$CH_3CH_2CH_2CH_2\overset{\displaystyle O}{\overset{\displaystyle \|}{C}}-OH$$

   (b) The reduction yields 2-octanol:

$$CH_3CH_2CH_2CH_2CH_2CH_2\underset{\underset{\displaystyle OH}{|}}{C}HCH_3$$

53. The following equations show the preparation of propyl propanoate:

$$CH_3CH_2CH_2OH \xrightarrow[\text{H}^+]{\text{KMnO}_4} CH_3CH_2CO_2H$$

$$CH_3CH_2CO_2H + HOCH_2CH_2CH_3 \xrightarrow{\text{H}^+} CH_3CH_2CO_2CH_2CH_2CH_3$$

55. The products of the hydrolysis of the ester are 1-butanol and sodium acetate:

$$CH_3CH_2CH_2CH_2OH \qquad\qquad CH_2CO_2Na$$

57. Regarding phenylalanine:

   (a) Carbon 3 has three groups attached, and a trigonal planar geometry.

   (b) The O-C-O bond angle would be 120°.

   (c) Carbon 2 has four different groups attached, so the molecule is chiral.

   (d) The H attached to the carboxylic acid group (Carbon 3) is acidic.

59. The reaction of a carboxylic acid with an amine is a condensation reaction, forming the amide, so important in proteins!

$$CH_3CH_2CH_2\overset{\displaystyle O}{\overset{\displaystyle \|}{C}}OH + H_2NCH_3 \longrightarrow CH_3CH_2CH_2\overset{\displaystyle O}{\overset{\displaystyle \|}{C}}NHCH_3 + H_2O$$

Functional Groups

61. Functional groups present:

   (a) the –OH group makes this molecule an *alcohol*.

   (b) the carbonyl group (C=O) adjacent to the N-H makes this molecule an *amide*.

   (c) the carbonyl group (C=O) adjacent to the O-H makes this molecule a *carboxylic acid*.

   (d) the carbonyl group (C=O) with an attached O-C makes this molecule an *ester*.

Polymers

63. (a) An equation for the formation of polyvinyl acetate from vinyl acetate:

n CH2CHOCOCH3 ⟶ [-CH2CH(OCOCH3)-]n

(b) The structure for polyvinylacetate:

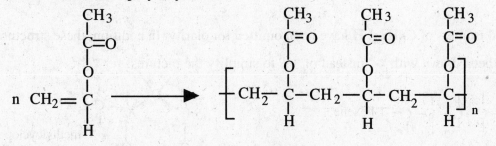

(c) Prepare polyvinyl alcohol from polyvinyl acetate:

Polyvinyl acetate is an ester. Hydrolysis of the ester (structure shown above) with NaOH will produce the sodium salt, $NaC_2H_3O_2$, and polyvinyl alcohol. Acidification with a strong acid (e.g. HCl) will produce polyvinyl alcohol (structure below).

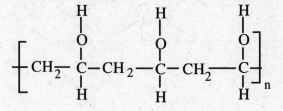

65.

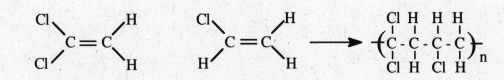

1,1-dichloroethene   chloroethene

The reaction proceeds with the free radical addition of the copolymers:

General Questions on Organic Chemistry

67. (a) geometric isomers of $C_2H_2Cl_2$:

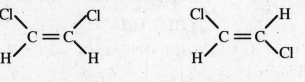

cis isomer   trans isomer

(b) structural isomer of $C_2H_2Cl_2$:

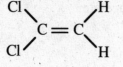

69. Structural isomers of C_6H_{12}: **H** have been omitted for clarity. In addition, these structures have been drawn with "-" instead of "··" to simplify the pictures.

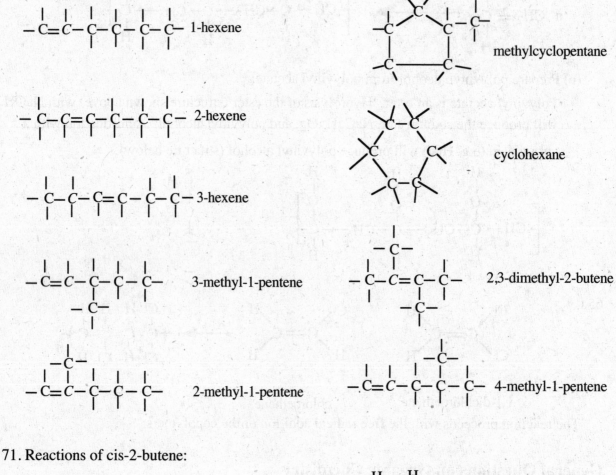

71. Reactions of cis-2-butene:

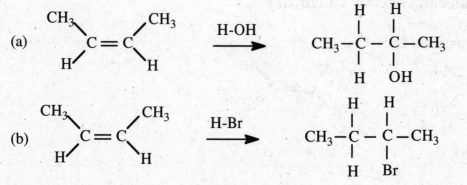

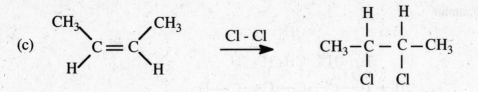

73. In (a) acetic acid reacts with NaOH (a base) to form a salt (an ionic compound). In (b) there is also an acid-base reaction. The amine (base) reacts with HCl (acid) to form a salt.

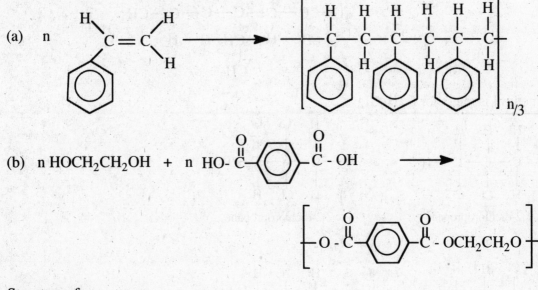

75. The equation for the formation of:

77. Structures for:
(a) 2,2-dimethylpentane

(b) 3,3-diethylpentane

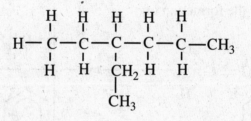

(c) 3-ethyl-2-methylpentane

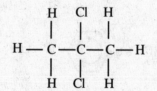

(d) 3-ethylhexane

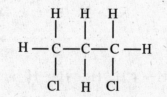

79. Structural isomers for $C_3H_6Cl_2$:

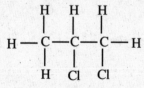

1,2-dichloropropane

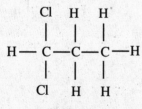

2,2-dichloropropane

1,1-dichloropropane

1,3-dichloropropane

178

81. Structural isomers for trimethylbenzene:

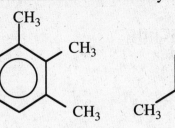
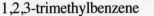
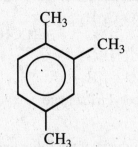

    1,2,3-trimethylbenzene   1,3,5-trimethylbenzene   1,2,4-trimethylbenzene

83. The decarboxylase enzyme would remove the COOH functionality-- releasing CO_2 and
    appending the H in the former location of the COOH group.

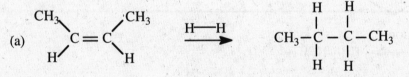

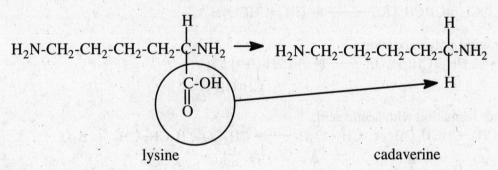

                    lysine   cadaverine

85. Reaction of cis-2-butene:

(a)

$CH_3 - \overset{\overset{\displaystyle H}{|}}{\underset{\underset{\displaystyle H}{|}}{C}} - \overset{\overset{\displaystyle H}{|}}{\underset{\underset{\displaystyle H}{|}}{C}} - CH_3$

butane, not chiral

(b)   an isomer of butane

$CH_3 - \overset{\overset{\displaystyle CH_3}{|}}{\underset{\underset{\displaystyle H}{|}}{C}} - \overset{\overset{\displaystyle H}{|}}{\underset{\underset{\displaystyle H}{|}}{C}} - H$

87. (a) Equation for the saponification of glyceryl trilaurate.

$CH_2O_2CC_{11}H_{23}$
$|$
$CHO_2CC_{11}H_{23}$ $\xrightarrow{\text{NaOH}}$
$|$
$CH_2O_2CC_{11}H_{23}$

CH_2OH
$|$
$CHOH$ $+ \ 3 \ Na^+ \ ^-O_2CC_{11}H_{23}$
$|$
CH_2OH

    glyceryl trilaureate   glycerol

(b) Prepare biodiesel fuel from the fat:

$$\begin{array}{c} CH_2O_2C_{11}H_{23} \\ | \\ CHO_2C_{11}H_{23} \\ | \\ CH_2O_2C_{11}H_{23} \end{array} \xrightarrow{CH_3OH} \begin{array}{c} CH_2OH \\ | \\ CHOH \\ | \\ CH_2OH \end{array} + 3\ CH_3O_2C_{11}H_{23}$$

glyceryl trilaureate   glycerol

89. Show products of the following reactions of CH_2CHCH_2OH:

(a) Hydrogenation of the compound:

$$CH_2{=}CHCH_2OH \longrightarrow CH_3CH_2CH_2OH$$

(b) Oxidation of the compound:

$$CH_2{=}CHCH_2OH \longrightarrow CH_2{=}CHCO_2H$$

(c) Addition polymerization:

$$n\ CH_2{=}CHCH_2OH \longrightarrow \begin{array}{c} [-CH_2{-}CH{-}]_n \\ | \\ CH_2OH \end{array}$$

(d) Ester formation with acetic acid:

$$CH_2{=}CHCH_2OH\ +\ CH_3CO_2H \longrightarrow CH_3CO_2CH_2CH{=}CH_2\ +\ H_2O$$

91. For the reactions in question:

(a) $CH_3CH{=}CH_2\ +\ HBr \longrightarrow \begin{array}{c} CH_3CHCH_3 \\ | \\ Br \end{array}$

2-bromopropane

(b) $\begin{array}{c} CH_3 \\ \diagdown \\ \quad C{=}C \\ \diagup \qquad \diagdown \\ CH_3CH_2 \qquad\qquad H \end{array} \begin{array}{c} H \\ \diagup \\ \\ \\ \end{array} + \ H{-}OH \longrightarrow \begin{array}{c} CH_3 \\ | \\ CH_3CH_2C{-}CH_2 \\ | \quad | \\ HO\ \ H \end{array}$

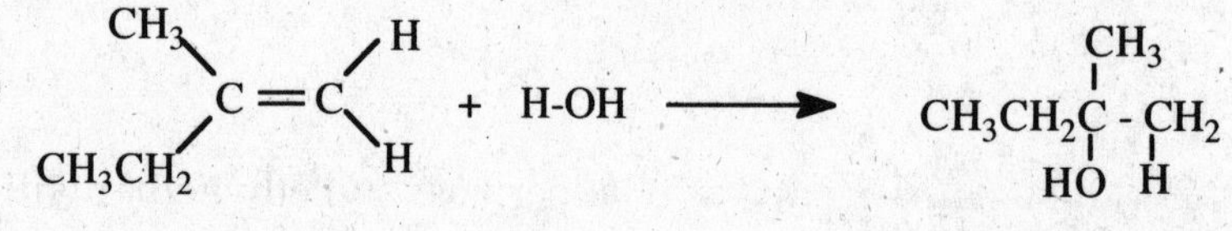

2-methyl-2-butanol

(c) $\begin{array}{c} H \\ \diagdown \\ \quad C{=}C \\ \diagup \qquad \diagdown \\ CH_3 \qquad\qquad CH_3 \end{array} \begin{array}{c} CH_3 \\ \diagup \\ \\ \\ \end{array} + \ H{-}OH \longrightarrow \begin{array}{c} CH_3 \\ | \\ CH_3CH{-}CCH_3 \\ | \qquad | \\ H \qquad OH \end{array}$

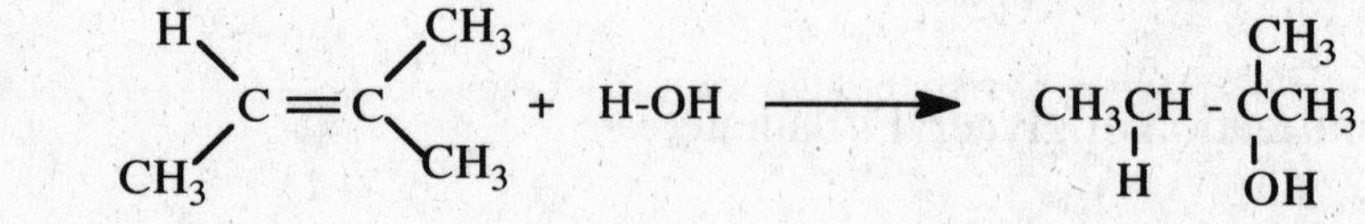

2-methyl-2-butanol

Given the identical names for the product formed in (b) and (c), the products of these two reactions are the same.

IN THE LABORATORY

93. Which of the following produce acetic acid when reacted with $KMnO_4$:

Both (b) and (c) produce acetic acid. The alcohol (c) is oxidized to the aldehyde (b) which subsequently is oxidized to acetic acid.

$$H_3C-\overset{\overset{\displaystyle OH}{|}}{\underset{\underset{\displaystyle H}{|}}{C}}-H \xrightarrow{\quad KMnO_4 \quad} H_3C-\overset{\overset{\displaystyle O}{\|}}{C}-H \xrightarrow{\quad KMnO_4 \quad} CH_3-\overset{\overset{\displaystyle O}{\|}}{C}-OH$$

$\qquad\qquad$ (c) $\qquad\qquad\qquad\qquad\qquad\qquad$ (b)

95. The reaction between bromine and cyclohexene:

The double bond in cyclohexene absorbs elemental bromine—adding across the double bond so that the adjacent carbon atoms (originally participating in the double bond) each have one Br atom bound to them. Benzene does not add Br_2 under these conditions.

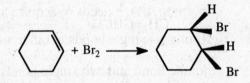

97. (a) Two structures with formula C_3H_6O:

$$CH_3\overset{\overset{\displaystyle O}{\|}}{C}CH_3 \qquad\qquad CH_3CH_2\overset{\overset{\displaystyle O}{\|}}{C}H$$

$\quad$ propanone (acetone) $\qquad\quad$ propanal

(b) Oxidation of the compound gives an acidic solution (a carboxylic acid, perhaps?), indicating that the unknown is propanal.

(c) Oxidation of propanal gives propanoic acid: $\quad CH_3CH_2\overset{\overset{\displaystyle O}{\|}}{C}OH$

99. Test to distinguish between 2-propanol and methyl ethyl ether: Methyl ethyl ether will not react with an oxidizing agent like $KMnO_4$. The alcohol will react with $KMnO_4$, forming a ketone.

101. Addition of water to an alkene, X, gives an alcohol, Y. Oxidation of Y

gives 3,3-dimethyl-2-pentanone. The structure for X and Y:

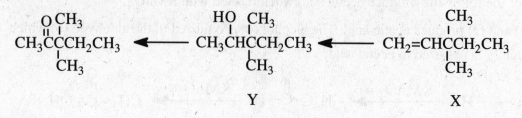

SUMMARY AND CONCEPTUAL QUESTIONS:

103. Modes by which C can achieve an octet of electrons. Since C has 4 valence electrons (it is
in group 4A), it needs to acquire an octet by gaining 4 electrons. This can be done by
forming 4 single bonds to other atoms (e.g. 4H atoms in CH_4). It can also participate in a
double bond and two single bonds (e.g. ethylene), two double bonds (as in allene), OR a
triple bond and one single bond (e.g. ethyne—or acetylene).

105. Properties imparted by the following characteristics:

(a) Cross-linking in polyethylene: Brings structural integrity to the polymeric chain. This
also increases the rigidity of the chain. CLPE (the material of soda bottle caps) is a good
example of a cross-linked polymer.

(b) OH groups in polyvinyl alcohol: These OH groups increase water solubility as hydrogen-
bonding between polyvinyl alcohol and water is possible.
Additionally these OH group provide a locus for cross-linking agents (e.g. borax is a
good cross-linking agent in the commercial polymer called Slime™).

(c) Hydrogen bonding in a polyamide: Causes polyamides (e.g. peptides) to form coils and
sheets.

107. (a) Combustion of ethane and ethanol:
$$2 \ CH_3CH_3(g) + 7 \ O_2(g) \rightarrow 4 \ CO_2(g) + 6 \ H_2O(g)$$
$$CH_3CH_2OH(l) + 3 \ O_2(g) \rightarrow 2 \ CO_2(g) + 3 \ H_2O(g)$$
$$\Delta_r H° \ (C_2H_6) = [4 \bullet \Delta_f H° \ (CO_2) + 6 \bullet \Delta_f H° \ (H_2O)] - [2 \bullet \Delta_f H° \ (CH_3CH_3) + 7 \bullet \Delta_f H° \ (O_2)]$$
$$= [\ (4 \ mol \bullet -393.509 \ kJ/mol) + (6 \ mol \bullet -241.83 \ kJ/mol)]$$
$$- [2 \ mol \bullet -83.85 kJ/mol) + 7 \ mol \bullet 0]$$

$$= [-1574.036 \text{ kJ} + -1450.98 \text{ kJ}] - [-167.7 \text{ kJ}] = -2857.32 \text{ kJ for 2 moles of}$$

ethane or -1428.66 kJ/mol

$$\Delta_r H^\circ (C_2H_5OH) = [2 \cdot \Delta_f H^\circ (CO_2) + 3 \cdot \Delta_f H^\circ (H_2O)] - [1 \cdot \Delta_f H^\circ (CH_3CH_3) + 3 \cdot \Delta_f H^\circ (O_2)]$$

$$= [(2 \text{ mol} \cdot -393.509 \text{ kJ/mol}) + (3 \text{ mol} \cdot -241.83 \text{ kJ/mol})]$$

$$- [1 \text{ mol} \cdot -277.0 \text{ kJ/mol}) + 3 \text{ mol} \cdot 0]$$

$$= [-787.018 \text{ kJ} + -725.49 \text{ kJ}] - [-277.0 \text{ kJ}] = -1235.5 \text{ kJ/mol ethanol}$$

On a per gram basis:

-1428.66 kJ/mol • 1 mol/30.0694 g or approximately - 47.51 kJ/g of ethane and

-1235.5 kJ/mol ethanol • 1 mol/46.0688 g or approximately – 26.82 kJ/g ethanol

(b) If ethanol is partially oxidized ethane, the ΔH of reaction is **less negative**. So partially oxidized ethane has less energy to release during the combustion process.

109. (a) Combustion of 0.125 g (125 mg) of maleic acid gives 0.190 g (190. mg) of CO_2 and 0.0388 g (38.8 mg) of H_2O. The empirical formula of maleic acid:

The combustion of maleic acid can be represented: $C_xH_yO_z + O_2 \rightarrow CO_2 + H_2O$

It's important to note that while **all** the C in CO_2 originated in the maleic acid and **all** the H in H_2O originated in the maleic acid—**not all** of the oxygen originates in the maleic acid.

So we begin by calculating the masses of C and H, and subtracting those masses from the 125 mg of compound to determine the mass of O present in maleic acid.

$$190. \text{ mg } CO_2 \cdot \frac{12.01 \text{ g C}}{44.02 \text{ g } CO_2} = 51.8 \text{ mg C}$$

$$38.8 \text{ mg } H_2O \cdot \frac{2.02 \text{ g H}}{18.02 \text{ g } H_2O} = 4.35 \text{ mg H}$$

Mass O = 125 mg - (51.8 mg C + 4.35 mg H) = 68.8 mg O

Now we can calculate the # of moles of each of these atoms:

$$51.8 \times 10^{-3} \text{ g C} \cdot \frac{1 \text{ mol C}}{12.01 \text{ g C}} = 4.32 \times 10^{-3} \text{ mol C}$$

$$4.35 \times 10^{-3} \text{ g H} \cdot \frac{1 \text{ mol H}}{1.008 \text{ g H}} = 4.31 \times 10^{-3} \text{ mol H}$$

$$68.8 \times 10^{-3} \text{ g O} \cdot \frac{1 \text{ mol O}}{16.00 \text{ g O}} = 4.30 \times 10^{-3} \text{ mol O}$$

Note that the **ratio** of C, H, and O present in maleic acid indicates equal numbers of C, H, and O atoms. The empirical formula is then $C_1H_1O_1$—or CHO.

(b) 0.261 g of maleic acid requires 34.60 mL of 0.130 M NaOH. The molecular formula of maleic acid:

The text indicates that there are 2 carboxylic acid groups per molecule of maleic acid.

The titration indicates: $\dfrac{0.130 \text{ mol NaOH}}{1 \text{ L}} \cdot \dfrac{0.03460 \text{ L}}{1} = 0.004498$ mol NaOH.

Since 1 mol of NaOH reacts with 1 mol of an acidic group (a –COOH group), we know that there are 0.002249 mol of maleic acid present.

The molecular weight of maleic acid is: 0.261 g maleic acid/0.002249 mol maleic acid or 116 g/mol. The empirical formula (CHO) would have a formula weight of (12 + 1 + 16) 29 g. So the molecular formula is: $\dfrac{116 \text{ g}}{1 \text{ mol}} \cdot \dfrac{1 \text{ empirical formula}}{29 \text{ g}} = 4$ empirical

formulas/mol or a molecular formula of $C_4H_4O_4$

(c) A Lewis structure for maleic acid:

(d) Hybridization used by C atoms: All four C atoms are attached to three other groups (and a double bond), making the hybridization of all sp^2.

(e) Bond angles around each carbon will be 120°.

Applying Chemical Principles

1. A balanced chemical equation for the reaction when 1 mol of methyl myristate, $C_{13}H_{27}CO_2CH_3(l)$, is burned:

$$2\, C_{13}H_{27}CO_2CH_3(l) + 43\, O_2(g) \rightarrow 30\, CO_2(g) + 30\, H_2O\,(g)$$

3. $\Delta_rH = \Sigma\, \Delta_fH_{products} - \Sigma\, \Delta_fH_{reactants}$

$$\Delta_rH = [(30\ mol)(-393.509\ \tfrac{kJ}{mol}) + (30\ mol)(-241.83\ \tfrac{kJ}{mol})] -$$

$$[(2\ mol)(-771.0\ \tfrac{kJ}{mol}) + (43\ mol)(0\ \tfrac{kJ}{mol})]$$

$\Delta_rH = -19060.17\,kJ - (-1542.0\ kJ) = -17518.2\,kJ$ —for 2 mol of $C_{13}H_{27}CO_2CH_3(l)$, so

the energy change **per mole** of $C_{13}H_{27}CO_2CH_3$ is -8759.1 kJ.

Now for hexadecane, the calculation:
The balanced equation: $2\, C_{16}H_{34}(l) + 49\, O_2(g) \rightarrow 32\, CO_2(g) + 34\, H_2O\,(g)$

and

$\Delta_rH = \Sigma\, \Delta_fH_{products} - \Sigma\, \Delta_fH_{reactants}$

$$\Delta_rH = [(32\ mol)(-393.509\ \tfrac{kJ}{mol}) + (34\ mol)(-241.83\ \tfrac{kJ}{mol})] -$$

$$[(2\ mol)(-456.1\ \tfrac{kJ}{mol}) + (49\ mol)(0\ \tfrac{kJ}{mol})]$$

$\Delta_rH = -20814.5\,kJ - (-912.2\ kJ) = -19902.3\,kJ$ —for 2 mol of $C_{16}H_{34}(l)$, so
the energy change **per mole** is -9951.2 kJ.

The **greater energy per mole** is realized with hexadecane.

To determine the greater energy per liter, we need to convert the molar masses of each
into mass (g) and calculate the volume associated with 1 mol of each fuel.
The molar mass of methyl myristate is 242.4 g/mol and that of hexadecane is 226.4 g/mol.

$$\frac{-8759.1\ kJ}{1\ mol} \cdot \frac{1\ mol}{242.4\ g} \cdot \frac{0.86\ g}{1\ mL} \cdot \frac{1000\ mL}{1\ L} = -31075.436\ kJ/L\ or\ -3.1 \times 10^4\ kJ/L\ (2\ sf)\ for$$

methyl myristate and for hexadecane:

$$\frac{-9951.2\ kJ}{1\ mol} \cdot \frac{1\ mol}{226.4\ g} \cdot \frac{0.77\ g}{1\ mL} \cdot \frac{1000\ mL}{1\ L} = -33837.451\ kJ/L\ or\ -3.4 \times 10^4\ kJ/L\ (2\ sf)$$

Chapter 11
Gases & Their Properties

PRACTICING SKILLS
Pressure

1. (a) $440 \text{ mm Hg} \cdot \dfrac{1 \text{ atm}}{760 \text{ mm Hg}} = 0.58 \text{ atm}$ (2 significant figures in 440)

   (b) $440 \text{ mm Hg} \cdot \dfrac{1.013 \text{ bar}}{760 \text{ mm Hg}} = 0.59 \text{ bar}$

   (c) $440 \text{ mm Hg} \cdot \dfrac{101.325 \text{ kPa}}{760 \text{ mm Hg}} = 59 \text{ kPa}$

3. The higher pressure in each of the pairs: (using appropriate conversion factors)
   (a) 534 mm Hg or 0.754 bar

   $534 \text{ mm Hg} \cdot \dfrac{1.013 \text{ bar}}{760 \text{ mm Hg}} = 0.712 \text{ bar}$ so 0.754 bar is higher

   (b) 534 mm Hg or 650 kPa

   $534 \text{ mm Hg} \cdot \dfrac{101.325 \text{ kPa}}{760 \text{ mm Hg}} = 71.2 \text{ kPa}$ so 650 kPa is higher

   (c) 1.34 bar or 934 kPa

   $1.34 \text{ bar} \cdot \dfrac{101.325 \text{ kPa}}{1.01325 \text{ bar}} = 134 \text{ kPa}$ so 934 kPa is higher

Boyle's Law and Charles's Law

5. **Boyle's law** states that the pressure a gas exerts is inversely proportional to the volume it occupies, or for a given amount of gas--PV = constant . We can write this as: $P_1V_1 = P_2V_2$

   So $(67.5 \text{ mm Hg})(500. \text{ mL}) = (P_2)(125 \text{ mL})$

   and $\dfrac{(67.5 \text{ mmHg})(500. \text{mL})}{125 \text{mL}} = 270. \text{ mm Hg}$

   Note that **volume decreased** by a factor of 4, and the **pressure increased** by a factor of 4.

7. **Charles' law** states that $V \alpha T$ (in Kelvin) or $\dfrac{V_1}{T_1} = \dfrac{V_2}{T_2}$

   $\dfrac{3.5 \text{ L}}{295 \text{ K}} = \dfrac{V_2}{310 \text{ K}}$ and rearranging to solve for V_2 yields:

   $V_2 = \dfrac{(3.5 \text{ L})(310 \text{ K})}{(295 \text{ K})} = 3.7 \text{ L}$

   Note that with the **increase in T** there has been an **increase in volume**.

The General Gas Law

9. Pressure of 3.6 L of H_2 at 380 mm Hg and 25 °C, if transferred to a 5.0 L flask at 0.0 °C:

Substituting into $\dfrac{P_1V_1}{T_1} = \dfrac{P_2V_2}{T_2}$, will permit us to calculate P_2. Rearranging the

equation:

$$P_2 = P_1 \cdot \dfrac{T_2}{T_1} \cdot \dfrac{V_1}{V_2} = 380 \text{ mm Hg} \cdot \dfrac{273 \text{ K}}{298 \text{ K}} \cdot \dfrac{3.6 \text{ L}}{5.0 \text{ L}} = 250 \text{ mm Hg}$$

11. Using the general gas law we can write: $\dfrac{P_1V_1}{T_1} = \dfrac{P_2V_2}{T_2}$

and for a fixed volume: $\dfrac{P_1}{T_1} = \dfrac{P_2}{T_2}$ or if we rearrange, we obtain $P_2 = P_1 \cdot \dfrac{T_2}{T_1}$

$$P_2 = 360 \text{ mm Hg} \cdot \dfrac{268.2 \text{ K}}{298.7 \text{ K}} = 320 \text{ mm Hg} \text{ (2 sf)}$$

13. Using the general gas law we can write: $\dfrac{P_1V_1}{T_1} = \dfrac{P_2V_2}{T_2}$

The volume is changing from 400. cm^3 to 50.0 cm^3, and the temperature from 15°C to 77 °C.
We can rearrange the equation to solve for the new pressure:

$$P_2 = P_1 \cdot \dfrac{T_2}{T_1} \cdot \dfrac{V_1}{V_2} = 1.00 \text{ atm} \cdot \dfrac{350. \text{ K}}{288 \text{ K}} \cdot \dfrac{400. \text{ cm}^3}{50.0 \text{ cm}^3} = 9.72 \text{ atm}$$

Avogadro's Hypothesis

15. (a) The balanced equation indicates that 1 O_2 is needed for 2 NO. At the same conditions of

T and P, the amount of O_2 is 1/2 the amount of NO, so 75 mL of O_2 is required.

(b) The amount of NO_2 produced will have the same volume as the amount of NO (since

their coefficients are equal in the balanced equation)—150 mL of NO_2.

Ideal Gas Law

17. The pressure of 1.25 g of gaseous carbon dioxide may be calculated with the ideal gas law:

$$1.25 \text{ g CO}_2 \cdot \dfrac{1 \text{ mol CO}_2}{44.01 \text{ g CO}_2} = 0.0284 \text{ mol CO}_2. \text{ Rearranging PV = nRT to solve for P:}$$

$$P = \dfrac{nRT}{V} = \dfrac{(0.0284 \text{ mol})(0.082057 \frac{L \cdot atm}{K \cdot mol})(295.7 \text{ K})}{0.750 \text{ L}} = 0.919 \text{ atm}$$

19. The volume of the flask may be calculated by realizing that the gas will expand to fill the flask.

$$2.2 \text{ g } CO_2 \cdot \frac{1 \text{ mol } CO_2}{44.0 \text{ g } CO_2} = 0.050 \text{ mol } CO_2$$

$$P = 318 \text{ mm Hg} \cdot \frac{1 \text{ atm}}{760 \text{ mm Hg}} = 0.418 \text{ atm}$$

$$V = \frac{(0.050 \text{ mol})(0.082057 \frac{L \cdot atm}{K \cdot mol})(295 \text{ K})}{0.418 \text{ atm}} = 2.9 \text{ L}$$

21. Rearranging $PV = nRT$ to solve for n, we obtain: $n = \dfrac{PV}{RT}$

Converting 737 mm Hg to atmospheres, we obtain

$$737 \text{ mm Hg} \cdot \frac{1 \text{ atm}}{760 \text{ mm Hg}} = 0.970 \text{ atm}$$

$$n = \frac{(0.970 \text{ atm})(1.2 \times 10^7 \text{ L})}{(0.082057 \frac{L \cdot atm}{K \cdot mol})(298.2 \text{ K})} = 4.8 \times 10^5 \text{ moles of He}$$

and since each mole of He has a mass of 4.00 g,

$$4.8 \times 10^5 \text{ mol He} \cdot \frac{4.00 \text{ g He}}{1 \text{ mol He}} = 1.9 \times 10^6 \text{ g He}$$

Gas Density and Molar Mass

23. Write the ideal gas law as: Molar Mass $= \dfrac{dRT}{P}$ where d = density in grams per liter.

Solving for d, we obtain: $\dfrac{(\text{Molar Mass}) \cdot P}{R \cdot T} = d$

The average molar mass for air is approximately 28.96 g/mol.

$$\frac{(28.96 \text{ g/mol})(0.20 \text{ mm Hg} \cdot \frac{1 \text{ atm}}{760 \text{ mm Hg}})}{(0.082057 \frac{L \cdot atm}{K \cdot mol})(250 \text{ K})} = 3.7 \times 10^{-4} \text{ g/L} = d$$

25. Molar mass $= \dfrac{(0.355 \text{ g/L})(0.082057 \frac{L \cdot atm}{K \cdot mol})(290. \text{K})}{(189 \text{ mmHg})\left(\frac{1 atm}{760 \text{ mmHg}}\right)} = 34.0 \text{ g/mol}$

27. Rearranging $PV = nRT$ to solve for n, we obtain: $n = \dfrac{PV}{RT}$

Converting 715 mm Hg to atmospheres, we obtain:

$$715\,mm\,Hg \cdot \dfrac{1\,atm}{760\,mm\,Hg} = 0.941\,atm$$

$$n = \dfrac{(0.941\,atm)(0.452\,L)}{(0.082057\frac{L \bullet atm}{K \bullet mol})(296.2\,K)} = 0.0175\,\text{moles of unknown gas.}$$

Since this number of moles of the gas has a mass of 1.007 g, we can calculate the molar mass.

$$\dfrac{1.007\,g\,\text{of unknown gas}}{0.0175\,\text{moles of unknown gas}} = 57.5\,g/mol$$

29. To calculate the molar mass:

$$\text{Molar mass} = \dfrac{(0.0125\,g/0.125L)(0.082057\frac{L \bullet atm}{K \bullet mol})(298.2\,K)}{(24.8\,mmHg \bullet 1atm/760mmHg)} = 74.9\,g/mol$$

B_6H_{10} has a molar mass of 74.9 grams.

Gas Laws and Chemical Reactions

31. Determine the amount of H_2 generated when 2.2 g Fe reacts:

$$2.2\,g\,Fe \cdot \dfrac{1\,mol\,Fe}{55.85\,g\,Fe} \cdot \dfrac{1\,mol\,H_2}{1\,mol\,Fe} = 0.0394\,mol\,H_2\,(0.039\,\text{to 2 sf})$$

Note that the latter factor is achieved by examining the balanced equation!

The pressure of this amount of H_2 is:

$$P = \dfrac{nRT}{V} = \dfrac{(0.039\,mol\,H_2)(62.4\frac{L \bullet torr}{K \bullet mol})(298\,K)}{10.0\,L} = 73\,torr\,\text{or}\,73\,mm\,Hg$$

Using the factor 1 atm/760 mm Hg, this P corresponds to 0.096 atm.

33. Calculate the moles of N_2 needed:

$$n = \dfrac{PV}{RT} = \dfrac{(1.3\,atm)(75.0\,L)}{(0.082057\frac{L \bullet atm}{K \bullet mol})(298\,K)} = 3.99\,mol\,N_2$$

The mass of NaN_3 needed is obtained from the stoichiometry of the equation:

$$3.99\,mol\,N_2 \cdot \dfrac{2\,mol\,NaN_3}{3\,mol\,N_2} \cdot \dfrac{65.0\,g\,NaN_3}{1\,mol\,NaN_3} = 170\,g\,NaN_3\,(\,2\,sf)$$

35. $N_2H_4 (g) + O_2 (g) \rightarrow N_2 (g) + 2 H_2O (l)$

$$1.00 \text{ kg } N_2H_4 \cdot \frac{1.0 \times 10^3 \text{ g } N_2H_4}{1.0 \text{ kg } N_2H_4} \cdot \frac{1 \text{ mol } N_2H_4}{32.0 \text{ g } N_2H_4} \cdot \frac{1 \text{ mol } O_2}{1 \text{ mol } N_2H_4}$$

$$= 3.13 \times 10^1 \text{ mole } O_2$$

$$P(O_2) = \frac{n(O_2) \cdot R \cdot T}{V} = \frac{(3.13 \times 10^1 \text{mol})(0.082057 \text{ L} \cdot \text{atm/K} \cdot \text{mol})(296 \text{ K})}{450 \text{ L}}$$

$$P(O_2) = 1.69 \text{ atm or } 1.7 \text{ atm (2 sf)}$$

Gas Mixtures and Dalton's Law

37. We know that the total pressure will be equal to the sum of the pressure of each gas (also called the *partial pressure* of each gas).

$$1.0 \text{ g } H_2 \cdot \frac{1 \text{ mol } H_2}{2.02 \text{ g } H_2} = 0.50 \text{ mol } H_2 \text{ and } 8.0 \text{ g Ar} \cdot \frac{1 \text{ mol Ar}}{39.9 \text{ g Ar}} = 0.20 \text{ mol Ar}$$

We can calculate the **total pressure** using the ideal gas law:

$$P = \frac{n \cdot R \cdot T}{V} = \frac{(0.70 \text{ mol})(0.082057 \text{ L} \cdot \text{atm/K} \cdot \text{mol})(300 \text{ K})}{3.0 \text{ L}} = 5.7 \text{ atm}$$

The pressure of **each gas** can be calculated by multiplying the total pressure (5.7 atm) by the mole fraction of the gas.

$$\text{Pressure of } H_2 = \frac{0.50 \text{ mol } H_2}{0.70 \text{ mol } H_2 + Ar} \cdot 5.7 = 4.1 \text{ atm and the}$$

$$\text{Pressure of Ar} = \frac{0.20 \text{ mol Ar}}{0.70 \text{ mol } H_2 + Ar} \cdot 5.7 = 1.6 \text{ atm}$$

Note that the total pressure is indeed (4.1+1.6) or 5.7 atm

39. $P_{total} = P_{halothane} + P_{oxygen} = 170 \text{ mm Hg} + 570 \text{ mm Hg} = 740 \text{ mm Hg}$

(a) Since we know that the pressure a gas exerts is **proportional** to the # of moles of gas present we can calculate the ratio of moles by using their partial pressures:

$$\frac{\text{moles of halothane}}{\text{moles of oxygen}} = \frac{170 \text{ mm Hg}}{570 \text{ mm Hg}} = 0.30$$

(b) $160 \text{ g oxygen} \cdot \dfrac{1 \text{ mol oxygen}}{32.0 \text{ g oxygen}} \cdot \dfrac{0.30 \text{ mol halothane}}{1 \text{ mol oxygen}} \cdot \dfrac{197.38 \text{ g halothane}}{1 \text{ mol halothane}} =$

$$3.0 \times 10^2 \text{ g halothane (2 sf)}$$

Kinetic-Molecular Theory

41. (a) The average kinetic energy depends on the temperature so the average kinetic *energies of the molecules of CO_2 will be greater than those of H_2.*

    (b) The average molecular velocity, v, is related to the kinetic energy by the expression: KE = $1/2\ mv^2$ and KE is related to the temperature by the expression KE = $(3/2)RT$. Setting the two quantities that are equal to KE gives $1/2\ mv^2 = (3/2)RT$ Multiplying both sides of the equation by 2 gives $mv^2 = 3RT$, and solving for v gives $v = \sqrt{\dfrac{3RT}{m}}$

    If we think in terms of moles of the gas, we can replace m with M (the molar mass).

    For CO_2 this would be $v_{CO2} = \sqrt{\dfrac{3R \cdot 298K}{44.0\ {}^{g}\!/_{mol}}}$ and for H_2 the expression would be:

    $v_{H2} = \sqrt{\dfrac{3R \cdot 273K}{2.0\ {}^{g}\!/_{mol}}}$. Note that 3R is equal in the two expressions, so we treat 3R as a

    constant (which won't affect the relative velocities).

    The two velocities are then proportional to the term $\sqrt{\dfrac{298}{44}}$ or 2.6 for CO_2

    And for H_2 $\sqrt{\dfrac{273}{2}}$ or 11.7 so the hydrogen molecules have an average velocity greater

    than the average velocity of the $CO2$ molecules.

    (c) Since the volumes are equal for these two gas samples, the pressure is proportional to the amount of gas present.

    $$V_A = \frac{n_A RT_A}{P_A} \quad\text{and}\quad V_B = \frac{n_B RT_B}{P_B} \text{ and } V_A = V_B \text{ so}$$

    $$\frac{n_A RT_A}{P_A} = \frac{n_B RT_B}{P_B} \text{ and rearranging to solve for the ratio of molecules present}$$

    $$\frac{P_B T_A}{T_B P_A} = \frac{n_B}{n_A} \text{ substituting yields: } \frac{2\text{ atm} \bullet 273\text{ K}}{298\text{ K} \bullet 1\text{ atm}} = \frac{n_B}{n_A} = 1.8$$

    There are 1.8 times as many moles (and molecules) of gas in Flask B ($CO2$) as there are in Flask A ($H2$).

    (d) Since Flask B contains 1.8 times as many moles of $CO2$ as Flask A contains of $H2$, the *ratio* of masses of gas present are:

    $$\frac{\text{Mass (Flask B)}}{\text{Mass (Flask A)}} = \frac{(1.8\text{ mole } CO_2)(44\text{ g } CO_2/\text{mol } CO_2)}{(1\text{ mol } H_2)(2\text{ g } H_2/\text{mol } H_2)} = \frac{40}{1}$$

    Note that any number of moles of $CO2$ and $H2$ (in the ratio of 1.8:1) would provide the same answer.

43. Since two gases at the same temperature have the same kinetic energy:

$$KE_{O_2} = KE_{CO_2}$$

and since the average $KE = 1/2\, M\bar{U}^2$

where $\bar{U}$ is the average speed of a molecule, we can write.

$$1/2\, M_{O_2}\bar{U}^2_{O_2} = 1/2\, M_{CO_2}\bar{U}^2_{CO_2} \quad \text{or} \quad M_{O_2}\bar{U}^2_{O_2} = M_{CO_2}\bar{U}^2_{CO_2}$$

and

$$\frac{M_{O_2}}{M_{CO_2}} = \frac{\bar{U}^2_{CO_2}}{\bar{U}^2_{O_2}}$$

and solving for the average velocity of CO_2 :

$$\bar{U}^2_{CO_2} = \frac{M_{O_2}}{M_{CO_2}} \cdot \bar{U}^2_{O_2}$$

Taking the square root of both sides

$$\bar{U}_{CO_2} = \sqrt{\frac{M_{O_2}}{M_{CO_2}}} \cdot \bar{U}_{O_2} = \sqrt{\frac{32.0 \text{ g } O_2 / \text{mol } O_2}{44.0 \text{ g } CO_2 / \text{mol } CO_2}} \cdot 4.28 \times 10^4 \text{cm/s}$$

$$= 3.65 \times 10^4 \text{ cm/s}$$

45. The species will have average molecular speeds which inversely proportional to their molar

masses.	Slowest			Fastest	
	CH_2F_2	< Ar	< N_2	< CH_4	
Molar masses	54	40	28	16	(integral values)

Diffusion and Effusion

47. Relative rates of effusion for the following pairs of gases:

    (a) CO_2 or F_2 : *Fluorine* effuses faster, since the molar mass of F_2 is 38 g/mol and that of

       CO_2 is 44 g/mol.

    (b) O_2 or N_2: *Nitrogen* effuses faster. (MM N_2 = 28 g/mol; for O_2 = 32 g/mol)

    (c) C_2H_4 or C_2H_6: *Ethylene* effuses faster. (MM C_2H_4 = 28 g/mol; for C_2H_6 = 30 g/mol)

    (d) $CFCl_3$ or $C_2Cl_2F_4$: *$CFCl_3$* effuses faster. (MM $CFCl_3$ = 137 g/mol;

       for $C_2Cl_2F_4$ = 171 g/mol)

49. Determine the molar mass of a gas which effuses at a rate 1/3 that of He:

$$\frac{\text{Rate of effusion of He}}{\text{Rate of effusion of unknown}} = \sqrt{\frac{\text{M of unknown}}{\text{M of He}}}$$

$$\frac{3}{1} = \sqrt{\frac{\text{M of unknown}}{4.0 \text{ g/mol}}}$$

Squaring both sides gives: $9 = \dfrac{\text{M}}{4.0}$ or M = 36 g/mol

Nonideal Gases

51. According to the Ideal Gas Law, the pressure would be:

$$P = \frac{n \bullet R \bullet T}{V} = \frac{(4.00\text{mol})(0.082057\frac{\text{Latm}}{\text{Kmol}})(373\text{K})}{4.00\text{L}} = 30.6 \text{ atm}$$

The van der Waal's equation is: $\left[P + a\left(\dfrac{n}{V}\right)^2 \right][V - bn] = nRT$. Substituting, we get:

$$\left[P + 6.49\frac{\text{atm}\bullet\text{L}^2}{\text{mol}^2}\left(\frac{4.00 \text{ mol}}{4.00 \text{ L}}\right)^2 \right]\left[4.00\text{L}-0.0562\frac{\text{L}}{\text{mol}}\bullet 4.00\text{mol} \right] = 4.00\text{mol}\bullet 0.082057\frac{\text{atm}\bullet\text{L}}{\text{K}\bullet\text{mol}}\bullet 373\text{K}$$

Simplifying: $[P + 6.49 \text{ atm}][4.00\text{L}-0.22\text{L}] = 122.43\text{atm}\bullet\text{L}$ and

$$P = \frac{122.43 \text{ atm}\bullet\text{L}}{3.78 \text{ L}} - 6.49 \text{ atm} = 25.9 \text{ atm}$$

GENERAL QUESTIONS

53.	atm	mm Hg	kPa	bar
Standard atmosphere:	1	1 atm • $\dfrac{760.\text{ mm Hg}}{1 \text{ atm}}$ = 760. mm Hg	1 atm • $\dfrac{101.325 \text{ kPa}}{1 \text{ atm}}$ = 101.325 kPa	1 atm • $\dfrac{1.013 \text{ bar}}{1 \text{ atm}}$ = 1.013 bar
Partial pressure of N_2 in the atmosphere	593 mm Hg • $\dfrac{1 \text{ atm}}{760. \text{ mm Hg}}$ = 0.780 atm	**593 mm Hg**	0.780 atm • $\dfrac{101.3 \text{ kPa}}{1 \text{ atm}}$ = 79.1 kPa	0.780 atm • $\dfrac{1.013 \text{ bar}}{1 \text{ atm}}$ = 0.791 bar
Tank of compressed H_2	133 bar • $\dfrac{1 \text{ atm}}{1.013 \text{ bar}}$ = 131 atm	131 atm • $\dfrac{760. \text{ mm Hg}}{1 \text{ atm}}$ = 9.98 x 10^4 mm Hg	131 atm • $\dfrac{101.3 \text{ kPa}}{1 \text{ atm}}$ = 1.33 x 10^4 kPa	**133 bar**
Atmospheric pressure at top of Mt. Everest	33.7 kPa • $\dfrac{1 \text{ atm}}{101.3 \text{ kPa}}$ = 0.333 atm	0.333 atm • $\dfrac{760. \text{ mm Hg}}{1 \text{ atm}}$ = 253 mm Hg	**33.7 kPa**	0.333 atm • $\dfrac{1.013 \text{ bar}}{1 \text{ atm}}$ = 0.337 bar

55. To increase the average speed of helium atoms by 10.0%, we must know the average speed initially.

$$\sqrt{u^2} = \sqrt{\frac{3RT}{M}} \quad \text{and substituting for He:} = \sqrt{\frac{3 \bullet (8.314 \text{ J/K} \bullet \text{mol}) \bullet 240\text{K}}{4.00 \times 10^{-3} \text{ kg/mol}}} = 1220 \text{ m/s}$$

NOTE: A Joule is a kg•m/s, so it's necessary to express the molar mass of helium (in kg/mol).

The new average speed = 110%(1220 m/s) = 1350 m/s. Substituting this value as u:

$$\frac{Mu^2}{3R} = T; \quad \frac{4.00 \times 10^{-3} \text{ kg/mol} \bullet (1350 \text{ m/s})^2}{3 \bullet 8.314 \text{ J/K} \bullet \text{mol}} = 290.\text{ K} \quad \text{or } (290.-273)°\text{C or } 17°\text{C}.$$

57. The balanced equation for the combustion of C_4H_9SH is:

2 C_4H_9SH (g) + 15 O_2 (g) $\rightarrow$ 2 SO_2 (g) + 8 CO_2 (g) + 10 H_2O (g).

Note that this stoichiometry indicates that 2 mol of C_4H_9SH produces a total of (2+8+10) or

20 mol of product gases.

The amount of C_4H_9SH present in 95.0 mg is:

$$\frac{95.0 \text{ mg } C_4H_9SH}{1} \bullet \frac{1 \text{ g } C_4H_9SH}{1000 \text{ mg } C_4H_9SH} \bullet \frac{1 \text{ mol } C_4H_9SH}{90.189 \text{ g } C_4H_9SH} \bullet \frac{10 \text{ mol gas}}{1 \text{ mol } C_4H_9SH} = 0.01045 \text{ mol gas}$$

(0.0105 mol to 3 sf)

The total pressure is:

$$P = \frac{n \bullet R \bullet T}{V} = \frac{(0.0105 \text{ mol})(0.082057 \frac{\text{Latm}}{\text{Kmol}})(298\text{K})}{5.25 \text{ L}} = 0.0489 \text{ atm or } 37.3 \text{ mm Hg}$$

As water is 10/20 (or 0.50) mol of the product gases, $P(H_2O)$ = 18.6 mm Hg

SO_2 is responsible for 2/20 (or 0.10) mol of the product gases, $P(SO_2)$ = 3.73 mm Hg

CO_2 is responsible for 8/20 (or 0.40) mol of the product gases, $P(CO_2)$ = 14.9 mm Hg

Note alternatively that the pressure of carbon dioxide may be calculated:
37.3 mm Hg = 18.6 mm Hg + 3.7.3 mm Hg + $P(CO_2)$ or

37.3 mm Hg – (18.6 mm Hg + 3.73 mm Hg) = 14.9 mm Hg.

59. Using the Ideal Gas Law,

$$n = \frac{P \bullet V}{R \bullet T} = \frac{\left(8 \text{ mm Hg} \bullet \frac{1 \text{ atm}}{760 \text{ mm Hg}}\right) \bullet \left(10.\text{ m}^3 \bullet \frac{1000 L}{1 \text{ m}^3}\right)}{0.082057 \frac{L \bullet \text{atm}}{K \bullet \text{mol}} \bullet 300.\text{ K}} = 4 \text{ mol (1 sf)}$$

194

61. This problem has two parts: 1) How many moles of Ni are present?

                                     2) How many moles of CO are present?

1) # moles of Ni present:

$$0.450 \text{ g Ni} \cdot \frac{1 \text{ mol Ni}}{58.693 \text{ g Ni}} = 7.67 \times 10^{-3} \text{ mol Ni and since 1 mol Ni(CO)}_4$$

is formed for **each mol of Ni**, one can form 7.67×10^{-3} mol Ni(CO)$_4$

2) # moles of CO present:

Using the Ideal Gas Law:

$$n = \frac{\frac{418}{760} \text{ atm} \cdot (1.50\text{L})}{(0.082057 \frac{\text{L} \cdot \text{atom}}{\text{K} \cdot \text{mol}})(298\text{K})} = 0.0337 \text{ mol CO which would be capable of}$$

forming 0.0337 mol CO $\cdot \dfrac{1 \text{ mol Ni(CO)}_4}{4 \text{ mol CO}} = 8.43 \times 10^{-3}$ mol Ni(CO)$_4$

Since the amount of *nickel limits the maximum amount of Ni(CO)$_4$ that can be formed*, the maximum mass of Ni(CO)$_4$ is then:

7.67×10^{-3} mol Ni(CO)$_4 \cdot \dfrac{170.7 \text{ g Ni(CO)}_4}{1 \text{ mol Ni(CO)}_4} = 1.31$ g Ni(CO)$_4$

63. For the four samples given:
    (1) 1.0 L of H$_2$ at STP   (2) 1.0 L of Ar at STP
    (3) 1.0 L of H$_2$ at 27°C and 760 mm Hg   (4) 1.0 L of He at 0°C and 900 mm Hg

For samples (1) and (2), the calculation is identical for # of particles:

Rearranging the Ideal Gas Law: $n = \dfrac{PV}{RT} = \dfrac{(1 \text{ atm})(1.0 \text{ L})}{(0.082057 \frac{\text{L} \cdot \text{atm}}{\text{K} \cdot \text{mol}})(273 \text{ K})} = 4.4 \times 10^{-2}$ mol

Note that we can alternatively use the factor: 22.4 L = 1 mol (since samples 1 and 2 are at

STP) For sample (3) $n = \dfrac{(1 \text{ atm})(1.0 \text{ L})}{(0.082057 \frac{\text{L} \cdot \text{atm}}{\text{K} \cdot \text{mol}})(300 \text{ K})} = 4.1 \times 10^{-2}$ mol of H$_2$

For sample (4) $n = \dfrac{(\frac{900}{760} \text{ atm})(1.0 \text{ L})}{(0.082057 \frac{\text{L} \cdot \text{atm}}{\text{K} \cdot \text{mol}})(273 \text{ K})} = 5.3 \times 10^{-2}$ mol of He

(a) Which sample has greatest number of gas particles?   Sample 4
(b) Which sample has the fewest number of gas particles? Sample 3

(c) Which sample represents the largest mass?

Given the number of moles of each gas, we can calculate the masses:

Sample (1): 4.4×10^{-2} mol $H_2 \cdot 2.0$ g H_2/mol $H_2 = 8.8 \times 10^{-2}$ g H_2

Sample (2): 4.4×10^{-2} mol Ar $\cdot 39.95$ g Ar /mol Ar $= 1.7$ g Ar

Sample (3): 4.1×10^{-2} mol $H_2 \cdot 2.0$ g H_2 /mol $H_2 = 8.2 \times 10^{-2}$ g H_2

Sample (4): 5.3×10^{-2} mol of He $\cdot 4.00$ g He/mol He $= 0.21$ g He

Sample 2 has the greatest mass.

65. To determine the theoretical yield of $Fe(CO)_5$, we need to know the # of moles of both Fe and of CO. Use the Ideal Gas Law to calculate moles of CO.

$$n = \frac{(\frac{732}{760}\text{ atm})(5.50\text{ L})}{(0.082057\frac{L \bullet atm}{K \bullet mol})(296\text{ K})} = 0.218 \text{ mol CO}$$

For Fe: $\dfrac{3.52 \text{ g Fe}}{1} \bullet \dfrac{1 \text{ mol Fe}}{55.85 \text{ g Fe}} = 0.0630$ mol Fe

The equation is: $Fe + 5\ CO \rightarrow Fe(CO)_5$, indicating that we need 5 mol of CO per mol of Fe.

Moles available $= \dfrac{0.218 \text{ mol CO}}{0.0630 \text{ mol Fe}} = 3.46$, indicating that CO is the Limiting Reagent.

The theoretical yield of $Fe(CO)_5$ will be:

$$\dfrac{0.218 \text{ mol CO}}{1} \bullet \dfrac{1 \text{ mol Fe(CO)}_5}{5 \text{ mol CO}} \bullet \dfrac{195.90 \text{ g Fe(CO)}_5}{1 \text{ mol Fe(CO)}_5} = 8.54 \text{ g Fe(CO)}_5$$

67. Determine the empirical formula of the S_xF_y compound:

Given the data, let's calculate the # of mol of the compound:

Note that we convert 89 mL to the volume in liters (0.089 L). We also need to express 83.8 mm Hg in units of atmospheres. With those conversions, we can calculate the # mol of the compound.

$$n = \frac{(\frac{83.8}{760}\text{ atm})(0.089\text{ L})}{(0.082057\frac{L \bullet atm}{K \bullet mol})(318\text{ K})} = 3.8 \times 10^{-4} \text{ mol compound (2 sf)}$$

The molecular weight is then: 0.0955 g/ 3.8×10^{-4} mol $= 253.9$ g/mol or 250 (to 2 sf).

The formula is 25.23% S (and therefore 74.77 % F). So we can anticipate that of the 250 g, 25.23% is S $(0.2523 \bullet 250) = 64.1$ g S. The amount of F is $(0.7477 \bullet 250) = 190$ g F.

Now we can calculate the # of moles of each element present:

$$\frac{64.1 \text{ g S}}{1} \cdot \frac{1 \text{ mol S}}{32.07 \text{ g S}} = 2.0 \text{ mol (to 2 sf)}$$

and for F: $\dfrac{190 \text{ g F}}{1} \cdot \dfrac{1 \text{ mol F}}{19.00 \text{ g}} = 10. \text{ mol (2 sf)}$. We note the formula is then S_2F_{10}.

69. (a) Average molar mass of air at 20 km above the earth's surface:

Converting volume to L, we note the conversion factor, $1 \text{ L} = 1 \times 10^{-3} \text{ m}^3$ so 1 m^3 is 1000 L. Noting also that $-63°C$ will be 210K, we can substitute:

$$n = \frac{(\frac{42}{760}\text{atm})(1000 \text{ L})}{(0.082057\frac{\text{L} \bullet \text{atm}}{\text{K} \bullet \text{mol}})(210 \text{ K})} = 3.2 \text{ mol air. Given the density as 92 g/m}^3\text{, we can}$$

calculate the molar mass: 92 g/3.2 mol = 29 g/mol. (to 2 sf)

Calculations for part (b) will be made using 2 sf. Using the un-rounded molar mass in part (a), the answers in part (b) will be different from those shown here.

(b) If the atmosphere is only O_2 and N_2, what is the mole fraction of each gas?

We know that the mf O_2 + mf N_2 = 1. We know that the mixture of oxygen and nitrogen has a molar mass of 29 g/mol. We can then calculate the % (or mole fraction) of this weighted average.

(mf O_2) • 32.0 g/mol + (mf N_2) • 28.0g/mol = 28.7 g/mol.

Using the mf relationship from above, we get:

(mf O_2) • 32.0 g/mol + (1 – mf O_2) • 28.0g/mol = 28.7 g/mol.

For simplicity let's use x to represent mf O_2.

x • 32.0 g/mol + (1 – x) • 28.0g/mol = 28.7 g/mol.

32.0x + 28.0 –28.0x = 28.7 and

4.0 x = (28.7 - 28.0) and x = (28.7 - 28.0)/4.0 or 0.17

The mf O_2 = 0.17 and the mf N_2 = 0.83

71. Calculate the # of moles of gas present in the flask:

$$n = \frac{(\frac{17.2}{760}\text{atm})(1.850 \text{ L})}{(0.082057\frac{\text{L} \bullet \text{atm}}{\text{K} \bullet \text{mol}})(294 \text{ K})} = 1.74 \times 10^{-3} \text{ mol ClxOyFz}$$

This amount of the gas has a mass of 0.150 g, so the molar mass is 0.150 g/1.74 x 10^{-3} mol or 86.4 g/mol. If the gas contains Cl,O, and F, the "molar mass" of ClOF = (35.5 + 16 + 19)= 70.5 g/mol. With our calculated value of 86.4, we hypothesize the formula to be ClO_2F.

73. Which of the following is not correct?

(a) Diffusion of gases occurs more rapidly at higher temperatures.

Correct. Since the kinetic energy of the gases will be greater at higher temperatures, the gases will diffuse more rapidly.

(b) Effusion of H_2 is faster than effusion of He (assume similar conditions and a rate expressed in units of mol/h).

Correct. Graham's Law tells us that effusion occurs at a rate that is inversely proportional to the square root of the molar masses of gases involved. H_2 is less massive than He (by a factor of 2), so should effuse faster than He.

(c) Diffusion will occur faster at low pressure than at high pressure.

Correct. At lower pressures, the concentration of gas is lower than at higher pressures. Collisions will be less frequent, so diffusion should occur faster.

(d) The rate of effusion of a gas (mol/h) is directly proportional to molar mass.

Incorrect. See part (b).

75. Carbon dioxide, CO_2, was shown to effuse through a porous plate at the rate of 0.033 mol/min. The same quantity of an unknown gas effuses through the same porous barrier in 104 seconds. Calculate the molar mass of the unknown gas.

$$\frac{\text{Rate of effusion of } CO_2}{\text{Rate of effusion of unknown}} = \sqrt{\frac{\text{MM of unknown}}{\text{MM of } CO_2}}$$

Note that the rates should be expressed in the same units of time, so convert 104 seconds into minutes (1.733 min). The rate of effusion of the unknown is then 0.033mol/1.733 minutes or 0.0190 mol/min. Substituting:

$$\frac{0.033 \text{ mol/min}}{0.0190 \text{ mol/min}} = \sqrt{\frac{\text{MM of unknown}}{44.01 \text{ g/mol}}} \text{ and squaring both sides of the equation yields:}$$

$$3.0166 = \frac{\text{MM of unknown}}{44.01 \text{ g/mol}} \text{ and (3.0166)(44.01 g/mol) = 130 g/mol (2 sf)}$$

77. The number of moles of He in the balloon can be calculated with the Ideal Gas Law. First calculate the P of He in the balloon:

gauge = total - barometric and rearranging:

gauge + barometric = total pressure = 22 mm Hg + 755 mm Hg = 777 mm Hg (777 torr)

$$n = \frac{PV}{RT} = \frac{(777 \text{ torr})(0.305 \text{ L})}{(62.4 \frac{L \cdot torr}{K \cdot mol})(298 \text{ K})} = 0.0128 \text{ mol He}$$

79. Calculate moles of O_2 to determine the amount of $KClO_3$ originally present.

$$n = \frac{(\frac{735}{760} \text{ atm})(0.327 \text{ L})}{(0.082057 \frac{L \cdot atm}{K \cdot mol})(292 \text{ K})} = 1.32 \times 10^{-2} \text{ mol } O_2$$

The mole ratio from the equation tells us that for each mol of O_2, we had 2/3 mole of $KClO_3$. The number of moles of $KClO_3$ will be 1.32×10^{-2} mol $O_2 \cdot$ (2 mol $KClO_3$/3 mol O_2)= 8.80×10^{-3} mol of $KClO_3$. The mass to which this corresponds is:

$$\frac{8.80 \times 10^{-3} \text{ mol } KClO_3}{1} \cdot \frac{122.55 \text{ g } KClO_3}{1 \text{mol } KClO_3} = 1.08 \text{ g } KClO_3.$$

The percent perchlorate in the original mixture:1.08 g $KClO_3$/1.56 g mixture $\cdot$ 100 = 69.1 %

81. (a) NO, O_2 and NO_2 in increasing velocity at 298 K. We know that the velocity of gases is inversely related to the molar masses. So the NO molecules would be moving fastest, and the NO_2 molecules slowest, with O_2 molecules intermediate in velocity.

(b) Partial pressure of O_2 when mixed in the appropriate ratio with NO: The equation shows that 2 molecules of NO are needed per molecule of O_2. With the partial pressure of NO = 150 mm Hg, the partial pressure of O_2 would be 150/2 or 75 mm Hg.

(c) The partial pressure of NO_2 after reaction should be equal to that of NO before reaction (150 mm Hg), since the partial pressures are proportional to the number of moles, 2 moles of NO (and 1 mole of O_2) would form 2 moles of NO_2.

83. For the process: 4 NH_3(g) + 3 F_2(g) $\longrightarrow$ 3 NH_4F(s) + NF_3(g), the correct mix of NH_3 with F_2 is a 4:3 mole ratio. This will provide (4+3) or 7 total moles of gas with a total pressure of 120 mm Hg, The gases will exert a pressure in proportion to the mole fraction of each gas.

$$P_{F_2} = \frac{3 \text{ mol } F_2}{7 \text{ mol gas}} \bullet 120 \text{ mm Hg} = 51 \text{ mm Hg}.$$ While one can perform a similar calculation

for NH_3, one can also note that the two gases provide a total 120 mm Hg pressure, so NH_3

exerts a pressure of (120-51) or 69 mm Hg.

One mole of gas will be produced for each 3 mol of F_2, so we anticipate a pressure that is 1/3

of that exerted by the F_2, or 1/3(51) = 17 mm Hg. (Assuming T is constant.)

85. Mass of water per liter of air under the conditions:

(a) At 20 °C and 45% relative humidity. Recalling the definition of Relative Humidity (RH):

$$RH = \frac{P (H_2O) \text{ in air}}{\text{Vapor Pressure(VP) of water}}$$

We can calculate the $P(H_2O)$ in air by multiplying the VP of water by the RH.

At 20 °C, the VP of water is 17.5 torr (Data from Appendix at back of textbook)

$$P (H_2O) = RH \bullet VP = 0.45 \bullet 17.5 \text{ torr}$$

Recalling that water vapor is a gas, we can use the Ideal Gas Law to calculate the # of

moles—and the mass of water present in 1 L of air. Substituting the $P (H_2O)$ that we

calculated above into "P", rearranging the Ideal Gas Law: $\dfrac{P \bullet V}{R \bullet T} = n$, and recalling that

$n = mass(g)/MW$ we can further rearrange to produce: $\dfrac{P \bullet V \bullet MW}{R \bullet T} = mass$.

Given that VP is given in torr, use the value of $R = 62.4 \dfrac{L \bullet torr}{K \bullet mol}$.

[Alternatively convert 17.5 torr to atm, by dividing by 760, and using the value

of $R = 0.082057 \dfrac{L \bullet atm}{K \bullet mol}$]

$$\frac{(0.45 \bullet 17.5 \text{ torr}) \bullet 1 \text{ L} \bullet 18.02 \text{ g/mol}}{62.4 \dfrac{L \bullet torr}{K \bullet mol} \bullet 293 \text{ K}} = mass = 7.8 \times 10^{-3} \text{ g per liter. (2 sf)}$$

(b) At 0 °C and 95% relative humidity: (VP of water at 0°C = 4.6 torr)

$$\frac{(0.95 \bullet 4.6 \text{ torr}) \bullet 1 \text{ L} \bullet 18.02 \text{ g/mol}}{62.4 \dfrac{L \bullet torr}{K \bullet mol} \bullet 273 \text{ K}} = mass = 4.6 \times 10^{-3} \text{ g (2 sf)}$$

IN THE LABORATORY

87. Calculate the pressure of O_2 in the tank:

Mass of O_2 = 0.0870 g or (0.0870 g O_2 • 1 mol O_2/32.00 g O_2) = 0.00272 mol O_2

$$P = \frac{n \cdot R \cdot T}{V} = \frac{(2.72 \times 10^{-3}\ \text{mol})\left(0.082057 \frac{L \cdot atm}{K \cdot mol}\right)(297\ K)}{0.550\ L} = 0.12\ \text{atm (2 sf)}$$

With the total pressure (of O_2, CO, and CO_2) = 1.56 atm, the pressure of CO and CO_2 will

be 1.56 atm = $P(O_2)$ + $P(CO)$ + $P(CO_2)$ = 0.12 atm + $P(CO)$ + $P(CO_2)$.

Then 1.44 atm = $P(CO)$ + $P(CO_2)$.

We know that $P(CO)$ + $P(O_2)$ = 1.34 atm. So $P(CO)$ + 0.12 = 1.34 atm and

$P(CO)$ = 1.22 atm.

Since 1.44 atm = $P(CO)$ + $P(CO_2)$, and $P(CO)$ = 1.22 atm, then $P(CO_2)$ = 0.22 atm.

The masses of CO and CO_2 can be found by substitution into the Ideal Gas Law:

$$n = \frac{(1.22\ atm)(0.550\ L)}{(0.082057 \frac{L \cdot atm}{K \cdot mol})(297\ K)} = 2.75 \times 10^{-2}\ \text{mol CO and (• 28 g CO/mol = 0.77 g CO}$$

$$n = \frac{(0.22\ atm)(0.550\ L)}{(0.082057 \frac{L \cdot atm}{K \cdot mol})(297\ K)} = 4.96 \times 10^{-3}\ \text{mol } CO_2 \text{ and (• 44 g/mol = 0.22 g of } CO_2)$$

89. The CO_2 evolved is:

$$n = \frac{(\frac{44.9}{760}\ atm)(1.50\ L)}{(0.082057 \frac{L \cdot atm}{K \cdot mol})(298\ K)} = 3.62 \times 10^{-3}\ \text{mol } CO_2$$

Note that the CO in the compound is oxidized to CO_2, so we know the original number of

moles of CO in the compound (3.62 x 10^{-3} mol).

The mass of this amount of CO would be: (3.62 x 10^{-3} mol • 28.0 g CO/mol) = 0.101 g CO.

From 0.142 g sample of the compound, the mass of Fe is (0.142- 0.101) or 4.0 x 10^{-2}

grams Fe, and corresponding to (4.0 x 10^{-2} g Fe • 1 mol Fe/55.85 g Fe) = 7.26 x 10^{-4} mol

Fe. The ratio of Fe: CO would be: 7.26 x 10^{-4} mol Fe: 3.62 x 10^{-3} mol CO or 1:5.

The empirical formula would be $Fe(CO)_5$.

91. (a) Pressure of B_2H_6 formed:

$$\frac{0.136 \text{ g NaBH}_4}{1} \bullet \frac{1 \text{ mol NaBH}_4}{37.83 \text{ g NaBH}_4} = 3.60 \times 10^{-3} \text{ mol NaBH}_4.$$

The equation shows that we get 1 mol of B_2H_6 for each 2 mol of $NaBH_4$.

Using the mol of $NaBH_4$, we can calculate the pressure of the B_2H_6.

$$P = \frac{\left(1.80 \times 10^{-3} \text{ mol}\right)\left(0.082057 \frac{L \bullet atm}{K \bullet mol}\right)(298 \text{ K})}{2.75 \text{ L}} = 0.0160 \text{ atm (to 3 sf)}$$

(b) The total pressure is that of B_2H_6 and H_2: The balanced equation shows 2 mol of H_2 gas for each 1 mol of B_2H_6. If the pressure of B_2H_6 = 0. 0160 atm, the pressure of H_2 will be twice that pressure (0.0320 atm); total pressure = 0.0160 + 0.0320 = 0.0480 atm.

93. The amount of HCl used is: (1.50 mol HCl/L $\bullet$ 0.0120 L) = 0.0180 mol HCl.

Let's use x = mass of Na_2CO_3, and (1.249 – x) = mass of $NaHCO_3$.

We know that each mol of Na_2CO_3 consumes 2 mol HCl and each mol of $NaHCO_3$

consumes 1 mol of HCl---and that the total number of mol of HCL consumed is 0.0180 mol.

mol HCl (consumed by Na_2CO_3) + mol HCl (consumed by $NaHCO_3$) = 0.0180 mol [1]

Express the mol HCl consumed by each substance:

$$\frac{x}{1} \bullet \frac{1 \text{ mol Na}_2CO_3}{106 \text{ g Na}_2CO_3} \bullet \frac{2 \text{ mol HCl}}{1 \text{ mol Na}_2CO_3} = \text{mol HCl consumed by Na}_2CO_3 \text{ and}$$

$$\frac{(1.249 - x)}{1} \bullet \frac{1 \text{ mol NaHCO}_3}{84.0 \text{ g NaHCO}_3} \bullet \frac{1 \text{ mol HCl}}{1 \text{ mol NaHCO}_3} = \text{mol HCl consumed by NaHCO}_3$$

Substituting into equation [1]:

$$\frac{x}{1} \bullet \frac{1 \text{ mol Na}_2CO_3}{106 \text{ g Na}_2CO_3} \bullet \frac{2 \text{ mol HCl}}{1 \text{ mol Na}_2CO_3} + \frac{(1.249 - x)}{1} \bullet \frac{1 \text{ mol NaHCO}_3}{84.0 \text{ g NaHCO}_3} \bullet \frac{1 \text{ mol HCl}}{1 \text{ mol NaHCO}_3} = 0.0180 \text{ mol}$$

Simplifying yields:

$$\frac{2x}{106} + \left[\frac{1.249}{84} - \frac{x}{84}\right] = 0.0180 \text{ and}$$

$$0.01887 \, x + 0.01487 - 0.01190x = 0.0180 \text{ and}$$

$$0.006965 \, x = 0.00313 \text{ and } x = 0.449 \text{ g (Na}_2CO_3\text{), indicating that mass of NaHCO}_3$$
$$= (1.249 - 0.449) = 0.800 \text{ g NaHCO}_3.$$

Converting to moles of each substance:

$$\frac{0.449 \text{ g Na}_2\text{CO}_3}{1} \cdot \frac{1 \text{ mol Na}_2\text{CO}_3}{106 \text{ g Na}_2\text{CO}_3} = 0.00424 \text{ mol Na}_2\text{CO}_3.$$

$$\frac{0.800 \text{ g NaHCO}_3}{1} \cdot \frac{1 \text{ mol NaHCO}_3}{84.0 \text{ g NaHCO}_3} = 0.00951 \text{ mol NaHCO}_3$$

The two equations tell us that 1 mol of either substance produces 1 mol of CO_2.

The amount of CO_2 produced is then: 0.00424 mol Na_2CO_3 + 0.00951 mol $NaHCO_3$ or 0.0138 mol CO_2.

This gas would occupy a volume of:

$$V = \frac{(0.0138 \text{ mol})(0.082057 \frac{\text{L} \cdot \text{atm}}{\text{K} \cdot \text{mol}})(298 \text{ K})}{\left(\frac{745}{760} \text{ atm}\right)} = 0.343 \text{ L or } 343 \text{ mL}$$

95. For the decomposition of copper(II) nitrate one may write the equation:

$$2 \text{ Cu(NO}_3)_2 \text{ (s)} \rightarrow 2 \text{ CuO (s)} + 4 \text{ NO}_2 \text{ (g)} + \text{O}_2 \text{ (g)}.$$

The ideal gas law provides the information:

$$n = \frac{(\frac{725}{760} \text{ atm})(0.125 \text{ L})}{(0.082057 \frac{\text{L} \cdot \text{atm}}{\text{K} \cdot \text{mol}})(308 \text{ K})} = 4.72 \text{x } 10^{-3} \text{ mol of gas}$$

The average molar mass is: 0.195 g gas/4.72x 10^{-3} mol of gas = 41.3g/mol

The mole fractions of the two gases:

We need to solve a system of 2 equations:

(1) # mol NO_2 + # mol O_2 = 0.00472 mol

(2) mass NO_2 + mass O_2 = 0.195 g   Let's represent the mass of NO_2 as x.

Substitute equation (2) into equation (1), expressing the mass of O_2 in terms of the mass of NO_2

$$\frac{\text{x g}}{46.00 \text{ g/mol}} + \frac{(0.195 - \text{x})\text{g}}{32.00 \text{ g/mol}} = 0.00472$$

$$\frac{(32.00\text{g/mol x}) + (46.01\text{g/mol})(0.195 - \text{x})}{(46.01\text{g/mol})(32.00\text{g/mol})} = 0.00472 \text{ mol}$$

And (omitting units for clarity of the mathematics):

32.00x - 46.01x + (0.195)(46.01)= 0.00472(46.01)(32.00)

Solving for x gives x = 0.144 g NO_2 and (0.195-0.144)g O_2 or 0.050 g O_2

Calculating the amounts (moles) of each substance:

$$\frac{0.144 \text{ g NO}_2}{1} \bullet \frac{1 \text{ mol NO}_2}{46.01 \text{ g NO}_2} = 3.13 \times 10^{-3} \text{ mol NO}_2$$

$$\frac{0.050 \text{ g O}_2}{1} \bullet \frac{1 \text{ mol O}_2}{32.00 \text{ g O}_2} = 1.57 \times 10^{-3} \text{ mol O}_2$$

Calculate the mol fractions of each substance:

$$\frac{0.00313}{\left(0.00313 + 0.00157\right)} = 0.666 \text{ mf NO}_2 \text{ and mf O}_2 = 1.000 - 0.666 = 0.334$$

Noting that the balanced equation would make one anticipate a 4mol:1mol ratio of NO_2:O_2 the amount of NO_2 is **not** as expected. Clearly the dimerization of NO_2 plays a role in this reaction.

97. What is the molar mass of the phosphorus–fluorine compound?

(a) What is the molar mass of the gas whose density is 5.60 g/L at a pressure of 0.971 atm and a T= 18.2 °C.

$$\text{Molar mass} = \frac{(5.60 \text{ g/L})(0.082057 \frac{\text{L} \bullet \text{atm}}{\text{K} \bullet \text{mol}})([18.2 + 273.15]\text{K})}{(0.971 \text{ atm})} = 138 \text{ g/mol}$$

(b) If the unknown phosphorus fluoride effuses at a rate of 0.028 mol/min gas and CO_2 effuses at a rate of 0.050 mol/min, what is the molar mass of the unknown gas?

$$\frac{\text{Rate of effusion of CO}_2}{\text{Rate of effusion of unknown}} = \sqrt{\frac{\text{MM of unknown}}{\text{MM of CO}_2}}$$

$$\frac{0.050 \text{ mol/min}}{0.028 \text{ mol/min}} = \sqrt{\frac{\text{MM of unknown}}{44.01 \text{ g/mol}}} \text{ and squaring both sides gives:}$$

$$\left(\frac{0.050}{0.028}\right)^2 \bullet 44.01 \text{ g/mol } = \text{MM of unknown} = 140 \text{ g/mol (2 sf)}$$

The molar masses of the gases given are: PF_3 (87.9), PF_5 (125.9) and P_2F_4 (137.9), so the unknown gas is most likely P_2F_4.

SUMMARY AND CONCEPTUAL QUESTIONS

99. In a 1.0-L flask containing 10.0 g each of O_2 and CO_2 at 25 °C.

(a) The gas with the greater partial pressure:

Partial pressure is a relative measure of the number of moles of gas present. The molar mass of oxygen is approximately 32 g/mol while that of CO_2 is approximately 44 g/mol.

There will be a greater number of moles of oxygen in the flask—hence the **partial pressure of O_2 will be greater**.

(b) The gas with the greater average speed:

The kinetic energy of each gas is given as: $KE = 1/2\ mu^2$, where u is the average speed of the molecules. Since the average KE of both gases are the same (They are at the same T), The lighter of the gases **(O_2) will have a greater average speed**.

(c) The gas with the greater average kinetic energy:

The KE depends upon temperature, and since both gases are at the same T, the **average KE of the two gases is the same**.

101. Two containers with 1.0 kg of CO and 1.0 kg C_2H_2

(a) Cylinder with the greater pressure:

Since P is proportional to the amount of substance present, let's calculate the # of moles of each gas.

Molar Mass of CO = 28 g/mol   Molar Mass of C_2H_2 = 26 g/mol

While we could calculate the # of moles, note that since acetylene has a small molar mass, 1.0 kg of acetylene would have more moles of gas (and *a greater pressure*) than 1.0 kg of CO.

(b) Cylinder with the greater number of molecules:

Since we know that the cylinder with acetylene has more moles of gas, the cylinder *with the acetylene* will have the greater number of molecules--since to convert between moles of gas and molecules of gas, we multiply by the constant, Avogadro's number.

103. Which of the following samples is a gas?

(a) Material expands 1% when a sample of gas originally at 100 atm is suddenly allowed to exist at one atmosphere pressure: This sample is **not a gas**, since a gas would expand in volume by 100-fold (to exist at 1 atm).

(b) A 1.0-ml sample of material weighs 8.2 g: This sample is **not a gas** since the density of the sample (at 8.2 g/mL) is too great.

(c) Material is transparent and pale green in color: **Insufficient information** to tell. Liquids could also be pale green and transparent.

(d) One cubic meter of material contains as many molecules as an equal volume of air at the same temperature and pressure: This material **is a gas**, since one cubic meter of a liquid or solid would contain a greater number of molecules than a cubic meter of air.

105. (a) To determine the balloon containing the greater number of molecules we need only to calculate the ratio of the amounts of gas present in the balloons. Using the Ideal Gas Law allows us to solve for the number of moles of each gas:

Pick a volume of gas—say 5L for He. Since the hydrogen balloon is twice the size of the He balloon, we'll use 10 L for the volume of H. We can then establish a ratio of this

expression for the two gases.
$$\frac{n_{He}}{n_{H_2}} = \frac{\dfrac{P_{He} \bullet V_{He}}{R \bullet T_{He}}}{\dfrac{P_{H_2} \bullet V_{H_2}}{R \bullet T_{H_2}}} = \frac{\dfrac{2atm \bullet 5L}{296K}}{\dfrac{1atm \bullet 10 L}{268K}} = \frac{268 \text{ K}}{296 \text{ K}} = \frac{0.9 \text{ mol He}}{1 \text{ mol H}_2}$$

Note that **R** cancels in the numerator and denominator, simplifying the calculation.

(b) Two calculate which balloon contains the greater mass of gas, we can use the ratio found:
0.9 mol He • 4.0 g/mol He = 3.6 g He; 1 mol H_2 • 2.0 g/mol H_2 = 2 g H_2

The balloon containing the *HELIUM* has the greater mass. Note that the *ratio* of moles of helium and hydrogen are **independent** of the sizes of the balloons.

107. The change of average speed of a gaseous molecules when T doubles:

Since the speed of a gaseous molecule is related to the square root of the *absolute* T, a change by a factor of 2 in the absolute T will lead to a change of $\sqrt{2}$ in the speed.

Applying Chemical Principles

1. What is the density (in g/L) of helium at 1.00 atm pressure and 25 °C?

Solving the Ideal Gas Law for d (which is n/V), we obtain: $\dfrac{(\text{Molar Mass}) \cdot P}{R \cdot T} = d$

$$\dfrac{(4.0026 \text{ g/mol})(1.00 \text{ atm})}{(0.082057 \dfrac{L \cdot atm}{K \cdot mol})(298.2 \text{ K})} = 0.164 \text{ g/L (3 sf)} = \text{density}$$

3. The density of the blimp is its total weight (blimp, helium and air, passengers, and ballast) divided by its volume. Gross weight of the blimp includes the blimp's structure and the helium, but does not include the air in the ballonets or the weights of the passengers and ballast. If the ballonets are filled with 340 m³ of air at 1.00 atm and 25 °C, what additional weight (of passengers and ballast) is required for neutral buoyancy?

We need the density of dry air at 1.00 atm and 25 deg C, so solve as in (1) above:

From Table 11.1 we find the average molar mass of dry air to be 2.8960 g/mol.

$$\dfrac{(28.960 \text{ g/mol})(1.00 \text{ atm})}{(0.082057 \dfrac{L \cdot atm}{K \cdot mol})(298.2 \text{ K})} = 1.18 \text{ g/L (3sf)}$$

Density is calculated in g/L, so express the masses in units of g, and the volumes in liters!

[From the Conversion factors table: 1 L = 1.00 x 10⁻³ m³]

For neutral buoyancy: Density of the blimp = Density of the air

D(blimp) = 1.18 g/L

D(blimp) = [(weight of blimp + weight of He) + weight of air + weight of passengers + weight of ballast]/volume

$$D(\text{blimp}) = \dfrac{[(5820 \text{kg}) + (1.18 \text{ g/L})(340\text{m}^3)(\dfrac{1L}{1 \times 10^{-3}\text{m}^3}) + \text{weight of passengers} + \text{ballast}]}{(5740 \text{ m}^3)(\dfrac{1L}{1 \times 10^{-3}\text{m}^3})}$$

Simplifying gives:

$$D(\text{blimp}) = \dfrac{(5.820 \times 10^6 \text{g} + 4.012 \times 10^5 \text{g} + (\text{weight of passengers} + \text{ballast})}{5.740 \times 10^6 \text{ L}}$$

Substituting this expression into the expression: D(blimp) = 1.18 g/L gives

$$\dfrac{(5.820 \times 10^6 \text{g} + 4.012 \times 10^5 \text{g} + (\text{weight of passengers} + \text{ballast})}{5.740 \times 10^6 \text{ L}} = 1.18 \text{ g/L}$$

6.2212 x 10⁶ g + (weight of passengers + ballast) = (1.18 g/L)(5.740 x 10⁶L)

6.2212 x 10⁶ g + (weight of passengers + ballast) = 6.7732 x 10⁶ g

(weight of passengers + ballast) = 6.7732 x 10⁶ g - 6.2212 x 10⁶ g = 552,000 g or 552 kg.

Chapter 12
Intermolecular Forces and Liquids

PRACTICING SKILLS

Intermolecular Forces

1. The <u>intermolecular</u> forces one must overcome to:

<u>change</u>	<u>intermolecular force</u>
(a) melt ice	hydrogen bonds and dipole-dipole (molecule with OH bonds)
(b) sublime solid I_2	induced dipole-induced dipole (nonpolar molecule)
(c) convert $NH_3(l)$ to $NH_3(g)$	hydrogen bonds and dipole-dipole (molecule with NH bonds)

3. To convert <u>species</u> from a liquid to a gas the <u>intermolecular</u> forces one must overcome:

<u>species</u>	<u>intermolecular force</u>
(a) liquid O_2	induced dipole-induced dipole (nonpolar molecule)
(b) mercury	induced dipole-induced dipole (nonpolar atom)
(c) methyl iodide	dipole-dipole (polar molecule)
(d) ethanol	hydrogen bonding and dipole-dipole (polar molecule with OH bonds)

5. Increasing strength of intermolecular forces:

$$CH_4 < Ne < CO < CCl_4$$

Neon and methane are nonpolar species and possess only induced dipole-induced dipole interactions. Methane has a smaller molar mass than neon, and therefore weaker London (dispersion) forces. Carbon monoxide is a polar molecule. Molecules of CO would be attracted to each other by dipole-dipole interactions, but the CO molecule is not a very strong dipole. The CCl_4 molecule is a non-polar molecule, but very heavy (when compared to the other three). Hence the greater London forces that accompany larger molecules would result in the strongest attractions of this set of molecules. The lower molecular weight molecules with weaker interparticle forces should be gases at 25 °C and 1 atmosphere: CH_4, Ne, CO.

7. Compounds which are capable of forming hydrogen bonds with water are those containing polar O-H bonds and lone pairs of electrons on N, O, or F.

(a) CH_3-O-CH_3 no; no "polar H's" and the C-O bond is not very polar

(b) CH_4 no

(c) HF   yes: lone pairs of electrons on F and a "polar hydrogen".

(d) CH_3COOH yes: lone pairs of electrons on O atoms, and a "polar hydrogen"

    attached to one of the oxygen atoms

(e) Br_2 no; nonpolar molecules

(f) CH_3OH yes: "polar H" and lone pairs of electrons on O

9. For each pair, which has the greater heat of hydration (more negative)?

(a) LiCl–since Li^{1+} is a smaller cation than Cs^{1+}.

(b) $Mg(NO_3)_2$; two effects here. Mg^{2+} is a smaller cation than Na^{1+}, and secondly magnesium has a 2+ charge while Na only a 1+ charge.

(c) $NiCl_2$ – same effects here as in (b) above. Ni^{2+} is a smaller cation than Rb^{1+}, and is doubly charged.

Liquids

11. Heat required: $125 \text{ mL} \cdot \dfrac{0.7849 \text{ g}}{1 \text{ mL}} \cdot \dfrac{1 \text{ mol}}{46.07 \text{ g}} \cdot \dfrac{42.32 \text{ kJ}}{1 \text{ mol}} = 90.1 \text{ kJ}$

13. Using Figure 12.17:

(a) The equilibrium vapor pressure of water at 60 °C is approximately 150 mm Hg.
    Appendix G lists this value as 149.4 mm Hg.
(b) Water has a vapor pressure of 600 mm Hg at 93 °C.
(c) At 70 °C the vapor pressure of water is approximately 225 mm Hg while that of ethanol is approximately 520 mm Hg.

15. The vapor pressure of $(C_2H_5)_2O$ at 30°C is **590 mm Hg**.

Calculate the amount of $(C_2H_5)_2O$ to furnish this vapor pressure at 30 °C (303K).

$$n = \frac{PV}{RT} = \frac{590 \text{ mm} \cdot \dfrac{1 \text{ atm}}{760 \text{ mm}} \cdot 0.100 \text{ L}}{0.082057 \dfrac{\text{L} \cdot \text{atm}}{\text{K} \cdot \text{mol}} \cdot 303 \text{ K}} = 3.1 \times 10^{-3} \text{ mol}$$

The total mass of $(C_2H_5)_2O$ [FW = 74.1 g] needed to create this pressure is about 0.23 g.

Since there is adequate ether to provide this pressure, we anticipate the pressure in the flask to be approximately 590 mm Hg. As the flask is cooled from 30.°C to 0 °C, **some of the gaseous ether will condense** to form liquid ether.

17. Member of each pair with the higher boiling point:
   (a) O_2 would have a higher boiling point than N_2 owing to its greater molar mass.
   (b) SO_2 would boil higher since SO_2 is a polar molecule while CO_2 is non-polar.
   (c) HF would boil higher since strong hydrogen bonds exist in HF but not in HI.
   (d) GeH_4 would boil higher. While both molecules are non-polar, germane has the greater molar mass and therefore stronger London forces.

19. (a) From the figure, we can read the vapor pressure of CS_2 as approximately 620 mmHg and for nitromethane as approximately 80 mm Hg.
   (b) The principle intermolecular forces for CS_2 (a non-polar molecule) are **induced dipole-induced dipole**; for nitromethane (polar molecule)--**dipole-dipole**.
   (c) The normal boiling point from the figure for CS_2 is 46 °C and for CH_3NO_2, 100 °C.
   (d) The temperature at which the vapor pressure of CS_2 is 600 mm Hg is about 39 °C.
   (e) The vapor pressure of CH_3NO_2 is 60 mm Hg at approximately 34 °C.

21. Regarding benzene:
   (a) Using the data provided, note that the vapor pressure of benzene is 760 mm Hg at 80.1°C
   (b) The graph of the data:
      The temperature at which the liquid has a vapor pressure of 250 mm Hg is about 47°C. The temperature at which the vapor pressure is 650 mm Hg is about 77 °C.

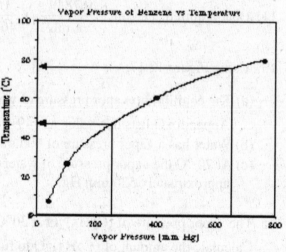

   (c) The molar enthalpy of vaporization from the Clausius-Clapeyron equation:

   The Clausius-Clapeyron equation is: $\ln\left(\dfrac{P_2}{P_1}\right) = \dfrac{\Delta H}{R}\left(\dfrac{1}{T_1} - \dfrac{1}{T_2}\right)$

   From the graph using (arbitrarily) two of the data points:

   $T_1 = 47$ °C and $P_1 = 250.$ mm Hg and $T_2 = 77$ °C and $P_2 = 650.$ mm Hg.

Recall however that the units of R include temperature units of K, So convert the two

temperatures to the K scale and substitute: $\ln\left(\dfrac{650.}{250.}\right)=\dfrac{\Delta H}{8.3145\ \dfrac{J}{K\bullet mol}}\left(\dfrac{1}{320\ K}-\dfrac{1}{350\ K}\right)$

Simplifying the equation:

$8.3145\ \dfrac{J}{K\bullet mol}\bullet \ln(2.60)=\Delta H\left(2.180\ x\ 10^{-4}\right)$ and $\Delta H = 29660$ J/mol or 30.0 kJ/mol

Using the graph, or two other data points than those used here, one may obtain a slightly different value.

23. A gas can be liquefied at or below its critical temperature. The critical temperature for CO is 132.09K (or –140° C) so CO can **not** be liquefied at room temperature (25 °C or –248K).

25. Surface tension is the energy needed to break through the surface of a liquid. Surface tension is manifest in many areas—the spherical form of a soap bubble is one. The ability of the "water strider" insect to "walk" across the surface of a pond is another. Surface tension is the consequence of intermolecular forces, since—at the surface, the molecules (e.g. of water) are being tightly held together to neighboring molecules on all sides, and to molecules *below* them, but **not** to molecules above them. This "unbalanced" force results in a taut surface.

27. The climbing of water upwards into the paper is an example of capillary action, brought about by the attraction of the water molecules in the container to the -OH bonds on the cellulose.

GENERAL QUESTIONS

29. Increasing strength of intermolecular forces: Ar < CO_2 < CH_3OH
    Argon and CO_2 are nonpolar species and possess only induced dipole-induced dipole (London) interactions. Ar has a smaller mass than CO_2, so London forces are expected to be weaker for Ar than for CO_2. The polar molecular CH_3OH is capable of forming the stronger hydrogen-bonds (with an O-H bond in the molecule).

31. Which salt has the *more exothermic* Enthalpy of Hydration? Between the two salts, the LITHIUM cation—being the smaller of the two—will have greater interaction with the water—and a resulting greater negative (exothermic) enthalpy of hydration.

33. Using the vapor pressure curves:

(a) The vapor pressure of ethanol at 60 °C is: 350 mm Hg (to the limits of this reader's ability to read the graph)

(b) The stronger intermolecular forces in the liquid state are those of: Ethanol has a lower vapor pressure than carbon disulfide at every temperature—and hence stronger intermolecular forces. This is quite expected since ethanol has hydrogen bonding as the intermolecular force while CS_2 has only induced dipole forces.

(c) The temperature at which heptane has a vapor pressure of 500 mm Hg is: 84 °C

(d) The approximate normal boiling points of the three substances are:

Carbon disulfide   bp = 46°C (literature value 46.5 °C)

Ethanol   bp = 78°C (literature value 78.5 °C)

Heptane   bp = 99°C (literature value 98.4 °C)

(e) At a pressure of 400 mm Hg and 70 °C, the state of the three substances is:

Carbon disulfide   state = gas

Ethanol   state = gas

Heptane   state = liquid

35. Increasing molar enthalpy of vaporization: CH_3OH, C_2H_6, HCl.

Owing to the H-bonding of CH_3OH, we predict that this substance will have the greatest molar enthalpy of vaporization. Ethane, on the other extreme, being a nonpolar compound, and of the 3 substances here the lowest molecular mass, will have the lowest. HCl—with a intermediate molecular mass and being polar—will have the intermediate molar enthalpy of vaporization.

$$C_2H_6 < HCl < CH_3OH$$

37. Calculate the number of atoms represented by the vapor:

The pressure of Hg is $0.00169 \text{ mm Hg} \cdot \dfrac{1 \text{ atm}}{760 \text{ mm Hg}} = 2.22 \times 10^{-6} \text{ atm}$

The number of moles/L is: $\dfrac{P}{R \cdot T} = \dfrac{n}{V} = \dfrac{2.22 \times 10^{-6} \text{ atm}}{(0.082057 \frac{L \cdot atm}{K \cdot mol})(297K)} = 9.12 \times 10^{-8} \text{ mol/L}$

Converting this to atoms/m^3:

$9.12 \times 10^{-8} \dfrac{mol}{L} \cdot \dfrac{1000 \text{ L}}{1 \text{ m}^3} \cdot \dfrac{6.022 \times 10^{23} \text{ atoms}}{1 \text{ mol}} = 5.49 \times 10^{19} \dfrac{atoms}{m^3}$

Note that the information that the air was saturated with mercury vapor obviates the need to calculate the volume of the room.

IN THE LABORATORY

39. Regarding dichlorodimethylsilane, and given the data:

    (a) The normal boiling point is easily determined, since it is *defined* as the temperature at which the vapor pressure of a substance is equal to 760 mm Hg. The normal boiling point is 70.3 °C.

    (b) Plotting the data:

    Vapor pressure of dichlorodimethylsilane

    $\ln P = 17.949 - 3885(1/T)$

    Using the equation for the straight-line, we can calculate T at P values of 250 and 650 mm Hg. Substituting P = 250 mm Hg, $\ln P = 5.52$ and T = 312.6 K (39.5 °C).

    Similarly, for P = 650 mm Hg, $\ln P = 6.48$ and T = 338.7 K (65.5 °C).

    (c) The Molar Enthalpy of vaporization for dichlorodimethylsilane from the Clausius-Clapeyron equation:

    $$\ln\left(\frac{P_2}{P_1}\right) = \frac{\Delta H_{vap}}{R}\left[\frac{1}{T_1} - \frac{1}{T_2}\right]$$ Using the T,P data from part (b):

    $$\ln\left(\frac{650. \text{ mm Hg}}{250. \text{ mm Hg}}\right) = \frac{\Delta_{vap}H}{8.3145 \times 10^{-3} kJ / K \bullet mol}\left[\frac{1}{312.6K} - \frac{1}{338.7K}\right]$$

    $$\ln 2.60 \bullet 8.3145 \times 10^{-3} kJ/K\bullet mol = \Delta_{vap}H\left[\frac{1}{312.6 \text{ K}} - \frac{1}{338.7 \text{ K}}\right]$$

    $$\ln 2.60 \bullet 8.3145 \times 10^{-3} \frac{kJ}{K\bullet mol} = \Delta_{vap}H (2.46 \times 10^{-4} \text{ K}^{-1})$$

    $$\frac{7.9445 \times 10^{-3} \text{ kJ/K}\bullet mol}{2.46 \times 10^{-4} \text{ K}^{-1}} = 32.3 \text{ kJ/mol}$$

41. (a) The can collapses as a result of the condensation of the gas in the can—which has filled the heated can—to a liquid. The distances between the particles of liquid are **much less**

than the distances between the particles of gas. The resulting decrease in pressure inside the can causes the greater pressure outside the can to crush it.

(b) Before the can is heated, most of the water exists in the liquid form, while some exists as the vapor (above the liquid surface). After heating, the liquid has been converted into the gaseous form, filling the volume of the can.

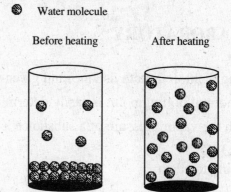

SUMMARY AND CONCEPTUAL QUESTIONS

43. Acetone readily absorbs water owing to *hydrogen bonding* between the C = O oxygen atom and the O—H bonds of water.

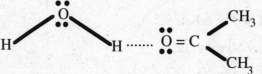

45. The **viscosity of ethylene glycol would be predicted to be greater** than that of ethanol since the glycol possesses two O-H groups per molecule while ethanol possesses one. Two OH groups/molecule would provide more hydrogen bonding!

47. Explain the fact that:

(a) Ethanol has a lower boiling point than water. Both molecules are polar, and both can form hydrogen bonds. Water has **two** polar H atoms, and two lone pairs of electrons on the O atom. Ethanol has only **one** polar H atom, with the accompanying two lone pairs on the O atom. Given the increased ability of water to form hydrogen bonds with other water molecules, one would expect that water would have a higher boiling point.

(b) Mixing 50 mL of water with 50 mL of ethanol results in less than 100 mL of solution. The H-bonding that occurs not only between (water molecules and other water molecules) and (ethanol molecules and other ethanol molecules) also occurs between ethanol and water molecules. The attraction results in the molecules occupying less space than one would anticipate—and a non-additive volume.

49. Evidence that water molecules in the liquid state exert attractive forces on one another:

a) Water has a relatively large specific heat capacity (4.18 J/g •K). This large value is a reflection of the strong forces that hold water molecules together and require a large amount of energy to overcome.

b) Water is a liquid at room temperature—even though it has a molar mass of approximately 18 g/mol. With this molar mass, one would anticipate a boiling point well below 0°C (and a *gaseous* physical state).

51. Referring to Figure 12.12:

(a) The hydrogen halide with the *largest total intermolecular force* is HI.

(b) The dispersion forces are greater for HI than for HCl owing to the fact that HI (specifically I) has a larger volume than HCl (or Cl).

(c) Dipole-Dipole forces are greater for HCl than for HI owing to the more polar H-Cl bond. The electronegativities (H= 2.2; Cl = 3.2; I = 2.7) indicate this difference in polarity (3.2-2.2) versus (2.7-2.2).

(d) The HI molecule has the largest dispersion forces of the molecules shown in the Figure. This is quite reasonable given the large volume of the Iodine atom (see part b).

53. The critical temperature for CF_4 (-45.7 °C) is below room temperature (25 °C). Since room temperature is greater than $_{Tc}$ for CF_4, the substance **can not be** liquefied at room temperature.

55. The electrostatic potential surface is shown below:

Hydrogen bonding occurs between H-atoms and the **lone pairs** of electrons on very electronegative atoms (F, O, N). The electropotential surface indicates two oxygen atoms (red spheres in the original diagram—and indicated by arrows below), each of which will contain two lone pairs of electrons at which hydrogen bonding is likely to occur. Also the N atom of the amide group will also exhibit H-bonding.

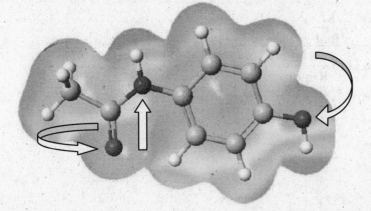

57. Four properties of liquids that are directly determined by intermolecular forces:

    a. vapor pressure—the pressure exerted by a liquid in equilibrium with its gaseous phase.

    b. boiling point—the temperature at which the vapor pressure of the liquid equals the ambient atmospheric pressure

    c. surface tension--energy required to break through the surface of a liquid

    d. viscosity--the resistance of liquids to flow

59. The branched chain isomers have a smaller exposed surface than their unbranched analogues. With decreased molecular contact, the dispersion forces are weakened, with the resulting lower boiling points.

61. Rank the halogens (F_2, Cl_2, Br_2, I_2) and the noble gases (He, Ne, Ar, Kr, Xe) in order of polarizability (from least polarizable to most polarizable):

We know that dispersion forces (and polarizability) increase *with increasing atomic or molecular volume*, so use the position of the elements in the Periodic Table (and hence their respective volumes) to rank them:

	Least polarizable						Most polarizable
Halogens:	F_2	$<$	Cl_2	$<$	Br_2	$<$	I_2
Noble gases:	He	$<$	Ne	$<$	Ar	$<$	Kr	$<$	Xe

63. Use the Clausius-Clapeyron equation to calculate the temperature at which water boils in the pressure cooker:

$$\ln\frac{P_2}{P_1} = -\frac{\Delta H^0_{vap}}{R}\left[\frac{1}{T_2}-\frac{1}{T_1}\right] \text{ and substituting}$$

$$\ln\frac{(14.70+15)\text{ psi}}{14.70\text{ psi}} = -\frac{40.7\frac{\text{kJ}}{\text{mol}}}{0.0083145\frac{\text{kJ}}{\text{K}\cdot\text{mol}}}\left[\frac{1}{T_2}-\frac{1}{373\text{K}}\right]$$

$$0.7033 = -4895\left[\frac{1}{T_2} - \frac{1}{373K}\right] \text{ and } \frac{0.7033}{-4895} = \left[\frac{1}{T_2} - \frac{1}{373K}\right]$$

$$-0.000143675 = \left[\frac{1}{T_2} - \frac{1}{373K}\right] \text{ and } -0.000143675 + \frac{1}{373K} = \frac{1}{T_2}; \text{ solving for T2 gives}$$

394 K or 121°C.

65. Using the data in Question 64, the following plot was obtained:

    The best straight line for the data is:

    y = -2808.123x + 11.70385

    Substituting 20°C (293.15K) into

    the equation gives a "y" value of:

    2.124716451 which is lnP.

    P is then: 8.4 atm

67. Water (10.0 g) is placed in a thick walled glass tube whose internal volume is 50.0 cm³. Then all the air is removed, the tube is sealed, and then the tube and contents are heated to 100 °C.

    (a) What is the appearance of the system at 100 °C.

    There will be liquid water in the tube, and some of the liquid will be converted to the gaseous state (water vapor)

    (b) What is the pressure inside the tube?

    Since the water is at 100°C, the vapor pressure of water is 1.00 atm (the temperature at which water boils at 1 atm is 100 °C, check?)

    (c) What volume of liquid water is in the tube at this temperature,

    [$D_{\text{liquid water}}$= 0.958 g/cm³] If all the water is in the liquid state, the volume of

    water would be: $10.0 \text{ g} \cdot \dfrac{1 \text{ cm}^3}{0.958 \text{ g}} = 10.4 \text{ cm}^3$ (to 3 sf)

    (d) Some of the water is in the vapor state. Determine the mass of water in the gaseous state.

    The volume of the tube is 50.0 cm³. The water (in the liquid state) would occupy

    10.4 cm³. So the gas would have a volume of (50.0-10.4) or 39.6 cm³.

Substituting into the Ideal Gas Law:

$$n = \frac{P \cdot V}{R \cdot T} = \frac{1.00 \text{ atm} \cdot 0.0396 \text{ L}}{0.082057 \dfrac{\text{L} \cdot \text{atm}}{\text{K} \cdot \text{mol}} \cdot 373.15 K} = 0.001293 \text{ mol}$$

Which corresponds to 18.02 g/mol x 0.001293 mol = 0.0233 g H_2O.

Applying Chemical Principles

1. Assume that a mixture of the three molecules below are separated on a C-18 column using a methanol/water mixture as the mobile phase.

   (a) The *most polar* of the 3 molecules, 1,5-pentanediol, will be most attracted to the polar mobile phase. The diol has two –OH groups, so hydrogen bonding will be a major intermolecular force. Since the molecule is polar, dipole-dipole forces will be present, as are the omnipresent London dispersion forces.

   (b) The molecule most attracted to the stationary phase will be the *least polar* of the three molecules—propyl ethyl ether. As in part (a), dipole-dipole forces and London dispersion forces will attract the ether to the stationary phase.

   (c) In what order will the three molecules exit (or "elute from") the column?
   The only question remaining is how 1-pentanol fits into the elution order. This alcohol possesses that important –OH functionality (but only one, unlike the pentanediol). We anticipate that the pentanol will be attracted to the mobile phase *to a greater extent* than the ether, so the elution order would be: 1,5-pentanediol first, followed by the 1-pentanol, with propyl ethyl ether eluting last.

Chapter 13
The Chemistry of Solids

PRACTICING SKILLS

Metallic and Ionic Solids

1. This compound would have the formula AB_8 since each black square (A) has

   eight corresponding white squares (B).

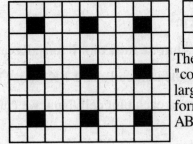 The unit shown above is the smallest "cookie cutter" that would reproduce the larger pattern shown at the left. The formula for such a compound would be AB_8.

3. To determine the perovskite formula, determine the number of each atom belonging **uniquely** to the unit cell shown. The Ca atom is wholly contained within the unit cell. There are Ti atoms at each of the eight corners. Since each of these atoms belong to eight unit cells, the portion of each Ti atom belonging to the pictured unit cell is 1/8 so 8 Ti atoms x 1/8 = 1 Ti atom. The O atoms on an edge belong to 4 unit cells, so the fraction contained within the pictured cell is 1/4. There are twelve such O atoms, leading to 12 x 1/4 = 3 O atoms— and a formula of $CaTiO_3$.

5. For cuprite:

   (a) Formula for cuprite: There are 8 oxygen atoms at the corner of the cell (each 1/8 within the cell) and 1 oxygen atom internal to the cell (wholly within the cell).

   The number of oxygen atoms is then [(8 • 1/8) +(1)] 2. There are 4 Cu atoms wholly within the cell, so the ratio was Cu_2O.

   (b) With a formula of Cu_2O, and the oxidation state of O = -2, the oxidation state of copper must be +1.

7. Radius of a calcium atom, given the density of solid Ca is 1.54 g/cm^3, and it crystallizes in fcc cell: Noting that calcium uses the fcc unit cell, we know that there are 4 calcium atoms in the unit cell: (6 faces • 1/2) + (8 corners • 1/8) = 4

   Using the density, let's calculate the number of calcium atoms in a unit cell:

$$\frac{4 \text{ atoms Ca}}{1 \text{ unit cell}} \bullet \frac{1 \text{mol Ca}}{6.022 \text{x} 10^{23} \text{atoms}} \bullet \frac{40.08 \text{ g Ca}}{1 \text{ mol Ca}} \bullet \frac{1 \text{ cm}^3}{1.54 \text{ g Ca}} = 1.73 \text{ x } 10^{-22} \text{ cm}^3$$

As volume is (length)^3, the length of the edge is $(1.73 \text{ x } 10^{-22} \text{ cm}^3)^{1/3}$ or $5.57 \text{ x } 10^{-8}$ cm

Referring to the diagram found in SQ13.39 in this manual, the face diagonal of the unit cell is 4• atomic radius, and the $\sqrt{2}$ • edge of the unit cell.

$$4 \bullet \text{radius} = \sqrt{2} \bullet 5.57 \text{ x } 10^{-8} \text{ cm}$$

$$\text{radius} = \frac{\sqrt{2} \bullet 5.57 \text{ x } 10^{-8} \text{ cm}}{4} = 1.97 \text{ x } 10^{-8} \text{cm or 197 pm}$$

9. Length of the side of a unit cell of KI, with density = 3.12 g/cm^3:

The face-centered cubic lattice with I$^-$ ions in the sites, would have a total of 4 I$^-$ ions. Recall that there are 8 ions at the corners (each contained 1/8 within the unit cell), and 4 K$^+$ ions (12 octahedral holes x 1/4 and 1 in center). So we have 4 KI pairs in each unit cell. Now we can calculate the volume of a unit cell:

$$\frac{1 \text{ cm}^3}{4.12 \text{ g KI}} \bullet \frac{166.0 \text{ g KI}}{1 \text{ mol KI}} \bullet \frac{1 \text{ mol KI}}{6.02214 \text{ x } 10^{23} \text{KI}} \bullet \frac{4 \text{ KI pairs}}{1 \text{ unit cell}} = 3.534 \text{x} 10^{-22} \text{cm}^3$$

The unit cell length is the cube root of the volume (definition of volume of a cube = l^3):

$$\text{Length} = \sqrt[3]{3.534 \text{ x } 10^{-22} \text{ cm}^3} = 7.07 \text{ x } 10^{-8} \text{ cm}$$

Expressed in units of picometers: $\dfrac{7.07 \text{x} 10^{-8} \text{cm}}{1} \bullet \dfrac{1 \text{m}}{100 \text{ cm}} \bullet \dfrac{1 \text{ pm}}{1 \text{ x } 10^{-12} \text{m}} = 707 \text{ pm}$

Metals and Semiconductors

11. If one uses the 2s atomic orbitals of 1000 Li atoms, one can form 1000 molecular orbitals (See Figure 13.11) In the lowest energy state, half of these orbitals will be populated by pairs of electrons and half will be empty.

13. How does the theory for metallic bonding explain conductivity? Metals have very closely spaced molecular orbitals. Small amounts of energy result in promoting electrons into higher (unoccupied) levels, creating a space (called a hole) that the electron had occupied. Electrical conductivity arises from the presence of the holes and "promoted" electrons in singly occupied states in the presence of an applied electric field. As the "promoted" electrons move toward the "+" side of the applied field, and the holes move toward the "-" side of the applied electric field—*current results*.

15. Diamond is an insulator and silicon is a semi-conductor. Why? This is easily understood if one recognizes that the band gap for silicon is small enough to permit electrons to be promoted from the valence band to the conduction band, while the band gap for carbon is too great to permit this to occur.

17. Intrinsic semiconductors are those in which the semi-conducting phenomenon is naturally occurring (e.g. silicon), while extrinsic semiconductors are those in which dopants are added. Extrinsic conductors are made by the addition of a Group III or Group V atom to a "base" material from Group IV.

Ionic Bonding and Lattice Energy

19. Arrange lattice energies from least negative to most negative:

Since CaO involves 2+ and 2- ions, the lattice energy for this compound is greater than the other compounds. Given that lattice energy is **inversely** related to the distance between the ions, the lattice energy for LiI is greater (more negative) than that for RbI ($Rb^+ > Li^+$). The small diameter of the fluoride ion (compared to iodide) indicates that the lattice energy for LiF would be more negative than for either LiI or RbI. The lattice energies are then:

least   -----   RbI   -----   LiI   -----   LiF   -----   CaO   **most**
negative   **negative**

21. Since melting a solid involves disassembling the crystal lattice of cations and anions, the lesser the distance between cations and anions—the greater the attraction between the cation and anion, and the harder it becomes to disassemble the lattice, and hence the **higher the melting point**.

23. The molar enthalpy of formation of solid lithium fluoride:

$1/2\ F_2(g) \longrightarrow F(g)$	$\Delta_f H =$ + 78.99 kJ/mol F
$F(g) + e^- \longrightarrow F^{-1}(g)$	$E\ A =$ -328.0 kJ/mol F
$Li\ (s) \longrightarrow Li\ (g)$	$\Delta_f H =$ + 159.37 kJ/mol Li
$Li(g) \longrightarrow Li^+(g) + 1\ e^-$	$IE =$ +520. kJ/mol Li
$\underline{Li^+(g) + F^-(g) \longrightarrow LiF(s)}$	$\underline{Lattice\ E = -1037.\ kJ/mol\ LiF}$
$Li(s) + 1/2\ F_2(g) \longrightarrow LiF(s)$	- 606.64 (-607 kJ/mol)

Other Types of Solids

25. For the unit cell of diamond:

(a) The unit cell has 8 corner atoms (1/8 in the cell), 6 face atoms (1/2 in the cell), and 4 atoms wholly within the cell, for a total of **8 carbon atoms**.

(b) Diamond uses a fcc unit cell (The structure shown also has 4 atoms occupying holes in the lattice). The holes are **tetrahedral**.

27. What particles make up the five types of solids? What are the forces of attraction between these particles?

Type	Particles	Attractive Forces
metallic	Metal atoms	Metallic bond
ionic	Cations and anions	Ion-ion attractions
molecular	Molecules	Intermolecular forces between the molecules;covalent bonds between the atoms in each molecule
network	Bonded atoms in long-range networks	Covalent bonds
amorphous	Bonded atoms in short-range networks	Covalent bonds

29.

	Classification	Particles	Attractive Forces	Physical Property
(a) gallium arsenide	Network	Atoms	Covalent bonds	semiconductor
(b) polystyrene	Amorphous	Atoms	Covalent bonds	Wide melting range
(c) silicon carbide	Network	Atoms	Covalent bonds	Very hard substance
(d) perovskite	Ionic	Cations and Anions	Ion-Ion attractions	Brittle; high-melting

Phase Changes for Solids

31. The heat evolved when 15.5 g of benzene freezes at 5.5 °C:

$$15.5 \text{ g benzene} \cdot \frac{1 \text{ mol benzene}}{78.1 \text{ g benzene}} \cdot \frac{9.95 \text{ kJ}}{1 \text{ mol benzene}} = -1.97 \text{ kJ}$$

Note once again the negative sign indicates that heat is evolved.

The quantity of heat needed to remelt this 15.5 g sample of benzene would be +1.97 kJ.

Phase Diagrams and Phase Changes

33. (a) The positive slope of the solid/liquid equilibrium line means the liquid CO_2 is **less dense** than solid CO_2.

(b) At 5 atm and 0 °C, CO_2 is in the **gaseous phase**.

(c) The phase diagram for CO_2 shows the critical pressure for CO_2 to be 73 atm, and the critical temperature to be +31 °C, so CO_2 cannot be liquefied at 45 °C.

35. The heat required is a summation of three "steps":

a. heat the liquid (at -50.0 °C) to its boiling point (-33.3 °C)

b. "boil" the liquid—converting it to a gas and

c. warm the gas from -33.3 °C to 0.0 °C

   1. To heat the liquid (at -50.0 °C) to its boiling point (-33.3 °C):

$$q_{liquid} = 1.2 \times 10^4 \text{ g} \cdot 4.7 \, \frac{J}{g \cdot K} \cdot (239.9 \text{ K} - 223.2 \text{ K}) = 9.4 \times 10^5 \text{ J}$$

   2. To boil the liquid:

$$23.3 \times 10^3 \, \frac{J}{mol} \cdot \frac{1 \text{ mol NH}_3}{17.03 \text{ g NH}_3} \cdot 1.2 \times 10^4 \text{ g} = 1.6 \times 10^7 \text{ J}$$

   3. To heat the gas from -33.3 °C to 0.0 °C:

$$q_{gas} = 1.2 \times 10^4 \text{ g} \cdot 2.2 \, \frac{J}{g \cdot K} \cdot (273.2 \text{ K} - 239.9 \text{ K}) = 8.8 \times 10^5 \text{ J}$$

The total heat required is:

$$9.4 \times 10^5 \text{ J} + 1.6 \times 10^7 \text{ J} + 8.8 \times 10^5 \text{ J} = 1.8 \times 10^7 \text{ J or } 1.8 \times 10^4 \text{ kJ}$$

GENERAL QUESTIONS

37. The estimated vapor pressure at 77 K (-196 °C) is between 150-200 mm Hg. The very slight positive slope of the solid/liquid equilibrium line indicates the **solid is more dense than the liquid**.

39. Silver crystallizes in the face-centered cubic cell, with a side of 409 pm. The radius of a silver atom can be found by examining the geometry of a face of the cell. The diagram shows such a face with an edge of 409 pm. A look at the diagram will reveal that the diagonal is the hypotenuse of a right triangle—two sides of which are 409 pm, and the hypotenuse^2 is $(409pm)^2 + (409pm)^2$. Let d be the hypotenuse. Then we have: $d^2 = 2 \cdot (409pm)^2$. The interest in d lies in the fact that d equals 1 diameter (of the center atom) and two-halves of two other atoms (or 4 radii). $d = \sqrt{2} \cdot 409pm$ or

Phase Diagram of Oxygen

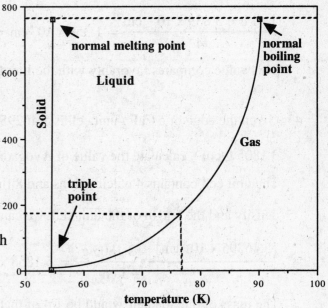

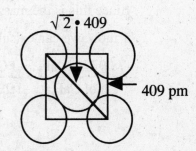

(1.414•409pm) = 578 pm (3sf) So 4 radii = 578 pm and 1 radius =145 pm (3sf). This radius compares favorably with the

reported radius of Ag (See the CD-ROM Periodic Table).

41. Density for Iridium = 22.56 g/cm^3. The radius of an iridium atom is:

Ir crystallizes in the fcc lattice. There are 4 atoms in the unit cell

(8 •1/8 corner atoms + 6•1/2 face atoms).

The mass/unit cell is: $\dfrac{4 \text{ Ir atoms}}{1 \text{ unit cell}} \bullet \dfrac{192.22 \text{ g Ir}}{6.0221 \times 10^{23} \text{ atoms}} = 1.277 \times 10^{-21}$ g/unit cell

The volume of the unit cell is:

$$\dfrac{1.277 \times 10^{-21}\text{g}}{1 \text{ unit cell}} \bullet \dfrac{1 \text{ cm}^3}{22.56 \text{ g}} = \dfrac{5.659 \times 10^{-23}\text{ cm}^3}{\text{unit cell}}$$

Since the cell is a cube, we determine an edge by taking the cube root of this volume:

Edge = 3.839 x 10^{-8} cm

The diagonal of the unit cell (which corresponds to 4•radii) = 1.414 • 3.839 x 10^{-8} cm

So 4•radii = 1.414 • 3.839 x 10^{-8} cm and the radius =

$$\dfrac{1.414 \bullet 3.839 \times 10^{-8} \text{ cm}}{4} = 1.356 \times 10^{-8}\text{cm} = \text{or } 135.6 \times 10^{-12} \text{ m or } 135.6 \text{ pm.}$$

This value compares favorably with the literature value of 136 pm.

43. Given the edge of a CaF$_2$ unit cell is 5.46295 x 10^{-8} cm, and the density of the solid is

3.1805 g/cm^3, calculate the value of Avogadro's number.

The unit cell contains 4 calcium ions and 8 fluoride ions (or 4 CaF$_2$ ion pairs). With the

density and the length of the unit cell, we can calculate the mass of these 4 ion pairs.

$$\dfrac{\left(5.46295 \times 10^{-8} \text{ cm}\right)^3}{1} \bullet \dfrac{3.1805 \text{ g}}{1 \text{ cm}^3} = 5.1853 \times 10^{-22} \text{ g.}$$

The mass of one ion pair would be 1/4 of that mass or 1.2963 x 10^{-22} g. If we add the

atomic masses of Ca and 2 F we get a molar mass of 78.077 g.

Since this is the mass corresponding to Avogadro's number of formula units (also known as

1 mol), we can calculate:

$$\dfrac{78.077 \text{ g CaF}_2}{1 \text{ mol CaF}_2} \bullet \dfrac{1 \text{ CaF}_2 \text{ formula unit}}{1.2963 \times 10^{-22} \text{ g CaF}_2} = 6.0230 \times 10^{23} \text{ formula units/ 1mol CaF}_2$$

45. For the two unit cells, determine the amount of filled space in each.

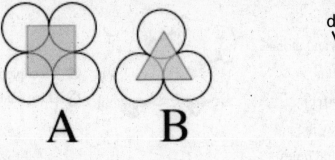

Assign a diameter of each of the atoms = d.

For unit cell A:

The length of the square inscribed is equal to 2 radii or 1 diameter (d). The area associated with the square is then $d \cdot d$ or d^2. Each circle has **one-fourth** of its area covered by the inscribed square. The area of a circle is Πr^2. So the r (the radius) = d/2. So the area of the circle is $\Pi(d/2)^2$ or $(\Pi d^2)/4$. Only one-fourth of each circle is covered by the portion of the inscribed square, so the area covered by the inscribed square is $1/4 \cdot (\Pi d^2)/4$. Noting that there are 4 circles, each of which has the area we've calculated, the **total** area covered by the inscribed square is $\Pi d^2/4$. The area **not covered** is the *difference* between the area of the square (d^2) and the area of the circles inside the square ($\Pi d^2/4$): $d^2 - \Pi d^2/4$. Arbitrarily assign a length to d (say 2 inches). Then the area not covered is $(2)^2 - 3.14(2)^2/4$ or $4 - 3.14 = 0.86$ sq. in. The amount of coverage is: 3.14/4.00 or 78.5 %.

For unit cell B:

Once again, assign a diameter of **d** to each circle. With the equilateral triangle shown, each interior angle (at the corners of the triangle) is 60 degrees. Since there are 360 degrees in the circle, this corresponds to 1/6 of each of the circles covered by the triangle. With three circles involved, $3 \cdot 1/6 = 1/2$ of the area of the 3 circles. Since the area of a circle (in terms of the diameter) is $(\Pi d^2)/4$, the total area of the 3 circles covered by the inscribed triangle is $1/2 \cdot (\Pi d^2)/4$ or $(\Pi d^2)/8$. The relationship for the area of the triangle is $= 1/2 \, b \cdot h$, where b = base and h = height. The base of the triangle is equal to 2r, as shown by the diagram, above right. The height may be calculated using the Pythagorean Theorem, since we know the length of the hypotenuse (d) and the base (of the triangle, r). So $d^2 = r^2 + h^2$ and rearranging, $d^2 - r^2 = h^2$. Noting that the radius is $1/2 \cdot d$, $d^2 - (d/2)^2 = h^2$ or $3/4(d)^2 = h^2$. Substituting the arbitrary value of 2 for d, gives $3/4(2)^2 = h^2$ or $3 = h^2$ and $1.732 = h$.

The area of the triangle is: $1/2 \cdot d \cdot h$, and A = $1/2 \cdot 2 \cdot 1.732 = 1.732$ sq. inches

The area of the circles (covered by the triangle) is $(\Pi d^2)/8$, and substituting the arbitrary value 2 for d gives: $(\Pi \cdot 4)/8$ or $3.14/2 = 1.57$ sq in.

The percent occupied is then $1.57/1.732 \cdot 100 = 90.7$ %.

47. The structure of silicon is shown here:

(a) Note that the unit cell has Si atoms at each corner, and atoms in each face. The cell is **face-centered cubic (fcc).** Highlighted in the drawing are the occupied tetrahedral holes

⬡ occupied tetrahedral sites

⊘ face centered atoms

(b) The density can be calculated by calculating the mass and volume of a unit cell. The volume is the (length of the unit cell)^3.

$$\frac{543.1 pm}{1} \cdot \frac{1 \times 10^{-12}\ m}{1\ pm} \cdot \frac{100\ cm}{1\ m} = 5.43 \times 10^{-8}\ cm\ ;$$

$$V = (5.43 \times 10^{-8}\ cm)^3 = 1.602 \times 10^{-22}\ cm^3$$

The mass of the unit cell: $\frac{28.08\ g}{1\ mol\ Si} \cdot \frac{1\ mol}{6.022 \times 10^{23}\ Si\ atoms} \cdot \frac{8\ atoms}{1\ unit\ cell} = 3.731 \times 10^{-22} g$

The density is: $3.731 \times 10^{-22}\ g / 1.602 \times 10^{-22}\ cm^3 = 2.329\ g/cm^3$

Note that the diagonal is equal to $\sqrt{2} \cdot 543.1 pm$ (768 pm).

Recognizing that the sine of the angle of 54.75 (half of 109.5) is equal to the ratio of 192/ Si-Si distance

$$\sin(54.75) = \frac{192\ pm}{Distance}\ \text{and solving}$$

Distance = 192 pm/0.817 = or 235 pm. Noting that this distance corresponds to two Si radii, the radius of one Si atom is 235/2 or 118 pm.

Si atom in tetrahedral hole

Si in center of face

1/2(109.5)

Si at cell corner

Distance = 1/2(384 pm) =192 pm

Distance = 1/2 cell diagonal = 384 pm

49. The mineral spinel is typical of the large class of compounds with the general formula AB_2X_4 and the unit cell contains 32 oxygen atoms in ccp array, $A_8B_{16}X_{32}$. In the normal spinel structure, 8 metal atoms occupy tetrahedral sites and 16 occupy octahedral sites.

(a) For $MgAl_2O_4$, Mg occupies 1/8 of the tetrahedral sites and Al 1/2 of the octahedral sites.

(b) For chromite, Fe(II) occupies 1/8 of the tetrahedral sites and Cr(III) occupies 1/2 of the octahedral sites.

[See Greenwood & Earnshaw, *Chemistry of the Element,* Pergamon, 1984, p. 279.]

51. The band gap in gallium arsenide is 140 kJ/mol. What is the maximum wavelength of light needed to excite an electron to move from the valence band to the conduction band?

Given energy, we can calculate λ: $E = h\dfrac{c}{\lambda}$

First we need to convert 140 kJ/mol into J/photon.

$$\dfrac{140 \times 10^3 J}{1\ mol} \cdot \dfrac{1\ mol}{6.02 \times 10^{23}\ photons} = (6.626 \times 10^{-34} J \bullet s)\dfrac{2.9979 \times 10^8\ m/s}{\lambda}$$

$$\dfrac{2.33 \times 10^{-19} J/photon}{(6.626 \times 10^{-34} J \bullet s)(2.9979 \times 10^8\ m/s)} = \dfrac{1}{\lambda} = 1.17 \times 10^6\ meters^{-1}\ so\ \lambda = 8.54 \times 10^{-7}\ m\ or$$

$(8.54 \times 10^{-7} m)(1 \times 10^9\ nm/m) = 8.54 \times 10^2\ nm$

53. Germanium has a greater metallic tendency, so it will show a higher conductivity at 298K.

55. Figure 13.14 helps with this question. Boron is a Group III element, and has 3 valence electrons, while Carbon has 4. Doping carbon with boron leads to a p-type semiconductor, since B would create a "hole"

IN THE LABORATORY

57. The unit cell for galena is shown here: This face-centered cubic structure has 4 lead ions (larger spheres) in the unit cell [(12 atoms with 1/4 in the cell)=3 ions and 1 ion in the center] and 4 sulfide ions (smaller spheres) [(8 atoms in the corners with 1/8 in the cell)= 1 ion and (6 atoms with 1/2 in the cell)= 3 ions], giving the formula PbS.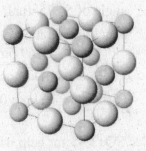

ZnS has a cubic close packed lattice, so it differs from that of galena. Since a crystal is formed by the repeated "stacking" of unit cells into a 3-dimensional lattice, the shape of the crystal is derived from the unit cell.

SUMMARY AND CONCEPTUAL QUESTIONS

59. (a) The balanced equation for the synthesis of BP:

$BBr_3(g) + PBr_3(g) + 3\ H_2(g) \longrightarrow BP(s) + 6\ HBr(g)$

(b) How many tetrahedral holes are filled with P in each cell? BP uses the zinc blende structure in which B atoms are in face-centered cubic pattern, and P atoms occupy half the tetrahedral sites. Note that in the fcc lattice, there are 4 B atoms [(8 x 1/8) + (6 x 1/2)]. With the formula BP, with 4B atoms, there must be 4 P atoms.

(c) With a length of 478 pm, what is the density (in g/cm³)?

First, determine the volume of the unit cell:

$V = l^3 = (478 \text{ pm})^3$ but converting the length to units of cm $\Rightarrow$ 478 x 10⁻¹²m = 4.78 x 10⁻⁸cm

so V = (4.78 x 10⁻⁸ cm)³ = 1.092 x 10⁻²² cm³

There are 4 BP pairs in each unit cell [(8B in each corner x 1/8) + (6B in each face x 1/2)]

Knowing that the molar mass of BP = 41.7847 g/mol, calculate the mass of BP/unit cell:

$$\frac{41.7848 \text{ g BP}}{1 \text{ mol BP}} \cdot \frac{1 \text{ mol BP}}{6.02214 \times 10^{23} \text{ BP pairs}} \cdot \frac{4 \text{ BP pairs}}{1 \text{ unit cell}} = 2.775 \times 10^{-22} \text{ g BP/unit cell}$$

$$\text{Density} = \frac{2.775 \times 10^{-22} \text{ g BP/unit cell}}{1.092 \times 10^{-22} \text{ cm}^3} = 2.541 \text{ g/cm}^3$$

(d) The closest distance between a B and a P atom in the unit cell:

With a unit cell length of 478 pm, the cell diagonal will be $\sqrt{2} \cdot 478\,pm = 676$ pm

The B-P distance corresponds to the hypotenuse of a right triangle.

Recognizing that the sine of the angle of 54.75 (half of 109.5) is equal to the ratio of 338pm/ BP distance

$\sin(54.75) = \dfrac{169 \text{ pm}}{\text{Distance}}$ and solving

Distance = 169 pm/0.817 = or 207 pm.

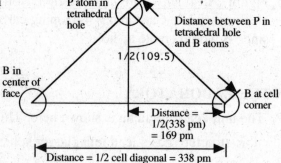

61. See SQ13.45 for help with this problem. If one assumes that the packing pattern used in the two pools is identical, note that the percent of space occupied—e.g. with unit cell A— would not be a function of the size of the spheres, so the two pools will have identical levels of water when the ice melts.

Applying Chemical Principles

1. How many tin atoms are contained in the tetragonal crystal lattice unit cell of β-tin?

   How many tin atoms are contained in the cubic crystal lattice unit cell of α-tin?

   The tetragonal crystal lattice unit cell for β-tin (Figure 2) shows:

8 corner atoms (shared by 8 other cells) x 1/8	= 1	
4 face atoms (each shared by 2 other cells) x ½	= 2	
1 center atom (not shared) x 1	= 1	
Total tin atoms	= 4	

   For the cubic crystal lattice of α-tin:

8 corner atoms (shared by 8 other cells) x 1/8	= 1	
6 face atoms (each shared by 2 other cells) x ½	= 3	
4 atoms (not shared) x 1	= 4	
Total tin atoms	= 8	

3. Dimensions of the unit cell of gray tin:

   Density for Gray tin = 5.769 g/cm^3. As noted in 1 above, there are 8 tin atoms in the unit cell of the cubic crystal lattice of grey tin.

   The mass/unit cell is: $\dfrac{8 \text{ Sn atoms}}{1 \text{ unit cell}} \bullet \dfrac{118.710 \text{ g Sn}}{6.0221 \times 10^{23} \text{atoms}} = 1.57699 \times 10^{-21} \text{g/unit cell}$

   The volume of the unit cell is:

   $\dfrac{1.57699 \times 10^{-21} \text{g}}{1 \text{ unit cell}} \cdot \dfrac{1 \text{ cm}^3}{5.769 \text{ g}} = 2.734 \times 10^{-22} \text{cm}^3$ and converting this volume to "atomic

   dimensions: $2.734 \times 10^{-22} \text{cm}^3 \cdot \dfrac{1 \times 10^{30} \text{pm}^3}{1 \text{cm}^3} = 2.734 \times 10^{8} \text{ pm}^3$

   Since the cell is a cube, we determine an edge by taking the cube root of this volume:

   Edge = 649 pm

Chapter 14
Solutions and Their Behavior

PRACTICING SKILLS

Concentration

1. For 2.56 g of succinic acid in 500. mL of water:

The molality of the solution:

Molality = #mole solute/kg solvent:

With a density of water of 1.00 g/cm^3, 500. mL = 0.500 kg

$$Molality = \frac{0.0217 \text{ mol}}{0.500 \text{ kg}} = 0.0434 \text{ molal}$$

The mole fraction of succinic acid in the solution:

For mole fraction we need *both* the # moles of solute *and* # moles of solvent.

$$\text{Moles of water} = 500. \text{ g } H_2O \cdot \frac{1 \text{ mol } H_2O}{18.02 \text{ g } H_2O} = 27.7 \text{ mol } H_2O$$

$$\text{The mf of acid} = \frac{0.0217 \text{ mol}}{(0.0217 \text{ mol} + 27.7 \text{ mol})} = 7.81 \times 10^{-4}$$

The weight percentage of succinic acid in the solution:

The fraction of *total* mass of solute + solvent which is solute:

$$\text{Weight percentage} = \frac{2.56 \text{ g succinic acid}}{502.56 \text{ g acid + water}} \cdot 100 = 0.509\% \text{ succinic acid}$$

3. Complete the following transformations for

NaI:

Weight percent:

$$\frac{0.15 \text{ mol NaI}}{1 \text{ kg solvent}} \cdot \frac{150 \text{ g NaI}}{1 \text{ mol NaI}} = \frac{22.5 \text{ g NaI}}{1 \text{ kg solvent}}$$

$$\frac{22.5 \text{ g NaI}}{1000 \text{ g solvent} + 22.5 \text{ g NaI}} \cdot 100 = 2.2 \% \text{ NaI}$$

Mole fraction:

1000 g H_2O = 55.51 mol H_2O

$$X_{NaI} = \frac{0.15 \text{ mol NaI}}{55.51 \text{ mol } H_2O + 0.15 \text{ mol NaI}} = 2.7 \times 10^{-3}$$

C₂H₅OH:

Molality:

$$\frac{5.0 \text{ g C}_2\text{H}_5\text{OH}}{100 \text{ g solution}} \cdot \frac{1 \text{ mol C}_2\text{H}_5\text{OH}}{46.07 \text{ g C}_2\text{H}_5\text{OH}} \cdot \frac{100 \text{ g solution}}{95 \text{ g solvent}} \cdot \frac{1000 \text{ g solvent}}{1 \text{ kg solvent}} = 1.1 \text{ molal}$$

Mole fraction:

$$\frac{5.0 \text{ g C}_2\text{H}_5\text{OH}}{1} \cdot \frac{1 \text{ mol C}_2\text{H}_5\text{OH}}{46.07 \text{ g C}_2\text{H}_5\text{OH}} = 0.11 \text{ mol C}_2\text{H}_5\text{OH}$$

and for water: $\dfrac{95 \text{ g H}_2\text{O}}{1} \cdot \dfrac{1 \text{ mol H}_2\text{O}}{18.02 \text{ g H}_2\text{O}} = 5.27 \text{ mol H}_2\text{O}$

$$X_{C_2H_5OH} = \frac{0.11 \text{ mol C}_2\text{H}_5\text{OH}}{5.27 \text{ mol H}_2\text{O} + 0.11 \text{ mol C}_2\text{H}_5\text{OH}} = 0.020$$

C₁₂H₂₂O₁₁:

Weight percent:

$$\frac{0.15 \text{ mol C}_{12}\text{H}_{22}\text{O}_{11}}{1 \text{ kg solvent}} \cdot \frac{342.3 \text{ g C}_{12}\text{H}_{22}\text{O}_{11}}{1 \text{ mol C}_{12}\text{H}_{22}\text{O}_{11}} = \frac{51.3 \text{ g C}_{12}\text{H}_{22}\text{O}_{11}}{1 \text{ kg solvent}}$$

$$\frac{51.3 \text{ g C}_{12}\text{H}_{22}\text{O}_{11}}{1000 \text{ g H}_2\text{O} + 51.3 \text{ g C}_{12}\text{H}_{22}\text{O}_{11}} \times 100 = 4.9 \text{ \% C}_{12}\text{H}_{22}\text{O}_{11}$$

Mole fraction:

$$X_{C_{12}H_{22}O_{11}} = \frac{0.15 \text{ mol C}_{12}\text{H}_{22}\text{O}_{11}}{55.51 \text{ mol H}_2\text{O} + 0.15 \text{ mol C}_{12}\text{H}_{22}\text{O}_{11}} = 2.7 \times 10^{-3}$$

5. To prepare a solution that is 0.200 m Na₂CO₃:

$$\frac{0.200 \text{ mol Na}_2\text{CO}_3}{1 \text{ kg H}_2\text{O}} \cdot \frac{0.125 \text{ kg H}_2\text{O}}{1} \cdot \frac{106.0 \text{ g Na}_2\text{CO}_3}{1 \text{ mol Na}_2\text{CO}_3} = 2.65 \text{ g Na}_2\text{CO}_3$$

$$\text{mol Na}_2\text{CO}_3 = \frac{0.200 \text{ mol Na}_2\text{CO}_3}{1 \text{ kg H}_2\text{O}} \cdot \frac{0.125 \text{ kg H}_2\text{O}}{1} = 0.025 \text{ mol}$$

The mole fraction of Na₂CO₃ in the resulting solution:

$$\frac{125. \text{ g H}_2\text{O}}{1} \cdot \frac{1 \text{ mol H}_2\text{O}}{18.02 \text{ g H}_2\text{O}} = 6.94 \text{ mol H}_2\text{O}$$

$$X_{Na_2CO_3} = \frac{0.025 \text{ mol Na}_2\text{CO}_3}{0.025 \text{ mol Na}_2\text{CO}_3 + 6.94 \text{ mol H}_2\text{O}} = 3.59 \times 10^{-3}$$

7. To calculate the number of mol of C3H5(OH)3:

$$0.093 = \frac{x \text{ mol } C_3H_5(OH)_3}{x \text{ mol } C_3H_5(OH)_3 + (425 \text{ g } H_2O \cdot \frac{1 \text{ mol } H_2O}{18.02 \text{ g } H_2O})}$$

$$0.093 = \frac{x \text{ mol } C_3H_5(OH)_3}{x \text{ mol } C_3H_5(OH)_3 + 23.58 \text{ mol } H_2O}$$

0.093(x + 23.58) = x and solving for x we get 2.4 mol C3H5(OH)3

Grams of glycerol needed: $2.4 \text{ mol } C_3H_5(OH)_3 \cdot \frac{92.1 \text{ g}}{1 \text{ mol}} = 220 \text{ g } C_3H_5(OH)_3$

The molality of the solution is (2.4 mol C3H5(OH)3/ 0.425 kg H2O)= 5.7 m

9. Concentrated HCl is 12.0 M and has a density of 1.18 g/cm^3.

(a) The molality of the solution:

Molality is defined as moles HCl/kg solvent, so begin by deciding the mass of 1 L, and the mass of water in that 1 L. Since the density = 1.18g/mL, then 1 L (1000 mL) will have a mass of 1180g.

The mass of HCl present in 12.0 mol HCl =

$$12.0 \text{ mol HCl} \cdot \frac{36.46 \text{ g HCl}}{1 \text{ mol HCl}} = 437.52 \text{ g HCl}$$

Since 1 L has a mass of 1180 g and 437.52 g is HCl, the difference (1180-437.52) is solvent. So 1 L has 742.98 g water.

$$\frac{12.0 \text{ mol HCl}}{1 \text{ L}} \cdot \frac{1 \text{ L}}{742.98 \text{ g } H_2O} \cdot \frac{1000 \text{ g } H_2O}{1 \text{ kg } H_2O} = 16.2 \text{ m}$$

(b) Weight percentage of HCl:

12.0 mol HCl has a mass of 437.52 g, and the 1 L of solution has a mass of 1180 g.

$$\%HCl = \frac{437.52 \text{ g HCl}}{1180 \text{ g solution}} \cdot 100 = 37.1 \%$$

11. The concentration of ppm expressed in grams is:

$$0.18 \text{ ppm} = \frac{0.18 \text{ g solute}}{1.0 \times 10^6 \text{ g solvent}} = \frac{0.18 \text{ g solute}}{1.0 \times 10^3 \text{ kg solvent}} \text{ or } \frac{0.00018 \text{ g solute}}{1 \text{ kg water}}$$

$$\frac{0.00018 \text{ g } Li^+}{1 \text{ kg water}} \cdot \frac{1 \text{ mol } Li^+}{6.939 \text{ g } Li^+} = 2.6 \times 10^{-5} \text{ molal } Li^+$$

The Solution Process

13. Pairs of liquids that will be miscible:

    (a) $H_2O/CH_3CH_2CH_2CH_3$

    Will **not** be miscible. Water is a polar substance, while butane is nonpolar.

    (b) C_6H_6/CCl_4

    Will **be** miscible. Both liquids are nonpolar and are expected to be miscible.

    (c) H_2O/CH_3CO_2H

    Will **be** miscible. Both substances can hydrogen bond, and we know that they mix—since a 5% aqueous solution of acetic acid is sold as "vinegar".

15. The enthalpy of solution for LiCl:

    The process can be represented as LiCl (s) $\rightarrow$ LiCl (aq)

    The $\Delta_r H° = \Sigma \Delta_f H°$ (product) - $\Sigma \Delta_f H°$ (reactant)

    $$= (-445.6 \text{ kJ/mol})(1 \text{mol}) - (-408.7 \text{ kJ/mol})(1\text{mol}) = -36.9 \text{ kJ}$$

    The similar calculation for NaCl is + 3.9 kJ. Note that the enthalpy of solution for NaCl is endothermic while that for LiCl is exothermic.

    Note the data (-408.7 kJ/mol) is from Table 14.1.

17. Raising the temperature of the solution will increase the solubility of NaCl in water. To increase the amount of dissolved NaCl in solution one must **(c) raise the temperature of the solution and add some NaCl.**

Henry's Law

19. Solubility of $O_2 = k_H \cdot P_{O_2}$

    $$= (1.66 \text{x } 10^{-6} \frac{M}{\text{mm Hg}}) \cdot 40 \text{ mm Hg} = 6.6 \text{ x } 10^{-5} \text{ M } O_2$$

    and $6.6 \text{ x } 10^{-5} \frac{\text{mol}}{L} \cdot \frac{32.0 \text{ g } O_2}{1 \text{ mol } O_2} = 2 \text{ x } 10^{-3} \frac{\text{g } O_2}{L}$

21. Solubility = $k_H \cdot P_{CO_2}$; $0.0506\ M = (4.48 \times 10^{-5}\dfrac{M}{mm\ Hg}) \cdot P_{CO_2}$

1130 mm Hg = P_{CO_2} or expressed in units of atmospheres:

1130 mm Hg $\cdot$ $\dfrac{1\ atm}{760\ mm\ Hg}$ = 1.49 atm and given the relationship of atm to bar: 1.49 bar

Raoult's Law

23. Since $P_{water} = X_{water}\ P^{\circ}_{water}$, to determine the vapor pressure of the solution (P_{water}), we need the mf of water.

35.0 g glycol $\cdot$ $\dfrac{1\ mol\ glycol}{62.07\ g\ glycol}$ = 0.564 mol glycol and

500.0 g H_2O $\cdot$ $\dfrac{1\ mol\ H_2O}{18.02\ g\ H_2O}$ = 27.75 mol H_2O . The mf of water is then:

$\dfrac{27.75\ mol\ H_2O}{(27.75\ mol\ +\ 0.564\ mol)}$ = 0.9801 and

P_{water} = $X_{water}\ P^{\circ}_{water}$ = 0.9801 $\cdot$ 35.7 mm Hg = 35.0 mm Hg

25. Using Raoult's Law, we know that the vapor pressure of pure water (P°) multiplied by the mole fraction (X) of the solute gives the vapor pressure of the solvent above the solution (P).

$$P_{water} = X_{water}\ P^{\circ}_{water}$$

The vapor pressure of pure water at 90 °C is 525.8 mmHg (from Appendix G).

Since the P_{water} is given as 457 mmHg, the mole fraction of the water is:

$$\dfrac{457\ mmHg}{525.8\ mmHg}\ =\ 0.869$$

The 2.00 kg of water correspond to a mf of 0.869. This mass of water corresponds to:

2.00 x 10^3 g H2O $\cdot$ $\dfrac{1 mol H_2O}{18.02 g H_2O}$ = 111 mol water.

Representing moles of ethylene glycol as x we can write:

$\dfrac{mol\ H_2O}{mol\ H_2O + mol\ C_2H_4(OH)_2}$ = $\dfrac{111}{111+x}$ = 0.869

$\dfrac{111}{0.869}$ = 111 + x; 16.7 = x (mol of ethylene glycol)

16.7 mol C2H4(OH)2 $\cdot$ $\dfrac{111}{111+x}$ = 1.04 x 10^3 g C2H4(OH)2

Boiling Point Elevation

27. Benzene normally boils at a temperature of 80.10 °C. If the solution boils at a temperature of 84.2 °C, the change in temperature is (84.2 - 80.10 °C) or 4.1 °C.

Calculate the Δt, using the equation $\Delta t = K_{bp} \bullet m_{solute}$:

The molality of the solution is $\dfrac{0.200 \text{ mol}}{0.125 \text{ kg solvent}}$ or 1.60 m

The K_{bp} for benzene is +2.53 °C/m

So $\Delta t = K_{bp} \bullet m_{solute} = +2.53$ °C/m $\bullet$ 1.60 m = +4.1 °C.

29. Calculate the molality of acenaphthene, $C_{12}H_{10}$, in the solution.

$$0.515 \text{ g } C_{12}H_{10} \bullet \frac{1 \text{ mol } C_{12}H_{10}}{154.2 \text{ g } C_{12}H_{10}} = 3.34 \times 10^{-3} \text{ mol } C_{12}H_{10}$$

and the molality is: $\dfrac{3.34 \times 10^{-3} \text{ mol acenaphthene}}{0.0150 \text{ kg } CHCl_3} = 0.223 \text{ molal}$

the boiling point *elevation* is: $\Delta t = m \bullet K_{bp} = 0.223 \text{ molal} \bullet \dfrac{+3.63 °C}{\text{molal}} = 0.808 °C$

and the boiling point will be 61.70 + 0.808 = 62.51 °C

Freezing Point Depression

31. The solution freezes 16.0 °C lower than pure water.

(a) We can calculate the molality of the ethanol:

$$\Delta t = mK_{fp}$$

$$-16.0 °C = m \,(-1.86 °C/molal)$$

$$8.60 = \text{molality of the alcohol}$$

(b) If the molality is 8.60 then there are 8.60 moles of C_2H_5OH

(8.60 mol x 46.07 g/mol = 396 g) in 1000 g of H_2O.

The weight percent of alcohol is $\dfrac{396 \text{ g}}{1396 \text{ g}}$ x 100 = 28.4 % ethanol

33. Freezing point of a solution containing 15.0 g sucrose in 225 g water:

   (1) Calculate the molality of sucrose in the solution:

   $$15.0 \text{ g } C_{12}H_{22}O_{11} \bullet \frac{1 \text{ mol } C_{12}H_{22}O_{11}}{342.30 \text{ g } C_{12}H_{22}O_{11}} = 0.0438 \text{ mol}$$

   $$\frac{0.0438 \text{ mol } C_{12}H_{22}O_{11}}{0.225 \text{ kg } H_2O} = 0.195 \text{ molal}$$

   (2) Use the Δt equation to calculate the freezing point change:

   $$\Delta t = mK_{fp} = 0.195 \text{ molal} \bullet (-1.86 \text{ }°C/molal) = -0.362 \text{ }°C$$

   The solution is expected to begin freezing at $-0.362 \text{ }°C$.

Osmosis

35. Assume we have 100 g of this solution, the number of moles of phenylalanine is

   $$3.00 \text{ g phenylalanine} \bullet \frac{1 \text{ mol phenylalanine}}{165.2 \text{ g phenylalanine}} = 0.0182 \text{ mol phenylalanine}$$

   The molality of the solution is: $\dfrac{0.0182 \text{ mol phenylalanine}}{0.09700 \text{ kg water}} = 0.187 \text{ molal}$

   (a) The freezing point:

   $$\Delta t = 0.187 \text{ molal} \bullet -1.86 \text{ }°C/molal = -0.348 \text{ }°C$$

   The new freezing point is $0.0 - 0.348 \text{ }°C = -0.348 \text{ }°C$.

   (b) The boiling point of the solution

   $$\Delta t = m K_{bp} = 0.187 \text{ molal} \bullet 0.5121 °C/molal = +0.0959 \text{ }°C$$

   The new boiling point is $100.000 + 0.0959 = +100.0959 \text{ }°C$

   (c) The osmotic pressure of the solution:

   If we assume that the **Molarity** of the solution is equal to the **molality**, then
   the osmotic pressure should be:

   $$\Pi = (0.187 \text{ mol/L})(0.0821 \frac{L \bullet atm}{K \bullet mol})(298 \text{ K}) = 4.58 \text{ atm}$$

   The osmotic pressure will be most easily measured, since the magnitudes of osmotic
   pressures (large values) result in decreased experimental error.

37. The molar mass of bovine insulin with a solution having an osmotic pressure of 3.1 mm Hg:

   $$3.1 \text{ mm Hg} \bullet \frac{1 \text{ atm}}{760 \text{ mm Hg}} = (M)(0.08205 \frac{L \bullet atm}{K \bullet mol})(298 \text{ K})$$

   $$1.67 \times 10^{-4} = \text{Molarity or } 1.7 \times 10^{-4} \text{ (to 2 sf)}$$

   The definition of molarity is #mol/L. Substituting into the definition we obtain:

   $$1.7 \times 10^{-4} \frac{\text{mol bovine insulin}}{L} = \frac{\frac{1.00 \text{ g bovine insulin}}{MM}}{1 \text{ L}} \text{ ; Solving for MM} = 6.0 \times 10^3 \text{ g/mol}$$

Colligative Properties and Molar Mass Determination

39. The change in the temperature of the boiling point is (80.26 - 80.10)°C or 0.16 °C.

Using the equation $\Delta t = m \cdot K_{bp}$; 0.16 °C = m • +2.53 °C/m, and the molality

is: $\dfrac{0.16\ °C}{+2.53\ °C/m} = 0.063$ molal

The solution contains 11.12 g of solvent (or 0.01112 kg solvent). We can calculate the # of

moles of the orange compound, since we know the molality:

0.063 molal = $\dfrac{x\ mol\ compound}{0.01112\ kg\ solvent}$ or 7.0 x 10^{-4} mol compound.

This number of moles of compound has a mass of 0.255 g, so 1 mol of compound is:

$\dfrac{0.255\ g\ compound}{7.0\ x\ 10^{-4}\ mol} = 360$ g/mol.

The empirical formula, $C_{10}H_8Fe$, has a mass of 184 g, so the # of "empirical formula units"

in one molecular formula is: $\dfrac{360\ g/mol}{184\ g\ empirical\ formula} = 2$ mol/empirical formula or a

molecular formula of $C_{20}H_{16}Fe_2$.

41. The change in the temperature of the boiling point is (61.82 - 61.70)°C or 0.12 °C.

Using the equation $\Delta t = m \cdot K_{bp}$; 0.12 °C = m • +3.63 °C/m, and the molality is:

$\dfrac{0.12\ °C}{+3.63\ °C/m} = 0.033$ molal

The solution contains 25.0 g of solvent (or 0.0250 kg solvent). We can calculate the # of

moles of benzyl acetate:

0.033 molal = $\dfrac{x\ mol\ compound}{0.0250\ kg\ solvent}$ or 8.3 x 10^{-4} mol compound.

This number of moles of benzyl acetate has a mass of 0.125 g, so 1 mol of benzyl acetate is:

$\dfrac{0.125\ g\ compound}{8.3 x 10^{-4}\ mol} = 150$ g/mol. (2 sf)

43. To determine the molar mass, first determine the molality of the solution.

-0.040 °C = m • -1.86 °C/molal = 0.0215 molal (or 0.022 to 2 sf)

and   0.022 molal = $\dfrac{\dfrac{0.180\ g\ solute}{MM}}{0.0500\ kg\ water}$ and solving for MM, MM = 167 or 170 (to 2 sf)

Colligative Properties of Ionic Compounds

45. The number of moles of LiF is: 52.5 g LiF $\bullet$ $\dfrac{1 \text{ mol LiF}}{25.94 \text{ g LiF}}$ = 2.02 mol LiF

So Δt_{fp} = $\dfrac{2.02 \text{ mol LiF}}{0.306 \text{ kg H}_2\text{O}}$ $\bullet$ -1.86 °C/molal $\bullet$ 2 = -24.6 °C

The anticipated freezing point is then 24.6 °C lower than pure water (0.0°C) or -24.6 °C

47. Solutions given in order of increasing freezing point (lowest freezing point listed first):

The solution with the **greatest number** of particles will have the lowest freezing point.

The total molality of solutions is:

	Solution	Particles / formula unit	Identity of particles	Total molality
(a)	0.1 m sugar	1	covalently bonded molecules	0.1 m $\bullet$ 1 = 0.1 m
(b)	0.1 m NaCl	2	Na^+, Cl^-	0.1 m $\bullet$ 2 = 0.2 m
(c)	0.08 m CaCl$_2$	3	Ca^{2+}, 2 Cl^-	0.08 m $\bullet$ 3 = 0.24 m
(d)	0.04 m Na$_2$SO$_4$	3	2 Na^+, SO_4^{2-}	0.04 m $\bullet$ 3 = 0.12 m

The freezing points would increase in the order: CaCl$_2$ < NaCl < Na$_2$SO$_4$ < sugar

Colloids

49. (a) $BaCl_2(aq)$ + $Na_2SO_4(aq)$ → $BaSO_4(s)$ + 2 $NaCl(aq)$

(b) The BaSO$_4$ initially formed is of a colloidal size — not large enough to precipitate fully.

(c) The particles of BaSO$_4$ grow with time, owing to a gradual loss of charge and become

large enough to have gravity affect them—and settle to the bottom.

GENERAL QUESTIONS

51. A solution of 0.52 g of phenylcarbinol in 25.0 g of water melts at – 0.36 °C.

Δt = m $\bullet$ -1.86 °C/molal

- 0.36 °C = m $\bullet$ -1.86 °C/molal so $\dfrac{-0.36°C}{-1.86 °C/molal}$ = m = 0.19 molal (0.194 to 3 sf)

0.19 m = $\dfrac{\text{mol phenylcarbinol}}{0.025 \text{ kg}}$ and (0.19 m $\bullet$ 0.025 kg) = 4.8 x 10^{-3} mol phenylcarbinol

The molar mass of phenylcarbinol is $\dfrac{0.52 \text{ g phenylcarbinol}}{4.8 \times 10^{-3} \text{ mol}}$ = 110 g/mol (2 sf)

53. Arranged the solutions in order of (i) increasing vapor pressure of water and (ii) increasing boiling points:

(i) The solution with the highest water vapor pressure would have the **lowest particle concentration**, since according to Raoult's Law, the vapor pressure of the water in the solution is directly proportional to the mole fraction of the water. The lower the number of particles, the greater the mf of water, and the greater the vapor pressure. Hence the order of *increasing* vapor pressure is:

$$Na_2SO_4 < sugar < KBr < glycol$$

(See part (ii) for particle concentrations—(m•i))

(ii) Recall that $\Delta t = m \cdot K_{fp} \cdot i$. The difference in these four solutions will be in the product (m • i). The products for these solutions are:

$$glycol \quad = 0.35 \cdot 1 = 0.35$$
$$sugar \quad = 0.50 \cdot 1 = 0.50$$
$$KBr \quad = 0.20 \cdot 2 = 0.40$$
$$Na_2SO_4 = 0.20 \cdot 3 = 0.60$$

Arranged in *increasing* boiling points: $glycol < KBr < sugar < Na_2SO_4$

55. For DMG, $(CH_3CNOH)_2$, the MM is 116.1 g/mol

So 53.0 g is: $53.0 \text{ g} \cdot \dfrac{1 \text{ mol DMG}}{116.1 \text{ g DMG}} = 0.456 \text{ mol DMG}$

525 g of C_2H_5OH is: $525 \text{ g} \cdot \dfrac{1 \text{ mol } C_2H_5OH}{46.07 \text{ g } C_2H_5OH} = 11.4 \text{ mol } C_2H_5OH$

(a) the mole fraction of DMG: $\dfrac{0.456 \text{ mol}}{(11.4 + 0.456) \text{ mol}} = 0.0385 \text{ mf DMG}$

(b) The molality of the solution: $\dfrac{0.456 \text{ mol DMG}}{0.525 \text{ kg}} = 0.869 \text{ molal DMG}$

(c) $P_{alcohol} = P^{\circ}_{alcohol} \cdot X_{alcohol}$

$= (760. \text{ mm Hg})(1 - 0.0385) = 730.7 \text{ mm Hg}$

(d) The boiling point of the solution:

$\Delta t = m \cdot K_{bp} \cdot i = (0.870)(+1.22 \text{ }^{\circ}C/molal)(1) = 1.06 \text{ }^{\circ}C$

The new boiling point is $78.4 \text{ }^{\circ}C + 1.06 \text{ }^{\circ}C = 79.46 \text{ }^{\circ}C$ or $79.5 \text{ }^{\circ}C$

57. Concentrated NH_3 is 14.8 M and has a density of 0.90 g/cm^3.

(1) The molality of the solution:

Molality is defined as moles NH_3/kg solvent, so begin by deciding the mass of 1 L, and the mass of water in that 1L. Since the density = 0.90 g/mL, then 1 L (1000 mL) will have a mass of 900 g.

The mass of NH_3 present in 14.8 mol NH_3=

$$14.8 \text{ mol } NH_3 \bullet \frac{17.03 \text{ g } NH_3}{1 \text{ mol } NH_3} = 252 \text{ g } NH_3$$

Since 1 L has a mass of 900 g and 252 g is NH_3, the difference (900-252) is solvent. So 1 L has 648 g water.

$$\frac{14.8 \text{ mol } NH_3}{1 \text{ L}} \bullet \frac{1 \text{ L}}{648 \text{ g } H_2O} \bullet \frac{1000 \text{ g } H_2O}{1 \text{ kg } H_2O} = 22.8 \text{ m} \text{ or } 23 \text{ m (2 sf)}$$

(2) The mole fraction of ammonia is:

Calculate the # of moles of water present:

$$648 \text{ g } H_2O \bullet \frac{1 \text{ mol } H_2O}{18.02 \text{ g } H_2O} = 35.96 \text{ mol } H_2O \text{ (retaining 1 extra sf)}$$

The mf NH_3 is: $\dfrac{14.8 \text{ mol } NH_3}{(14.8 \text{ mol } + 35.96 \text{ mol})} = 0.29$

(3) Weight percentage of NH_3:

14.8 mol NH_3 has a mass of 252.0 g, and 1 L of the solution has a mass of 900 g.

$$\% \ NH_3 = \frac{252.0 \text{ g } NH_3}{900 \text{ g solution}} \bullet 100 = 28 \% \text{ (2 sf)}$$

59. To make a 0.100 m solution, we need a ratio of #moles of ions/kg solvent that is 0.100.

$0.100 \text{ m} = \dfrac{\# \text{ mol ions}}{0.125 \text{ kg solvent}}$ and solving for # mol ions: $\quad$ 0.0125 mol ions

The salt will dissociate into 3 ions per formula unit (2 Na^+ and 1 SO_4^{2-}).

The amount of Na_2SO_4 is:

$$0.0125 \text{ mol ions} \bullet \frac{1 \text{ mol } Na_2SO_4}{3 \text{ mol ions}} \bullet \frac{142.04 \text{ g } Na_2SO_4}{1 \text{ mol } Na_2SO_4} = 0.592 \text{ g } Na_2SO_4$$

61. Solution properties:

   (a) The solution with the higher boiling point:

   Recall that $\Delta t = m \cdot Kfp \cdot i$. The difference in these solutions will be in the product $(m \cdot i)$.

   The products for these solutions are:

   sugar $= 0.30 \cdot 1 = 0.30$ (the sugar molecule remains as one unit)

   KBr $= 0.20 \cdot 2 = 0.40$ (KBr dissociates into K^+ and Br^- ions)

   KBr will provide the larger Δt.

   (b) The solution with the lower freezing point:

   Using the same logic as in part (a), NH_4NO_3 provides 2 ions/formula unit while Na_2CO_3

   provides 3. The product, $(m \cdot i)$, is larger for Na_2CO_3, so **Na_2CO_3** gives the greater Δt

   and the lower freezing point.

63. The change in temperature of the freezing point is: $\Delta t = m \cdot K_{fp} \cdot i$

   Calculate the molality:

   $$35.0 \text{ g CaCl}_2 \cdot \frac{1 \text{ mol CaCl}_2}{111.0 \text{ g CaCl}_2} = 0.315 \text{ mol CaCl}_2 \text{ in } 0.150 \text{ kg water.}$$

   $$m = \frac{0.315 \text{ mol CaCl}_2}{0.150 \text{ kg}} = 2.10 \text{ molal CaCl}_2$$

   $\Delta t = m \cdot K_{fp} \cdot i = (2.10 \text{ molal} \cdot -1.86 \text{ °C/molal} \cdot 2.7) = -10.6 \text{ °C.}(-11 \text{ to 2sf})$

   The freezing point of the solution is 0.0°C - 11 °C = -11 °C

65. The molar mass of hexachlorophene if 0.640 g of the compound in 25.0 g of $CHCl_3$ boils at 61.93 °C:

   Recalling the Δt equation: $\Delta t = m \cdot K_{bp} = m \cdot \dfrac{+3.63 \text{ °C}}{\text{molal}} = (61.93 - 61.70) \text{ °C}$

   Solving for m: $\dfrac{0.23 \text{ °C}}{3.63 \text{ °C/m}} = 0.0634 \text{ m}$

   Substitute into the definition for molality: m = #mol/kg solvent

   $$0.0634 \text{ molal} = \frac{\dfrac{0.640 \text{ g hexachloraphene}}{MM}}{0.025 \text{ kg}} \quad \text{and solving for MM; } 4.0 \times 10^2 \text{ g/mol} = MM$$

67. Solubility of $N_2 = k_H \cdot P_{N_2}$

   $$= (6.0 \times 10^{-4} \frac{\text{mol}}{\text{kg} \cdot \text{bar}}) \cdot 585 \text{ mmHg} \cdot \frac{1.01325 \text{ bar}}{760 \text{ mmHg}} = 4.7 \times 10^{-4} \text{ mol/kg N}_2$$

69. (a) *Average* MM of starch if 10.0 g starch/L has an osmotic pressure = 3.8 mm Hg at 25 °C.

$$3.8 \text{ mm Hg} \cdot \frac{1 \text{ atm}}{760 \text{ mm Hg}} = (M)(0.08205 \frac{L \cdot atm}{K \cdot mol})(298 \text{ K})$$

$$2.045 \times 10^{-4} = \text{Molarity or } 2.0 \times 10^{-4} \text{ (to 2 sf)}$$

The definition of molarity is #mol/L. Substituting into the definition we obtain:

$$2.0 \times 10^{-4} \frac{\text{mol bovine insulin}}{L} = \frac{\frac{10.0 \text{ g starch}}{MM}}{1L} ; \text{Solving for MM} = 4.9 \times 10^{4} \text{ g/mol}$$

(b) Freezing point of the solution:

$$\Delta t = m \cdot K_{fp} \cdot i \text{ (assume that i=1 and that Molarity = molality)}$$

$$\Delta t = m \cdot (-1.86 \text{ °C/molal})$$

and the $M = \dfrac{\dfrac{10.0 \text{ g starch}}{4.9 \times 10^{4} \text{ g/mol}}}{1 \text{ L}} = 2.0 \times 10^{-4}$

so the $\Delta t = 2.0 \times 10^{-4} \cdot (-1.86 \text{ °C/molal}) = -3.8 \times 10^{-4}$ °C. In essence the starch will boil at the temperature of pure water. From this data we can assume that *it will NOT be easy* to measure the molecular weight of starch using this technique.

71. The enthalpies of solution for Li_2SO_4 and K_2SO_4:

The process is MX(s) → MX(aq)

Using the data for **Li_2SO_4**:

$\Delta_{solution}H° = \Delta_fH°(aq) - \Delta_fH°(s) = (-1464.4 \text{ kJ/mol}) - (-1436.4 \text{ kJ/mol}) = -28.0 \text{ kJ/mol}$

Using the data for **K_2SO_4**:

$\Delta_{solution}H° = \Delta_fH°(aq) - \Delta_fH°(s) = (-1414.0 \text{ kJ/mol}) - (-1437.7 \text{ kJ/mol}) = 23.7 \text{ kJ/mol}$

Note that for Li_2SO_4 *the process is* **exothermic** *while for* K_2SO_4 *the process is endothermic.*

Similar data for LiCl and KCl:

For LiCl: $\Delta_fH°(aq) - \Delta_fH°(s) = (-445.6 \text{ kJ/mol}) - (-408.7 \text{ kJ/mol}) = -36.9 \text{ kJ/mol}$ and

for KCl: $\Delta_fH°(aq) - \Delta_fH°(s) = (-419.5 \text{ kJ/mol}) - (-436.7 \text{ kJ/mol}) = 17.2 \text{ kJ/mol}$

Note the similarities of the chloride salts, with the lithium salt being **exothermic** *while the potassium salt is* **endothermic**.

73. Graham's law says that the pressure of a mixture of gases (benzene and toluene) is the sum of the partial pressures. Using Raoult's Law:

$P_{benzene}$ = mf benzene • $P°_{benzene}$ and similarly for toluene.

The total pressure is:

$P_{total} = P_{benzene} + P_{toluene}$

$$= (\frac{2 \text{ mol benzene}}{3 \text{ mol}} \cdot 75 \text{ mm Hg}) + (\frac{1 \text{ mol toluene}}{3 \text{ mol}} \cdot 22 \text{ mm Hg}) = 57 \text{mm Hg}$$

What is the mole fraction of each component in the liquid and in the vapor?

The **mf of the components in the liquid** are: benzene: 2/3 and toluene: 1/3

The **mf of the components in the vapor** are proportional to their pressures in the vapor state.

The mf of benzene is: $\frac{50 \text{ mm Hg}}{57 \text{ mm Hg}}$ = 0.87; the mf of toluene would be (1 - 0.87) or 0.13.

75. A 2.0 % aqueous solution of novocainium chloride(NC) is also 98.0 % in water. Assume that we begin with 100 g of solution. The molality of the solution is:

$$\frac{2.0 \text{ g} \cdot \frac{1 \text{mol NC}}{272.8 \text{ g NC}}}{0.0980 \text{ kg water}} = 0.075 \text{ m}$$

Using the "delta T" equation:

$\Delta t = m \cdot K_{fp} \cdot i$, we can solve for i: $\frac{\Delta t}{m \cdot K_{fp}} = i$

$$\frac{-0.237°C}{0.075m \cdot -1.86°C/m} = 1.7$$

So approximately **2 moles of ions are present per mole of compound**.

77. (a) We can calculate the freezing point of sea water if we calculate the molality of the solution. Let's imagine that we have 1,000,000 (or 10^6) g of sea water. The amounts of the ions are then equal to the concentration (in ppm).

Cl^-: $1.95 \times 10^4 \text{ g Cl}^- \cdot \frac{1 \text{ mol Cl}^-}{35.45 \text{ g Cl}^-}$ = 550. mol Cl^-

Na^+: $1.08 \times 10^4 \text{ g Cl}^- \cdot \frac{1 \text{mol Na}^+}{22.99 \text{ g Na}^+}$ = 470. mol Na^+

Mg^{+2}: $1.29 \times 10^3 \text{ g Mg}^{+2} \cdot \frac{1 \text{mol Mg}^{+2}}{24.31 \text{ g Mg}^{+2}}$ = 53.1 mol Mg^{+2}

SO_4^{-2}: $9.05 \times 10^2 \text{ g SO}_4^{-2} \cdot \frac{1 \text{mol SO}_4^{-2}}{96.06 \text{ g SO}_4^{-2}}$ = 9.42 mol SO_4^{-2}

Ca^{+2}: 4.12×10^2 g $Ca^{+2} \cdot \dfrac{1 \text{mol Ca}^{+2}}{40.08 \text{ g Ca}^{+2}}$ = 10.3 mol Ca^{+2}

K^+: 3.80×10^2 g $K^+ \cdot \dfrac{1 \text{mol K}^+}{39.10 \text{ g K}^+}$ = 9.72 mol K^+

Br^-:   67 g $Br^- \cdot \dfrac{1 \text{mol Br}^-}{79.90 \text{ g Br}^-}$ = 0.84 mol Br^-

For a total of: 1103 mol ions

The concentration per gram is: $\dfrac{1103 \text{ mol ions}}{10^6 \text{ g H}_2\text{O}}$

The *change* in the freezing point of the sea water is:

$\Delta t = m \cdot K_{fp} = \dfrac{1103 \text{ mol ions}}{10^6 \text{ g H}_2\text{O}} \cdot \dfrac{1000 \text{ g H}_2\text{O}}{1 \text{ kg H}_2\text{O}} \cdot -1.86 \text{ °C/molal} = -2.05 \text{ °C}$

So we expect this sea water to begin freezing at -2.05 °C.

(b) The osmotic pressure (in atmospheres) can be calculated if *we assume the density of sea water is 1.00 g/mL.*

$\Pi = MRT = \dfrac{1.103 \text{ mol}}{1 \text{ L}} \cdot 0.082057 \dfrac{\text{L} \cdot \text{atm}}{\text{K} \cdot \text{mol}} \cdot 298 \text{ K} = 27.0 \text{ atm}$

The pressure needed to purify sea water by reverse osmosis would then be a pressure greater than 27.0 atm.

79. A 2.00 % aqueous solution of sulfuric acid is also 98.00 % in water.

Assume that we begin with 100 g of solution.

(a) We can calculate the van't Hoff factor by first calculating the molality of the solution:

$$\dfrac{2.00 \text{ g} \cdot \dfrac{1 \text{ mol H}_2\text{SO}_4}{98.06 \text{ g H}_2\text{SO}_4}}{0.09800 \text{ kg water}} = 0.208 \text{ m}$$

Using the "delta T" equation:

$\Delta t = m \cdot K_{fp} \cdot i$, we can solve for i: $\dfrac{\Delta t}{m \cdot K_{fp}} = i$

$\dfrac{-0.796 \text{°C}}{0.208 m \cdot -1.86 \text{°C/m}} = 2.06 = i$

(b) Given the van't Hoff factor of 2 (above), the best representation of a dilute solution of sulfuric acid in water has to be: $H^+ + HSO_4^-$.

81. The Henry's law constant for N_2O is 2.4×10^{-2} mol/kg · bar. What mass of N_2O will dissolve in 500. mL of water, under an N_2O pressure of 1.00 bar?

Solubility of $N_2O = k_H \bullet P_{N_2O}$

$= (2.4 \times 10^{-2} \text{ mol/kg} \cdot \text{bar}) \bullet 1.00 \text{ bar} = 2.4 \times 10^{-2} \text{ mol } N_2O/kg$

and $\dfrac{2.4 \times 10^{-2} \text{ mol } N_2O}{1 \text{ kg}} \bullet \dfrac{44.01 \text{ g } N_2O}{1 \text{ mol } N_2O} \bullet 0.500 \text{ kg} = 0.53 \text{ g } N_2O$

What is the concentration of N_2O in this solution, expressed in ppm (d H_2O = 1.00 g/mL)?
Since the density of water is 1.00 g/mL, 500. mL of water will have a mass of 500. grams.

The concentration of N_2O is: $\dfrac{0.53 \text{ g } N_2O}{500. \text{ g } H_2O} = \dfrac{x}{1,000,000 \text{ g } H_2O}$ or 1060 ppm

or 1.1×10^3 to 2 sf.
Note that the mass of N_2O is negligible and therefore ignored in the mass of the solution!

83. Tests to determine if the contents of a flask is a solution or a colloid:

One very simple test is to shine a beam of light through the liquid. A colloid will show the "path of the light" as the light is reflected off the particles of the colloid—the Tyndall effect. Another test is to attempt to pass the colloid through a membrane. The colloidal particles will diffuse through the membrane either slowly or not at all, owing to the sizes of the suspended medium.

IN THE LABORATORY

85. Using the freezing point depression and boiling point elevation equations, calculate the term (m • i). Since we have no quantitative information about the quantity of benzoic acid dissolved in the benzene, the term (m • i) will be the best metric by which we can judge the degree of dissociation of benzoic acid at the freezing point and boiling point of benzene.
At the freezing point: $\Delta t = m \bullet i \bullet K_{fp}$; (3.1 °C - 5.50 °C) = m • i • (-5.12 °C/molal) i

and $\dfrac{-2.4 \text{ °C}}{-5.12 \text{ °C/molal}}$ = m • i so 0.47 molal = m • i

At the boiling point: $\Delta t = m \bullet i \bullet K_{bp}$; (82.6 °C – 80.1 °C) = m • i • (+2.53 °C/molal) i

and $\dfrac{+2.5 \text{ °C}}{+2.53 \text{ °C/molal}}$ = m • i so 0.99 molal = m • i

If we assume the amount of benzoic acid dissolved in benzene is constant over the temperature range described, the conclusion one reaches is that *i* has a greater value at higher temperatures than at lower ones. Another way of expressing this is that *at higher temperatures*, the *degree of association* between benzoic acid molecules *decreases*.

87. The apparent molecular weight of acetic acid in benzene, determined by the depression of benzene's freezing point:

$$\Delta t = m \cdot K_{fp} \cdot i; \ (3.37\ °C - 5.50\ °C) = m(-5.12\ °C/molal)\ i$$

and $\dfrac{-2.13°C}{-5.12°C/molal} = m \cdot i$ so 0.416 molal $= m \cdot i$ (assume i = 1)

and the apparent molecular weight is:

$$0.416\ molal = \dfrac{\dfrac{5.00\ g\ acetic\ acid}{MM}}{0.100\ kg}$$ and solving for MM; 120 g/mol = MM

The apparent molecular weight of acetic acid in water:

$$\Delta t = m \cdot K_{fp} \cdot i; \ (-1.49\ °C - 0.00\ °C) = m(-1.86\ °C/molal)\ i$$

and $\dfrac{-1.49°C}{-1.86°C/molal} = m \cdot i$ so 0.801 molal $= m \cdot i$

(once again, i = 1) and the apparent molecular weight is:

$$0.801\ molal = \dfrac{\dfrac{5.00\ g\ acetic\ acid}{MM}}{0.100\ kg}$$ and solving for MM; 62.4 g/mol = MM

The accepted value for acetic acid's molecular weight is approximately 60.1 g/mol. Hence the value for i isn't much larger than 1, indicating that the degree of dissociation of acetic acid molecules in water is not great—a finding consistent with the designation of acetic acid as a weak acid. The apparently doubled molecular weight of acetic acid in benzene indicates that the acid must exist primarily as a dimer.

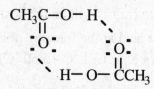

89. The vapor pressure data should permit us to calculate the molar mass of the boron compound.

$$P_{benzene} = X_{benzene} \cdot P°_{benzene}$$

94.16 mm Hg = $X_{benzene} \cdot$ 95.26 mm Hg, and rearranging: $X_{benzene} = \dfrac{94.16\ mm\ Hg}{95.26\ mm\ Hg}$

$$X_{benzene} = 0.9885$$

Now we need to know the # of moles of the boron compound, so let's use the mf of benzene

to find that: 10.0 g benzene $\cdot \dfrac{1\ mol\ benzene}{78.11\ g\ benzene}$ = 0.128 mol benzene

$$0.9885 = \frac{0.128 \text{ mol benzene}}{0.128 \text{ mol benzene} + x \text{ mol } B_xF_y}$$

$0.9885(0.128 + x) = 0.128 \, x$ and solving for $x = 0.001489$ mol B_xF_y

Knowing that this # of moles of compound has a mass of 0.146 g, we can calculate the molar mass:

$$\frac{0.146}{0.001489} = 98.0 \text{ g/mol}$$

We can calculate the empirical formula, since we know that the compound is 22.1% boron and 77.9% fluorine.

In 100 g of the compound there are $22.1 \text{ g B} \cdot \dfrac{1 \text{ mol B}}{10.81 \text{ g B}} = 2.11$ mol B and

$$77.9 \text{ g F} \cdot \frac{1 \text{ mol F}}{19.00 \text{ g F}} = 4.10 \text{ mol F}$$

The empirical formula is BF_2, which would have a formula weight of 48.8.

Dividing the molar mass (found from the vapor pressure experiment) by the mass of the empirical

formula, we get: $\dfrac{98.0}{48.8} = 2.00$.

(a) The molecular formula is then B_2F_4.

(b) A Lewis structure for the molecule:
  We know that the molecule is nonpolar (does not have

  a dipole moment). The F-B-F bond angles are 120°, as are the F-B-B bond angles, hence

  the molecule is planar (flat). The hybridization of the boron atoms is sp^2.

SUMMARY AND CONCEPTUAL QUESTIONS

91. Greater hydration of Be^{2+}, Mg^{2+}, and Ca^{2+}: In general, the smaller the ion, the greater the

    degree of hydration, so Be^{2+} will be most strongly hydrated, while Ca^{2+} will be least strongly

    hydrated.

93. Equimolar amounts of $CaCl_2$ and NaCl lower freezing points differently. The formulas tell

    us that $CaCl_2$ provides 3 particles per formula unit while NaCl provides only two. Hence we

    expect—given van't Hoff factors of 3 and 2 respectively that $CaCl_2$ should have a freezing

    point depression that is about 50% greater than that of NaCl.

95. Solutes likely to dissolve in water; and solutes likely to dissolve in benzene:

Substances likely to dissolve in water are polar (ionic compounds) and those polar substances capable of hydrogen bonding. Substances likely to dissolve in benzene are non-polar substances.

Likely to dissolve in water: (a) $NaNO_3$-ionic; (d) NH_4Cl

Likely to dissolve in benzene: (b) $CH_3CH_2OCH_2CH_3$- only slightly polar, with large fraction of the molecule being non-polar (C-C, and C-H bonds); (c) $C_{10}H_8$--nonpolar

97. Since hydrophilic colloids are those that "love water", we would expect starch to form a hydrophilic colloid since it contains the OH bonds that can hydrogen bond to water. Hydrocarbons on the other hand have non-polar bonds that should have little-to-no attraction to water molecules, and form a hydrophobic colloid.

99. Semipermeable membrane dividing container into two parts; one side containing 5.85 g NaCl in 100 mL solution, and the other side containing 8.88 g KNO_3 in 100 mL solution.

Calculate the osmotic pressure of both solutions:$\Pi = MRT$

Note that we don't actually have to calculate the osmotic pressures, only to note that the solution with the **greater molarity** will have the greater osmotic pressure.

M for NaCl: $\dfrac{5.85 \text{ g NaCl}}{1} \bullet \dfrac{1 \text{ mol NaCl}}{58.5 \text{ g NaCl}} \bullet \dfrac{1}{0.100\text{L}} = 1.00 \text{ M}$

M for KNO_3: $\dfrac{8.88 \text{ g } KNO_3}{1} \bullet \dfrac{1 \text{ mol } KNO_3}{101.1 \text{ g } KNO_3} \bullet \dfrac{1}{0.100\text{L}} = 0.878 \text{ M}$

So the osmotic pressure for NaCl will be greater than that for KNO_3, and the *solvent should flow from the KNO_3 to the NaCl*, reducing the osmotic pressure for NaCl.

101. (a) mole fraction of ethanol and water:

$\dfrac{12 \text{ g ethanol}}{100 \text{ g solutions}} \cdot \dfrac{100 \text{ g solutions}}{88 \text{ g water}}$ so we know there are 12 g ethanol and 88 g water.

$\dfrac{12 \text{ g ethanol}}{1} \cdot \dfrac{1 \text{ mol ethanol}}{46.07 \text{ g ethanol}} = 0.260 \text{ mol ethanol}$

and $\dfrac{88 \text{ g water}}{1} \cdot \dfrac{1 \text{ mol water}}{18.02 \text{ g water}} = 4.883 \text{ mol water}$

248

Mole fraction of ethanol $= \dfrac{0.260 \text{ mol ethanol}}{(0.260 + 4.883)\text{mol}} = 0.051$ mf ethanol

and 1.000-0.051 or 0.949 mf water.

(b) What are the equilibrium vapor pressures of ethanol and water at this temperature: Given that normal boiling point is defined as the temperature at which the vapor pressure of a substance is equal to atmospheric pressure, we have:

$\dfrac{760 \text{ mmHg}}{1} \cdot \dfrac{0.051}{1} = 38$ mmHg (to 2 sf)

We know, from Appendix G, that the vapor pressure of water at 78.5°C is 334.2 mmHg. Since the mixture has a mf of water that is 0.949, then the vapor pressure of water is 0.949 x 334.2 mmHg or 317 mmHg.

(c) The mole fractions can be found by using the equilibrium vapor pressures calculated above:

mf ethanol – (38mmHg)/(38mmHg + 217 mmHg) = 0.11 and the mf water = 0.89 (since the two mole fractions must equal 1.0)

(d) The mole fraction of ethanol originally was 0.51. The mole fraction of ethanol in the vapor (and hence in the condensate) is 0.11. The mf of ethanol has obviously been increased by more than 2!

The mass percent of ethanol can be calculated by asking "How much ethanol and how much water are present?"

$\dfrac{46.06 \text{ g ethanol}}{1 \text{ mol ethanol}} \cdot \dfrac{0.11 \text{ mol ethanol}}{1} = 4.93$ g ethanol and

$\dfrac{18.02 \text{ g water}}{1 \text{ mol water}} \cdot \dfrac{0.89 \text{ mol water}}{1} = 16.04$ g water so the mass % is

4.93/(4.93+16.04) x 100 or 23.5% (or 24% to 2 sf) of ethanol.

103. For NaCl, 5% NaCl contains 5 g NaCl and 95 g water (by definition). The molality of the solution is then: $\dfrac{5.0 \text{ g NaCl}}{1} \cdot \dfrac{1 \text{ mol NaCl}}{58.44 \text{ g NaCl}} = 0.08555$ mol NaCl and the molality is:

$\dfrac{0.08555 \text{ mol NaCl}}{0.0950 \text{ kg water}} = 0.901$ m and calculating the ΔT,

ΔT = (-1.86 °C/m)(0.901m)= -1.68°C. Given that the measured ΔT = -3.05°C, the van't Hoff factor is (-3.05°C/-1.68°C) = 1.82.

For Na_2SO_4: 5% Na_2SO_4 contains 5 g Na_2SO_4 and 95 g water (by definition). The

molality of the solution is: $\dfrac{5.0 \text{ g Na}_2\text{SO}_4}{1} \cdot \dfrac{1 \text{ mol Na}_2\text{SO}_4}{142.05 \text{ g Na}_2\text{SO}_4} = 0.03519$ mol Na_2SO_4 and

the molality is: $\dfrac{0.03519 \text{ mol Na}_2\text{SO}_4}{0.0950 \text{ kg water}} = 0.371$ m Na_2SO_4 and calculating the ΔT,

ΔT = (-1.86 °C/m)(0.371m)= -0.689°C.

Given that the measured ΔT = -1.36 °C. the van't Hoff factor is (-1.36/-0.689) = 1.97.

Looking at the values of the factor in Table 14.4, these values certainly are consistent with the values expected—being less than the 2% solutions of NaCl and Na_2SO_4.

105. This question really asks "Is $-16.46\,°C$ equivalent to $0\,°F$? So let's convert!

$$°F = (9/5)°C + 32 = (9/5)(-16.46) + 32 = -29.6 + 32 = 2.37\,°F$$

Not a very strong endorsement of the story!

Applying Chemical Principles

1. What is the approximate mole fraction of hexane in the vapor phase after two evaporation-condensation cycles?

A copy of Figure A is attached:
Letter a indicates the first evaporation-condensation cycle, and letter b the second evaporation-condensation cycle. So the mf of hexane in the 2nd evaporation-condensation cycle is approximately 0.6—perhaps around 0.59.

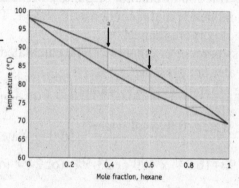

3. What is the mass percent of hexane in a mixture with heptane if the mole fraction of hexane is 0.20?

We need the masses of both hexane and heptane! We know BOTH mole fractions: since mf (hexane) = 0.20, and this is a binary mixture, the mf(heptane) is 1.00 - 0.20 = 0.80.

Calculate the mass of one mole of each compound:

Hexane: C_6H_{14} has a mass of 86.17 g/mol; Heptane: C_7H_{16} has a mass of 100.2 g/mol

The mass of hexane present would be: $0.20 \cdot 86.17$ g/mol= 17.23 g and the mass of heptane present would be: $0.80 \cdot 100.2$ g/mol= 80.16g. The total mass is: 17.23 g + 80.16g = 97.39 g, and the mass percent of hexane is $(17.23/97.39) \cdot 100 = 18\%$ (to 2 sf).

Chapter 15
Chemical Kinetics: The Rates of Chemical Reactions

PRACTICING SKILLS
Reaction Rates

1. (a) $2 O_3 (g) \rightarrow 3 O_2 (g)$

$$\text{Reaction Rate} = -\frac{1}{2} \cdot \frac{\Delta[O_3]}{\Delta t} = +\frac{1}{3} \cdot \frac{\Delta[O_2]}{\Delta t}$$

(b) $2 HOF (g) \rightarrow 2 HF (g) + O_2 (g)$

$$\text{Reaction Rate} = -\frac{1}{2} \cdot \frac{\Delta[HOF]}{\Delta t} = +\frac{1}{2} \cdot \frac{\Delta[HF]}{\Delta t} = +\frac{\Delta[O_2]}{\Delta t}$$

3. For the reaction, $2 O_3 (g) \rightarrow 3 O_2 (g)$, the rate of formation of O_2 is 1.5×10^{-3} mol/L•s. SQ15.1(a) offers a clear assist, indicating that O_2 forms at a rate 1.5 times the rate that ozone decomposes. (2 ozones produce 3 oxygens!) Hence the rate of decomposition of O_3 is -1.0×10^{-3} mol/L•s.

5. Plot the data for the hypothetical reaction $A \rightarrow 2 B$

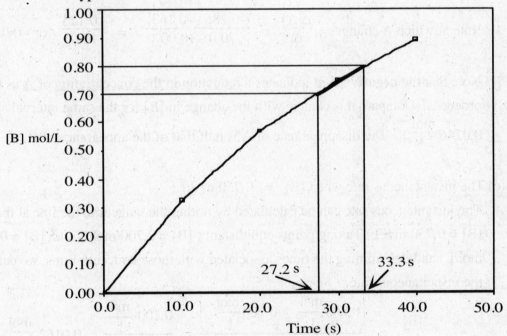

(a) Rate = $\dfrac{\Delta[B]}{\Delta t}$ = $\dfrac{(0.326 - 0.000)}{10.0 - 0.00}$ = $+\dfrac{0.326}{10.0}$ = $+0.0326\ \dfrac{mol}{L \bullet s}$

$\qquad\qquad\quad$ = $\dfrac{(0.572 - 0.326)}{20.0 - 10.00}$ = $+\dfrac{0.246}{10.0}$ = $+0.0246\ \dfrac{mol}{L \bullet s}$

$\qquad\qquad\quad$ = $\dfrac{(0.750 - 0.572)}{30.0 - 20.00}$ = $+\dfrac{0.178}{10.0}$ = $+0.0178\ \dfrac{mol}{L \bullet s}$

$\qquad\qquad\quad$ = $\dfrac{(0.890 - 0.750)}{40.0 - 30.00}$ = $+\dfrac{0.140}{10.0}$ = $+0.0140\ \dfrac{mol}{L \bullet s}$

The rate of change decreases from one time interval to the next *due to a continuing decrease* in the amount of reacting material (A).

(b) Since each A molecule forms 2 molecules of B, the concentration of A will decrease at a rate that is **half** of the rate at which B appears. The negative signs here indicate a **decrease in [A]--not a negative concentration of A!**

T	[B]	$[A] = 1/2([B]_0 - [B])$
10.0 s	0.326	$-1/2(0.326) = -0.163$
20.0 s	0.572	$-1/2(0.572) = -0.286$

Rate at which A changes = $\dfrac{\Delta[A]}{\Delta t}$ = $\dfrac{(-0.286 - -0.163)}{20.0 - 10.00}$ = $\dfrac{-0.123}{10.0}$ or $-0.0123\ \dfrac{mol}{L \bullet s}$

Note that the **negative sign** indicates a <u>reduction in the concentration of A</u> as the reaction proceeds. Compare this change with the change in [B] for the same interval above (+ $0.0246\ \dfrac{mol}{L \bullet s}$). The disappearance of A is half that of the appearance of B.

(c) The instantaneous rate when [B] = 0.750 mol/L:

The instantaneous rate can be calculated by noting the tangent to the line at the point , [B] = 0.750 mol/L. Taking points equidistant ([B] = 0.700 mol/L and [B] = 0.800 mol/L) and determining the times associated with those concentrations, we can calculate the instantaneous rate.

$$\dfrac{\Delta[B]}{\Delta t} = \dfrac{\left(0.800\,\dfrac{mol}{L} - 0.700\,\dfrac{mol}{L}\right)}{(33.3\ s - 27.2\ s)} = \dfrac{0.100\,\dfrac{mol}{L}}{6.1\ s} = 0.0163\ \dfrac{mol}{L \bullet s}$$

Concentration and Rate Equations

7. For the rate equation Rate $= k[A]^2[B]$, the reaction is 2nd order in A (superscript 2 with A), 1st order in B (implied superscript of 1 with B), and (2+1) or 3rd order overall.

9. (a) The rate equation :   Rate $= k[NO_2][O_3]$

   (b) Since k is constant, if $[O_3]$ is held constant, the rate would be tripled if the concentration of NO_2 is tripled. Let **C** represent the concentration of NO_2. Substituting into the rate equation:

   $$Rate_1 = k[C][O_3]$$
   $$Rate_2 = k[3C][O_3] = 3 \cdot k[C][O_3] \text{ or } 3 \cdot Rate_1$$

   (c) Halving the concentration of O_3—assuming $[NO_2]$ is constant, would halve the rate.

   $$Rate_1 = k[NO_2][C]$$
   $$Rate_2 = k[NO_2][1/2 \, C] = 1/2[NO_2][C] \text{ or } 1/2 \cdot Rate_1$$

11. (a) If we designate the three experiments (data sets in the table as i, ii, and iii respectively),

Experiment	[NO]	[O2]	$-\dfrac{\Delta[NO]}{\Delta t}$ $\left(\dfrac{mol}{L \cdot s}\right)$
i	0.010	0.010	2.5×10^{-5}
ii	0.020	0.010	1.0×10^{-4}
iii	0.010	0.020	5.0×10^{-5}

   Note that experiment ii proceeds at a rate four times that of experiment i.

   $$\frac{\text{experiment ii rate}}{\text{experiment i rate}} = \frac{1.0 \times 10^{-4} \frac{mol}{L \cdot s}}{2.5 \times 10^{-5} \frac{mol}{L \cdot s}} = 4$$

   This rate change was the result of doubling the concentration of NO. The *order of dependence of NO must be second order*. Comparing experiments i and iii, we see that changing the concentration of O_2 by a factor of two, also affects the rate by a factor of two. The *order of dependence of O_2 must be first order*.

   (b) Using the results above we can write the rate equation: Rate $= k[NO]^2[O_2]^1$

   (c) To calculate the rate constant we have to have a rate. Note the data provided gives the rate of disappearance of NO. The relation of this concentration to the rate is:

   Rate $= -1/2 \cdot \dfrac{\Delta[NO]}{\Delta t}$; Using experiment ii, dividing the disappearance of NO by 2

   gives a rate of $5.0 \times 10^{-5} \dfrac{mol}{L \cdot s}$ which we can substitute into the rate law:

   $$5.0 \times 10^{-5} \frac{mol}{L \cdot s} = k[0.020 \frac{mol}{L}]^2[0.010 \frac{mol}{L}] \text{ gives } 13 \frac{L^2}{mol^2 \cdot s} = k$$

(d) Rate when [NO] = 0.015 M and [O_2] = 0.0050 M

$$\text{Rate} = k[NO]^2[O_2]$$

$$= 12.5 \frac{L^2}{mol^2 \cdot s} (0.015 \frac{mol}{L})^2 (0.0050 \frac{mol}{L})$$

$$= 1.4 \times 10^{-5} \frac{mol}{L \cdot s}$$

(e) The relation between reaction rate and concentration changes:

$$\text{Rate} = -1/2 \cdot \frac{\Delta[NO]}{\Delta t} = -\frac{\Delta[O_2]}{\Delta t} = +1/2 \cdot \frac{\Delta[NO_2]}{\Delta t}$$

So when NO is reacting at $1.0 \times 10^{-4} \frac{mol}{L \cdot s}$ then O_2 will be reacting at

$5.0 \times 10^{-5} \frac{mol}{L \cdot s}$ and NO_2 will be forming at $1.0 \times 10^{-4} \frac{mol}{L \cdot s}$

13. For the reaction 2 NO(g) + O_2 (g) → 2 NO_2 (g):

(a) The rate law can be determined by examining the effect on the rate by changing the concentration of *either* NO *or* O_2.

In Data sets 1 and 2, the [O_2] doubles, and the rate doubles—a first-order dependence.

In Data sets 2 and 3, the [NO] is halved, and the rate is quartered —a second-order dependence.

The rate law will be: Rate = $k[O_2][NO]^2$

(b) To calculate the rate constant we have to have a rate. Note the data provided gives the rate of disappearance of NO. We need to adjust the rate by a factor of 1/2.

The relation of this concentration to the rate is:

$$\text{Rate} = -1/2 \cdot \frac{\Delta[NO]}{\Delta t} ; \text{ and } \frac{1}{2} (3.4 \times 10^{-8}) = 1.7 \times 10^{-8}$$

Rate: $1.7 \times 10^{-8} = k[5.2 \times 10^{-3} \frac{mol}{L}][3.6 \times 10^{-4} \frac{mol}{L}]^2$

Solving for k: k = 25.2 (or 25 $L^2/mol^2 \cdot h$ to 2sf).

Note that I selected the data from Experiment 1. Any of the data sets, (1, 2, or 3) would have provided the same value of k.

(c) The initial rate for Experiment 4 is determined by substitution into the rate law (with the value of k determined in (b):

$$\text{Rate} = 25 \, L^2/mol^2 \cdot h[5.2 \times 10^{-3} \frac{mol}{L}][1.8 \times 10^{-4} \frac{mol}{L}]^2 = 4.3 \times 10^{-9} \, mol/L \cdot h.$$

Concentration-Time Relationships

15. Note that the reaction is first order. We can write the rate expression:

$$\ln \left(\frac{[C_{12}H_{22}O_{11}]}{[C_{12}H_{22}O_{11}]_0} \right) = -kt$$

Substitute the concentrations of sucrose at $t = 0$ and $t = 27$ minutes into the equation:

$$\ln \left(\frac{[0.0132 \text{ mol/L}]}{[0.0146 \text{ mol/L}]_0} \right) = -k(27 \text{ min}) \text{ and solve for k to obtain: } k = 3.7 \times 10^{-3} \text{ min}^{-1}$$

17. Since the reaction is first order, we can write:

$$\ln \left(\frac{[SO_2Cl_2]}{[SO_2Cl_2]_0} \right) = -kt. \text{ Given the rate constant, } 2.8 \times 10^{-3} \text{ min}^{-1}, \text{ we can calculate the time}$$

required for the concentration to fall from 1.24×10^{-3} M to 0.31×10^{-3}M

$$\ln \left(\frac{0.31 \times 10^{-3}}{1.24 \times 10^{-3}} \right) = -(2.8 \times 10^{-3} \text{ min}^{-1})t.$$

$$\frac{\ln(0.25)}{(-2.8 \times 10^{-3} \text{ min}^{-1})} = t = 495 \text{ min or } 5.0 \times 10^{2} \text{ min } \text{(to 2 sf)}$$

19. (a) Since the reaction is first order, we can write:

$$\ln \left(\frac{[H_2O_2]}{[H_2O_2]_0} \right) = -kt. \text{ Given the rate constant, } 1.06 \times 10^{-3} \text{ min}^{-1}, \text{ we can calculate the time}$$

required for the concentration to fall from the original concentration to 85% of that value. Note that the concentrations *per se* are not that critical.

Let's assume the initial concentration is 100. M and after the passage of t time the concentration is 85.0M (that's 15% decomposed, yes?)

$$\frac{\ln \left(\frac{85.0}{100} \right)}{1.06 \times 10^{-3} \text{ min}^{-1}} = -t \text{ and solving for the fraction: } \frac{-0.163}{1.06 \times 10^{-3} \text{ min}^{-1}} = -t$$

and $t = 153$ min (to 3 sf).

(b) For 85% of the sample to decompose, we repeat the process, substituting 15.0 for the $[H_2O_2]$ remaining:

$$\frac{\ln \left(\frac{15.0}{100} \right)}{1.06 \times 10^{-3} \text{ min}^{-1}} = -t \text{ and solving for } t = 1790 \text{ min } \text{(to 3 sf).}$$

21. $[NO_2] = 2.8 \times 10^{-2}$ mol/L initially; Decomposition is 2nd order in NO_2; $k = 1.1$ L/mol•s

The integrated rate equation for a 2nd order process is: $\dfrac{1}{[NO_2]_t} - \dfrac{1}{[NO_2]_0} = kt$

We want to know the time, t, needed for 75% of the original NO_2 to decompose. That reduced concentration would be $(0.25)(2.8 \times 10^{-2}) = 7.0 \times 10^{-3}$.

Substituting into the integrated rate equation gives:

$$\dfrac{1}{[7.0 \times 10^{-3}]_t} - \dfrac{1}{[2.8 \times 10^{-2}]_0} = (1.1\ ^{L}\!/_{mol \cdot s})t$$

$142.8 - 35.7 = 1.1$ L/mol•s x t and solving for t gives: 97.4 s (or 97s to 2 sf)

23. Rate constant for decomposition of ammonia = 1.5×10^{-3} mol/L · s. What time will be required to completely decompose 1.0 g of NH_3 in a 1.0-L flask (by a 0 order process)?

The integrated rate equation for such a process is $[R]_0 - [R]_t = kt$. Before substituting into the rate equation, we need to express the amount of ammonia in mol/L!

$$\dfrac{1.0\ g\,NH_3}{1\ L} \cdot \dfrac{1\ mol\ NH_3}{17.03\ g\ NH_3} = 0.059\,mol/L$$

Now substituting we get: 0.059 M = 1.5×10^{-3} mol/L · s • t [Note the concentration of the ammonia at time t = 0, so the second term on the left side of the rate equation disappears!]

0.059 mol/L/1.5×10^{-3} mol/L · s = t and solving for t yields: 39 s (to 2sf)

Half-Life

25. Given that the reaction is first order we can use the integrated form of the rate law:

$$\ln\left(\dfrac{[N_2O_5]}{[N_2O_5]_0}\right) = -kt.$$

(a) Since the **definition of half-life** is "the time required for half of a substance to react", the fraction on the left side = 1/2, and $\ln(0.50) = -0.693$

Given the rate constant 6.7×10^{-5} s^{-1} we can solve for t:

$-0.693 = -(6.7 \times 10^{-5}$ s$^{-1})t$ and t = 1.0×10^4 seconds

(b) Time required for the concentration to drop to 1/10 of the original value:

Substitute the ratio 1/10 for the concentration of N_2O_5:

$\ln(0.10) = -(6.7 \times 10^{-5}$ s$^{-1})t$ and t = 3.4×10^4 seconds

27. Since the decomposition is first order: $\ln \dfrac{[\text{azomethane}]}{[\text{azomethane}]_0} = -kt$

The masses will be proportional to the number of moles, so we can simply substitute the mass (in g) of azomethane.

$$\ln \frac{[x \text{ g azomethane}]}{[2.00\text{g azomethane}]_0} = -(0.0216 \text{ min}^{-1})(0.0500 \text{ hr})(60\text{min/hr})$$

$$\ln \frac{[x \text{ g azomethane}]}{[2.00\text{g azomethane}]_0} = -0.0648$$

$$\frac{[x \text{ g azomethane}]}{[2.00\text{g azomethane}]_0} = e^{-0.0648} = 0.93725$$

and $0.9374 \times 2.00\text{g} = 1.87$ g azomethane remaining after 0.0500 hr.

Since 1 mol N_2 is produced when 1 mol of azomethane decomposes, the amount of N_2 formed is:

$$\frac{0.13 \text{ g azomethane decomposed}}{1} \cdot \frac{1 \text{ mol azomethane}}{58.08 \text{ g azomethane}} \cdot \frac{1 \text{ mol } N_2}{1 \text{ mol azomethane}} \cdot \frac{28.0 \text{ g } N_2}{1 \text{ mol } N_2}$$

$$= 0.063 \text{ g } N_2 \text{ formed}$$

29. Since this is a first-order process, $\quad \ln \dfrac{[Cu^{2+}]}{[Cu^{2+}]_0} = -kt$ and $k = -\dfrac{0.693}{12.70 \text{ hr}}$

What fraction of the copper remains after time, t = 64 hr?

Radioactive decay is a first-order process so we use the equation:

$$\ln \frac{[Cu^{2+}]}{[Cu^{2+}]_0} = -\frac{0.693}{12.70 \text{ hr}} \cdot 64 \text{ hr}$$

$\ln \dfrac{[Cu^{2+}]}{[Cu^{2+}]_0} = -3.49$ and $\dfrac{[Cu^{2+}]}{[Cu^{2+}]_0} = e^{-3.49}$ or 0.030 so 3.0 % remains (to 2 sf)

Graphical Analysis:Rate Equations and k

31. For the decomposition of N_2O:

Since **ln[N₂O] vs t** gives a **straight line**, we know that the reaction is first order with respect to N_2O, and the line has a slope = - k. Taking the natural log (ln) of the concentrations at t=120 min and t =15.0 min gives ln(0.0220) = -3.8167; ln(0.0835) = -2.4829.

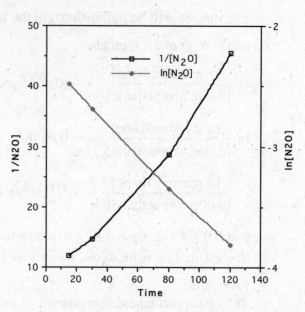

$$\text{slope} = -k = \frac{(-3.8167) - (-2.4829)}{(120.0 - 15.0)\text{min}} =$$

$$\frac{1.3338}{105.0 \text{ min}} = 0.0127 \text{ min}^{-1}$$

The rate equation is: Rate = k [N₂O].

The rate of decomposition when

$$[N_2O] = 0.035 \text{ mol/L} : \text{Rate} = (0.0127 \text{ min}^{-1})(0.035 \frac{\text{mol}}{\text{L}}) = 4.4 \times 10^{-4} \frac{\text{mol}}{\text{L} \cdot \text{min}}$$

33. Since the graph of reciprocal concentration gives a straight line, we know the reaction is **second-order** with respect to NO_2. Equation 15.2 indicates that the **slope** of the line is k, so k = 1.1 L/mol • s. The rate law is Rate = k[NO₂]²

35. The straight line obtained when the reciprocal concentration of C_2F_4 is plotted vs t indicates that the reaction is second-order in C_2F_4.

The rate expression is: Rate = $\dfrac{-\Delta[C_2F_4]}{\Delta t} = 0.04 \dfrac{L}{\text{mol} \cdot s}[C_2F_4]^2$

Kinetics and Energy

37. The E_a for the reaction N_2O_5 (g) → 2 NO₂ (g) + 1/2 O₂ (g)

Given k at 25 °C = 3.46×10^{-5} s⁻¹ and k at 55 °C = 1.5×10^{-3} s⁻¹

The rearrangement of the Arrhenius equation (in your text as Equation 15.7) is helpful here.

$$\ln \frac{k_2}{k_1} = -\frac{E_a}{R}\left(\frac{1}{T_2} - \frac{1}{T_1}\right) \; ; \ln \frac{1.5 \times 10^{-3} \text{ s}^{-1}}{3.46 \times 10^{-5} \text{ s}^{-1}} = -\frac{E_a}{8.31 \times 10^{-3} \text{kJ/mol} \cdot \text{K}}\left(\frac{1}{328 \text{ K}} - \frac{1}{298 \text{ K}}\right)$$

and solving for E_a yields a value of 102 kJ/mol for E_a.

39. Using the Arrhenius equation: $\ln \dfrac{k_2}{k_1} = -\dfrac{E_a}{R}\left(\dfrac{1}{T_2} - \dfrac{1}{T_1}\right)$, $T_1 = 800K$, and $T_2 = 850$ K

Given $E_a = 260$ kJ/mol and $k_1 = 0.0315$ s^{-1}, we can calculate k_1.

$$\ln \frac{k_2}{0.0315 \text{ s}^{-1}} = -\frac{260 \text{ kJ/mol}}{8.3145 \times 10^{-3} \text{ kJ/mol} \cdot \text{K}}\left(\frac{1}{850 \text{ K}} - \frac{1}{800 \text{ K}}\right)$$

$$\ln \frac{k_2}{0.0315 \text{ s}^{-1}} = 2.30 = \ln k_2 - \ln(0.0315 \text{ s}^{-1})$$

$$2.30 + \ln(0.0315 \text{ s}^{-1}) = \ln k_2 = 2.30 - 3.458 = -1.16$$

$$k_2 = e^{-1.16} = 0.3 \text{ s}^{-1} \text{ (1 sf owing to a temperature (800K) with 1 sf)}$$

41. Energy progress diagram:

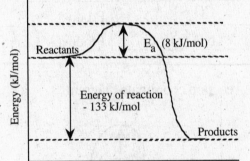

Reaction Mechanisms

43. Elementary Step

    (a) NO (g) + NO$_3$ (g) $\rightarrow$ 2 NO$_2$ (g)

    (b) Cl (g) + H$_2$ (g) $\rightarrow$ HCl (g) + H (g)

    (c) (CH$_3$)$_3$CBr (aq) $\rightarrow$ (CH$_3$)$_3$C$^+$ (aq) + Br$^-$ (aq)

Rate law

Rate = k[NO][NO$_3$]

Reaction is bimolecular

Rate = k[Cl][H$_2$]

Reaction is bimolecular

Rate = k[(CH$_3$)$_3$CBr]

Reaction is unimolecular

45. For the reaction reflecting the decomposition of ozone:

(a) The second step is the slow step, and therefore rate-determining.

(b) The rate equation involves *only* these substances that affect the rate, (since they
participate in the rate determining step), Rate = k[O₃][O].

47. For the reaction of NO₂ and CO:

Slow NO₂ + ~~NO₂~~ → NO + ~~NO₃~~

Fast ~~NO₃~~ + CO → ~~NO₂~~ + CO₂
_____

Net NO₂ + CO → NO + CO₂

Note that when the two steps are added, the desired overall equation results.

(a) Classify the species:

NO₂(g)	Reactant (step 1); Product (step 2)
CO (g)	Reactant (step 2)
NO₃ (g)	Intermediate (produced & consumed subsequently)
CO₂ (g)	Product
NO (g)	Product

(b) A reaction coordinate diagram

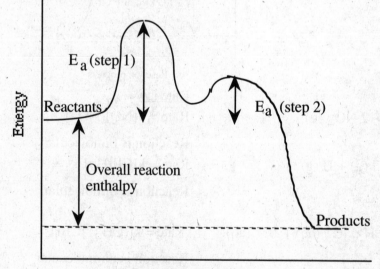

GENERAL QUESTIONS

49. What happens to the reaction rate for a reaction with the rate equation: Rate = $k[A]^2[B]$

   $Rate_1 = k[A]^2[B]$. Double the concentration of A—call it 2A, and halve the concentration of

   B—call it B/2.

   $Rate_2 = k[2A]^2[B/2]$. Reducing the concentrations gives $Rate_2 = (2)^2 \bullet (1/2) \, k[A]^2[B]$

   $Rate_2 = 2 \bullet k \, [A]^2[B]$. So $Rate_2$ is **two times** that of $Rate_1$.

51. To determine second-order dependence, after acquisition of the pH vs time data,

   plot 1/[OH$^-$] versus time. The reaction is second-order in OH$^-$ if a straight line is obtained.

53. For first-order kinetics, we know that $\ln\left(\dfrac{[HCO_2H]}{[HCO_2H]_0}\right)$ = -kt.

   Substituting into the equation, we solve for **k**.

   $\ln\left(\dfrac{[25]}{[100]_0}\right)$ = - k(72 s) and rearranging to solve for k gives $\dfrac{-1.386}{-72 \text{ s}}$ = k = 0.01925 s^{-1}

   and since we're pursing the $t_{1/2}$, $t_{1/2}$ = 0.693/k so $t_{1/2}$ = 0.693/0.01925 s^{-1} = 36 s.

   A **much simpler** route to this answer is to recognize that 1 half-life would consume 50% of

   the original sample, and the 2nd half-life would consume half of the remaining amount (25%).

   So 2 half-lives would result in the consumption of 75% of the original sample—or 1/2(72 s)!

55. For the dimerization to form octafluorocyclobutane, the following plot is obtained:

   (a) The plot of reciprocal concentration
      vs time gives a straight line. Such
      behavior is indicative of
      a second-order process.

      The rate law is: Rate = $k[C_2F_4]^2$

   (b) The rate constant is equal to the
      slope of the line: Using a graphical
      package (I used Graph Sketcher)
      the equation for the line is:

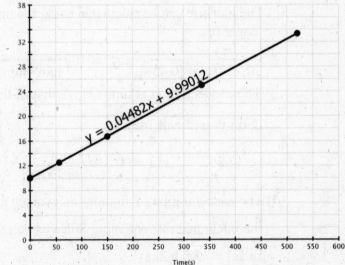

y = 9.990 + 0.04482x.

So k = 0.045 L/mol•s

(c) The concentration after 600 s is found by using the integrated rate equation for second-order processes:

$$\frac{1}{[C_2F_4]_t} - \frac{1}{[(0.100)]_0} = 0.045 \frac{L}{mol \bullet s}(600 \ s).$$ Rearranging gives

$$\frac{1}{[C_2F_4]_t} = 0.045 \frac{L}{mol \bullet s}(600 \ s) + \frac{1}{0.100}$$ and $[C_2F_4]_{600} = 0.027$ (or 0.03 M to 1sf)

(d) Time required for 90% completion: Using the same equation as in part (c), and substituting 10% of the initial concentration as our "concentration at time t", we can solve for t.

$$\frac{1}{[(0.010)]_t} - \frac{1}{[(0.100)]_0} = 0.045 \frac{L}{mol \bullet s}(t)$$

$$(100 - 10) = 0.045 \frac{L}{mol \bullet s}(t)$$ and t = 2000 s

57. For the formation of urea from ammonium cyanate:

(a) Plot the data to determine the order.

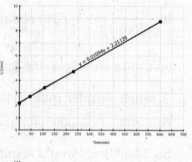

The upper graph shown here has a plot of the

1/[NH$_4$NCO] vs time, and the lower graph ln[NH$_4$NCO]

versus time. Note that the plot of reciprocal

concentration gives a straight line, indicating the reaction

is *second-order in ammonium cyanate*.

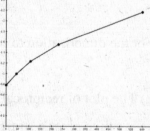

(b) k is the slope of the line, which is 0.0109 L/mol•min

(c) The half-life can be calculated using the integrated rate equation

$$\frac{1}{[0.229]_t} - \frac{1}{[0.458]_0} = 0.0109 \frac{L}{mol \bullet min}(t)$$

The concentration at time t is 1/2 that of the original concentration of NH$_4$NCO (the

definition of half-life).

$$4.367 - 2.183 = 0.0109 \frac{L}{mol \bullet min}(t)$$ and solving for t= 200. minutes.

(d) The concentration of ammonium cyanate after 12.0 hours (720. min)is found by using

the integrated rate equation. Since we know k and t, we can solve for "concentration at

time t =(12.0 hours)".

262

$$\frac{1}{[NH_4NCO]_t} - \frac{1}{[0.458]_0} = 0.0109 \frac{L}{mol \cdot min}(t)$$

$$\frac{1}{[NH_4NCO]_t} = 0.0109 \frac{L}{mol \cdot min}(720.\ min) + \frac{1}{[0.458]_0}$$

Solving for [NH$_4$NCO] we obtain [NH$_4$NCO] = 0.0997 M

59. The reaction between carbon monoxide and nitrogen dioxide has a rate equation that is second-order in NO$_2$ This means that the *slowest step* in the mechanism involves **2** molecules of nitrogen dioxide. Mechanism 2 has a SLOW step that fulfills this requirement. Note that Mechanisms 1 and 3 are only 1st order in nitrogen dioxide.

61. The decomposition of dinitrogen pentaoxide has a first-order rate equation. Determine the rate constant and the half-life by substitution into the integrated rate equation for first-order reactions .

$$\ln\left(\frac{[N_2O_5]_t}{[N_2O_5]_0}\right) = -k \cdot t \quad \text{The decomposition is 20.5\% complete in 13.0 hours at 298K.}$$

The amount of N$_2$O$_5$ remaining after 13.0 h is 79.5% of the original concentration. The left hand term is then 79.5/100.

$$\ln\left(\frac{79.5}{100}\right) = -k \cdot 13.0\ h \quad \text{and } k = 0.0176\ h^{-1}$$

Calculation of the half-life is accomplished by noting that the left-hand side of the equation has a value of 50% --and ln(0.50) = - 0.693. Substituting the value of k from above:

$$\frac{-0.693}{-0.0176\ hr^{-1}} = 39.3\ h$$

63. For the decomposition of dimethyl ether:

(a) The mass of dimethyl ether remaining after 125 min and after 145 min:

The half-life is 25.0 min. A period of 125 minutes is 5 half-lives. The fraction remaining after n half-lives is $\left(\frac{1}{2}\right)^n$ and with n = 5, the fraction remaining is 0.03125

(1/32 of the original amount). The *mass remaining is (0.03125)(8.00 g) = 0.251 g dimethyl ether*. Note that 145 minutes is *almost* 6 half-lives. So you should be able to "guess" at a value for the amount remaining. The mass should be slightly greater than 1/64 of the original amount. The exact amount can be found by substitution into the first-order rate equation to solve for the rate constant:

$$\frac{-0.693}{25.0 \text{ min}} = -k \text{ and } k = 0.0277 \text{ min}^{-1}$$

Note that the "ln term" is simplified by remembering that a "half-life" is a time for which 50% decomposes. Then $\ln(0.50) = -0.693$ {A HANDY THING TO REMEMBER}

Substituting our value of k into the equation and solving for the fraction remaining:

$$\ln\left(\frac{[\text{dimethyl ether}]_t}{[\text{dimethyl ether}]_0}\right) = -0.0277 \text{ min}^{-1} \bullet 145 \text{ min} = -4.020$$

For simplicity, let's represent the fraction remaining as x. Solving for the ratio of concentrations gives $\ln(x) = -4.020$, for which $x = 0.0179$

The mass of dimethyl ether remaining is $(0.0179)(8.00 \text{ g}) = 0.144 \text{ g}$

(b) The time required for 7.60 ng of ether to be reduced to 2.25 ng:

Substitute into the first order equation (as we've done above). Now we know the value for the "left side" and we know k; we can solve for t

$$\frac{\ln\left(\frac{[2.25 \text{ ng}]_t}{[7.60 \text{ ng}]_0}\right)}{-0.0277 \text{ min}^{-1}} = t = \frac{\ln(0.296)}{-0.0277 \text{ min}^{-1}} = \frac{-1.217}{-0.0277 \text{ min}^{-1}} = 43.9 \text{ min}$$

(c) The fraction remaining after 150 minutes is easily calculated by noting that 150 minutes is *exactly* 6 half-lives. The fraction remaining is 1/64 or 0.016 (to 2 sf).

65. Show consistency of mechanism with the rate law: Rate = $k[O_3]^2/[O_2]$:

The rate law for the **slow** step is Rate=$k[O_3][O]$, but the concentration of O is affected by the preceding equilibrium step. Solve for the [O] in that equilibrium step:

The equilibrium constant expression for the fast step is:

$$K = \frac{[O_2][O]}{O_3}; \text{solving for } [O] = \frac{K \bullet [O_3]}{[O_2]}$$

Substitute this concentration into the Rate law for the slow step above:

$$\text{Rate=}k[O_3][O] ; \text{ Rate} = \frac{K \bullet k[O_3][O_3]}{[O_2]}$$

Combining $K \bullet k$ as k', $\text{Rate} = \frac{k'[O_3]^2}{[O_2]}$

67. We can calculate Ea if we know values of rate constants at two temperatures:

$$\ln \frac{k2}{k1} = -\frac{Ea}{R}\left(\frac{1}{T_2} - \frac{1}{T_1}\right)$$

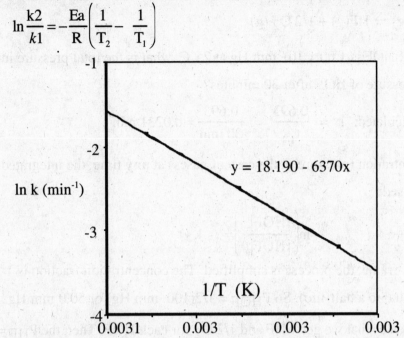

The slope was determined to be –6370. Since slope = -Ea/R,

$$-6370 = -\frac{Ea}{8.314 \times 10^{-3} \dfrac{kJ}{K \bullet mol}} = 53.0 \text{ kJ/mol}$$

69. Since the time required to prepare the egg will be inversely proportional to the rate constant,

we can calculate the ratio of the time, by calculating the ratio of the rate constants. Using the

Arrhenius equation, $\ln \dfrac{k_2}{k_1} = -\dfrac{E_a}{R}\left[\dfrac{1}{T_2} - \dfrac{1}{T_1}\right]$

and recalling the reciprocal relationship between k and t, we write:

$$\ln \frac{t_{90}}{t_{100}} = -\frac{E_a}{R}\left[\frac{1}{T_2} - \frac{1}{T_1}\right] \text{ or } \ln \frac{t_{90}}{t_{100}} = -\frac{52.0 \dfrac{kJ}{mol}}{8.3145 \times 10^{-3} \dfrac{kJ}{K \bullet mol}}\left[\frac{1}{373K} - \frac{1}{363K}\right]$$

$$\ln \frac{t_{90}}{t_{100}} = -6254 \text{ K}\left(-7.385 \times 10^{-5} \text{K}^{-1}\right) = 0.4619$$

$$\frac{t_{90}}{t_{100}} = 1.59 \text{ or } t_{90} = t_{100} \bullet 1.59 \text{ so } t_{90} = (3 \text{ minutes}) \bullet 1.59 = 4.76 \text{ minutes or}$$

5 minutes (to 1 sf)

71. The decomposition of HOF proceeds in a first-order reaction with a half-life of 30 minutes. The equation is: $HOF(g) \rightarrow HF(g) + 1/2\ O_2\ (g)$

If the initial pressure of HOF is 1.00×10^2 mm Hg at 25°C, what is the total pressure in the flask and the partial pressure of HOF after 30 minutes?

The rate constant is calculated: $k = \dfrac{0.693}{t_{1/2}} = \dfrac{0.693}{30\ min} = 0.0231\ min^{-1}$

To calculate the concentration (pressure in the case of gases) at **any** time, the integrated first-order rate law can be used:

$$\ln\left(\frac{[HOF]_t}{[HOF]_0}\right) = -k \bullet t$$

For the 30 minute time-frame, the process is simplified. The concentration fraction is 1/2 (since 30 minutes is equal to a half-life). So $P_{HOF} = 1/2(100.\ mm\ Hg)$ or 50.0 mm Hg.

The stoichiometry indicates that we get 1 HF and $1/2\ O_2$ for each HOF. Then the $P_{HF} = 50.0$ mm Hg, and $P_{O2} = 25.0$ mm Hg. The **total** pressure is then $(50.0 + 50.0 + 25.0)$ or 125.0 mm Hg.

The corresponding values after 45 minutes:

$$\ln\left(\frac{[HOF]_t}{[HOF]_0}\right) = -k \bullet t \quad \text{becomes} \quad \ln\left(\frac{[HOF]_t}{[100.\ mm\ Hg]_0}\right) = -0.0231\,min^{-1} \bullet 45\,min$$

$$\frac{[HOF]_t}{100.\ mmHg} = e^{-(0.0231min^{-1} \bullet 45\ min)} \quad \text{and} \quad [HOF]_t = 0.3536 \bullet 100.\ mmHg = 35.4\ mm\ Hg$$

$P_{HOF} = 35.4$ mm Hg; $P_{HF} = (100.0-35.4)$ mm Hg or 65 mm Hg ; $P_{O2} = 32$ mm Hg

The **total** pressure is then (35.4 mm Hg + 65 mm Hg + 32 mm Hg) or 132 mm Hg.

73. (a) Given that the experimental rate law is $\text{Rate} = \dfrac{k[NO_2NH_2]}{[H_3O^+]}$, the apparent order will be first order in NO_2NH_2 and –1 in hydronium ion.

(b) Note that the rate expression shows that H_3O^+ as a factor in the rate determining step.

Examine **mechanism 3**. Note that the rate determining step is has the apparent rate law:

$$\text{Rate} = k_5[NO_2NH^-]$$

As this specie (NO_2NH^-)is an intermediate, we can—using a steady state approximation—express the concentration as:

$[NO_2NH^-] = \dfrac{k_4 [NO_2NH_2]}{[H_3O^+]}$ and substituting into the apparent rate law:

Rate $= \dfrac{k_5 k_4 [NO_2NH_2]}{[H_3O^+]}$ and since both k_4 and k_5 are constants, we can combine them into

a k' or Rate $= \dfrac{k' [NO_2NH_2]}{[H_3O^+]}$ as expressed by the experimental rate law.

(c) Note that $k_4 \cdot k_5$ is another constant k' as explained in part b

(d) Note that in the mechanism , in the "very fast reaction" step, hydronium ion reacts

with OH⁻. Speeding up the production of water. The consumption of hydronium ion

would accelerate the overall rate (since the hydronium ion concentration—in the rate law)

and increase the $[NO_2NH]^-$

IN THE LABORATORY

75. The data are plotted as follows:

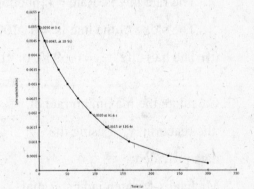

(a) For the instantaneous rate initially, we have:

$\dfrac{0.0050 - 0.0045M}{0 - 10.5s} = - 4.7 \times 10^{-5}$ M/s

For the time from 100 to 125 s

(we have data at 91.6 and 120.4s)

$\dfrac{0.0020 - 0.0015M}{91.6 - 120.4\ s} = - 1.6 \times 10^{-5}$ M/s

Note that the rate *decreases* with time as expected, since we anticipate rate to be

proportional to the concentration of the phenolphthalein.

(b) The instantaneous rate at 50 s is:

$\dfrac{0.0035 - 0.0030}{35.7 - 51.1\ s} = \dfrac{0.0005}{-15.4} = - 3.0 \times 10^{-5}$ M/s . While the order of magnitude (10^{-5} M/s)

will be stable, the value (3.0) will vary considerably depending on the points one

chooses.

Note that we calculated the instantaneous rate by drawing a tangent at 50 s. We have

used the data at 35.7 s and 51.1 s for this calculation.

(c) To determine the order graphically, we can plot [phenolphthalein] vs time (above),

1/[phenolphthalein] vs time, and ln[phenolphthalein]vs time.

The latter plots are below:

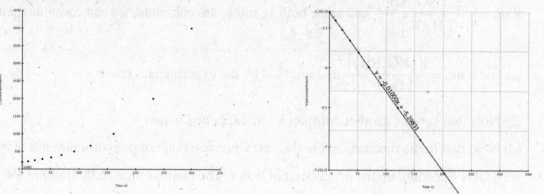

The plot of ln[phenolphthalein] vs time provides a straight line indicating that the

reaction is **1st order.**

The rate law is: Rate = k[phenolphthalein].

The slope of the line is -=0.0100, so $k = 1 \times 10^{-2}$ M/s

(d) The half-life, $t_{1/2} = 0.693/k$ or $0.693/0.010 = 69.3$ s

77. Calculate the maximum rate of

the reaction, v_{max}, using the

data provided:

Michaelis-Menten indicates that

a plot of 1/[concentration] vs 1/rate

will have an intercept that is

equal to $1/v_{max}$. The intercept

7.5×10^4, is $1/v_{max}$. and

$V_{max} = 1/7.5 \times 10^4$

$= 1.3 \times 10^{-5}$ M/min

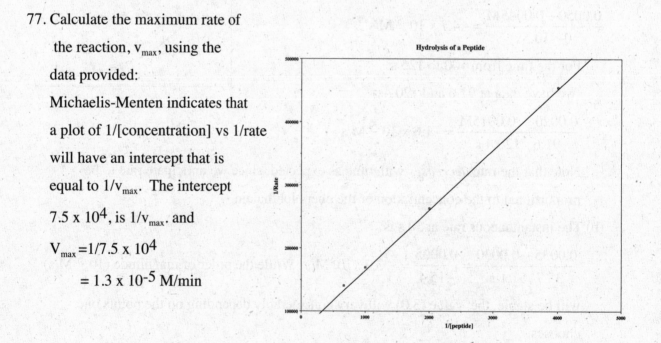

Hydrolysis of a Peptide

SUMMARY AND CONCEPTUAL QUESTIONS

79. Finely divided rhodium has a larger surface area than a block of the metal with the same mass. Since hydrogenation reactions depend on adsorption of H_2 on the catalyst surface, the greater the surface area, the greater the locations for such adsorption to occur.

81. For the reaction: H_2 (g) + I_2 (g) → 2HI(g), the rate law is Rate = k[H_2][I_2], which of the following statements is false—and why it is incorrect.

    (a) The reaction **must** occur in a single step. **False**. Only an elaboration of the mechanism can clarify this fact. While it might occur in a single step, it *need not occur in a single step*.

    (b) This is a 2nd order reaction overall. **True**. Addition of the order of H_2 and I_2 (1 +1) gives an overall order of 2.

    (c) Raising the temperature will cause the value of k to decrease. **False**. Increases in T lead to increases in the value of k.

    (d) Raising the temperature lowers the activation energy for this reaction. **False**. Temperature does not affect activation energy.

    (e) If the concentrations of both reactants are doubled the rate will double. **False**. Doubling the concentration of *either reactant* wil double the rate. Doubling the concentration of both reactants will increase the rate by a factor of 4.

    (f) Adding a catalyst in the reaction will cause the initial rate to increase. **True**. A catalyst should increase the rate at which the reaction occurs.

83. (a) True—the rate-determining step in a mechanism **is the slowest step**.

    (b) True—the rate constant is a proportionality constant that relates the concentration and rate *at a given temperature* and varies with temperature.

    (c) False—As a reaction proceeds the concentration of reactants diminshes and the rate of the reaction slows as fewer collisions occur.

    (d) False—If the slow (single) step involves a termolecular process, then only one step is necessary. However termolecular collisions are **not highly likely**.

85. For the reaction: cyclopropane → propene, describe the **change** on the following quantities.

(a) [cyclopropane] – Concentrations of reactants *decrease* as reactions proceed.

(b) [propene] – Concentrations of products *increase* as reactions proceed.

(c) [catalyst]—As catalysts are not consumed during a process, the concentration of the catalyst *will not change*.

(d) rate constant—The rate constant *will not change* as the reaction proceeds.

(e) The order of the reaction—*will not change* as the reaction proceeds.

(f) the half-life of cyclopropane—Since the half-life of cyclopropane is a function of the rate constant (which doesn't change—see (d) above), the half-life *does not change*.

87. For the reaction coordinate diagram shown:

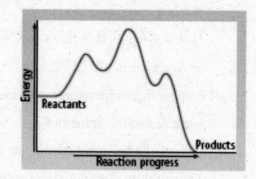

(a) Number of steps: **3**- Note that each "peak" indicates a transition between a "reactant" and a "product".

(b) Is the reaction exo- or endothermic? Noting that the energy of the products is *lower* than that of the reactants, the reaction is **exothermic**.

Applying Chemical Principles

1. Sullivan used 578 nm light to dissociate I_2 molecules to I atoms.

a) What is the energy (in kJ/mol) of 578 nm light?

$E = hc/\lambda = (6.626 \times 10^{-34}\ J \cdot s)(2.998 \times 10^8\ m/s)/578 \times 10^{-9}\ m) = 3.44 \times 10^{-19}\ J/photon$

Now we need this energy in kJ/mol of photons, so multiply by Avogadro's number, and divide by 1000 (to convert J to kJ):

$(3.44 \times 10^{-19}\ J/photon)(6.022 \times 10^{23}\ photon/mol)(1\ kJ/1000\ J) = 207\ kJ/mol$

b) Breaking an I_2 bond requires 151 kJ/mol of energy. What is the longest wavelength of light that has enough energy to dissociate I_2?

In part (a) we solved for the energy of light with a specific wavelength. In part (b) we reverse the procedure, and ask what wavelength will have at least 151 kJ/mol! If we divide the required energy (151 kJ/mol) by the 2nd and 3rd terms in the expression above, we'll have the energy in J/photon

$$\frac{151\ kJ/mol}{6.022 \times 10^{23}} \cdot \frac{1000J}{1\ kJ} = 2.51 \times 10^{-19}\ J/photon$$

Now we can use the energy equation ($E = hc/\lambda$) to solve for λ

Note that we'll divide hc/E to give the wavelength:

$$\lambda = \frac{(6.626 x 10^{-34}\ J \bullet s)(2.998 \times 10^8\ m/s)}{2.51 x 10^{-19}\ J/photon} = 792 x 10^{-9}\ m\ \ or\ 792\ nm$$

3. Why is a termolecular elementary step likely to be the slowest in a mechanism?

The probability of a 3-body collision is low, making a termolecular elementary step slow!

Chapter 16
Principles of Reactivity: Chemical Equilibria

PRACTICING SKILLS
Writing Equilibrium Constant Expressions

1. Equilibrium constant expressions:

(a) $K = \dfrac{[H_2O]^2[O_2]}{[H_2O_2]^2}$ (b) $K = \dfrac{[CO_2]}{[CO][O_2]^{1/2}}$ (c) $K = \dfrac{[CO]^2}{[CO_2]}$ (d) $K = \dfrac{[CO_2]}{[CO]}$

Note that **solids** in the equations are *omitted* in the equilibrium constant expressions.

The Equilibrium Constant and Reaction Quotient

3. The equilibrium expression for the reaction: $I_2\,(g) \Leftrightarrow 2\,I\,(g)$ has a $K = 5.6 \times 10^{-12}$

Substituting the molar concentrations into the equilibrium expression:

$Q = \dfrac{[I]^2}{[I_2]} = \dfrac{(2.00 \times 10^{-8})^2}{(2.00 \times 10^{-2})} = 2.0 \times 10^{-14}$

Since $Q < K$, the system **is not at equilibrium** and will move to **the right to make more product (and reach equilibrium)**.

5. Is the system at equilibrium?

$[SO_2]$	5.0×10^{-3} M
$[O_2]$	1.9×10^{-3} M
$[SO_3]$	6.9×10^{-3} M

Substituting into the equilibrium expression:

$\dfrac{[SO_3]^2}{[SO_2]^2[O_2]} = \dfrac{\left[6.9 \times 10^{-3}\right]^2}{\left[5.0 \times 10^{-3}\right]^2\left[1.9 \times 10^{-3}\right]} = 1000$

Since $Q > K$, the system is **not** at equilibrium, and will move "to the left" to make more reactants to reach equilibrium.

Calculating an Equilibrium Constant

7. For the equilibrium: PCl_5 (g) $\Leftrightarrow$ PCl_3 (g) + Cl_2 (g), calculate K

The equilibrium concentrations are: $[PCl_5]$ = 4.2 x 10^{-5} M

$[PCl_3]$ = 1.3 x 10^{-2} M

$[Cl_2]$ = 3.9 x 10^{-3} M

The equilibrium expression is: $\dfrac{[PCl_3][Cl_2]}{[PCl_5]} = \dfrac{\left[1.3 \times 10^{-2}\right]\left[3.9 \times 10^{-3}\right]}{\left[4.2 \times 10^{-5}\right]} = 1.2$

9. The equilibrium expression is: $K = \dfrac{[CO]^2}{[CO_2]}$

The quantities given are **moles**, so first calculate the **concentrations at equilibrium**.

$[CO] = \dfrac{0.10 \text{ moles}}{2.0 \text{ L}} = 0.050$ M $[CO_2] = \dfrac{0.20 \text{ moles}}{2.0 \text{ L}} = 0.10$ M

(a) $K = \dfrac{[CO]^2}{[CO_2]} = \dfrac{[0.050]^2}{[0.10]} = 2.5 \times 10^{-2}$

(b & c) The only change here is in the amount of carbon. Since C does not appear in the equilibrium expression, K would not change.

11. To keep track of concentrations, construct a table:

The reaction is CO (g) + Cl_2 (g) $\Leftrightarrow$ $COCl_2$(g)

	CO	Cl_2	$COCl_2$
Initial	0.0102 M	0.00609 M	0 M
Change	- 0.00308 M	- 0.00308 M	+ 0.00308 M
Equilibrium	0.0071 M	0.00301M	+ 0.00308 M

(a) Equilibrium concentrations of CO and $COCl_2$ are found by noting that 0.00308 M Cl_2 is

consumed (to leave 0.00301M at equilibrium), an equimolar amount of CO is consumed

and an equal amount of $COCl_2$ is produced. So $[COCl_2]$ = 0.00308 M and

$[CO]$ = 0.0071 M.

(b) $K = \dfrac{[+0.00308]}{[0.0071][0.00301]} = 140$ (2 sf)

Using Equilibrium Constants

13. For the system butane $\Leftrightarrow$ isobutene, K = 2.5

Equilibrium concentrations may be found using a table. First note that the amount of butane must be converted to molar concentrations. So 0.017mol butane/0.50 L = 0.034 M.

	butane	isobutane
Initial	0.034	0
Change	- x	+ x
Equilibrium	0.034 - x	x

Substituting these equilibrium concentrations into the equilibrium expression:

$$K = \frac{[\text{isobutene}]}{[\text{butane}]}; K = \frac{x}{0.034 - x} = 2.5 \text{ and } x = 2.5(0.034 - x);$$

multiplying: $x = 0.085 - 2.5\,x$ and solving for x gives $x = 2.4 \times 10^{-2}$

and [isobutane] = 2.4×10^{-2} M, [butane] = 0.034 - x = 1.0×10^{-2} M

15. For the equilibrium, the equilibrium constant is $\dfrac{[I]^2}{[I_2]} = 3.76 \times 10^{-3}$

The initial concentration of I_2 is $\dfrac{0.105 \text{ mol}}{12.3 \text{ L}}$ or 8.54×10^{-3} M

The equation indicates that 2 mol of I form for each mol of I_2 that reacts.

If some amount, say x M, of I_2 reacts, then the amount of I that forms is 2x, and the amount of I_2 remaining at equilibrium is (8.54×10^{-3} - x). The equilibrium concentrations can then be substituted into the equilibrium expression.

$$\frac{(2x)^2}{8.54 \times 10^{-3} - x} = 3.76 \times 10^{-3}$$

or $\quad 4\,x^2 = 3.76 \times 10^{-3}(8.54 \times 10^{-3} - x)$ rearranging we get:

$4\,x^2 + 3.76 \times 10^{-3}\,x - 3.21 \times 10^{-5} = 0.$

Solve this, using the quadratic equation. Using the positive solution to that equation we find that $x = 2.40 \times 10^{-3}$.

So at equilibrium:

$[I_2] = 8.54 \times 10^{-3} - 2.40 \times 10^{-3} = 6.14 \times 10^{-3}$ M

$[I] = 2(2.40 \times 10^{-3}) = 4.80 \times 10^{-3}$ M

17. Given at equilibrium: $K = \dfrac{[CO][Br_2]}{[COBr_2]} = 0.190$ at 73°C

First, calculate concentrations: $[COBr_2] = \dfrac{0.500 \text{ mol}}{2.00 \text{ L}} = 0.250$ M

Substituting the $[COBr_2]$ into the equilibrium expression we get: $\dfrac{[CO][Br_2]}{(0.250)} = 0.190$

Note that the stoichiometry of the equation tells us that for **each CO, we obtain 1 Br$_2$**.

	COBr$_2$	CO	Br2
Initial	0.250 M	0 M	0 M
Change	-x	+ x	+ x
Equilibrium	0.250 -x	+ x	+ x

We can rewrite the expression to read: $\dfrac{[CO][Br_2]}{0.250 - x} = \dfrac{[x]^2}{0.250 - x} = 0.190$

Solve this using the quadratic equation. $x^2 + 0.190x - 0.0475 = 0$

Using the positive solution to that equation, $x = 0.143$ M $= [CO]$ and $[COBr_2] =$

$(0.250-0.143) = 0.107$ M.

The percentage of COBr$_2$ that has decomposed is: $\dfrac{0.143 \text{ M}}{0.250 \text{ M}} \bullet 100 = 57.1$ %

Manipulating Equilibrium Constant Expressions

19. To compare the two equilibrium constants, write the equilibrium expressions for the two.

$$\dfrac{[C]^2}{[A][B]} = K_1 \quad \text{and} \quad \dfrac{[C]^4}{[A]^2[B]^2} = K_2$$

Note that the first expression *squared* is equal to the second expression, so **(b) $K_2 = K_1^2$**

21. Comparing the two equilibria:

(1) $SO_2(g) + \frac{1}{2} O_2(g) \Leftrightarrow SO_3(g)$ K_1

(2) $2 SO_3(g) \Leftrightarrow 2 SO_2(g) + O_2(g)$ K_2

The expression that relates K_1 and K_2 is (e): $K_2 = \dfrac{1}{K_1^2}$

Reversing equation 1 gives: $SO_3(g) \Leftrightarrow SO_2(g) + \frac{1}{2} O_2(g)$ with the eq. constant $\dfrac{1}{K_1}$

Multiplying the (reversed equation 1) x 2 gives: $2 SO_3(g) \Leftrightarrow 2 SO_2(g) + O_2(g)$ and the

modified equilibrium constant $\left(\dfrac{1}{K_1}\right)^2$.

23. Calculate K for the reaction:

$$SnO_2 (s) + 2 CO (g) \Leftrightarrow Sn (s) + 2 CO_2 (g) \quad \text{given:}$$

(1) $SnO_2 (s) + 2 H_2 (g) \Leftrightarrow Sn (s) + 2 H_2O (g)$ K = 8.12

(2) $H_2 (g) + CO_2 (g) \Leftrightarrow H_2O (g) + CO (g)$ K = 0.771

Take equation 2 and reverse it, then multiply by 2 to give:

$$2H_2O (g) + 2 CO (g) \Leftrightarrow 2 H_2 (g) + 2 CO_2 (g) \qquad K = (\dfrac{1}{0.771})^2 = 1.68$$

add equation 1 : $SnO_2 (s) + 2 H_2 (g) \Leftrightarrow Sn (s) + 2 H_2O (g)$ K = 8.12

$$SnO_2 (s) + 2 CO (g) \Leftrightarrow Sn (s) + 2 CO_2 (g) \qquad K_{net} = 8.12 \cdot 1.68 = 13.7$$

Disturbing a Chemical Equilibrium

25. The equilibrium may be represented : $N_2O_3 (g) + \text{heat} \Leftrightarrow NO_2 (g) + NO (g)$

Since the process is **endothermic** ($\Delta H = +$), heat is absorbed in the "left to right" reaction.
The effect of:

(a) Adding more $N_2O_3(g)$: An increase in the pressure of N_2O_3 (adding more N_2O_3) will

shift the equilibrium to the **right**, producing more NO_2 and NO.

(b) Adding more $NO_2(g)$: An increase in the pressure of NO_2 (adding more NO_2) will shift

the equilibrium to the **left**, producing more N_2O_3.

(c) Increasing the volume of the reaction flask: If the volume of the flask is increased, the

pressure will drop. (Remember that P for gases is inversely related to volume.) A drop in

pressure will favor that side of the equilibrium with the "larger total number of moles of

gas"—so this equilibrium will shift to the **right**, producing more NO_2 and NO.

(d) Lowering the temperature: You should note that a change in T (up or down) will result in

a **change in the equilibrium constant**. (None of the three changes mentioned above

change K!) However, the same principle applies. The removal of heat (a decrease in T)

favors the exothermic process (shifts the equilibrium to the **left**) producing more N_2O_3.

27. K for butane $\Leftrightarrow$ iso-butane is 2.5.

(a) Equilibrium concentrations if 0.50 mol/L of isobutane is added:

	butane	iso-butane
Original concentration	1.0	2.5
Change immediately after addition	1.0	2.5 + 0.50
Change (going to equilibrium)	+ x	- x
Equilibrium concentration	1.0 + x	3.0 - x

Substituting into the equilibrium expression: $K = \dfrac{[\text{iso-butane}]}{[\text{butane}]} = \dfrac{3.0 - x}{1.0 + x} = 2.5$

$$3.0 - x = 2.5\,(1.0 + x) \quad \text{and } 0.14 = x$$

The equilibrium concentrations are:

[butane] = 1.0 + x = 1.1 M and [iso-butane] = 3.0 - x = 2.9 M

(b) Equilibrium concentrations if 0.50 mol /L of butane is added:

	butane	iso-butane
Original concentration	1.0	2.5
Change immediately after addition	1.0 + 0.50	2.5
Change (going to equilibrium)	- x	+ x
Equilibrium concentration	1.5 - x	2.5 + x

$K = \dfrac{[\text{iso-butane}]}{[\text{butane}]} = 2.5$ and solving: $2.5 + x = 2.5\,(1.5 - x)$ and $x = 0.36$

The eq. concentrations are:[butane] = 1.5 - x = 1.1 M and [iso-butane] = 2.5 + x = 2.9 M

GENERAL QUESTIONS

29. For the equilibrium: Br_2 (g) $\Leftrightarrow$ 2Br (g), we must first calculate the concentration of Br_2

$$[Br_2] = \frac{0.086 \text{ mol}}{1.26 \text{ L}} = 0.068 \text{ M}$$

Now we can complete an equilibrium table:

	Br_2	Br
Initial	0.068 M	0
Change	- (0.037)(0.068) M	+2 (0.037)(0.068) M
Equilibrium	0.066 M	+0.0051 M

Substituting into the K expression: $K = \dfrac{[Br]^2}{[Br_2]} = \dfrac{(0.0051)^2}{(0.066)} = 3.9 \times 10^{-4}$

31. K_p is 6.5×10^{11} at 25°C for CO (g) + Cl_2 (g) $\Leftrightarrow$ $COCl_2$ (g)

What is the value of K_p for $COCl_2$ (g) $\Leftrightarrow$ CO(g) + Cl_2 (g)?

Writing the equilibrium expressions for these two processes will show that one is the *reciprocal* of the other. Then K_p for the two will be mathematically related by the reciprocal relationship. So K_p for the second reaction is $1/K_p$ for the first reaction or $1/6.5 \times 10^{11}$ or 1.5×10^{-12}.

33. Calculate K for the system CS_2 (g) + 3 Cl_2 $\Leftrightarrow$ S_2Cl_2 (g) + CCl_4 (g)

Note that the reaction occurs in a 1.00 L flask, so the **molar concentrations** and the **number of moles** will be numerically identical.

	CS_2	Cl_2	S_2Cl_2	CCl_4
Initial	1.2	3.6	0	0
Change	- 0.90	- 3 • 0.90	+0.90	+0.90
Equilibrium	0.3	0.9	0.90	0.90

Equilibrium concentrations are easily determined by noting the stoichiometry of the reaction. If 0.90 mol of CCl_4 are formed (to reach equilibrium), an equal amount of S_2Cl_2 is formed and an equal amount of CS_2 reacted. The amount of Cl_2 that reacts is found by noting the 1:3 ratio of CS_2 to Cl_2 . If 0.90 mol of CS_2 react, then 3 x 0.90 mol of Cl_2 react.

Substituting into the equilibrium expression gives: $\dfrac{[S_2Cl_2][CCl_4]}{[CS_2][Cl_2]^3} = \dfrac{[0.90][0.90]}{[0.3][0.9]^3} = 4$ (1 sf)

35. K for butane $\Leftrightarrow$ isobutane is 2.5. Is system at equilibrium if 1.75 mol of butane and 1.25 mol of isobutane are mixed?

Substituting into the equilibrium expression: $K = \dfrac{[\text{isobutane}]}{[\text{butane}]} = \dfrac{1.25 \text{ mol}}{1.75 \text{ mol}} = 0.714.$

The system is not at equilibrium (since Q is not equal to K). Since Q is less than K, the system will need to shift to the right (producing more isobutane) to reach equilibrium.

	butane	isobutane
Original concentration	1.75	1.25
Change (going to equilibrium)	- x	+ x
Equilibrium concentration	1.75 - x	1.25 + x

$$K = \frac{[\text{isobutane}]}{[\text{butane}]} = \frac{1.25 + x}{1.75 - x} = 2.5 \text{ and solving for x: } 2.5(1.75 - x) = (1.25 + x)$$

Expanding: $4.375 - 2.5x = 1.25 + x$; $(4.375 - 1.25) = 3.5\,x$; $3.125 = 3.5\,x$; $x = 0.89$

The equilibrium concentrations are: [butane] $= 1.75 - x = 0.86$ M and

[isobutane] $= 1.25 + x = 2.14$ M

37. For the two equilibria shown, which correctly relates the two equilibrium constants?

K_1 $\quad$ $NOCl(g) \Leftrightarrow NO(g) + 1/2\,Cl_2(g)$

K_2 $\quad$ $2\,NO(g) + Cl_2(g) \Leftrightarrow 2\,NOCl(g)$

One way to answer the question is to write the equilibrium expressions for the two equilibria:

$$K_1 = \frac{[NO][Cl_2]^{1/2}}{[NOCl]} \text{ and } K_2 = \frac{[NOCl]^2}{[NO]^2[Cl_2]}.$$

Note two differences between the K expressions: (1) they are reciprocals of each other—in one case with NOCl in the denominator, and in the other with NOCl in the numerator. Mentally invert K_2. Note the 2nd difference. K_2 is $(K_1)^2$. So the relationship between the two is: (c) $K_2 = \dfrac{1}{K_1^2}$

39. For the equilibrium: $BaCO_3(s) \Leftrightarrow CO_2(g) + BaO(s)$, the effect of:

(a) adding $BaCO_3(s)$: (i) no effect. Since the solid does not appear in the equilibrium expression, the addition of more solid has no effect on the position of equilibrium.

(b) adding CO_2: Increasing the concentration of CO_2 would shift the equilbrium (ii) to the left—consuming CO_2 until the system returned to equilibrium.

(c) Adding BaO: As in (a) above, since the solid, BaO, does not appear in the equilibrium expression, adding BaO would have (i) no effect.

(d) Raising the temperature: Raising T would have the effect of accelerating the decomposition of the carbonate, producing more CO_2—shifting the equilibrium (iii) to the right.

(e) Increasing the volume of the flask: This decreases the concentration of CO_2. and (iii) shifts the equilibrium to the right—just as in (b) increasing the CO_2 concentration shifted the equilibrium to the left.

41. For the equilibrium: PCl_5 (g) $\Leftrightarrow$ PCl_3 (g) + Cl_2 (g), calculate K

	[PCl5]	[PCl3]	[Cl2]
Equilibrium (expressed as mol)	3.120g/208.24g/mol = 0.01498 mol	3.845g/137.33 g/mol = 0.02800 mol	1.787g/70.91 g/mol = 0.02520 mol
Concentration immediately after addition	0.01498 mol	0.02800 mol	0.02520 mol + 0.02000 mol (corresponds to adding 1.418g/70.91g/mol)
Change (going to equilibrium)	+x	-x	-x
New equilibrium	0.01498 + x	0.02800 - x	0.04520 - x

(a) The addition of the chlorine will shift the equilibrium to the *left*

(b) The new equilibrium concentrations will be found using the values from the table, but we must know the value of K, which we can get by using the equilibrium concentrations given in the problem.

$$K = \frac{[PCl_3][Cl_2]}{[PCl_5]} = \frac{[0.02800][0.02520]}{[0.01498]} = 0.0470$$

Substituting the values from our table above: $\dfrac{[0.02800 - x][0.04520 - x]}{[0.01498 + x]} = 0.0470$

The resulting quadratic equation can be solved for x.

The "sensible" root for x = 0.00485 M.

The resulting equilibrium concentrations are:

$[PCl_5] = 0.01498 + x = 0.01498 + 0.00485 = 0.0199$ M

$[PCl_3] = 0.02800 - x = 0.02800 - 0.00485 = 0.0231$ M

$[Cl_2] = 0.04520 - x = 0.04520 - 0.00485 = 0.0403$ M

43. For the system: $NH_4I(s) \Leftrightarrow NH_3 (g) + HI (g)$, what is K_p ?

Given that the total pressure is attributable **only** to NH_3 + HI, **and** that the stoichiometry of the equation tells us that equal amounts of the two substances are formed, we can easily determine the pressure of **each** of the gases. Since we desire K_p in atmospheres, we must first convert 705 mm Hg to units of atmospheres:

$$705 \text{ mm Hg} \cdot \frac{1 \text{ atm}}{760 \text{ mm Hg}} = 0.928 \text{ atm}$$

$P_{total} = P(NH_3) + P(HI) = 0.928$ atm so the pressure of each gas is 1/2(0.928) or 0.464 atm.

$K_p = P(NH_3) \cdot P(HI) = (0.464 \text{ atm})^2 = 0.215$

45. Given Kp for $N_2O_4(g) \Leftrightarrow 2 NO_2 (g)$ is 0.148 at 25 °C.

(a) If total P is 1.50 atm, what fraction of N_2O_4 has dissociated?

$$P_{total} = P_{NO_2} + P_{N_2O_4} = 1.50 \text{ atm} ; \quad P_{N_2O_4} = 1.50 - P_{NO_2}$$

$$K_p = \frac{P^2_{NO_2}}{P_{N_2O_4}} = 0.148 \text{ and } K_p = \frac{P^2_{NO_2}}{1.50 - P_{NO_2}} = 0.148 \text{ or } 0.148(1.50 - P_{NO_2}) = P^2_{NO_2}$$

and rearranging, $0.222 - 0.148P_{NO_2} = P^2_{NO_2}$

Solving for P_{NO_2} with the quadratic equations yields:

$$P_{NO_2} = 0.403 \text{ atm and } P_{N_2O_4} = 1.10 \text{ atm}$$

To determine the fraction of N_2O_4 that has dissociated, we need to know the amount of N_2O_4 that was originally present. The stoichiometry tells us that we get 2 NO_2 for each N_2O_4 that dissociates. The 0.403 atm of NO_2 that exist indicate that (0.403/2) atm of N_2O_4 dissociated. The **original pressure** of N_2O_4 will then be 1.10 atm + 0.20 atm or 1.30 atm. The fraction that has dissociated is then: (1.30 - 1.10)/1.30 or 0.15.

(b) The fraction dissociated if the total equilibrium pressure falls to 1.00 atm:

$$P_{total} = P_{NO_2} + P_{N_2O_4} = 1.00 \text{ atm} ; \quad P_{N_2O_4} = 1.00 - P_{NO_2}$$

$$K_p = \frac{P^2_{NO_2}}{P_{N_2O_4}} = 0.148 = \frac{P^2_{NO_2}}{1.00 - P_{NO_2}} \; ; \; 0.148(1.00 - P_{NO_2}) = P^2_{NO_2}$$

Solving via the quadratic equation: $P_{NO_2} = 0.318$ atm and $P_{N_2O_4} = 0.682$ atm

The equilibrium pressure of 0.682 atm for N_2O_4 tells us that the **original pressure of**

N_2O_4 was 0.682 + 1/2(0.318) or 0.841 atm. [Recall that 0.318 atm of NO_2 represents

0.159 atm of N_2O_4 decomposing].

The fraction dissociated is: 0.159 atm/0.841 atm = 0.189 or approximately 19%.

47. For the decomposition of ammonia at 450 °C, K = 6.3.

What are equilibrium concentrations of NH_3, N_2, and H_2 and what is the **total** pressure in

the flask?

3.60 mol NH_3 in a 2.00 L vessel is 1.80 M NH_3. Establish a reaction table:

	$[NH_3]$	$[N_2]$	$[H_2]$
Initial concentrations (mol/L)	1.80	0	0
Change (going to equilibrium)	- x	+ x/2	+3x/2
New equilibrium	1.80 - x	x/2	3x/2

Substituting into the K expression:

$$6.3 = \frac{\left[\dfrac{x}{2}\right]\left[\dfrac{3x}{2}\right]^3}{[1.80 - x]^2} \quad \text{and simplifying gives: } 6.3 = \frac{\dfrac{27x^4}{16}}{[1.80 - x]^2}$$

Simplifying further: $6.3 = \dfrac{1.6875x^4}{[1.80 - x]^2}$, taking the square root of both sides will simplify

the math: $2.51 = \dfrac{1.299x^2}{[1.80 - x]}$ and $2.51[1.80 - x] = 1.299x^2$.

Expanding gives: $4.4519 - 2.51x = 1.299x^2$, and using the quadratic equation to solve

gives x = 1.13.

Using the table above, we get:

$[NH_3] = 1.80 - x = 0.67$ M; $[N_2] = x/2 = 0.57$ M; $[H_2] = 3x/2 = 1.7$ M

Total pressure in flask: $P = nRT/V$. and since $M = n/M$, we can use $P = MRT$

$$P = (0.67 + 0.57 + 1.7) \bullet 0.082057 \frac{L \bullet atm}{K \bullet mol} \bullet (723K) = 180 \text{ atm (2 sf)}$$

49. K_c for the decomposition of NH_4HS into ammonia and hydrogen sulfide gases is 1.8×10^{-4}.

   (a) When pure salt decomposes, the equilibrium concentrations of $[NH_3]$ and $[H_2S]$:

   $K_p = [NH_3][H_2S] = 1.8 \times 10^{-4}$. Noting the stoichiometry of the reaction, the two

   concentrations will be equal. $[NH_3]^2 = [H_2S]^2 = 1.8 \times 10^{-4}$ and

   $[NH_3] = [H_2S] = 1.3 \times 10^{-2}$ M.

   (b) If NH_4HS is placed into a flask containing $[NH_3] = 0.020$ M,

   when system achieves equilibrium, what are equilibrium concentrations?

   The equilibrium table:

	$[NH_3]$	$[H_2S]$
Initial concentration	0.020	0
Change (going to equilibrium).	+ x	+ x
New equilibrium	0.020 + x	+ x

   $K_p = [NH_3][H_2S] = 1.8 \times 10^{-4}$ and $[0.020 + x][+ x] = 1.8 \times 10^{-4}$

   Simplifying gives: $0.020x + x^2 = 1.8 \times 10^{-4}$. This will require the quadratic equation to

   resolve the equation: $x^2 + 0.020x - 1.8 \times 10^{-4} = 0$. Solving gives $x = 6.7 \times 10^{-3}$ M.

   The equilibrium concentrations are then $[NH_3] = 0.020 + 6.7 \times 10^{-3}$ or 0.0267M

   (or 0.027 M to 2 sf) and for $[H_2S] = 6.7 \times 10^{-3}$M

51. What is P_{total} for a mixture of NO_2 and N_2O_4 (total mass = 64.4g) in a 15 L flask at 300K?

Note that Kp for $2 NO_2 (g) \Leftrightarrow N_2O_4 (g)$ is 7.1 at 300K so $K_p = \dfrac{P_{N_2O_4}}{P^2_{NO_2}} = 7.1$ (eq 1)

Mass of NO_2 + Mass of N_2O_4 = 64.4 g (eq 2) and $P_{total} = P_{NO_2} + P_{N_2O_4}$ (eq 3)

Begin by substituting into equation 2:

Mass of NO_2 + Mass of N_2O_4 = 64.4 g. We can calculate mass of each by noting that the product of #mol • MM = mass (in grams).

#mol • MM (for NO_2) +#mol • MM (for N_2O_4) = 64.4

[#mol NO_2 • 46.01g/mol NO_2] + [#mol N_2O_4 • 92.01 g/mol N_2O_4] = 64.4

Recall that $P = \dfrac{nRT}{V} = \dfrac{\text{\# mol} \bullet 0.082 \frac{L \bullet atm}{K \bullet mol} \bullet 300 K}{15 L}$

Using equation 1: $P_{N2O4} = 7.1 \cdot P^2_{NO2}$

$$\dfrac{\text{\# mol } N_2O_4 \bullet 0.082 \frac{L \bullet atm}{K \bullet mol} \bullet 300 K}{15 L} = 7.1 \cdot \left(\dfrac{\text{\#mol } NO_2 \cdot 0.082 \frac{L \cdot atm}{K \cdot mol} \cdot 300 K}{15 L} \right)^2$$

Cancelling similar terms on both sides gives:

$$\text{\# mol } N_2O_4 = \dfrac{7.1 \cdot \left(\text{\#mol } NO_2 \right)^2 \cdot 0.082 \frac{L \cdot atm}{K \cdot mol} \cdot 300 K}{15 L}$$

Multiplying all the constants on the right hand side gives:

mol N_2O_4 = 11.644 (#mol $NO_2)^2$ and substituting into the equation below:

[#mol NO_2 • 46.01g/mol NO_2] + [#mol N_2O_4 • 92.01 g/mol N_2O_4] = 64.4 g gives

[#mol NO_2 • 46.01g/mol NO_2] + [11.644 (#mol $NO_2)^2$ • 92.01] = 64.4 g and leaving off all units (for clarity) and multiplying (11.644 • 92.01) gives

46.01 n + 1071.36444 • n^2 = 64.4 ,which can be solved by the quadratic equation to give n = 0.225 mol NO_2. Since each mol of NO_2 has a mass of 46.01 g, the mass of NO_2 = 0.225 • 46.01 = 10.4 g. From equation 2: 64.4 g – 10.4 g = 54.0 g of N_2O_4.

The pressure of each gas is:

$$P = \dfrac{nRT}{V} = \dfrac{(0.225 \text{mol } NO_2) \cdot 0.082 \frac{L \cdot atm}{K \cdot mol} \cdot 300 K}{15 L} = 0.369 \text{ atm}$$

Substituting into equation 1, we can solve for the pressure of N_2O_4.

$P_{N2O4} = 7.1 \cdot P_{NO2}^2 = 7.1(0.369)^2 = 0.967$ atm (0.97 to 2sf)

The total pressure is: $0.37 + 0.97 = 1.34$ atm

53. (a) For the reaction of hydrogen and iodine to give HI, if Kc = 56 at 435 °C, what is Kp ?

The relation between the equilibrium constants is: $K_p = K_c(RT)^{\Delta n}$. The equation is:

$$H_2(g) + I_2(g) \Leftrightarrow 2\,HI(g)$$

Note that the **total** number of moles of gas is 2 on both sides of the equilibrium, so that $\Delta n = 0$. So Kp = 56.

(b) Mix 0.45 mol of each gas in a 10.0 L flask at 435 °C, what is the total pressure of the mixture before and after equilibrium?

Before equilibrium: P = nRT/V or

$$\frac{0.90 \text{ mol} \bullet 0.082057\frac{L \bullet atm}{K \bullet mol} \bullet 708 \text{ K}}{10.0 \text{ L}} = 5.2 \text{ atm} \quad (\text{Note that we can simply add the}$$

amount of each gas (obviously the P of either gas will be 2.6 atm).

At equilibrium:

Substituting into the equilibrium expression gives:

$$K_p = \frac{P^2(HI)}{P(H_2) \bullet P(I_2)} = \frac{(2x)^2}{(2.6 - x)^2} = 56 \quad \text{Simplify the expression by taking the square root}$$

of both sides to give: $\dfrac{(2x)}{(2.6 - x)} = 7.48$. Solving for x gives x = 2.06 atm, so

$P_{HI} = 2 \bullet 2.06$ or 4.1 atm (to 2 sf), $P(H_2) = P(I_2) = (2.6 - 2.06) = 0.54$ atm

The total pressure is : 4.1 atm + 0.54 atm + 0.54 atm = 5.2 atm

(c) The partial pressures, as noted above, are: $P_{HI} = 4.1$ atm , $P(H_2) = P(I_2) = 0.54$ atm

55. For the reaction of hemoglobin (Hb) with CO we can write: $K = \dfrac{[HbCO][O_2]}{[HbO_2][CO]} = 2.0 \times 10^2$

If $\dfrac{[HbCO]}{[HbO_2]} = 1$, substitution into the K expression shows that $\dfrac{[O_2]}{[CO]} = 2.0 \times 10^2$

If $[O_2] = 0.20$ atm, then $\dfrac{0.20 \text{ atm}}{[CO]} = 2.0 \times 10^2$ and solving for $[CO] = 1 \times 10^{-3}$ atm.

So a partial pressure of $[CO] = 1 \times 10^{-3}$ atm would likely be fatal.

57. How many O atoms are present if 1.0 mol of O_2 is placed in a 10.L vessel at 1800 K?

Given Kp for $O_2 (g) \Leftrightarrow 2 O (g) = 1.2 \times 10^{-10}$

Since we need to express the # of O atoms in a 10. L vessel, let's convert Kp into Kc:

Since $Kp = Kc(RN)^{\Delta n}$ then $Kc = Kp/(RN)^{\Delta n} = 1.2 \times 10^{-10}/(0.082057 \frac{L \bullet atm}{K \bullet mol} \bullet 1800K)^1$

and solving, $Kc = 8.12 \times 10^{-13}$.

Substituting into the Kc expression, we get: $\frac{[O]^2}{[O_2]} = 8.12 \times 10^{-13}$

If we let a mol/L of O_2 dissociate into atoms, we get: $\frac{[2a]^2}{[0.10-a]} = 8.12 \times 10^{-13}$

So $4a^2 = 8.12 \times 10^{-13}(0.10-a)$ and solving for a: $a = 1.4 \times 10^{-7}$. Note here that with the small size of K, the denominator could be simplified to 0.10, and an approximate value of a calculated. The [O] is then 2a or 2.8×10^{-7} M. To calculate the # of O atoms:

10. L $\bullet$ 2.8×10^{-7} mol/L $\bullet$ 6.02×10^{23} atoms/mol $= 1.7 \times 10^{18}$ O atoms.

59. For the equilibrium with boric acid (BA) + glycerin(gly), K = 0.90.
    $B(OH)_3 (aq) + glycerin(aq) \Leftrightarrow B(OH)_3 \bullet glycerin(aq)$

	BA	gly	BA•gly
Initial	0.10 M	?	0
Change	- 0.60 • 0.10 M		+ 0.60 • 0.10M
Equilibrium	0.040M	x	0.060 M

Substituting into the equilibrium expression yields:

$\frac{[BA \bullet gly]}{[BA][gly]} = 0.90$, and substituting from our table gives: $\frac{[0.060]}{[0.040][gly]} = 0.90$

Rearranging:

$\frac{[0.060]}{[0.040][0.90]} = [gly] = 1.67$ M (or 1.7 to 2 sf)

Solving for [gly] gives the *equilibrium* amount of glycerin.

Recall, however that the equilibrium amount of glycerin represents the amount of glycerin remaining uncomplexed from the original amount. Recall that *some* glycerin is consumed in making the complex (in this case 0.060M). So the initial amount of glycerin present would be 1.67 + 0.060 or 1.73M—which is 1.7 M (to 2 sf).

61. For the reaction of N_2O_4 decomposing into NO_2, a sample of N_2O_4 at a pressure of 1.00 atm reaches equilibrium with 20.0% of the N_2O_4 having been converted into NO_2.

(a) What is K_p?

	N_2O_4	NO_2
Initial	1.00 atm	0 atm
Change	$- 0.200 \cdot 1.00$ atm	$+ 2 \cdot 0.200 \cdot 1.00$ atm
Equilibrium	0.80 atm	0.400 atm

Substituting into the equilibrium expression gives:

$$K_p = \frac{P^2(NO_2)}{P(N_2O_4)} = \frac{(0.400)^2}{(0.80)} = 0.20$$

(b) If the initial pressure is 0.10 atm, the percent dissociation:

Now we know that $K_p = 0.20$. Substituting into the equilibrium expression:

$$K_p = \frac{P^2(NO_2)}{P(N_2O_4)} = \frac{(2x)^2}{(0.10 - x)} = 0.20 \text{ and rearranging: } 4x^2 = (0.10-x)(0.20)$$

or $4x^2 = 0.020 - 0.20x$, and solving for the quadratic equation, $x = 0.050$ atm.

This represents half (or 50%) of the original N_2O_4.

Does this agree with LeChatelier's Principle? YES. We began with a lower concentration (pressure), so we expect that a larger percentage of the N_2O_4 will dissociate (50% compared to 20%), and it does.

IN THE LABORATORY

63. For the equilibrium: $(NH_3)[B(CH_3)_3] \Leftrightarrow B(CH_3)_3 + NH_3$ $K_p = 4.62$

Substituting $(CH_3)_3P$ for NH_3 gives an equilibrium with $K_p = 0.128$

while substituting $(CH_3)_3N$ gives an equilibrium with $K_p = 0.472$.

(a) To determine which of the three systems would provide the largest concentration of $B(CH_3)_3$, one needs to ask "What does K tell me?"

One answer to that question is "the extent of reaction—that is the larger the value of K, the more products are formed. Hence **the system with the largest K_p would give the largest partial pressure of $B(CH_3)_3$ at equilibrium.**

(b) Given $(NH_3)[B(CH_3)_3] \Leftrightarrow B(CH_3)_3 + NH_3$ $K_p = 4.62$

Since we have a K_p, we need to express "concentrations" as Pressures:

$$P = \frac{0.010 \text{ mol} \cdot 0.082057 \frac{L \cdot atm}{K \cdot mol} \cdot 373K}{0.100L} = 3.1 \text{ atm}$$

	$(NH_3)[B(CH_3)_3]$	$B(CH_3)_3$	NH_3
Initial:	3.1 atm	0 M	0 M
Change	- x	+ x	+ x
Equilibrium	3.1 - x	+x	+x

Substituting into the equilibrium expression we have: $K_p = \frac{(x)(x)}{3.1-x} = 4.62$

and simplifying gives : $x^2 = 4.62(3.1 - x)$ and $x^2 + 4.62x - 14.1 = 0$

Solving via the quadratic formula gives x = 2.1 atm

So the **concentrations at equilibrium** are:

$[B(CH_3)_3] = [NH_3] = 2.1$ atm and $[(NH_3)[B(CH_3)_3]] = 1.0$ atm

The percent dissociation of $(NH_3)[B(CH_3)_3]$ is:

$\frac{\text{amount changed}}{\text{original amount}} = \frac{2.1 \text{ atm}}{3.1 \text{ atm}}$ x 100 = 69% dissociated

65. (a) The addition of KSCN results in the color becomes even more red. Why? This is to be expected if the solution had not yet reached equilibrium prior to the addition of the KSCN. As more Fe^{2+} is added, more of the $(Fe(H_2O)_5SCN)^+$ complex ion would form—making the solution more red (in accordance with LeChatelier's principle).

$$[Fe(H_2O)_6]^{2+} + SCN^- \rightarrow [(Fe(H_2O)_5SCN)]^+$$

(b) Once again, LeChatelier's principle would answer the question. The addition of silver ion would initiate formation of the white AgSCN solid, removing SCN^- ions from solution—and reducing the concentration of the red $(Fe(H_2O)_5SCN)^+$ complex ion.

$Ag^+ + SCN^- \rightarrow AgSCN(s)$ resulting in $[Fe(H_2O)_6]^{2+} + SCN^- \leftarrow [(Fe(H_2O)_5SCN)]^+$

SUMMARY AND CONCEPTUAL QUESTIONS

67. Decide upon the truth of each of the statements:

(a) The magnitude of the equilibrium constant is always independept of T. **false**- K is always a function of the temperature.

(b) The equilibrium constant for the net equation is the produce of the equilibrium constants of the summed equations—**true**.

(c) The equilibrium constant for a reactions has the same value as K for the reverse reaction. **False**. The reverse reaction would have an equilibrium constant that is 1/K(orig).—that is the reciprocal of the original K.

(d) For an equilibrium with only one non-solid reactant or product, only that concentration (in this case, CO_2) appears in he equilibrium constant expression—**true**.

(e) For the equilibrium involving the decomposition of $CaCO_3$, the value of K is independent of the expression of the amount of CO_2, --**false; $Kp = Kc(RT)^{\Delta n}$**. with Δn equal to a non-zero number, Kp and Kc will differ by the factor $(RT)^{\Delta n}$.

69. Characterize each of the following as product- or reactant-favored:

(a) with $Kp = 1.2 \times 10^{45}$; A large K indicates the equilibrium would be **product-favored**.

(b) with $Kp = 9.1 \times 10^{-41}$; A small K indicates the equilibrium would "lie to the left"—that is would be **reactant-favored**.

(c) $Kp = 6.5 \times 10^{11}$; equilibrium would "lie to the right"—that is would be **product-favored**.

71. There are several ways to prove the dynamic nature of this equilibrium. Begin the experiment with a mixture of (a) deuterated H_2 (represented as D_2), (b) elemental N_2, and (c) NH_3. Let the mixture reach equilibrium. Separate the three different gases and measure the amount of D_2. If the system is dynamic, the amount of elemental D_2 will be reduced (as D is incorporated into the NH_3 molecules), AND the NH_3 present will consist of varying amounts of $NH_{(3-x)}D_x$ as a result of the incorporation of D atoms into the NH_3 molecular species.

Applying Chemical Principles

1. Freezing point depression is one means of determining the molar mass of a compound. The freezing point depression constant of benzene is $-5.12 \, °C/m$.

   (a) When a 0.503 g sample of the white crystalline dimer is dissolved in 10.0 g benzene, the freezing point of benzene is decreased by 0.542 °C. Verify that the molar mass of the dimer is 475 g/mol when determined by freezing point depression. Assume no dissociation of the dimer occurs.

   If we write the freezing point expression as $\Delta T = i \bullet m \bullet K_{fp}$

   $$0.542°C = i \cdot \frac{\dfrac{0.503g}{475g/mol}}{0.010 \, kg} \cdot 0.512 \, °C/m$$

   and solving, $i = 1$. So the molar mass is verified as being 475 g/mol and no dissociation is assumed ($i = 1$).

   (b) The correct molar mass of the dimer is 487 g/mol. Explain why the dissociation equilibrium causes the freezing point depression calculation to yield a lower molar mass for the dimer. K for this equilibrium is **much much** smaller than 1, so there is a greater abundance of the monomer (with molar mass much less than 487)—giving rise to the lower calculated molar mass.

3. A 0.64 g sample of the white crystalline dimer (4) is dissolved in 25.0 mL of benzene at 20 °C. Use the equilibrium constant to calculate the concentrations of monomer (2) and dimer (4) in this solution.

   $$K = \frac{[monomer]^2}{[dimer]} = 4.1 \times 10^{-4}$$ Assuming that the correct molar mass of the dimer is 487 (as

   in b above), we can calculate the [dimer] as $\dfrac{\dfrac{0.64g \, dimer}{487 \, g/mol \, dimer}}{0.025 \, L} = 0.053M$ (2 sf)

   Knowing the [dimer], we can substitute into the K expression:

   $$\frac{[2x]^2}{[0.053 - x]} = 4.1 \times 10^{-4}$$ and multiplying both sides by the denominator:

   $4x^2 = 4.1 \times 10^{-4}(0.053 - x); \quad 4x^2 = 2.16 \times 10^{-5} - 4.1 \times 10^{-4}x$

   Solving via the quadratic equation gives x = $2.27 \times 10^{-3}M$

The [monomer] = 2 x $2.27 \times 10^{-3}M = 4.5 \times 10^{-3}M$ (2 sf).

The concentration of the dimer at equilibrium would be $(0.053 - 0.00227) = 0.050M$ (2 sf).

5. Which of the organic species mentioned in this story is paramagnetic?

    (a) triphenylmethyl chloride

    (b) triphenylmethyl radical

    (c) the triphenylmethyl dimer

This one's easy! A radical **by definition** has an unpaired electron, so (b) is paramagnetic!

Chapter 17
Principles of Reactivity: Chemistry of Acids and Bases

PRACTICING SKILLS
The Brönsted Concept

1. Conjugate Base of: Formula   Name

   (a) HCN CN^- cyanide ion

   (b) HSO_4^- SO_4^{2-} sulfate ion

   (c) HF F^- fluoride ion

3. Products of acid-base reactions:

   (a) $HNO_3(aq)$ + $H_2O(l)$ → $H_3O^+(aq)$ + $NO_3^-(aq)$

       acid   base   conjugate acid   conjugate base

   (b) $HSO_4^-(aq)$ + $H_2O(l)$ → $H_3O^+(aq)$ + $SO_4^{2-}(aq)$

       acid   base   conjugate acid   conjugate base

   (c) $H_3O^+(aq)$ + $F^-(aq)$ → $HF(aq)$ + $H_2O(l)$

       acid   base   conjugate acid   conjugate base

5. Hydrogen oxalate acting as Brönsted acid and Brönsted base:

   Brönsted acid: $HC_2O_4^-(aq)$ + $H_2O(l)$ ⇔ $C_2O_4^{2-}(aq)$ + $H_3O^+(aq)$

   Brönsted base: $HC_2O_4^-(aq)$ + $H_2O(l)$ ⇔ $H_2C_2O_4(aq)$ + $OH^-(aq)$

   Hydrogen oxalate ion is an amphoteric (amphiprotic) substance. Note the characteristic of

   many such substances: (a) a negative charge—making it attractive to positively charged

   hydronium ions, and the presence of an acidic H, making it capable of donating a proton—

   and acting as a Brönsted acid.

7. (a) $HCO_2H(aq)$ + $H_2O(l)$ ⇔ $HCO_2^-(aq)$ + $H_3O^+(aq)$

       acid   base   conjugate   conjugate
                                of HCO_2H of H_2O

   (b) $NH_3(aq)$ + $H_2S(aq)$ ⇔ $NH_4^+(aq)$ + $HS^-(aq)$

       base   acid   conjugate   conjugate
                                of NH_3 of H_2S

(c) $HSO_4^-(aq) + OH^-(aq) \Leftrightarrow SO_4^{2-}(aq) + H_2O(l)$
    acid   base   conjugate   conjugate
                                of HSO_4^- of OH^-

pH Calculations

9. Since pH = 3.75, $[H_3O^+] = 10^{-pH}$ or $10^{-3.75}$ or 1.8×10^{-4} M

Since the $[H_3O^+]$ is greater than 1×10^{-7} (pH < 7), the solution is acidic.

11. pH of a solution of 0.0075 M HCl:

Since HCl is considered a strong acid, a solution of 0.0075 M HCl has

$[H_3O^+] = 0.0075$ or 7.5×10^{-3}; pH = $-\log[H_3O^+] = -\log[7.5 \times 10^{-3}] = 2.12$

The hydroxide ion concentration is readily determined since $[H_3O^+] \cdot [OH^-] = 1.0 \times 10^{-14}$

$$[OH^-] = \frac{1.0 \times 10^{-14}}{7.5 \times 10^{-3}} = 1.3 \times 10^{-12} \text{ M}$$

13. pH of a solution of 0.0015 M $Ba(OH)_2$:

Soluble metal hydroxides are strong bases. To the extent that it dissolves ($Ba(OH)_2$ is not that soluble), $Ba(OH)_2$ gives two OH^- for each formula unit of $Ba(OH)_2$. 0.0015 M $Ba(OH)_2$ would provide 0.0030 M OH^-. Since $[H_3O^+] \cdot [OH^-] = 1.0 \times 10^{-14}$

$[H_3O^+]$ would then be:

$$[H_3O^+] = \frac{1.0 \times 10^{-14}}{3.0 \times 10^{-3}} = 3.3 \times 10^{-12} \text{ M} \text{ and pH} = -\log[3.3 \times 10^{-12}] = 11.48.$$

Equilibrium Constants for Acids and Bases

15. Concerning the following acids:

Phenol		Formic acid		Hydrogen oxalate ion	
C_6H_5OH	1.3×10^{-10}	HCO_2H	1.8×10^{-4}	$HC_2O_4^-$	6.4×10^{-5}

(a) The strongest acid is formic. The weakest acid is phenol. Acid strength is proportional to the magnitude of K_a.

(b) Recall the relationship between acids and their conjugate base: $K_a \cdot K_b = K_w$

Since K_W is a constant, the greater the magnitude of K_a, the smaller the value of the K_b for the conjugate base. The strongest acid (formic) has the weakest conjugate base.

(c) The weakest acid (phenol) has the strongest conjugate base.

17. The substance which has the smallest value for K_a will have the strongest conjugate base. One can prove this quantitatively with the relationship: $K_a \cdot K_b = K_W$. An examination of Appendix H shows that--of these three substances--HClO has the smallest K_a, and ClO^- will be the strongest conjugate base.

19. The equation for potassium carbonate dissolving in water:

$$K_2CO_3 \text{ (aq)} \rightarrow 2\,K^+ \text{ (aq)} + CO_3^{2-} \text{ (aq)}$$

Soluble salts--like K_2CO_3--dissociate in water. The carbonate ion formed in this process is a base, and reacts with the acid, water.

$$CO_3^{2-} \text{ (aq)} + H_2O \text{ (l)} \Leftrightarrow HCO_3^- \text{ (aq)} + OH^- \text{ (aq)}$$

The production of the hydroxide ion, a strong base, in this second step is responsible for the basic nature of solutions of this carbonate salt.

21. Most of the salts shown are sodium salts. Since Na^+ does not hydrolyze, we can estimate the acidity (or basicity) of such solutions by looking at the extent of reaction of the anions with water (hydrolysis).

The Al^{3+} ion is acidic, as is the $H_2PO_4^-$ ion. From Table 17.3 we see that the K_a for the hydrated aluminum ion is greater than that for the $H_2PO_4^-$ ion, making the Al^{3+} solution more acidic —lower pH than that of $H_2PO_4^-$. All the other salts will produce basic solutions and since the S^{2-} ion has the largest K_b, we will anticipate that the Na_2S solution will be most basic—i.e. have the highest pH.

pKa: A Logarithmic Scale of Acid Strength

23. The pK_a for an acid with a K_a of 6.5×10^{-5}. $pK_a = -\log(6.5 \times 10^{-5}) = 4.19$

25. K_a for epinephrine, whose $pK_a = 9.53$. $K_a = 10^{-pK_a}$ so $K_a = 10^{-9.53}$ or 3.0×10^{-10}

From Table 17.3, we see that epinephrine belongs between:

Hexaaquairon(II) ion $Fe(H_2O)_6^{2+}$ 3.2×10^{-10}

Hydrogen carbonate ion HCO_3^{1-} 4.8×10^{-11}

27. *2-chlorobenzoic* acid has a smaller pK_a than benzoic acid, so it *is the stronger acid*.

Ionization Constants for Weak Acids and Their Conjugate Bases

29. The K_b for the chloroacetate ion:

Recall the relationship between acids and their conjugate bases: $K_a \cdot K_b = K_w$

K_b for the chloroacetate ion will be $\dfrac{1.00 \times 10^{-14}}{1.41 \times 10^{-3}} = 7.09 \times 10^{-12}$

31. The K_a for $(CH_3)_3NH^+$ is 10^{-pK_a} or $10^{-9.80} = 1.6 \times 10^{-10}$

Then $K_b = \dfrac{1.0 \times 10^{-14}}{1.6 \times 10^{-10}} = 6.3 \times 10^{-5}$

Predicting the Direction of Acid-Base Reactions

33. CH_3CO_2H (aq) + HCO_3^- (aq) $\Leftrightarrow$ $CH_3CO_2^-$ (aq) + H_2CO_3 (aq)

Since acetic acid is a stronger acid than carbonic acid, the equilibrium lies predominantly to the right.

35. Predict whether the equilibrium lies predominantly to the left or to the right:

(a) NH_4^+ (aq) + Br^- (aq) $\Leftrightarrow$ NH_3 (aq) + HBr (aq)

HBr is a stronger acid than NH_4^+; equilibrium lies to left.

(b) HPO_4^{2-} (aq) + $CH_3CO_2^-$ (aq) $\Leftrightarrow$ PO_4^{3-} (aq) + CH_3CO_2H (aq)

CH_3CO_2H is a stronger acid than HPO_4^{2-} ; equilibrium lies to left.

(c) $Fe(H_2O)_6^{3+}$ (aq) + HCO_3^- (aq) $\Leftrightarrow$ $Fe(H_2O)_5(OH)^{2+}$ (aq) + H_2CO_3 (aq)

$Fe(H_2O)_6^{3+}$ is a stronger acid than H_2CO_3; equilibrium lies to right.

Types of Acid-Base Reactions

37. (a) The net ionic equation for the reaction of NaOH with Na_2HPO_4:

$$OH^-(aq) + HPO_4^{2-}(aq) \Leftrightarrow H_2O(l) + PO_4^{3-}(aq)$$

(b) The equilibrium lies to the right (since HPO_4^{2-} is a stronger acid than H_2O and OH^- is a stronger base than PO_4^{3-}), but phosphate and hydroxide ions aren't too different in basic strength, so the position of equilibrium does not lie very far to the right. The result is that remaining OH^- will result in a basic solution.

39. (a) The net ionic equation for the reaction of CH_3CO_2H with Na_2HPO_4:

$$CH_3CO_2H\ (aq) + HPO_4^{2-}(aq) \Leftrightarrow CH_3CO_2^-(aq) + H_2PO_4^-(aq)$$

(b) The equilibrium lies to the right (since CH_3CO_2H ⁻is a stronger acid than $H_2PO_4^-$ and HPO_4^{2-} is a stronger base than $CH_3CO_2^-$). The result is that the weak acetic acid and the $H_2PO_4^-$ ion will result in a weakly acidic solution.

Using pH to Calculate Ionization Constants

41. For a 0.015 M HOCN, pH = 2.67:

(a) Since pH = 2.67, $[H_3O^+] = 10^{-pH}$ or $10^{-2.67}$ or 2.1×10^{-3} M

(b) The Ka for the acid is in general: $K_a = \dfrac{[H^+][A^-]}{[HA]}$

The acid is monoprotic, meaning that for each H^+ ion, one also gets a OCN^-
The two numerator terms are then equal. The concentration of the molecular acid is the (original concentration – concentration that dissociates).

$$K_a = \frac{[2.1 \times 10^{-3}][2.1 \times 10^{-3}]}{[0.015 - 2.1 \times 10^{-3}]} = 3.5 \times 10^{-4}$$

43. With a pH = 9.11, the solution has a pOH of 4.89 and $[OH^-] = 10^{-4.89} = 1.3 \times 10^{-5}$ M.

The equation for the base in water can be written:

$$H_2NOH\ (aq) + H_2O\ (l) \Leftrightarrow H_3NOH^+\ (aq) + OH^-\ (aq)$$

At equilibrium, $[H_2NOH] = [H_2NOH] - [OH^-] = (0.025 - 1.3 \times 10^{-5})$

or approximately 0.025 M.

$$K_b = \frac{[H_3NOH^+][OH^-]}{[H_2NOH]} = \frac{(1.3 \times 10^{-5})^2}{0.025} = 6.6 \times 10^{-9}$$

45. (a) With a pH = 3.80 the solution has a $[H_3O^+] = 10^{-3.80}$ or 1.6×10^{-4} M

(b) Writing the equation for the unknown acid, HA, in water we obtain:

$$HA\ (aq) + H_2O\ (l) \Leftrightarrow H_3O^+\ (aq) + A^-\ (aq)$$

$[H_3O^+] = 1.6 \times 10^{-4}$ implying that $[A^-]$ is also 1.6×10^{-4}. Therefore the equilibrium

concentration of acid, HA, is $(2.5 \times 10^{-3} - 1.6 \times 10^{-4})$ or $\approx 2.3 \times 10^{-3}$.

$$K_a = \frac{[H_3O^+][A]}{[HA]} = \frac{(1.6 \times 10^{-4})^2}{2.3 \times 10^{-3}} = 1.1 \times 10^{-5}$$

We would classify this acid as a moderately weak acid.

Using Ionization Constants

47. For the equilibrium system: $CH_3CO_2H\ (aq) + H_2O(l) \Leftrightarrow CH_3CO_2^-(aq) + H_3O^+(aq)$

Initial	0.20 M		0	0
Change	- x		+ x	+ x
Equilibrium	0.20 - x		+ x	+ x

Substituting into the K_a expression:

$$K_a = \frac{[H_3O^+][A]}{[HA]} = \frac{(x)^2}{0.20 - x} = 1.8 \times 10^{-5} \text{ and solving for } x = 1.9 \times 10^{-3}$$

So $[CH_3CO_2H] \div 0.20$ M; $[CH_3CO_2^-] = [H_3O^+] = 1.9 \times 10^{-3}$ M

49. Using the same logic as in question SQ17.47, we can write:

$$HCN\ (aq)\ +\ H_2O\ (l)\ \Leftrightarrow\ H_3O^+\ (aq)\ +\ CN^-\ (aq)$$

	HCN		H_3O^+	CN^-
Initial	0.025M		0	0
Change	- x		+ x	+ x
Equilibrium	0.025 - x		+ x	+ x

Substituting these values into the K_a expression for HCN:

$$K_a = \frac{[CN^-][H_3O^+]}{[HCN]} = \frac{(x)^2}{(0.025-x)} = 4.0\ x\ 10^{-10}$$

Assuming that the denominator may be approximated as 0.025 M, we obtain:

$$\frac{x^2}{0.025} = 4.0\ x\ 10^{-10}\ \text{and}\ x = 3.2\ x\ 10^{-6}.$$

The equilibrium concentrations of $[H_3O^+] = [CN^-] = 3.2\ x\ 10^{-6}$ M.

The equilibrium concentration of $[HCN] = (0.025 - 3.2\ x\ 10^{-6})$ or 0.025 M.

Since x represents $[H_3O^+]$ the pH = $- \log(3.2\ x\ 10^{-6})$ or 5.50.

51. The equilibrium of ammonia in water can be written:

$$NH_3\ (aq)\ +\ H_2O(l)\ \Leftrightarrow\ NH_4^+\ (aq)\ +\ OH^-\ (aq)\quad K_b = 1.8\ x\ 10^{-5}$$

	NH_3		NH_4^+	OH^-
Initial	0.15 M		0	0
Change	- x		+ x	+ x
Equilibrium	0.15 - x		+ x	+ x

Substituting into the K_b expression:

$$K_b = \frac{[NH_4^+][OH^-]}{[NH_3]} = \frac{(x)^2}{(0.15 - x)} = 1.8\ x\ 10^{-5}$$

With K_b small, the extent to which ammonia reacts with water is slight.

We can approximate the denominator (0.15 - x) as 0.15 M, and solve the equation:

$$\frac{(x)^2}{0.15} = 1.8\ x\ 10^{-5}\ \text{and}\ x = 1.6\ x\ 10^{-3}$$

So $[NH_4^+] = [OH^-] = 1.6\ x\ 10^{-3}$ M and $[NH_3] = (0.15 - 1.6\ x\ 10^{-3}) \approx 0.15$ M

The pH of the solution is then:

$$pOH = - \log(1.6\ x\ 10^{-3}) = 2.78\ \text{and the pH} = (14.00-2.78) = 11.22$$

53. For CH_3NH_2 (aq) + H_2O(l) $\Leftrightarrow$ $CH_3NH_3^+$(aq) + OH^-(aq) $K_b = 4.2 \times 10^{-4}$

Using the approach of SQ17.51, the equilibrium expression can be written:

$$K_b = \frac{x^2}{0.25 - x} = 4.2 \times 10^{-4} \approx \frac{x^2}{0.25}$$

Solving the expression **with approximations** gives $x = 1.0 \times 10^{-2}$; $[OH^-] = 1.0 \times 10^{-2}$ M

pOH would then be 1.99 and pH 12.01 (2 sf)

55. pH of 1.0×10^{-3} M HF; $K_a = 7.2 \times 10^{-4}$

Using the same method as in question 49, we can write the expression:

$$K_a = \frac{x^2}{(1.0 \times 10^{-3} - x)} = 7.2 \times 10^{-4}$$

The concentration of the HF and K_a preclude the use of our usual approximation

$(1.0 \times 10^{-3} - x \approx 1.0 \times 10^{-3})$

So we multiply both sides of the equation by the denominator to get:

$$x^2 = 7.2 \times 10^{-4}(1.0 \times 10^{-3} - x)$$

$$x^2 = 7.2 \times 10^{-7} - 7.2 \times 10^{-4} x$$

Using the quadratic equation we solve for x:

and $x = 5.6 \times 10^{-4}$ M = $[F^-]$ = $[H_3O^+]$ and pH = 3.25

Acid-Base Properties of Salts

57. Hydrolysis of the NH_4^+ produces H_3O^+ according to the equilibrium:

$$NH_4^+ \text{ (aq)} + H_2O \text{ (l)} \Leftrightarrow H_3O^+ \text{ (aq)} + NH_3 \text{ (aq)}$$

With the ammonium ion acting as an acid, to donate a proton, we can write the K_a

expression:

$$K_a = \frac{[NH_3][H_3O^+]}{[NH_4^+]} = \frac{K_w}{K_b} = \frac{1.0 \times 10^{-14}}{1.8 \times 10^{-5}} = 5.6 \times 10^{-10}$$

The concentrations of both terms in the numerator are equal, and the concentration of

ammonium ion is 0.20 M. (Note the approximation for the **equilibrium** concentration of

NH_4^+ to be equal to the **initial** concentration.) Substituting and rearranging we get

$$[H_3O^+] = \sqrt{0.20 \cdot 5.6 \times 10^{-10}} = 1.1 \times 10^{-5} \text{M and the pH} = 4.98.$$

59. The hydrolysis of CN^- produces OH^- according to the equilibrium:

$$CN^- \text{ (aq)} + H_2O \text{ (l)} \Leftrightarrow HCN \text{ (aq)} + OH^- \text{ (aq)}$$

Calculating the concentrations of Na^+ and CN^- :

$$[Na^+]_i = [CN^-]_i = \frac{10.8 \text{ g NaCN}}{0.500 \text{ L}} \cdot \frac{1 \text{ mol NaCN}}{49.01 \text{ g NaCN}} = 4.41 \times 10^{-1} \text{ M}$$

$$\text{Then } K_b = \frac{K_w}{K_a} = \frac{1.0 \times 10^{-14}}{4.0 \times 10^{-10}} = 2.5 \times 10^{-5} = \frac{[HCN][OH^-]}{[CN^-]}$$

Substituting the $[CN^-]$ concentration into the K_b expression and noting that:

$[OH^-]_e = [HCN]_e$ we may write

$$[OH^-]_e = [(2.5 \times 10^{-5})(4.41 \times 10^{-1})]^{1/2} = 3.3 \times 10^{-3} \text{ M}$$

$$[H_3O^+] = \frac{1.0 \times 10^{-14}}{3.3 \times 10^{-3}} = 3.0 \times 10^{-12} \text{ M}$$

pH After an Acid-Base Reaction

61. The net reaction is:

$$CH_3CO_2H \text{ (aq)} + NaOH \text{ (aq)} \Leftrightarrow CH_3CO_2^- \text{(aq)} + Na^+ \text{(aq)} + H_2O \text{(l)}$$

The addition of 22.0 mL of 0.15 M NaOH (3.3 mmol NaOH) to 22.0 mL of 0.15 M CH_3CO_2H (3.3 mmol CH_3CO_2H) produces water and the soluble salt, sodium acetate (3.3 mmol $CH_3CO_2^- Na^+$). The acetate ion is the anion of a weak acid and reacts with water according to the equation: $CH_3CO_2^- \text{ (aq)} + H_2O \text{ (l)} \Leftrightarrow CH_3CO_2H \text{ (aq)} + OH^- \text{ (aq)}$

The equilibrium constant expression is: $K_b = \dfrac{[CH_3CO_2H][OH^-]}{[CH_3COO^-]} = 5.6 \times 10^{-10}$

The concentration of acetate ion is: $\dfrac{3.3 \text{ mmol}}{(22.0 + 22.0) \text{ ml}} = 0.075 \text{ M}$

	$CH_3CO_2^-$	CH_3CO_2H	OH^-
Initial concentration	0.075	0	0
Change	-x	+x	+x
Equilibrium	0.075 - x	+x	+x

$$K_b = \frac{[CH_3CO_2H][OH^-]}{[CH_3CO_2^-]} = \frac{x^2}{0.075-x} = 5.6 \times 10^{-10}$$

Simplifying $(100 \cdot K_b \ll 0.075)$ we get $\dfrac{x^2}{0.075} = 5.6 \times 10^{-10}$; $x = 6.5 \times 10^{-6} = [OH^-]$

The hydrogen ion concentration is related to the hydroxyl ion concentration by the equation:

$$K_W = [H_3O^+][OH^-] = 1.0 \times 10^{-14}$$

$$[H_3O^+] = \dfrac{1.0 \times 10^{-14}}{[OH^-]} \quad \dfrac{1.0 \times 10^{-14}}{6.5 \times 10^{-6}} = 1.5 \times 10^{-9} \text{ and pH} = 8.81$$

The pH is greater than 7, as we expect for a salt of a strong base and weak acid.

63. Equal numbers of moles of acid and base are added in each case, leaving only the salt of the acid and base. The reaction (if any) of that salt with water (hydrolysis) will affect the pH.

pH of solution	Reacting Species	Reaction controlling pH
(a) >7	CH_3CO_2H/KOH	Hydrolysis of $CH_3CO_2^-$
(b) <7	HCl/NH_3	Hydrolysis of NH_4^+
(c) $=7$	$HNO_3/NaOH$	No hydrolysis

Polyprotic Acids and Bases

65. (a) pH of 0.45 M H_2SO_3: The equilibria for the diprotic acid are:

$$K_{a1} = \dfrac{[HSO_3^-][H_3O^+]}{[H_2SO_3]} = 1.2 \times 10^{-2} \text{ and } K_{a2} = \dfrac{[SO_3^{2-}][H_3O^+]}{[HSO_3^-]} = 6.2 \times 10^{-8}$$

For the first step of dissociation:

	H_2SO_3	HSO_3^-	H_3O^+
Initial concentration	0.45 M	0	0
Change	-x	+x	+x
Equilibrium	0.45 - x	+x	+x

Substituting into the K_{a1} expression: $K_{a1} = \dfrac{(x)(x)}{(0.45-x)} = 1.2 \times 10^{-2}$

We must solve this expression with the quadratic equation since $(0.45 < 100 \cdot K_{a1})$.

The equilibrium concentrations for HSO_3^- and H_3O^+ ions are found to be 0.0677 M.

The further dissociation is indicated by K_{a2}.

Using the equilibrium concentrations from the first step, substitute into the K_{a2} expression.

	HSO_3^-	SO_3^{2-}	H_3O^+
Initial concentration	0.0677	0	0.0677
Change	-x	+x	+x
Equilibrium	0.0677 - x	+x	0.0677 + x

$$K_{a2} = \frac{[SO_3^{2-}][H_3O^+]}{[HSO_3^-]} = \frac{(+x)(0.0677 + x)}{(0.0677 - x)} = 6.2 \times 10^{-8}$$

We note that x will be small in comparison to 0.0677, and we simplify the expression:

$$K_{a2} = \frac{(+x)(0.0677)}{(0.0677)} = 6.2 \times 10^{-8}$$

In summary, the concentrations of HSO_3^- and H_3O^+ ions have been virtually unaffected

by the second dissociation. So $[H_3O^+] = 0.0677$ M and pH = 1.17

(b) The equilibrium concentration of SO_3^{2-} :

From the K_{a2} expression above: $[SO_3^{2-}] = 6.2 \times 10^{-8}$ M

67. (a) Concentrations of OH^-, $N_2H_5^+$, and $N_2H_6^{2+}$ in 0.010 M N_2H_4:

The K_{b1} equilibrium allows us to calculate $N_2H_5^+$ and OH^- formed by the reaction of

N_2H_4 with H_2O. $K_{b1} = \dfrac{[N_2H_{5^+}^+][OH^-]}{[N_2H_4]} = 8.5 \times 10^{-7}$

	N_2H_4	$N_2H_5^+$	OH^-
Initial concentration	0.010	0	0
Change	-x	+x	+x
Equilibrium	0.010 - x	+x	+x

Substituting into the K_{b1} expression: $\dfrac{(x)(x)}{0.010 - x} = 8.5 \times 10^{-7}$

We can simplify the denominator $(0.010 > 100 \cdot K_{b1})$.

$\dfrac{(x)(x)}{0.010} = 8.5 \times 10^{-7}$ and $x = 9.2 \times 10^{-5}$ M $= [N_2H_5^+] = [OH^-]$

The second equilibrium (K_{b2}) indicates further reaction of the $N_2H_5^+$ ion with water.

The step should consume some $N_2H_5^+$ and produce more OH^-.

The magnitude of K_{b2} indicates that the equilibrium "lies to the left" and we anticipate that not much $N_2H_6^{2+}$ (or additional OH^-) will be formed by this interaction.

	$N_2H_5^+$	$N_2H_6^{2+}$	OH^-
Initial concentration	9.2×10^{-5}	0	9.2×10^{-5}
Change	$-x$	$+x$	$+x$
Equilibrium	$9.2 \times 10^{-5} - x$	$+x$	$9.2 \times 10^{-5} + x$

$$K_{b2} = \frac{[N_2H_6^{2+}][OH^-]}{[N_2H_5^+]} = 8.9 \times 10^{-16} = \frac{x \cdot (9.2 \times 10^{-5} + x)}{(9.2 \times 10^{-5} - x)}$$

Simplifying yields $\dfrac{x(9.2 \times 10^{-5})}{(9.2 \times 10^{-5})} = 8.9 \times 10^{-16}$; $x = 8.9 \times 10^{-16}\ M = [N_2H_6^{2+}]$

In summary, the second stage produces a negligible amount of OH^- and consumes very little $N_2H_5^+$ ion. The equilibrium concentrations are:

$[N_2H_5^+]$: $9.2 \times 10^{-5}\ M$; $[N_2H_6^{2+}]$: $8.9 \times 10^{-16}\ M$; $[OH^-]$: $9.2 \times 10^{-5}\ M$

(b) The pH of the 0.010 M solution: $[OH^-] = 9.2 \times 10^{-5}\ M$ so pOH = 4.04
   and pH = 14.0 - 4.04 = 9.96

Molecular Structure, Bonding, and Acid-Base Behavior

69. HOCN will be the stronger acid. In HOCN the proton is attached to the very electronegative O atom. This great electronegativity will provide a very polar bond, weakening the O-H bond, and making the H more acidic than in HCN.

71. Benzenesulfonic acid is a Brönsted acid owing to the inductive effect of three oxygen atoms attached to the S (and through the S to the benzene ring). These very electronegative O atoms will—through the inductive effect—remove electron density between the O and the H—weakening the OH bond, and making the H acidic.

Lewis Acid and Bases

73. (a) H_2NOH electron rich (accepts H^+)   Lewis base

    (b) Fe^{2+} electron poor   Lewis acid

    (c) CH_3NH_2 electron rich (accepts H^+)   Lewis base

75. CO is a Lewis base (donates electron pairs) in complexes with nickel and iron.

GENERAL QUESTIONS ON ACIDS AND BASES

77. For the equilibrium: $HC_9H_7O_4$ (aq) + H_2O (l) $\Leftrightarrow$ $C_9H_7O_4^-$ (aq) + H_3O^+ (aq)

we can write the K_a expression:

$$K_a = \frac{[C_9H_7O_4^-][H_3O^+]}{[HC_9H_7O_4]} = 3.27 \times 10^{-4}$$

The initial concentration of aspirin is:

$$2 \text{ tablets} \cdot \frac{0.325 \text{ g}}{1 \text{ tablet}} \cdot \frac{1 \text{ mol } HC_9H_7O_4}{180.2 \text{ g } HC_9H_7O_4} \cdot \frac{1}{0.225 \text{ L}} = 1.60 \times 10^{-2} \text{ M } HC_9H_7O_4$$

Substituting into the K_a expression:

$$K_a = \frac{[H_3O^+]^2}{1.60 \times 10^{-2} - x} = 3.27 \times 10^{-4} = \frac{x^2}{1.60 \times 10^{-2} - x}$$

Since $100 \cdot K_a \approx$ [aspirin], the quadratic equation will provide a "good" value.

Using the quadratic equation, $[H_3O^+] = 2.13 \times 10^{-3}$ M and pH = 2.671.

79. Calculate the molar mass of each base: $Ba(OH)_2$ = 171.3 g/mol and $Sr(OH)_2$ = 121.6 g/mol.

The pH gives us information about the $[OH^-]$, since pH + pOH = 14.00.

With pH= 12.61, pOH = 14.00-12.61 = 1.39 and [OH-] = $10^{-1.39}$ or 0.0407M (0.041 to 2 sf).

Note that two bases, as they totally dissolve, two moles of hydroxide ion per mol of the base.

So we can calculate the # of moles of base by dividing the hydroxide ion concentration by 2.
[Base] = 0.0407M/2 = 0.0204 or 0.021M (to 2sf) .

Recall that 2.50g of the solid sample provided this concentration of base, so we can calculate

the molar mass of the base: $\frac{2.50 \text{ g base}}{0.0204 \text{ mol base}}$ = 120 g/mol (2 sf). Comparing the molar

masses of the two possible bases, this solid sample is most likely $Sr(OH)_2$.

81. The reaction between H_2S and $NaCH_3CO_2$:

H_2S (aq) + $CH_3CO_2^-$ (aq) $\Leftrightarrow$ HS^- (aq) + CH_3CO_2H (aq)

An examination of Table 17.3 reveals that CH_3CO_2H is a stronger acid than H_2S, so the

equilibrium will lie *to the left (reactants)*.

83. Monoprotic acid has $K_a = 1.3 \times 10^{-3}$. Equilibrium concentrations of HX, H_3O^+ and pH for

0.010 M solution of HX:

	HX	H$^+$	X$^-$
Initial concentration	0.010	0	0
Change	-x	+x	+x
Equilibrium	0.010 - x	+x	+ x

$$K_a = \frac{[H_3O^+][A]}{[HA]} = \frac{(x)^2}{0.010 - x} = 1.3 \times 10^{-3}$$

Since Ka and the concentration are of the same order of magnitude, the quadratic

equation will help: $x^2 + 1.3 \times 10^{-3}x - 1.3 \times 10^{-5} = 0$.

The "reasonable solution" is $x = 3.0 \times 10^{-3}$

Concentrations are then: $[HX] = 0.010 - x = 0.010-0.0030 = 7.0 \times 10^{-3}$ M

$$[H^+] = +x = 3.0 \times 10^{-3} \text{ M} \quad [X^-] = +x = 3.0 \times 10^{-3} \text{M}$$

$$pH = -\log(3.0 \times 10^{-3}) = 2.52$$

85. The pKa of m-nitrophenol:

The Ka expression for this weak acid is: $K_a = \dfrac{[H^+][A^-]}{[HA]}$

The pH of a 0.010 M solution of nitrophenol is 3.44, so we know: $[H^+] = 3.63 \times 10^{-4}$

Since we get one " A^- " for each " H^+ ", the concentrations of the two are equal.

The $[HA] = (0.010 - 3.63 \times 10^{-4})$. Substituting those values into the K_a expression yields:

$$K_a = \frac{[H^+][A^-]}{[HA]} = \frac{[3.63 \times 10^{-4}][3.63 \times 10^{-4}]}{[0.010 - 3.63 \times 10^{-4}]} = 1.4 \times 10^{-5} \text{ (to 2sf)}$$

$$pK_a = -\log(1.4 \times 10^{-5}) = 4.86$$

87. For Novocain, the $pK_a = 8.85$. The pH of a 0.0015M solution is:

First calculate K_a: $K_a = 10^{-8.85}$ so $K_a = 1.41 \times 10^{-9}$

Now treat Novocain as any weak acid, with the appropriate K_a expression:

$$K_a = \frac{[H^+][A^-]}{[HA]} = \frac{[x][x]}{[0.0015 - x]} = 1.4 \times 10^{-9}$$

Since Ka is so small, we can assume that the denominator is approximated by 0.0015 M.

$$\frac{[x]^2}{0.0015} = 1.4 \times 10^{-9} \text{ and } x^2 = 1.45 \times 10^{-6}$$

Since $[H^+] = 1.45 \times 10^{-6}$; pH = -log (1.45×10^{-6}) = 5.84

89. Regarding ethylamine and ethanolamine:

(a) Since ethylamine has the larger K_b, ethylamine is the stronger base.

(b) The pH of 0.10 M solution of ethylamine:

$$C_2H_5NH_2 + H_2O \Leftrightarrow C_2H_5NH_3^+ + OH^- \qquad K_b = 4.3 \times 10^{-4}$$

Initial	0.10 M		0	0
Change	- x		+ x	+ x
Equilibrium	0.10 - x		+ x	+ x

Substituting into the K_b expression:

$$K_b = \frac{[NH_4^+][OH^-]}{[NH_3]} = \frac{x^2}{(0.10-x)} = 4.3 \times 10^{-4}$$

So x = 6.56×10^{-3} so pOH = -log(6.56×10^{-3}) = 2.18 and pH = 14.00-2.18 = 11.82

91. With a pKa = 2.32, the Ka for saccharin is $10^{-2.32} = 4.8 \times 10^{-3}$

As we are dealing with the conjugate base of saccharin, $C_7H_4NO_3S^-$, we need to use the Kb of the conjugate base. Recalling that $K_a \cdot K_b = K_w$, $K_b = K_w/K_a$ or $1.0 \times 10^{-14}/4.8 \times 10^{-3}$ or 2.1×10^{-12}. The equation for the K_b is:

$$C_7H_4NO_3S^- + H_2O \Leftrightarrow HC_7H_4NO_3S + OH^-$$

The equilibrium constant expression would be:

$$\frac{[HC_7H_4NO_3S][OH^-]}{[C_7H_4NO_3S^-]} = 2.1 \times 10^{-12}.$$

If we let the $[OH^-]$ and $[C_7H_4NO_3S^-] = x$, the equilibrium concentration of the molecular acid would be 0.10 – x. Substituting into the expression above:

$$\frac{[x][x]}{[0.10 - x]} = 2.1 \times 10^{-12} \cong \frac{x^2}{0.10}; \; x^2 = 2.1 \times 10^{-13} \text{ and } x \text{ (or } [OH^-]) = 4.6 \times 10^{-7} \text{ so}$$

$pOH = -\log[4.6 \times 10^{-7}]$ or 6.34 and $pH = 14.00 - 6.34 = 7.66$.

93. pH of aqueous solutions of

		reaction	pH
(a)	$NaHSO_4$	hydrolysis of HSO_4^- produces H_3O^+	< 7
(b)	NH_4Br	hydrolysis of NH_4^+ produces H_3O^+	< 7
(c)	$KClO_4$	no hydrolysis occurs	$= 7$
(d)	Na_2CO_3	hydrolysis of CO_3^{2-} produces OH^-	> 7
(e)	$(NH_4)_2S$	hydrolysis of S^{2-} produces OH^-	> 7
(f)	$NaNO_3$	no hydrolysis occurs	$= 7$
(g)	Na_2HPO_4	hydrolysis of HPO_4^{2-} produces OH^-	> 7
(h)	$LiBr$	no hydrolysis occurs	$= 7$
(i)	$FeCl_3$	hydrolysis of Fe^{3+} produces H_3O^+	< 7

95. For oxalic acid $K_{a1} = 5.9 \times 10^{-2}$ and $K_{a2} = 6.4 \times 10^{-5}$

Representing oxalic as H_2A. The first step can be written as:

$$H_2A + H_2O \Leftrightarrow HA^- + H_3O^+ \qquad K_{a1} = 5.9 \times 10^{-2}$$

We can write the second step:

$$HA^- + H_2O \Leftrightarrow A^{2-} + H_3O^+ \qquad K_{a2} = 6.4 \times 10^{-5}$$

If we add the two equations we get:

$$H_2A + 2 H_2O \Leftrightarrow A^{2-} + 2 H_3O^+ \quad K_{net} = (5.9 \times 10^{-2})(6.4 \times 10^{-5})$$

$$K_{net} = 3.8 \times 10^{-6}$$

97. Confirm the fact that 1.8×10^{10} is the value of the equilibrium constant for the reaction of formic acid and sodium hydroxide :

To do this, we need two equations that have a **net** equation that corresponds to:

$$HCO_2H(aq) + OH^-(aq) \Leftrightarrow HCO_2^-(aq) + H_2O \, (l)$$

Begin with equilibria associated with the weak acid in water, and the Kw for water:

$HCO_2H(aq) + H_2O(l) \Leftrightarrow HCO_2^-(aq) + H_3O^+(aq)$ $K_a = 1.8 \times 10^{-4}$

$H_3O^+(aq) + OH^-(aq) \Leftrightarrow 2 H_2O(l)$ $1/K_W = 1.0 \times 10^{14}$

The net equation is: $HCO_2H(aq) + OH^-(aq) \Leftrightarrow HCO_2^-(aq) + H_2O(l)$ and

the net K is: $K_a \cdot 1/K_W = (1.8 \times 10^{-4})(1.0 \times 10^{14}) = 1.8 \times 10^{10}$

99. Volume to which 1.00×10^2 mL of a 0.20 M solution of a weak acid, HA, should be diluted
to result in a doubling of the percent ionization:

Begin by writing the Ka expression: $K = \dfrac{[H^+][A^-]}{[HA]}$ (1)

The fraction of dissociation (which we want to double) is $\dfrac{[A^-]}{[HA]}$ or equivalently: $\dfrac{[H^+]}{[HA]}$ (2)

Knowing that for a monoprotic acid, $[H^+] = [A^-]$, we can rearrange equation (1) to yield:

$[H^+]^2 = Ka \cdot [HA-H^+]$, where $[HA-H^+]$ represents the concentration of molecular acid.

Expanding gives: $[H^+]^2 = Ka \cdot [HA] - Ka \cdot [H^+]$, and setting this up as a quadratic equation

gives: $[H^+]^2 + Ka \cdot [H^+] - Ka \cdot [HA] = 0$.

For the sake of argument, let's pick a value for Ka, say 1.0×10^{-5}.

Given the original concentration of acid is 0.2 M, let's substitute the values into the equation.

$1 \cdot [H^+]^2 + (1.0 \times 10^{-5}) \cdot [H^+] - (1.0 \times 10^{-5}) \cdot [0.2] = 0$, and solve the equation--call it (3).

This gives $[H^+] = 1.41 \times 10^{-3}$ M. The fraction of dissociation is (from equation (2) above)

$\dfrac{[H^+]}{[HA]}$ or 0.00705. Now we want this **fraction to double**. So substitute differing

concentrations of acid (increasingly dilute) into equation (3), solve the quadratic equation

that results, and calculate the fraction dissociated.

The table following shows such a series of calculations:

Acid conc	fraction	Relative dissociation
0.20	0.00704611	1.0
0.15	0.0081317	1.2
0.10	0.00995012	1.4
0.05	0.01404249	2.0

Note that when the acid concentration falls to 1/4 of the original value, the relative dissociation has doubled. So 100 mL of the acid would need to be diluted to 400 mL to double the fraction (or percent) of dissociation.

101. Oxalic acid, $H_2C_2O_4$, is diprotic, with K values: $Ka_1 = 5.9 \times 10^{-2}$ and $Ka_2 = 6.4 \times 10^{-5}$.

The K_{a1} equilibrium allows us to calculate $HC_2O_4^-$ and H_3O^+ formed by the reaction of

$H_2C_2O_4$ with H_2O. $K_{a1} = \dfrac{\left[HC_2O_4^-\right]\left[H_3O^+\right]}{\left[H_2C_2O_4\right]} = 5.9 \times 10^{-2}$

	$H_2C_2O_4$	$HC_2O_4^-$	H_3O^+
Initial concentration	0.10	0	
Change	-x	+x	+x
Equilibrium	0.10 - x	+x	+x

Substituting into the K_{a1} expression we obtain: $\dfrac{(x)(x)}{0.10 - x} = 5.9 \times 10^{-2}$

We cannot simplify the denominator, since $(0.10 < 100 \cdot K_{a1})$.

Solving via the quadratic equation, we obtain: $x = 9.8 \times 10^{-3}$.

$[H_2C_2O_4] = 0.10 - x = 9.0 \times 10^{-2}M$; $[HC_2O_4^-] = 9.8 \times 10^{-3}M$ & $[H_3O^+] = 9.8 \times 10^{-3}M$

Very little is formed as a result of the second ionization, so $[H_3O^+] = [HC_2O_4^-]$ and

$[C_2O_4^{2-}] = K_{a2} = 6.4 \times 10^{-5}$. The major species present, in decreasing concentration are:

$$H_2O > H_2C_2O_4 > H_3O^+ = HC_2O_4^- > C_2O_4^{2-} > OH^-$$

103. To determine the relative basic strengths of the substances, one can measure the pH of aqueous solutions of known concentration (e.g. 0.50M) of each solute. Remember the equilibrium associated with bases in water:

$$B + H_2O \Leftrightarrow BH^+ + OH^- \qquad\qquad K_b = y.y \times 10^{-z}$$

The strongest base will have an equilibrium lying farther "to the right" than the weaker bases-resulting in a solution with the *greatest* concentration of hydroxyl ions (and the greatest pH). The weakest base will obviously result in a solution with the *lowest* concentration of hydroxyl ions (and the lowest pH).

105. This puzzle is a great way to test your chemical knowledge. Organize what we know.

Cations	Anions
Na^+	Cl^-
NH_4^+	OH^-
H^+	

Experimentally we observe:

B + Y → acidic solution   B + Z → basic solution   A + Z   → neutral solution

If A + Z give a neutral solution, then A + Z must be acid and base (in some order).

Since B + Z gives a basic solution, B + Z cannot be acid and base (as A + Z), otherwise B + Z would be neutral (as A + Z). Z must be basic (OH^-), meaning Y must be neutral (Cl^-).

Since B + Y gives an acid solution, then B must contain NH_4^+. If B contains ammonium, then A must contain H^+, and C must contain Na^+.

In summary then: $A = H^+$; $B = NH_4^+$; $C = Na^+$; $Y = Cl^-$; $Z = OH^-$.

Pairing these cations with chloride and potassium ions, we have:

$A = HCl$; $B = NH_4Cl$; $C = NaCl$; $Y = KCl$; $Z = KOH$. The pH = 14.00-6.03 or 7.97.

SUMMARY AND CONCEPTUAL QUESTIONS:

107. The Ka expression for nicotinic acid in water is:

$$C_6H_5NO_2 + H_2O \Leftrightarrow C_6H_4NO_2^- + H_3O^+$$

The pH allows us to determine the $[H_3O^+]$; pH = $10^{-2.70}$ = 2.0 x 10^{-3}M. As this is a monoprotic acid, we also know that, at equilibrium the $[C_6H_4NO_2^-]$ = 2.0 x 10^{-3}M. The molecular acid concentration is then: Initial = 2.0 x 10^{-3}M.

Calculating the initial concentration:

For $C_6H_5NO_2$ the molar mass = 123.1 g, so

$$[C_6H_5NO_2] = \frac{1.00 \text{ g}}{0.060 \text{ L}} \cdot \frac{1 \text{ mol}}{123.1 \text{ g}} = 0.14 \text{ M (2 sf)}$$

Substituting into the Ka expression, we obtain:

$$\frac{\left[C_6H_4NO_2^-\right]\left[H_3O^+\right]}{\left[C_6H_5NO_2\right]} = \frac{\left(2.0 \text{ x } 10^{-3}\right)^2}{\left(0.14 - 2.0 \text{ x } 10^{-3}\right)} = 3.0 \text{ x } 10^{-5}$$

109. (a) Aniline can serve as **both** a Brönsted base **and** a Lewis base:

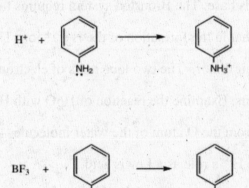

(b) The pH of a solution of 1.25 g sodium sulfanilate in 125 mL: Sulfanilic acid is a weak

acid, so its conjugate base should provide a slightly basic solution.

The pK_a = 3.23 so K_a = $10^{-3.23}$ = 5.9 x 10^{-4}.

The formula for sulfanilic acid is complex ($H_2NC_6H_4SO_3H$), so let's use HSA to

represent the acid, and SA to represent the anion.

The concentration of the sodium salt (represented as NaSA) will be necessary:

$$\frac{1.25 \text{ g}}{0.125 \text{ L}} \cdot \frac{1 \text{ mol NaSA}}{195.15 \text{ g NaSA}} = 0.0512 \text{ M NaSA}$$

In water, the NaSA salt will exist predominantly as sodium cations and SA anions, so we can ignore the sodium cations.

The reaction that occurs with the conjugate base and water is: $SA + H_2O \Leftrightarrow HSA + OH^-$

(Note the production of hydroxyl ion –the reason we expect the solution to be basic.)

As the anion is acting as a base, we'll need to calculate an appropriate K for the anion.

K_b (conjugate) $= K_w/K_a$ so $\dfrac{1.0 \times 10^{-14}}{5.9 \times 10^{-4}} = 1.7 \times 10^{-11}$

The K_b expression is: $\dfrac{[HSA][OH^-]}{[SA]} = \dfrac{[x][x]}{0.0512 - x} = 1.7 \times 10^{-11}$

The size of K ($100 \cdot K < 0.0512$) tells us that we can safely approximate the denominator of the fraction as 0.0512. The resulting equation is: $x^2 = (0.0512)(1.7 \times 10^{-11})$ and

$x = 9.33 \times 10^{-7}$ Since x represents the concentration of OH^- ion,

the pOH $= -\log(9.33 \times 10^{-7}) = 6.03$, and the pH $= 7.97$.

111. Water can be both a Brönsted base and a Lewis base. The Brönsted system requires that a base be able to accept a H^+ ion. Water does that in the formation of the H_3O^+ ion. The Lewis system defines a base as an electron pair donor. The two lone pairs of electrons on the O atom provide a source for those electrons. Examine the reaction of H_2O with H^+. The hydrogen ion accepts the electron pairs from the O atom of the water molecule— fulfilling water's role as a Lewis base and the H^+'s role as a Lewis acid.

A Brönsted acid furnishes H^+ to another specie. The autoionization of water provides just one example of water acting as a Brönsted acid. Water however cannot function as a Lewis acid, as it has no capacity to accept an electron pair.

113. (a) The strongest acid of the three oxyacids is HOCl, with the pKa = 7.46. Recall that the magnitude of the Ka indicates the degree of ionization for the acid. The greater the ionization, the greater the Ka, and the smaller the pKa.

(b)The change in acid strength is understandable if you recall that in changing from I to Br to Cl, the electronegativity of the halogen increases, strengthening the O-X bond (where

X represents the halogen) and reducing the electron density (and weakening) the H-O bond, making the acid stronger.

115.(a) The reaction of perchloric acid with sulfuric acid:

$HClO_4 + H_2SO_4 \Leftrightarrow ClO_4^- + H_3SO_4^+$

(b) Lewis dot structure for sulfuric acid:

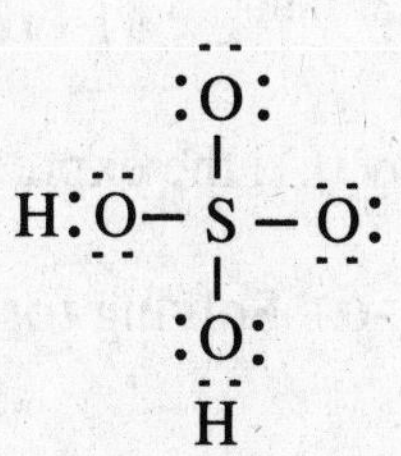

Sulfuric acid has two oxygen atoms capable of "donating" an electron pair, e.g. to a proton, H^+, thereby acting as a base.

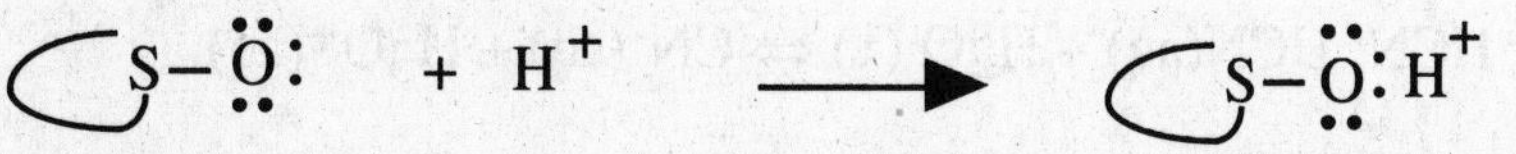

117. (a) Electron dot structure for I_3^- :

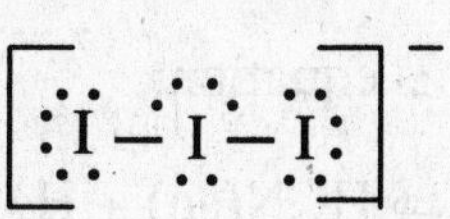

(b) $I^- + I_2 \longrightarrow I_3^-$ in which I^- donates the electron pair (Lewis base), and I_2 accepts the electron pair (Lewis acid).

119. (a) If the *degree of ionization* (α) is viewed as the part of the molecular specie (e.g. acid) that dissociates (or ionizes), then ($1-\alpha$) is the part of the molecular specie remaining at equilibrium. At equilibrium, for the hypothetical acid, HA, then concentrations are:

$[H^+] = \alpha Co$ $[A^-] = \alpha Co$ $[HA] = (1-\alpha)Co$

Substituting into an equilibrium constant expression:

$$K = \frac{[H^+][A^-]}{[HA]} = \frac{[\alpha Co][\alpha Co]}{[(1-\alpha)Co]} \text{ and cancelling a Co term} = \frac{[\alpha Co][\alpha]}{[(1-\alpha)]} = \frac{[\alpha^2 Co]}{[(1-\alpha)]}$$

(b) The degree of ionization for ammonium ion in 0.10 M NH_4Cl:

The Ka expression for the ammonium ion is:

$$Ka_{(NH_4^+)} = \frac{[H^+][NH_3]}{[NH_4^+]} \text{ or } \frac{1.0 \times 10^{-14}}{1.0 \times 10^{-5}} = 5.6 \times 10^{-10}$$

Using our equation: $Ka_{(NH_4^+)} = \frac{\alpha^2[0.1]}{(1-\alpha)} = 5.6 \times 10^{-10}$

Dividing both sides by 0.1, the expression becomes: $\frac{\alpha^2}{(1-\alpha)} = 5.6 \times 10^{-11}$

and $\alpha^2 = 5.6 \times 10^{-11}(1-\alpha)$. Solving for α using the quadratic equation, $\alpha = 7.5 \times 10^{-5}$.

121. (a) Determine the form for the equilibrium constant for the reaction:

$$NH_4^+ (aq) + CN^-(aq) \Leftrightarrow NH_3(aq) + HCN (aq)$$

Ka for HCN: $HCN(aq) + H_2O (l) \Leftrightarrow CN^-(aq) + H_3O^+(aq)$.

Noting that the HCN and its conjugate are on the opposite sides for our desired equation, we "swap sides" for this equation:

$CN^-(aq) + H_3O^+(aq) \Leftrightarrow HCN(aq) + H_2O (l)$ and K = 1/Ka

Similarly we can write the K_b expression for NH_3, but we note that NH_3 is on the "right side" of our desired expression, so let's write the "reversed equation", for which K would be 1/Kb: $NH_4^+(aq) + OH^-(aq) \Leftrightarrow NH_3(aq) + H_2O(l)$.

Finally note that the components of water, $H_2O(l)$ and $OH^-(aq)$ and $H_3O^+(aq)$, are missing from our desired equation: Recall that equilibrium may be written:

$$2 H_2O(l) \Leftrightarrow H_3O^+(aq) + OH^-(aq) \text{ with the K = Kw.}$$

Summarizing:

$CN^-(aq) + H_3O^+(aq) \Leftrightarrow HCN(aq) + H_2O (l)$	K = 1/Ka
$NH_4^+(aq) + OH^-(aq) \Leftrightarrow NH_3(aq) + H_2O(l)$	K = 1/Kb
$2 H_2O(l) \Leftrightarrow H_3O^+(aq) + OH^-(aq)$	K = Kw
$NH_4^+ (aq) + CN^-(aq) \Leftrightarrow NH_3(aq) + HCN (aq)$	$K_{net} = Kw \cdot 1/Kb \cdot 1/Ka$

(b) Calculate Knet values for each of the following: NH_4CN, $NH_4CH_3CO_2$, and NH_4F.

Which salt has the largest value of Knet and why?

$K_{net} = K_w \cdot 1/K_b \cdot 1/K_a$

For NH_4CN: $K_{net} = 1.0 \times 10^{-14} \cdot (1/4.0 \times 10^{-10}) \cdot (1/1.8 \times 10^{-5}) = 1.4$

For $NH_4CH_3CO_2$: $K_{net} = 1.0 \times 10^{-14} \cdot (1/1.8 \times 10^{-5}) \cdot (1/1.8 \times 10^{-5}) = 3.1 \times 10^{-5}$

For NH_4F: $K_{net} = 1.0 \times 10^{-14} \cdot (1/7.2 \times 10^{-4}) \cdot (1/1.8 \times 10^{-5}) = 7.7 \times 10^{-7}$

The base, cyanide, is the strongest of the three bases, and therefore the base most capable of extracting the hydrogen ion, producing the greatest amount of product—and the largest K_{net}.

(c) No calculation is really needed. If the K's of the respective conjugate acid and base are equal, each reacts to an equivalent degree, producing a neutral solution—as is the case for $NH_4C_2H_3O_2$. If the Kb for the base is greater than the Ka for the acid—as is the case for NH_4CN, the solution will be basic, while an acidic solution will result if the Ka > Kb.

Applying Chemical Principles

1. Convert the pK values to K values for the dissociations of HCl, $HClO_4$, and H_2SO_4 in glacial acetic acid. Rank these acids in order from strongest to weakest.

pK for HCl =8.8   so K = $10^{-8.8}$ and K = 2 x 10^{-9}

pK for $HClO_4$ = 5.3 so K = $10^{-5.3}$ and K = 5 x 10^{-6}

pK for H_2SO_4 = 6.8 so K = $10^{-6.8}$ and K= 1.6 x 10^{-7}

Ranking: $HClO_4$ > H_2SO_4 > HCl

3. Write an equation for the reaction of the amide ion (a stronger base than OH^-) and water. Does the equilibrium favor products or reactants?

NH_2^- (aq) + H_2O (l) $\Leftrightarrow$ NH_3 (aq) + OH^-(aq); the stronger base extracts H^+ from H_2O.

As the amide ion is a much stronger base than the hydroxide ion, the equilibrium favors the products.

5. To measure the relative strengths of bases stronger than OH^-, it is necessary to choose a solvent that is a weaker acid than water. One such solvent is liquid ammonia.

(a) Write a chemical equation for the autoionization of ammonia.

NH_3 (aq) + NH_3 (aq) $\Leftrightarrow$ NH_2^- (aq) + NH_4^+ (aq)

(b) What is the strongest acid and base that can exist in liquid ammonia?

The strongest acid that can exist will be NH_4^+ ; the strongest base will be NH_2^-.

(c) Will a solution of HCl in liquid ammonia be a strong electrical conductor, a weak conductor, or a nonconductor?

The HCl will react with the NH_4^+ ions to form the soluble salt, NH_4Cl, and make a **strongly conducting** solution.

(d) Oxide ion (O^{2-}) is a stronger base than the amide ion (NH_2^-).

Write an equation for the reaction of O^{2-} with NH_3 in liquid ammonia.

O^{2-} (aq) + NH_3 (aq)$\Leftrightarrow$ NH_2^- (aq) + OH^-(aq)

Since we have a stronger base (O^{2-}) forming a weaker base (NH_2^-), the equilibrium will favor the products.

Chapter 18
Principles of Reactivity: Other Aspects of Aqueous Equilibria

PRACTICING SKILLS
The Common Ion Effect and Buffer Solutions
1. To determine how pH is expected to change, examine the equilibria in each case:

(a) NH_3 (aq) + H_2O (l) $\Leftrightarrow$ NH_4^+ (aq) + OH^- (aq)

As the added NH_4Cl dissolves, ammonium ions are liberated—increasing the ammonium

ion concentration and shifting the position of equilibrium to the left—reducing OH^-, and

decreasing the pH.

(b) CH_3CO_2H (aq) + H_2O(l) $\Leftrightarrow$ $CH_3CO_2^-$ (aq) + H_3O^+(aq)

As sodium acetate dissolves, the additional acetate ion will shift the position of

equilibrium to the left—reducing H_3O^+, and *increasing* the pH.

(c) $NaOH$ (aq) $\rightarrow$ Na^+ (aq) + OH^- (aq)

NaOH is a strong base, and as such is totally dissociated. Since the added NaCl does not

hydrolyze to any appreciable extent—*no change* in pH occurs.

3. The pH of the buffer solution is:

$$K_b = \frac{[NH_4^+][OH^-]}{[NH_3]} = \frac{(0.20)[OH^-]}{(0.20)} = 1.8 \times 10^{-5}$$

solving for hydroxyl ion yields: $[OH^-] = 1.8 \times 10^{-5}$ M ; pOH = 4.75   pH = 9.25

5. pH of solution when 30.0 mL of 0.015 M KOH is added to 50.0 mL of 0.015 M benzoic acid:

The equilibrium affected is that of benzoic acid in water:

$C_6H_5CO_2H$ (aq) + H_2O (l) $\Leftrightarrow$ $C_6H_5CO_2^-$ (aq) + H_3O^+ (aq) $K_a = 6.3 \times 10^{-5}$

$$K_a = \frac{[C_6H_5CO_2^-][H_3O^+]}{[C_6H_5CO_2H]} = 6.3 \times 10^{-5}$$

The KOH will *consume benzoic acid* and *produce* the conjugate *benzoate anion*.

The base and benzoic acid react:

KOH(aq) + $C_6H_5CO_2H$(aq) $\rightarrow$ $C_6H_5CO_2^-K^+$(aq)+ H_2O(l)

The pH of the solution will depend on the ratio of the conjugate pairs (acid and anion).

Determine the concentrations of the two:

Amount of benzoic acid present: 50 mL • 0.015 mol/L = 0.75 mmol of benzoic acid.

Amount of KOH added: 30 mL • 0.015 mol/L = 0.45 mmol of KOH

Amount of benzoic acid remaining: (0.75 mmol - 0.45 mmol of KOH)= 0.30 mmol benzoic acid

Amount of benzoate anion produced: *0.45 mmol of benzoate anion.*

We can rearrange the K_a expression to solve for H_3O^+:

$$[H_3O^+] = 6.3 \times 10^{-5} \cdot \frac{[C_6H_5CO_2H]}{[C_6H_5CO_2^-]} = 6.3 \times 10^{-5} \cdot \frac{0.30 \text{ mmol}}{0.45 \text{ mmol}} = 4.2 \times 10^{-5}$$

$$pH = -\log(4.2 \times 10^{-5}) = 4.38$$

7. The original pH of the 0.12M NH_3 solution will be:

$$\frac{[NH_4^+][OH^-]}{[NH_3]} = 1.8 \times 10^{-5} \text{ and } \frac{x^2}{0.12 - x} = 1.8 \times 10^{-5}$$

Assuming $(0.12 - x \approx 0.12)$, $x = 1.5 \times 10^{-3}$

$[OH^-] = 1.5 \times 10^{-3}$ and $pOH = 2.83$ with $pH = 11.17$

Adding 2.2 g of NH_4Cl (0.041 mol) to 250 mL will produce an immediate increase of

0.16 M NH_4^+ (0.041 mol/0.250 L).

Substituting into the equilibrium expression as we did earlier, we get:

$$\frac{(x + 0.16)(x)}{0.12 - x} = 1.8 \times 10^{-5}$$

Assuming ($x + 0.16 \approx 0.16$) and $(0.12 - x \approx 0.12)$

$$\frac{0.16(x)}{0.12} = 1.8 \times 10^{-5} \quad x = 1.35 \times 10^{-5} \text{ M} = [OH^-]$$

Note the hundred-fold decrease in $[OH^-]$ over the initial ammonia solution, as predicted by LeChatelier's principle. So $pOH = 4.88$ and $pH = 9.12$ (lower than original).

9. Mass of sodium acetate needed to change 1.00 L solution of 0.10 M CH3CO2H to

pH = 4.50: The equilibrium affected is that of acetic acid in water:

$$CH_3CO_2H + H_2O \Leftrightarrow CH_3CO_2^- + H_3O^+ \quad K_a = 1.8 \times 10^{-5}$$

The equilibrium expression is: $K_a = \dfrac{[CH_3CO_2^-][H_3O^+]}{[CH_3CO_2H]} = 1.8 \times 10^{-5}$

We know the concentration of acetic acid (0.10M), and we know the desired $[H_3O^+]$:

pH = 4.50 so $[H_3O^+] = 10^{-4.50} = 3.2 \times 10^{-5}$

Substituting these values into the equilibrium expression gives:

$$\dfrac{[CH_3CO_2^-][H_3O^+]}{[CH_3CO_2H]} = 1.8 \times 10^{-5} = \dfrac{[CH_3CO_2^-][3.2 \times 10^{-5}]}{[0.10]}$$

We can solve for $[CH_3CO_2^-] = 0.057$ M

What mass of $NaCH_3CO_2$ would give this concentration of $CH_3CO_2^-$?

$$\dfrac{0.057 \text{ mol } C_2H_3O_2^-}{1 \text{ L}} \cdot \dfrac{1 \text{ mol } NaCH_3CO_2}{1 \text{ mol } CH_3CO_2^-} \cdot \dfrac{1 \text{ L}}{1} \cdot \dfrac{82.0 \text{ g } NaCH_3CO_2}{1 \text{ mol } NaCH_3CO_2} = 4.7 \text{ g } NaCH_3CO_2$$

Using the Henderson-Hasselbalch Equation

11. The pH of a solution with 0.050 M acetic acid and 0.075 M sodium acetate:

$$pH = pK_a + \log \dfrac{[\text{conjugate base}]}{[\text{acid}]}$$

The pK_a for acetic acid is $= -\log(K_a) = -\log(1.8 \times 10^{-5})$ or 4.74

$$pH = 4.74 + \log \dfrac{[0.075]}{[0.050]} = 4.74 + 0.176 = 4.92$$

13. (a) The pK_a for formic acid is $= -\log(K_a) = -\log(1.8 \times 10^{-4})$ or 3.74

$$pH = pK_a + \log \dfrac{[\text{conjugate base}]}{[\text{acid}]}$$

$$= 3.74 + \log \dfrac{0.035}{0.050} = 3.74 - 0.15 \text{ or } 3.59$$

(b) Ratio of conjugate pairs to increase pH by 0.5 (to pH= 4.09)

Substituting into the Henderson-Hasselbalch equation

319

$$4.09 = 3.74 + \log \frac{\text{[conjugate base]}}{\text{[acid]}}$$

$$+0.345 = \log \frac{\text{[conjugate base]}}{\text{[acid]}} \quad \text{so} \quad -0.345 = \log \frac{\text{[acid]}}{\text{[conjugate base]}}$$

$$0.45 = \frac{\text{[acid]}}{\text{[conjugate base]}} \quad \text{(2sf)}$$

Preparing a Buffer Solution

15. The best combination to provide a buffer solution of pH 9 is (b) the NH_3/NH_4^+ system.

    Note that K_a (for NH_4^+) is approximately 10^{-10}. Buffer systems are good when the desired pH is ± 1 unit from pK_a (10 in this case). The HCl and NaCl don't form a buffer. The acetic acid/sodium acetate system would form an acidic buffer ($pK_a \approx 5$) in the pH range 4 - 6.

17. Preparation of a buffer of NaH_2PO_4 and Na_2HPO_4 for a pH = 7.5

    The principle species present (other than Na^+ ions) will be $H_2PO_4^-$ and HPO_4^{2-}
    So let's use the Ka2 expression for H_3PO_4.

    $$\frac{[HPO_4^{2-}][H_3O^+]}{[H_2PO_4^-]} = 6.2 \times 10^{-8}$$

    For a pH = 7.5, $[H_3O^+] = 3.2 \times 10^{-8}$ M. So the ratio of the conjugate pairs is:

    $$\frac{[HPO_4^{2-}]}{[H_2PO_4^-]} = \frac{6.2 \times 10^{-8}}{3.2 \times 10^{-8}} = 1.94 \text{ (or 2 to 1 sf)}. \text{ The reciprocal of this value is } 1/2 = 0.5.$$

    So mixing 0.5 moles of NaH_2PO_4 with 1 mol Na_2HPO_4 would provide the desired pH.

Adding an Acid or Base to a Buffer Solution

19. (a) Initial pH

    Need to know the concentrations of the conjugate pairs:

    The equilibrium expression shows the *ratio of the conjugate pairs*, we can calculate

    moles of the conjugate pairs, and know that the ratio of the # of moles of the species will

    have the same value as the ratio of their concentrations!

    $$CH_3CO_2H = 0.250 \text{ L} \cdot 0.150 \text{ M} = 0.0375 \text{ mol}$$

    $$NaCH_3CO_2 = 4.95 \text{ g} \cdot \frac{1 \text{ mol}}{82.07 \text{ g}} = 0.0603 \text{ mol}$$

    Substituting into the K_a expression:

$$\frac{[CH_3CO_2^-][H_3O^+]}{[CH_3CO_2H]} = 1.8 \times 10^{-5} = \frac{[0.0603][H_3O^+]}{[0.0375]}$$

and solving for $[H_3O^+] = 1.1 \times 10^{-5}$ M ; pH = 4.95

(b) pH after 82. mg NaOH is added to 100. mL of the buffer. The amount of the conjugate

pairs in 100/250 of the buffer is $(100/250)(0.0375 \text{ mol}) = 0.0150$ mol CH_3CO_2H

and $(100/250)(0.0603 \text{ mol}) = 0.0241$ mol $CH_3CO_2^-$

$$82 \text{ mg NaOH} \cdot \frac{1 \text{ mmol NaOH}}{40.0 \text{ mg NaOH}} = 2.05 \text{ mmol NaOH or } 0.00205 \text{ mol NaOH}$$

or 2.1 mmol NaOH (2 sf)

This base would consume an equivalent amount of CH_3CO_2H and produce an equivalent

amount of $CH_3CO_2^-$.

After that process: $(0.0150- 0.0021)$ or 0.0129 mol CH_3CO_2H and $(0.0241 + 0.0021)$ or

0.0262 mol $CH_3CO_2^-$ are present. Substituting into the K_a expression as in part (a)

$$\frac{(0.0262)[H_3O^+]}{(0.0129)} = 1.8 \times 10^{-5} \text{ so } [H_3O^+] = 8.9 \times 10^{-6} \text{ and pH} = 5.05$$

21. (a) The pH of the buffer solution is:

$$K_b = \frac{[NH_4^+][OH^-]}{[NH_3]} = \frac{(0.250)[OH^-]}{(0.500)} = 1.8 \times 10^{-5}$$

Note : Here the data presented are given as moles (in the case of ammonium chloride)

and molar concentration (in the case of ammonia). In SQ18.19 we substituted the #

moles of the conjugate pairs into the K expression. Here we must first *decide* whether to

substitute # moles or molar *concentrations* into the K_b expression. Either would work!

What is critical to remember is that we have to have both species expression in one **or** the

other form—not a mix of the two. Here I chose to convert moles of NH_4Cl into molar

concentrations, and substitute.

Solving for hydroxyl ion in the K_b expression above yields: $[OH^-] = 3.6 \times 10^{-5}$ M

and pOH = 4.45 so pH = 9.55

(b) pH after addition of 0.0100 mol HCl:

The basic component of the buffer (NH_3) will react with the HCl, producing more

ammonium ion.

The composition of the solution is:	NH_3	NH_4Cl
Moles present (before HCl added)	0.250	0.125
Change (reaction)	- 0.0100	+ 0.0100
Following reaction	0.240	+ 0.135

The amounts of NH_3 and NH_4Cl following the reaction with HCl are only slightly different from the amounts prior to reaction. Converting these numbers into molar concentrations (Volume is 500. mL) and substituting the concentrations into the K_b expression yields:

$$K_b = \frac{[NH_4^+][OH^-]}{[NH_3]} = \frac{(0.270)[OH^-]}{(0.480)} = 1.8 \times 10^{-5}$$

$[OH^-] = 3.2 \times 10^{-5}$; pOH = 4.50, and the new pH = 9.50.

More About Acid-Base Reactions: Titrations

23. We can calculate the amount of phenol present by converting mass to moles:

$$0.515 \text{ g } C_6H_5OH \cdot \frac{1 \text{mol } C_6H_5OH}{94.11 \text{ g } C_6H_5OH} = 5.47 \times 10^{-3} \text{ mol phenol}$$

(a) The pH of the solution containing 5.47×10^{-3} mol phenol in 125 mL water:

With a $K_a = 1.3 \times 10^{-10}$, the K_a expression is: $K_a = \frac{[C_6H_5O^-][H_3O^+]}{[C_6H_5OH]} = 1.3 \times 10^{-10}$

The stoichiometry of the compound indicates one phenoxide ion:one hydronium ion. The initial concentration is 5.47×10^{-3} mol/ 0.125 L = 0.0438 M

The K_a expression becomes: $\frac{[x][x]}{[0.0438 - x]} = 1.3 \times 10^{-10}$

Given the magnitude of the K_a, we can safely approximate the denominator as 0.0438M.

$x^2 = 1.3 \times 10^{-10} \cdot 0.0438$ and $x = 2.4 \times 10^{-6}$ M and pH = $-\log(2.4 \times 10^{-6})$ = 5.62.

(b) At the equivalence point 5.47×10^{-3} mol of NaOH will have been added. Phenol is a monoprotic acid, so one mol of phenol reacts with one mol of sodium hydroxide. The volume of 0.123 NaOH needed to provide this amount of base is: moles = M x V

5.47×10^{-3} mol NaOH = $\frac{0.123 \text{ mol NaOH}}{L}$ x V or 44.5 mL of the NaOH solution.

The total volume would be (125 + 44.5) or 170. mL solution.

Sodium phenoxide is a soluble salt hence the initial concentration of both sodium and phenoxide ions will be equal to:

$$\frac{5.47 \times 10^{-3} \text{ mol}}{0.170 \text{ L}} = 3.23 \times 10^{-2} \text{ M}$$

The phenoxide ion however is the conjugate ion of a weak acid and undergoes hydrolysis.

$$C_6H_5O^- \text{ (aq)} + H_2O \text{ (l)} \Leftrightarrow C_6H_5OH \text{ (aq)} + OH^- \text{ (aq)}$$

	$C_6H_5O^-$	C_6H_5OH	OH^-
Initial	3.23×10^{-2} M		
Change	-x	+x	+x
Equilibrium	3.23×10^{-2} M - x	x	x

$$K_b = \frac{[C_6H_5OH][OH^-]}{[C_6H_5O^-]} = \frac{1.0 \times 10^{-14}}{1.3 \times 10^{-10}} = \frac{(x)(x)}{3.23 \times 10^{-2} - x} = 7.7 \times 10^{-5}$$

Since $7.7 \times 10^{-5} \cdot 100 < 3.23 \times 10^{-2}$ we simplify: $\frac{(x)(x)}{3.23 \times 10^{-2}} = 7.7 \times 10^{-5}$

and $x = 1.5 \times 10^{-3}$ M.

(c) At the equivalence point: $[OH^-] = 1.5 \times 10^{-3}$ M

and $[H_3O^+] = \frac{1.0 \times 10^{-14}}{1.5 \times 10^{-3}} = 6.5 \times 10^{-12}$ and pH = 11.19

While the phenoxide ion reacts with water (hydrolyzes) to some extent, the **Na$^+$** is a "spectator ion" and its concentration remains unchanged at 3.23×10^{-2} M.

The concentration of phenoxide **is** reduced (albeit slightly) and at equilibrium is $(3.23 \times 10^{-2}$ M - x) or 3.08×10^{-2} M.

25. (a) At the equivalence point the moles of acid = moles of base.

(0.03678 L) (0.0105 M HCl) = 3.86×10^{-4} mol HCl

If this amount of base were contained in 25.0 mL of solution, the concentration of NH_3 in the original solution was 0.0154 M.

(b) At the equivalence point NH_4Cl will hydrolyze according to the equation:

$$NH_4^+ \text{ (aq)} + H_2O \text{ (l)} \Leftrightarrow NH_3 \text{ (aq)} + H_3O^+ \text{ (aq)}$$

$$K_a = \frac{[NH_3][H_3O^+]}{[NH_4^+]} = \frac{1.0 \times 10^{-14}}{1.8 \times 10^{-5}} = 5.6 \times 10^{-10}$$

The salt (3.86×10^{-4} mol) is contained in ($25.0 + 36.78$) 61.78 mL. Its concentration will

be $\dfrac{3.69 \times 10^{-4} \text{ mol}}{0.06178 \text{ L}}$ or 6.25×10^{-3} M. Substituting into the K_a expression :

$$\frac{[H_3O^+]^2}{6.25 \times 10^{-3}} = 5.6 \times 10^{-10}$$

Since $[H_3O^+][OH^-] = 1.0 \times 10^{-14}$ then $[OH^-] = \dfrac{1.0 \times 10^{-14}}{1.9 \times 10^{-6}} = 5.3 \times 10^{-9}$ M

and $[NH_4^+] = 6.25 \times 10^{-3}$ M (i.e. the salt concentration)

(c) With $[H_3O^+] = 1.9 \times 10^{-6}$ the pH = 5.73.

Titration Curves and Indicators

27. The titration of 0.10 M NaOH with 0.10 M HCl (a strong base vs a strong acid)

The initial pH of a 0.10 M NaOH would be

pOH = - log[0.10] so pOH = 1.00 and

pH = 13.00

When 15.0 mL of 0.10 M HCl have been

added, one-half of the NaOH initially

present will be consumed, leaving

0.5 (0.030 L • 0.10 mol/L) or

1.50×10^{-3} mol NaOH in 45.0 mL—

therefore a concentration of 0.0333 M NaOH.

pOH = 1.48 and pH = 12.52.

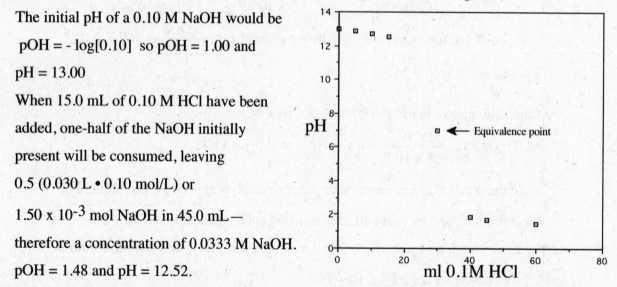

At the equivalence point (30.0 mL of the 0.10 M acid are added) there is only NaCl present.

Since this salt does not hydrolyze, the pH at that point is exactly 7.0. The total volume

present at this point is 60.0 mL.

Once a total of 60.0 mL of acid are added, there is an excess of 3.0×10^{-3} mol of HCl.

Contained in a total volume of 90.0 mL of solution, the [HCl] = 0.0333 M and the pH =1.5.

29. (a) pH of 25.0 mL of 0.11 M NH_3:

For the weak base, NH_3, the equilibrium in water is represented as:

$$NH_3 \text{ (aq)} + H_2O \text{ (l)} \Leftrightarrow NH_4^+ \text{ (aq)} + OH^- \text{ (aq)}$$

The slight dissociation of NH_3 would form equimolar amounts of NH_4^+ and OH^- ions.

$$K_b = \frac{[NH_4^+][OH^-]}{[NH_3]} = \frac{(x)(x)}{0.11 - x} = 1.8 \times 10^{-5}$$

Simplifying, we get : $\frac{x^2}{0.10} = 1.8 \times 10^{-5};$ $x = 1.4 \times 10^{-3}$ M = $[OH^-]$

$$pOH = 2.87 \quad \text{and pH} = 11.12$$

(b) Addition of HCl will consume NH_3 and produce NH_4^+ (the conjugate) according to the

net equation: $NH_3 \text{ (aq)} + H^+ \text{ (aq)} \Leftrightarrow NH_4^+ \text{ (aq)}$

The strong acid will drive this equilibrium to the right so we will assume this reaction to

be complete. Calculate the moles of NH_3 initially present:

$$(0.0250 \text{ L}) (0.10 \frac{\text{mol } NH_3}{\text{L}}) = 0.00250 \text{ mol } NH_3$$

Reaction with the HCl will produce the conjugate acid, NH_4^+. The task is two-fold. First

calculate the amounts of the conjugate pair present. Second substitute the concentrations

into the K_b expression. [One time-saving hint: The ratio of concentrations and the ratio of

the amounts (moles) will have the same numerical value. One can substitute the amounts

of the conjugate pair into the K_b expression.]

$$K_b = \frac{\left[NH_4^+\right][OH^-]}{[NH_3]} = 1.8 \times 10^{-5}$$

When 25.0 mL of the 0.10 M HCl has been added (total solution volume = 50.0 mL), the

reaction is at the equivalence point. All the NH_3 will be consumed, leaving the salt,

NH_4Cl. The NH_4Cl (2.50 millimol) has a concentration of 5.0×10^{-2} M.

This salt, being formed from a weak base and strong acid, undergoes hydrolysis.

$$NH_4^+(aq) + H_2O (l) \Leftrightarrow NH_3 (aq) + H_3O^+ (aq)$$

	NH_4^+	NH_3	H_3O^+
Initial concentration	$5.0 \times 10^{-2}M$	0	0
Change	-x	+x	+x
Equilibrium	5.0×10^{-2} - x	+x	+x

$$K_a = \frac{[NH_3][H_3O^+]}{[NH_4^+]} = 5.6 \times 10^{-10}$$

$$= \frac{x^2}{5.0 \times 10^{-2} - x} \approx \frac{x^2}{5.0 \times 10^{-2}} = 5.6 \times 10^{-10}$$

and $x = 5.3 \times 10^{-6} = [H_3O^+]$ and pH = 5.28 (equivalence point)

(c) The halfway point of the titration occurs when 12.50 mL of the acid have been added. At that point the amount of base and salt present are equal. An examination of the K_b expression will show that under these conditions the $[OH^-] = K_b$.

So pOH of 4.75 and pH = 9.25.

(d) From the table of indicators in your text, one indicator to use is Methyl Red. This indicator would be yellow prior to the equivalence point and red past that point. Bromcresol green would also be suitable, being blue prior to the equivalence point and yellow-green after the equivalence point.

(e)

mL of 0.10 M HCl added	millimol HCl added	millimol NH_3 after reaction	millimol NH_4^+ after reaction	$[OH^-]$ after reaction	pH
5.00	0.50	2.0	0.50	7.2×10^{-5}	9.85
15.0	1.5	1.0	1.5	1.2×10^{-5}	9.08
20.0	2.0	0.50	2.0	4.5×10^{-6}	8.65
22.0	2.2	0.30	2.2	2.5×10^{-6}	8.39

For the pH after 30.0 mL have been added: Addition of acid in excess of 25.00 mL will result in a solution which is essentially a strong acid. After the addition of 30.0 mL, substances present are: millimol HCl added: 3.00

millimol NH_3 present: 2.50

excess HCl present : 0.50 millimol

This HCl is present in a total volume of 55.0 mL of solution, hence the calculation for a strong acid proceeds as follows: $[H_3O^+] = 0.50$ mmol HCl/55.0 mL $= 9.1 \times 10^{-3}$ M and pH $= 2.04$.

A Summary

mL acid	pH
0.00	11.15
5.00	9.85
15.0	9.08
20.0	8.65
22.0	8.39
30.0	2.04

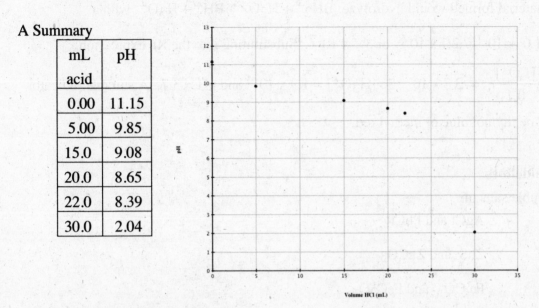

Volume HCl (mL)

31. Suitable indicators for titrations:

(a) HCl with pyridine: A solution of pyridinium chloride would have a pH of approximately 3. A suitable indicator would be thymol blue or bromphenol blue.

(b) NaOH with formic acid: The salt formed at the equivalence point is sodium formate. Hydrolysis of the formate ion would give rise to a basic solution (pH ≈ 8.5). Phenolphthalein would be a suitable indicator.

(c) Ethylenediamine and HCl: The base will have two endpoints—as it has two K's ($Kb_1 = 8.5 \times 10^{-5}$ and $Kb_2 = 2.7 \times 10^{-8}$). The first endpoint would contain, as the predominant specie, $H_2N(CH_2)_2NH_3^+$. This ion would hydrolyze with water. The Ka of this conjugate acid would be $1.0 \times 10^{-14}/ 8.5 \times 10^{-5}$ or 1.2×10^{-10}. Assume that we have a 0.1 M soln of the anion. The pH can be determined by the expression:

$$Ka = \frac{[H_3O^+]^2}{0.1} = 1.2 \times 10^{-10}.$$ [Recall that the concentration of the anion and the hydronium ion would be identical.][H_3O^+]$= 3.5 \times 10^{-6}$ and pH $= -\log(3.5 \times 10^{-6})$ or 5.5. Since the indicator would need to change colors +/- 1 on either side of the endpoint, methyl red would be a suitable indicator for this endpoint.

The 2nd endpoint would be reached when the monoprotonated base (let's call it BH^+) has accepted a 2nd proton (to form BH_2^+).

The material formed would hydrolyze: $BH_2^+ + H_2O \Leftrightarrow BH^+ + H_3O^+$, with a

$Ka = 1.0 \times 10^{-14}/ 2.7 \times 10^{-8}$ or 3.7×10^{-7}. Substituting into the Ka expression:

$$Ka = \frac{[H_3O^+]^2}{0.1} = 3.7 \times 10^{-7}.$$ So $[H_3O^+] = 1.9 \times 10^{-4}$ and pH = 3.7. A suitable indicator

would be thymol blue or methyl red.

Solubility Guidelines

33. Two insoluble salts of

    (a) Cl^- $AgCl$ and $PbCl_2$

    (b) Zn^{2+} ZnS and $ZnCO_3$

    (c) Fe^{2+} $Fe CO_3$ and FeC_2O_4

35. Using the table of solubility guidelines, predict water solubility for the following:

    (a) $(NH_4)_2CO_3$ Ammonium salts are **soluble**.

    (b) $ZnSO_4$ Sulfates are generally **soluble**.

    (c) NiS Sulfides are generally **insoluble**.

    (d) $BaSO_4$ Sr^{2+}, Ba^{2+}, and Pb^{2+} form **insoluble** sulfates.

Writing Solubility Product Constant Expressions

37.

Salt dissolving	Ksp expression	Ksp values
(a) $AgCN(s) \Leftrightarrow Ag^+(aq) + CN^-(aq)$	$Ksp = [Ag^+][CN^-]$	6.0×10^{-17}
(b) $NiCO_3(s) \Leftrightarrow Ni^{2+}(aq) + CO_3^{2-}(aq)$	$Ksp = [Ni^{2+}][CO_3^{2-}]$	1.4×10^{-7}
(c) $AuBr_3(s) \Leftrightarrow Au^{3+}(aq) + 3 Br^-(aq)$	$Ksp = [Au^{3+}][Br^-]^3$	4.0×10^{-36}

Calculating Ksp

39. Here we need only to substitute the equilibrium concentrations into the Ksp expression:

$$Ksp = [Tl^+][Br^-] = (1.9 \times 10^{-3})(1.9 \times 10^{-3}) = 3.6 \times 10^{-6}$$

41. What is K_{sp} for SrF_2?

The temptation is to calculate the number of moles of the solid that are added to water, but one must recall (a) not all that solid dissolves and (b) the concentration of the solid does not appear in the K_{sp} expression.

Note that the equilibrium concentration of $[Sr^{2+}] = 1.03 \times 10^{-3}$ M. The stoichiometry of the solid dissolving indicates two fluoride ions accompany the formation of one strontium ion. This tells us that $[F^-] = 2.06 \times 10^{-3}$ M. Substitution into the K_{sp} expression:

$$K_{sp} = [Sr^{+2}][F^-]^2 = (1.03 \times 10^{-3})(2.06 \times 10^{-3})^2 = 4.37 \times 10^{-9}.$$

43. For lead(II) hydroxide, the K_{sp} expression is $K_{sp} = [Pb^{+2}][OH^-]^2$.

Since we know the pH, we can calculate the $[OH^-]$.

pH = 9.15 and pOH = 14.00 - 9.15 = 4.85 so $[OH^-]= 1.4 \times 10^{-5}$.

For each mole of $Pb(OH)_2$ that dissolves, we get one mol of Pb^{+2} and two mol of OH^- .

Since $[OH^-]$ at equilibrium = 1.4×10^{-5} M, then $[Pb^{+2}] = 1/2 \cdot 1.4 \times 10^{-5}$ M

$K_{sp} = [Pb^{+2}][OH^-]^2 = K_{sp} = [7.0 \times 10^{-6}][1.4 \times 10^{-5}]^2 = 1.4 \times 10^{-15}$

Estimating Salt Solubility from K_{sp}

45. The solubility of AgI in water at 25°C in (a) mol/L and (b) g/L

(a) The K_{sp} for AgI (from Appendix J) is 8.5×10^{-17}. The K_{sp} expression is:

$$K_{sp} = [Ag^+][I^-] = 8.5 \times 10^{-17}$$

Since the solid dissolves to give one silver ion/one iodide ion, the concentrations of Ag+ and I⁻ will be equal. We can write the K_{sp} as: $[Ag^+]^2 = [I^-]^2 = 8.5 \times 10^{-17}$.

Taking the square root of both sides we obtain: $[Ag^+] = [I^-] = 9.2 \times 10^{-9}$ and recognizing that each mole of AgI that dissolves per liter gives one mol of silver ion, the solubility of AgI will be 9.2×10^{-9} mol/L.

(b) The solubility in g/L: $9.2 \times 10^{-9} \dfrac{\text{mol AgI}}{L} \cdot \dfrac{234.77 \text{ g AgI}}{1 \text{ mol AgI}} = 2.2 \times 10^{-6}$ g AgI/L

47. The K_{sp} for CaF_2 = 5.3×10^{-11}

   K_{sp} = $[Ca^{2+}][F^-]^2$ = 5.3×10^{-11}

   if a mol/L of CaF_2 dissolve, $[Ca^{2+}]$ = a and $[F^-]$ = 2a

   K_{sp} = $(a)(2a)^2 = 4a^3$ = 5.3×10^{-11} and a = 2.36×10^{-4}

   (a) The molar solubility is then 2.4×10^{-4} M (2 sf)

   (b) Solubility in g/L

$$2.4 \times 10^{-4} \frac{mol\ CaF_2}{L} \cdot \frac{78.07\ g\ CaF_2}{1\ mol\ CaF_2} = 1.8 \times 10^{-2}\ g\ CaF_2\ /L$$

49. K_{sp} = $[Ra^{2+}][SO_4^{2-}]$ = 4.2×10^{-11} so $[Ra^{2+}]$ = $[SO_4^{2-}]$ = 6.5×10^{-6} M

   $RaSO_4$ will dissolve to the extent of 6.5×10^{-6} mol/L

   Express this as grams in 100. mL (or 0.1 L)

$$\frac{6.5 \times 10^{-6}\ mol\ RaSO_4}{1\ L} \cdot \frac{0.1\ L}{1} \cdot \frac{322\ g\ RaSO_4}{1\ mol\ RaSO_4} = 2.1 \times 10^{-4}\ g$$

   Expressed as milligrams: 0.21 mg (2 sf) of $RaSO_4$ will dissolve!

51. Which solute in each of the following pairs is more soluble.

	Compound	Ksp
(a)	$PbCl_2$	1.7×10^{-5}
	$PbBr_2$	6.6×10^{-6}
(b)	HgS (red)	4×10^{-54}
	FeS	6×10^{-19}
(c)	$Fe(OH)_2$	4.9×10^{-17}
	$Zn(OH)_2$	3×10^{-17}

To compare relative solubilities of two compounds of the same general formula, one can examine the Ksp value. The larger the value of K_{sp}, the more soluble the compound.

The Common Ion Effect and Salt Solubility

53. The equilibrium for AgSCN dissolving is: AgSCN (s) $\Leftrightarrow$ Ag^+ (aq) + SCN^- (aq).

As x mol/L of AgSCN dissolve in pure water, x mol/L of Ag^+ and x mol/L of SCN^- are produced.

The expression would be: $K_{sp} = [Ag^+][SCN^-] = x^2 = 1.0 \times 10^{-12}$ and $x = 1.0 \times 10^{-6}$ M

So 1.0×10^{-6} mol AgSCN/L dissolve in pure water.

The equilibrium *for AgSCN dissolving in NaSCN* (0.010 M) is like that above. Equimolar amounts of Ag^+ and SCN^- ions are produced as the solid dissolves. However the $[SCN^-]$ is augmented by the soluble NaSCN.

$$K_{sp} = [Ag^+][SCN^-] = (x)(x + 0.010) = 1.0 \times 10^{-12}$$

We can simplify the expression by *assuming* that $x + 0.010 \approx 0.010$. Note that the value of x above (1.0×10^{-6}) lends credibility to this assumption.

$$(x)(0.010) = 1.0 \times 10^{-12} \quad \text{and} \quad x = 1.0 \times 10^{-10} \text{ M}$$

The solubility of AgSCN in 0.010 M NaSCN is 1.0×10^{-10} M -- reduced by four orders of magnitude from its solubility in pure water. LeChatelier strikes again!

55. Solubility in mg/mL of AgI in (a) pure water and (b) water that is 0.020 M in Ag^+ (aq).
$$K_{sp} = [Ag^+][I^-] = 8.5 \times 10^{-17}$$

(a) In SQ 18.45 we found the solubility of AgI to be 9.2×10^{-9} mol/L or 2.2×10^{-6} g/L
Expressing this in mg/mL = 2.2×10^{-6} mg/mL

(b) The difference in solubility arises because of the presence of the 0.020 M in Ag^+.
Recall that in SQ18.45 we knew that the silver and iodide ion concentrations were equal. With the addition of the 0.020 M silver solution, this is no longer a valid assumption. If we let x mol/L of the AgI dissolve, the concentrations of Ag^+ and I^- (from the salt dissolving) will be x mol/L. The $[Ag^+]$ will be amended by 0.020 M, so we write:

$K_{sp} = [x + 0.020][x] = 8.5 \times 10^{-17}$. This could be solved with the quadratic equation, but a bit of thought will simplify the process. In the (a) part we discovered that the $[Ag^+]$ was approximately 10^{-9} M. This value is small compared to 0.020. Let's use that approximation to convert the Ksp expression to:

$[0.020][x] = 8.5 \times 10^{-17}$ and $x = 4.25 \times 10^{-15}$ M.

This molar solubility translates into:
$$4.25 \times 10^{-15} \frac{\text{mol AgI}}{\text{L}} \cdot \frac{234.77 \text{ g AgI}}{1 \text{ mol AgI}} = 9.98 \times 10^{-13} \text{g AgI/L}$$

or 1.0×10^{-12} mg/mL (2 sf).

The Effect of Basic Anions on Salt Solubility

57. The salt that should be more soluble in nitric acid than in pure water from the pairs:

(a) $PbCl_2$ or PbS: PbS will be more soluble, since the S^{2-} ion will react with the nitric acid, reducing the $[S^{2-}]$, and increasing the amount of PbS that dissolves.

(b) Ag_2CO_3 or AgI: Ag_2CO_3 will be more soluble. The CO_3^{2-} ion will react with the nitric acid, and produce HCO_3^- and H_2CO_3 which will decompose to CO_2 and H_2O. The removal of carbonate shifts the equilibrium to the right, increasing the amount that dissolves.

(c) $Al(OH)_3$ or AgCl: $Al(OH)_3$ will be more soluble. As in (b) above, the OH^- will react with the H^+ to form water. The reduction in $[OH^-]$ will increase the amount of the salt that dissolves.

Precipitation Reactions

59. Given the equation for $PbCl_2$ dissolving in water: $PbCl_2 (s) \Leftrightarrow Pb^{+2} (aq) + 2 Cl^-$

we can write the K_{sp} expression : $K_{sp} = [Pb^{+2}][Cl^-]^2 = 1.7 \times 10^{-5}$

Substituting the ion concentrations into the Ksp expression we get:

$Q = [Pb^{+2}][Cl^-]^2 = (0.0012)(0.010)^2 = 1.2 \times 10^{-7}$

Since Q is less than K_{sp} , no $PbCl_2$ precipitates.

61. If $Zn(OH)_2$ is to precipitate, the reaction quotient (Q) must exceed the K_{sp} for the salt.

4.0 mg of NaOH in 10. mL corresponds to a concentration of:

$$[OH^-] = \frac{4.0 \times 10^{-3} \text{ g NaOH}}{0.0100 \text{ L}} \cdot \frac{1 \text{ mol NaOH}}{40.0 \text{ g NaOH}} = 0.010 \text{ M}$$

The value of Q is: $[Zn^{2+}][OH^-]^2 = (1.6 \times 10^{-4})(1.0 \times 10^{-2})^2 = 1.6 \times 10^{-8}$

The value of Q is greater than the K_{sp} for the salt (4.5×10^{-17}) , so $Zn(OH)_2$ precipitates.

63. The molar concentration of Mg^{2+} is:

$$\frac{1350 \text{ mg Mg}^{2+}}{1 \text{ L}} \cdot \frac{1 \text{ g Mg}^{2+}}{1000 \text{ mg Mg}^{2+}} \cdot \frac{1 \text{ mol Mg}^{2+}}{24.3050 \text{ g Mg}^{2+}} = 5.55 \times 10^{-2} \text{M}$$

For $Mg(OH)_2$ to precipitate, Q must be greater than the K_{sp} for the salt (5.6×10^{-12}).

$Q = [Mg^{2+}][OH^-]^2 = (5.55 \times 10^{-2})(OH^-)^2 = 5.6 \times 10^{-12}$ so

$(OH^-)^2 = 5.6 \times 10^{-12}/5.55 \times 10^{-2}$ or 1.01×10^{-10}, and $[OH^-] = 1.0 \times 10^{-5}$ M.

Equilibria Involving Complex Ions

65. An equilibrium to demonstrate that sufficient OH^- will dissolve $Zn(OH)_2$:

The equilibrium for the dissolution of $Zn(OH)_2$ is:

$Zn(OH)_2(s) \Leftrightarrow Zn^{2+}(aq) + 2OH^-(aq)$ $\qquad\qquad$ $K = 3 \times 10^{-17}$

$\underline{Zn^{2+}(aq) + 4\,OH^-(aq) \Leftrightarrow [Zn(OH)_4]^{2-}(aq) \qquad\qquad K = 4.6 \times 10^{17}}$

Net $\quad Zn(OH)_2(s) + 2OH^-(aq)) \Leftrightarrow [Zn(OH)_4]^{2-}(aq)$

$\qquad\qquad\qquad\qquad\qquad$ $K_{net} = (3 \times 10^{-17})(4.6 \times 10^{17}) = 10$ (1 sf)

Note that the magnitude of the net K indicates that sufficient OH^- will dissolve $Zn(OH)_2$ as

the K indicates the equilibrium is product favored.

67. Moles of ammonia to dissolve 0.050 mol of AgCl suspended in 1.0 L of water:

The equilibria involved are:

$AgCl(s) \Leftrightarrow Ag^+(aq) + Cl^-(aq)$ $\qquad\qquad\qquad$ $K = 1.8 \times 10^{-10}$

$\underline{Ag^+(aq) + 2\,NH_3(aq) \Leftrightarrow [Ag(NH_3)_2]^+(aq) \qquad\qquad K = 1.1 \times 10^7}$

$AgCl(s) + 2\,NH_3(aq) \Leftrightarrow [Ag(NH_3)_2]^+(aq) + Cl^-(aq)$

$\qquad\qquad$ $K_{net} = (1.8 \times 10^{-10})(1.1 \times 10^7) = 2.0 \times 10^{-3}$

The equilibrium expression for the net equation is:

$$\frac{[Ag(NH_3)_2]^+[Cl^-]}{[NH_3]^2} = 2.0 \times 10^{-3}$$

Note that if we dissolve 0.050 mol AgCl (in 1.0 L), the concentration of the complex ion

AND the chloride ion will be equal to 0.050 M. Substituting these values into the

expression above, and solving for the concentration of ammonia yields:

$$\frac{(0.050)(0.050)}{2.0 \times 10^{-3}} = [NH_3]^2 \text{ and } [NH_3] = 1.12 \text{ M (or 1.1 M to 2 sf)}$$

The equilibrium concentration of ammonia has to be 1.1M. Note that the stoichiometry of forming the complex ion requires (2 x 0.050 mol) or 0.10 mol NH_3.

The total amount of ammonia needed is then: 0.10 mol + 1.1 mol or 1.2 mol NH_3.

69. (a) Solubility of AgCl in water:

$$AgCl(s) \Leftrightarrow Ag^+(aq) + Cl^-(aq) \qquad\qquad K = 1.8 \times 10^{-10}$$

$[Ag^+][Cl^-] = 1.8 \times 10^{-10}$ and noting that the concentration of the two ions will be identical: $[Ag^+]^2 = 1.8 \times 10^{-10}$, and $[Ag^+] = 1.3 \times 10^{-5}$ so the solubility of AgCl is 1.3 x 10^{-5} M.

(b) Solubility of AgCl in 1.0 M NH_3:

The pertinent equilibrium is the same as that in SQ18.67 above:

$$AgCl(s) + 2\ NH_3\ (aq) \Leftrightarrow [Ag(NH_3)_2]^+(aq) + Cl^-\ (aq)\ \text{ with } K = 2.0 \times 10^{-3}$$

The equilbrium concentration of ammonia is 1.0 M. Substitute into the K expression:

$$\frac{[Ag(NH_3)_2]^+\left[Cl^-\right]}{\left[NH_3\right]^2} = 2.0 \times 10^{-3} = \frac{[Ag(NH_3)_2]^+\left[Cl^-\right]}{\left[1.0\right]^2} = 2.0 \times 10^{-3}$$

As the concentration of both the chloride and complex ion will be equal, the terms in the numerator can be represented as x^2: $x^2 = 2.0 \times 10^{-3}(1.0)^2$ and $x = 4.5 \times 10^{-2}$. The solubility of AgCl in 1.0 M NH_3 is 4.5×10^{-2} M.

GENERAL QUESTIONS

71. Solution producing precipitate:

For the solutes listed, note that each would exist as the individual cations and anions. For example, NaBr (aq) more accurately is represented as: $Na^+(aq)$ and $Br^-(aq)$.

The net equations that result are:

(a) $Ag^+\ (aq)\ +\ Br^-\ (aq)\ \rightarrow AgBr\ (s)$

(b) $Pb^{2+}\ (aq)\ +\ 2\ Cl^-\ (aq)\ \rightarrow PbCl_2\ (s)$

One can make these decisions in one of two ways: (1) Recalling the solubility tables— probably learned earlier or (2) Reviewing the data from the Ksp tables (which contains compounds that we normally classify as insoluble—and precipitate from a solution containing the appropriate pairs of cations and anions (Ag^+ and Br^- for example).

73. Will $BaSO_4$ precipitate?

Calculate the concentrations of barium and sulfate ions (after the solutions are mixed).

$$\frac{48\ mL}{72\ mL} \bullet 0.0012\ M\ Ba^{2+} = 0.00080\ M\ Ba^{2+} \text{ and}$$

$$\frac{24\ mL}{72\ mL} \bullet 1.0 \times 10^{-6}\ M\ SO_4^{2-} = 3.3 \times 10^{-7}\ M\ SO_4^{2-}$$

Substituting in the K_{sp} expression: $Q = [Ba^{+2}][SO_4^{2-}] = (8.00 \times 10^{-4})(3.3 \times 10^{-7})$

$Q = 2.7 \times 10^{-10}$; Q is larger than the Ksp for the solid (1.1×10^{-10}) so $BaSO_4$ precipitates.

Note that we are assuming that the sulfuric acid totally dissociates into protons and sulfate ions!

75. The pH and $[H_3O^+]$ when 50.0 mL of 0.40 M NH_3 is mixed with 25.0 mL of 0.20 M HCl:

The number of moles of each reactant:

(0.20 mol HCl/L)(0.025 L) = 0.0050 mol HCl and (0.40 mol NH_3/L)(0.050 L) =

0.020 mol NH_3. So 0.005 mol of NH_3 are consumed—and 0.005 mol NH_4^+ are produced.

	NH_3	H_2O	NH_4^+
Initial	0.020 mol NH_3		0
Change	- 0.0050 (consumed by HCl)		+ 0.0050 mol
Equilibrium	0.015 mol		0.0050 mol

So we have a buffer with a conjugate pair (NH_3 and NH_4^+) present.

The Kb expression for NH_3 can be written: $\dfrac{[NH_4^+][OH^-]}{[NH_3]} = 1.8 \times 10^{-5}$

Substituting the # of mol into the Kb expression yields:

$\dfrac{[0.005\ mol][OH^-]}{[0.015\ mol]} = 1.8 \times 10^{-5}$ or $[OH^-] = 5.4 \times 10^{-5}$ M.

So pOH = 4.27 and pH = 9.73, and a concomitant $[H_3O^+] = 10^{-9.73}$ or 1.9×10^{-10} M.

77. Compounds in order of increasing solubility in H_2O:

Compound	K_{sp}
$BaCO_3$	2.6×10^{-9}
Ag_2CO_3	8.5×10^{-12}
Na_2CO_3	not listed: Na_2CO_3 is very soluble in water and is the most soluble salt of the 3.

To determine the relative solubilities, find the molar solubilities.

For $BaCO_3$: $K_{sp} = [Ba^{2+}][CO_3^{2-}] = (x)(x) = 2.6 \times 10^{-9}$ and $x = 5.1 \times 10^{-5}$

The molar solubility of $BaCO_3$ is 5.1×10^{-5} M.

and for Ag_2CO_3: $K_{sp} = [Ag^+]^2[CO_3^{2-}] = (2x)^2(x) = 8.5 \times 10^{-12}$

$$4x^3 = 8.5 \times 10^{-12} \text{ and } x = 1.3 \times 10^{-4}$$

The molar solubility of Ag_2CO_3 is 1.3×10^{-4} M.

In order of increasing solubility: $BaCO_3 < Ag_2CO_3 < Na_2CO_3$

79. pH of a solution that contains 5.15 g NH_4NO_3 and 0.10 L of 0.15 M NH_3:

The concentration of the ammonium ion (from the nitrate salt) is:

$$5.15 \text{ g } NH_4NO_3 \cdot \frac{1 \text{ mol } NH_4NO_3}{80.04 \text{ g } NH_4NO_3} \cdot \frac{1}{0.10 \text{ L}} = 0.64 \text{ M}$$

We can calculate the pH using the K_b expression for ammonia:

$$K_b = \frac{[NH_4^+][OH^-]}{[NH_3]} = \frac{(0.64)(OH^-)}{(0.15)} = 1.8 \times 10^{-5} \text{ and solving for the hydroxide ion}$$

concentration, $x = \frac{(1.8 \times 10^{-5})(0.15)}{(0.64)} = 4.2 \times 10^{-6}$ M and pOH = 5.38 ; so pH = 8.62

Diluting the solution does not change the pH of the solution, since the dilution affects the concentration of both members of the conjugate pair. Since the pH of the buffer is a function of the *ratio* of the conjugate pair, the pH does not change.

81. The effect on pH of:
(a) Adding $CH_3CO_2^-Na^+$ to 0.100 M CH_3CO_2H:

The equilibrium affected is: $CH_3CO_2H + H_2O \Leftrightarrow CH_3CO_2^- + H_3O^+$

Addition of sodium acetate increases the concentration of acetate ion. The equilibrium will shift to the left—reducing hydronium ion, and increase the pH.

(b) Adding $NaNO_3$ to 0.100 M HNO_3: No effect on pH. Since nitric acid is a strong acid, the equilibrium lies very far to the right. Addition of NO_3^- will not shift the equilibrium. The effects differ owing to the nature of the two acids. Nitric acid (a strong acid) exists as hydronium and nitrate ions. Acetic acid exists as the molecular acid and acetate ion. The addition of conjugate base of both acids ($CH_3CO_2^-$ and NO_3^-) will affect the acetic acid equilibrium but not the nitric acid system.

83. Several approaches are valid. Perhaps the simplest is the use of the Henderson Hasselbalch equation:

(a) pH of buffer solution: $pH = pKa + \log\dfrac{\left[C_6H_5CO_2^-\right]}{\left[C_6H_5CO_2H\right]}$

$$\frac{1.50 \text{ g } C_6H_5CO_2^-}{1} \bullet \frac{1 \text{ mol } C_6H_5CO_2^-Na^+}{144.1 \text{ g } C_6H_5CO_2^-Na^+} = 0.0104 \text{ mol } C_6H_5CO_2^-Na^+$$

$$\frac{1.50 \text{ g } C_6H_5CO_2^-}{1} \bullet \frac{1 \text{ mol } C_6H_5CO_2H}{122.1 \text{ g } C_6H_5CO_2H} = 0.0123 \text{ mol } C_6H_5CO_2H$$

Note that these are NOT the concentrations, since these quantities are dissolved in 150.0 mL. Since we are using a **ratio** of the conjugate pairs, we can use the ratio of # mol exactly as we would have used the ratio of the molar concentrations:

$$pH = pKa + \log\frac{\left[C_6H_5CO_2^-\right]}{\left[C_6H_5CO_2H\right]} = -\log(6.3 \times 10^{-5}) + \log\frac{[0.0104]}{[0.0123]} = 4.20 + -0.07 = 4.13$$

(b) To reduce the pH to 4.00, we will need to add the "acidic" component of the buffer. Substituting into the equation, as before—solving for the amount of benzoic acid:

$$pH = pKa + \log\frac{\left[C_6H_5CO_2^-\right]}{\left[C_6H_5CO_2H\right]} \qquad 4.00 = 4.20 + \log\frac{[0.0104]}{\left[C_6H_5CO_2H\right]}$$

$$4.00 - 4.20 = \log\frac{[0.0104]}{\left[C_6H_5CO_2H\right]} \quad \text{and} \quad 10^{-0.20} = \frac{[0.0104]}{\left[C_6H_5CO_2H\right]}$$

Remembering that we're using # mol in the equation, mol $C_6H_5CO_2H$ = 0.017 mol (2 sf)

The amount of $C_6H_5CO_2H$ needed is (0.017 – 0.0123)mol x 122.1 g/mol = 0.57

(0.6g to1sf—subtraction!)

(c) Quantity of 2.0 M NaOH or HCl to change buffer pH to 4.00:

To reduce the pH of the buffer, we'll clearly need to add 2.0 M HCl. The question is "how much" of the 2.0 M HCl:

We'll need to add enough HCl to reduce the concentration of the benzoate ion AND increase the concentration of benzoic acid appropriately:

$$4.00 - 4.20 = \log\frac{\left[C_6H_5CO_2^-\,Na^+\right]}{\left[C_6H_5CO_2H\right]} \text{ so } 10^{-0.20} = \frac{\left[C_6H_5CO_2^-\,Na^+\right]}{\left[C_6H_5CO_2H\right]} = \frac{0.0104 - x}{0.0123 + x}$$

$$0.63 = \frac{0.0104 - x}{0.0123 + x} \text{ and } 0.63(0.0123 + x) = 0.0104 - x \text{ and } x = 0.00163 \text{ mol}$$

$$\text{The amount of 2.0M HCl} = \frac{0.00163 \text{ mol}}{1} \cdot \frac{1 \text{ L}}{2.0 \text{ mol HCl}} = 8.2 \times 10^{-4} \text{L}$$

85. The equation $AgCl(s) + I^-(aq) \Leftrightarrow AgI(aq) + Cl^-(aq)$

can be obtained by adding two equations:

$$1.\ AgCl(s) \Leftrightarrow Ag^+(aq) + Cl^-(aq) \qquad K_{sp1} = 1.8 \times 10^{-10}$$

$$2.\ Ag^+(aq) + I^-(aq) \Leftrightarrow AgI(s) \qquad \frac{1}{K_{sp2}} = 1.2 \times 10^{16}$$

The net equation: $AgCl(s) + I^-(aq) \Leftrightarrow AgI(aq) + Cl^-(aq)$

$$K_{net} = K_{sp1} \cdot \frac{1}{K_{sp2}} = (1.8 \times 10^{-10})(1.2 \times 10^{16}) = 2.2 \times 10^6$$

The equilibrium lies to the right. This indicates that *AgI will form* if I^- is added to a saturated solution of AgCl.

87. Regarding the solubility of barium and calcium fluorides:

(a) $[F^-]$ that will precipitate maximum amount of calcium ion:

BaF_2 begins to precipitate when the ion product just exceeds the K_{sp}.

$K_{sp} = [Ba^{2+}][F^-]^2 = 1.8 \times 10^{-7}$, and the concentration of barium ion $= 0.10$ M

$[F^-] = (1.8 \times 10^{-7}/0.10)^{0.5} = 1.3 \times 10^{-3}$ M

This would be the maximum fluoride concentration permissible.

(b) $[Ca^{2+}]$ remaining when BaF_2 just begins to precipitate:

The $[Ca^{2+}]$ ion remaining will be calculable from the K_{sp} expression for CaF_2.

$K_{sp} = [Ca^{2+}][F^-]^2 = 5.3 \times 10^{-11}$ and we know that $[F^-]$ is 1.3×10^{-3} M.

So $\dfrac{5.3 \times 10^{-11}}{\left(1.3 \times 10^{-3}\right)^2} = 2.9 \times 10^{-5}$ M

89. Regarding the precipitation of $CaSO_4$ and $PbSO_4$ from a solution 0.010 M in each metal:

(a) Examine the K_{sp} of the two salts:

K_{sp} for $CaSO_4 = 4.9 \times 10^{-5}$ and for $PbSO_4 = 2.5 \times 10^{-8}$

These data indicate that $PbSO_4$ will begin to precipitate first—as it is the less soluble of the two salts. This can be found by examining the general Ksp expression for 1:1 salts:

$K_{sp} = [M^{2+}][SO_4^{2-}]$. With both metal concentrations at 0.010 M, the sulfate ion for the less soluble sulfate will be exceeded first.

(b) When the calcium salt just begins to precipitate, the $[Pb^{2-}] = ?$

We know that the metal ion concentration is 0.010 M. So the sulfate ion concentration when the more soluble calcium sulfate begins to precipitate is:

$K_{sp} = [0.010][SO_4^{2-}] = 4.9 \times 10^{-5}$ and $[SO_4^{2-}] = (4.9 \times 10^{-5} / 0.010) = 4.9 \times 10^{-3}$ M

At that point the $[Pb^{2+}]$ would be: $K_{sp} = [Pb^{2+}][SO_4^{2-}] = 2.5 \times 10^{-8}$

Then $[Pb^{2+}][4.9 \times 10^{-3}] = 2.5 \times 10^{-8}$ and $[Pb^{2+}] = (2.5 \times 10^{-8} / 4.9 \times 10^{-3})$ or

5.1×10^{-6} M.

91. Percentage of calcium ions removed:

$[Ca^{2+}]_{initial} = 0.010$ M

At equilibrium, we can write: $[Ca^{2+}][CO_3^{2-}] = 3.4 \times 10^{-9}$

If the carbonate ion concentration is 0.050 M, then $[Ca^{2+}][0.050] = 3.4 \times 10^{-9}$ and

$[Ca^{2+}] = \dfrac{3.4 \times 10^{-9}}{[0.050]} = 6.8 \times 10^{-8}$ M. This represents the calcium ion concentration at

equilibrium. The amount of calcium ion removed per liter is: $(0.010 - 6.8 \times 10^{-8})$ or

9.99×10^{-3} M. In essence, ALL the calcium ion (> 99.9%) has been removed.

IN THE LABORATORY

93. Separate the following pairs of ions:

(a) Ba^{2+} and Na^+: Since most sodium salts are soluble, it is simple to find a barium salt which is not soluble--e.g. the sulfate. Addition of dilute sulfuric acid should provide a source of SO_4^{2-} ions in sufficient quantity to precipitate the barium ions, but not the sodium ions.
[Ksp $(BaSO_4)$ = 1.1 x 10^{-10} Ksp (Na_2SO_4) = not listed, owing to the large solubility of sodium sulfate]

(b) Ni^{2+} and Pb^{2+}: The carbonate ion will serve as an effective reagent for selective precipitation of the two ions. A solution of Na_2CO_3 can be added, and the less soluble $PbCO_3$ will begin precipitating first. The 7 orders of magnitude difference in their Ksp's should provide satisfactory separation.

[Ksp $(PbCO_3)$ = 7.4 x 10^{-14} Ksp $(NiCO_3)$ = 1.4 x 10^{-7}]

95. (a) These metal sulfates are 1:1 salts. The K_{sp} expression has the general form:
$$K_{sp} = [M^{2+}][SO_4^{2-}]$$
To determine the $[SO_4^{2-}]$ necessary to begin precipitation, we can divide the equation by the metal ion concentration to obtain:
$$\frac{K_{sp}}{[M^{2+}]} = [SO_4^{2-}]$$

The concentration of the three metal ions under consideration are each 0.10 M. Substitution of the appropriate K_{sp} for the sulfates and 0.10 M for the metal ion concentration yields the sulfate ion concentrations in the table below. As the soluble sulfate is added to the metal ion solution, the sulfate ion concentration increases from zero molarity. The lowest sulfate ion concentration is reached first, with higher concentrations reached later. The order of precipitation is listed in the last column of the table below.

Compound	K_{sp}	Maximum $[SO_4^{2-}]$	Order of Precipitation
$BaSO_4$	1.1 x 10^{-10}	1.1 x 10^{-9}	1
$SrSO_4$	3.4 x 10^{-7}	3.4 x 10^{-6}	2

(b) Concentration of Ba^{2+} when $SrSO_4$ begins to precipitate:

From (a) we note that $[SO_4^{2-}]$ will be 3.4×10^{-6} when $SrSO_4$ begins to precipitate.

So we calculate the concentration of Ba^{2+} by noting that since that barium salt has been precipitating, it is a **saturated solution of $BaSO_4$**.

$$K_{sp} = [Ba^{2+}][SO_4^{2-}] = 1.1 \times 10^{-10}.$$
$$\text{So } [Ba^{2+}] = 1.1 \times 10^{-10}/3.4 \times 10^{-6} = 3.2 \times 10^{-5}.$$

97. For the titration of aniline hydrochloride with NaOH:

(a) initial pH: Recall that the hydrochloride (represented in this example as HA) is a weak acid: $(K_a = 2.4 \times 10^{-5})$ Use a K_a expression to solve for H_3O^+:

$$K_a = \frac{[A^-][H_3O^+]}{[HA]} = 2.4 \times 10^{-5} \bullet 0.10 = x^2 = 2.4 \times 10^{-6}$$

$x = 1.5 \times 10^{-3} M = [H_3O^+]$ and pH = 2.81

(b) pH at the equivalence point:

At the equivalence point, we know that the number of moles of acid = # moles base (by definition). Since we have 0.00500 mol of HA (M $\bullet$ V), we must ask the question, "How much 0.185 M NaOH contains 0.00500 mol NaOH?" That volume is:

0.00500 mol OH^- $\bullet$ 1000 mL/ 0.185 mol NaOH = 27.0 mL of NaOH solution.

The total volume is 50.0 mL + 27.0 mL = 0.0770 L.

The [HA] = 0.00500 mol/0.0770 L = 0.0649 M

We can calculate the pH of the solution using the K_b expression for A:

$$K_b = \frac{[HA][OH^-]}{[A]} = \frac{1.0 \times 10^{-14}}{2.4 \times 10^{-5}} \text{ and knowing that the } [HA] = [OH^-]$$

$x^2 = 4.2 \times 10^{-12} \bullet 0.0649$ and $x = 5.2 \times 10^{-6}$ M; pOH = 5.28 and pH = 8.72

(c) pH at the midpont:

This portion is easy to solve if you examine the Ka expression or the Henderson-Hasselbalch expression for this system:

From the K_b expression above, note that the conjugate pairs are found in the numerator and denominator of the right-side of the K_b term (the same applies to Ka expressions).

At the mid-point you have reacted half the acid, (using 13.5 mL—see part (b)) forming its conjugate base. The result is that the concentrations of the conjugate pairs are equal and the pOH = pK_b and pH = pK_a So pH = $-\log(2.4 \times 10^{-5}) = 4.62$.

(d) With a pH= 8.72 at the equivalence point, o-cresolphthalein or phenolphthalein would serve as an adequate indicator.

(e) The pH after the addition of 10.0, 20.0 and 30.0 mL base:

The volumes of base correspond to

10 mL (0.185 mol/L • 0.0100 L)	0.00185 mol NaOH
20.0 mL (0.185 mol/L • 0.0200 L)	0.00370 mol NaOH
30.0 mL(0.185 mol/L • 0.0300L)	0.00555 mol NaOH

Each mol of NaOH consumes a mol of aniline hydrochloride(HA) and produces an equal number of mol of aniline(A).

Recall that we began with 0.00500 mol of the acid (which we're representing as HA)

Substitution into the Ka expression yields:

$$\frac{[A^-][H_3O^+]}{[HA]} = 2.4 \times 10^{-5} \quad \text{and rearranging them} \quad [H_3O^+] = 2.4 \times 10^{-5} \cdot \frac{[HA]}{[A]}$$

After 10 mL of base are added, 0.00185 mol HA are consumed, and 0.00185 mol A produced. The acid remaining is (000500 - 0.00185) mol.

Substituting into the rearranged equation:

$$[H_3O^+] = 2.4 \times 10^{-5} \cdot \frac{[0.00315 \text{ mol}]}{[0.00185 \text{ mol}]} = 4.1 \times 10^{-5} \quad \text{and pH} = -\log(4.1 \times 10^{-5}) = 4.39$$

After 20 mL of base are added, 0.00370 mol HA are consumed, and 0.00370 mol A produced. The acid remaining is (000500 - 0.00370)mol.

Substituting into the rearranged equation:

$$[H_3O^+] = 2.4 \times 10^{-5} \cdot \frac{[0.00130 \text{ mol}]}{[0.00370 \text{ mol}]} = 8.4 \times 10^{-6} \quad \text{and pH} = -\log(8.4 \times 10^{-6}) = 5.07$$

After 30 mL of base are added, all the acid is consumed, and excess strong base is present (0.00555 mol – 0.00500)mol. We can treat this solution as one of a strong base. The volume of the solution is (50.0 + 30.0 or 80.0 mL.).

The concentration of the NaOH is:

0.00055 mol/ 0.080 L = 0.006875 M so pOH = -log(0.006875) and pOH=2.16, the

pH=11.84.

(f)   The approximate titration curve:

Vol base	pH
0.0	2.81
10.0	4.39
13.5	4.62
20.0	5.07
27.0	8.72
30.0	11.84

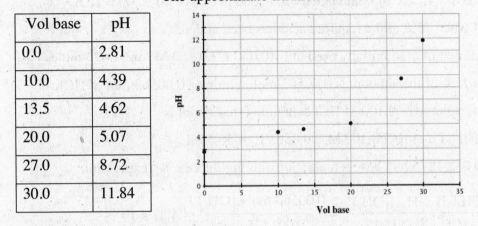

99. For the titration of 0.150 M ethylamine (K_b = 4.27 x 10^{-4}) with 0.100 M HCl:
    (a) pH of 50.0 mL of 0.150 M $CH_3CH_2NH_2$:

    For the weak base, $CH_3CH_2NH_2$, the equilibrium in water is represented as:

    $$CH_3CH_2NH_2 (aq) + H_2O (l) \Leftrightarrow CH_3CH_2NH_3^+ (aq) + OH^- (aq)$$

    Note : The K_b for ethylamine is given as 4.27 x 10^{-4}.

    The slight dissociation of $CH_3CH_2NH_2$ would form equimolar amounts of

    $CH_3CH_2NH_3^+$ and OH^- ions.

    $$K_b = \frac{[CH_3CH_2NH_3^+][OH^-]}{[CH_3CH_2NH_2]} = \frac{(x)(x)}{0.150 - x} = 4.27 \times 10^{-4}$$

    The quadratic equation will be needed to find an exact solution.

    $x^2 = 6.405 - 4.27 \times 10^{-4}x$

    Rearranging: $x^2 + 4.27 \times 10^{-4}x - 6.405 \times 10^{-5} = 0$

    and solving for x = 7.79 x 10^{-3} ; pOH = -log(7.79 x 10^{-3}) = 2.11

    and pH = 11.89

(b) pH at the halfway point of the titration:

    The volume of acid added isn't that important since the amounts of base and conjugate

    acid will be equal. In part (d), we find that 75.0 mL of acid are required for the

    equivalence point, so the volume of acid at this point is (0.5 • 75.0 mL)

Substituting that fact into the equilibrium expression we obtain:

$$\frac{[CH_3CH_2NH_3^+][OH^-]}{[CH_3CH_2NH_2]} = \frac{(x)[OH^-]}{(x)} = 4.27 \times 10^{-4} \text{ so we can see that the}$$

$[OH^-] = 4.27 \times 10^{-4}$ and pOH = 3.37 and pH = (14.00-3.37) = 10.63

(c) pH when 75% of the required acid has been added:

Amount of base initially present is (0.050 L • 0.150 M) or 0.0075 moles base.

So 75% of this amount is 0.00563 mol. requiring 0.00563 mol of HCl.

The volume of 0.100 M HCl containing that # mol is

0.00563 mol HCl/0.100 M = 0.0563 L or 56.3 mL

So 0.001875 mol base remain. Substituting into the K_b expression:

$$\frac{[CH_3CH_2NH_3^+][OH^-]}{[CH_3CH_2NH_2]} = \frac{0.00563 \text{ mol} \bullet [OH^-]}{0.001875 \text{ mol}} = 4.27 \times 10^{-4}$$

solving for hydroxyl ion concentration yields:1.42×10^{-4} M and pOH = 3.85 and pH = 10.15.

(d) pH at the equivalence point:

At the equivalence point, there are equal # of moles of acid and base. The number of moles of ethylamine = (0.050 L • 0.150 M) or 0.00750 moles ethylamine .

That amount of HCl would be:

$$7.50 \times 10^{-3} \text{ mol HCl} \bullet \frac{1 \text{ L}}{0.100 \text{ mol HCl}} = 0.0750 \text{ L (or 75.0 mL)}$$

This total amount of solution would be: 75.0 + 50.0 = 125.0 mL

Since we have added equal amounts of acid and base, the reaction between the two will result in the existence of only the salt.

and the concentration of salt would be: $\frac{7.50 \times 10^{-3} \text{ mol}}{0.1250 \text{ L}} = 6.00 \times 10^{-2}$ M

Recall that the salt will act as a weak acid and we can calculate the K_a:

$$CH_3CH_2NH_3^+ (aq) + H_2O (l) \Leftrightarrow CH_3CH_2NH_2 (aq) + H_3O^+ (aq)$$

The equilibrium constant (K_a) would be $\frac{K_w}{K_b} = \frac{1.0 \times 10^{-14}}{4.27 \times 10^{-4}} = 2.3 \times 10^{-11}$

The equilibrium expression would be: $\dfrac{[CH_3CH_2NH_2][H_3O^+]}{[CH_3CH_2NH_3^+]} = 2.3 \times 10^{-11}$

Given that the salt would hydrolyze to form equal amounts of ethylamine and hydronium ion (as shown by the equation above). If we represent the concentrations of those species as x, then we can write (using our usual approximation):

$$\frac{x^2}{6.00 \times 10^{-2}} = 2.3 \times 10^{-11} \text{ and solving for } x = 1.19 \times 10^{-6}$$

Since x represents $[H_3O^+]$, then pH = $-\log(1.19 \times 10^{-6}) = 5.93$

(e) pH after addition of 10.0 mL of HCl more than required:

Past the equivalence point, the excess strong acid controls the pH. To determine the pH, we need only calculate the concentration of HCl. In part (d) we found that the total volume at the equivalence point was 125.0 mL. The addition of 10.0 mL will bring that total volume to 135.0 mL of solution. The # of moles of excess HCl is:

(0.100 mol HCl/L • 0.0100 L = 0.00100 mol HCl contained within 135.0 mL, for a

concentration of 7.41×10^{-3} M HCl. Since HCl is a strong acid, the

pH $= -\log(7.41 \times 10^{-3})$ or 2.13.

(f) The titration curve.

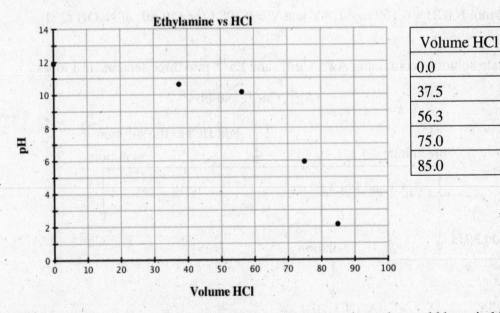

Volume HCl	pH
0.0	11.89
37.5	10.63
56.3	10.15
75.0	5.93
85.0	2.13

(g) Suitable indicator for endpoint: Alizarin or Bromcresol purple would be suitable indicators. (See Figure 18.10 for indicators.)

101. Make a buffer of pH= 2.50 from 100. mL of 0.230 M H_3PO_4 and 0.150 M NaOH.

For the first ionization of the acid, $K_a = 7.5 \times 10^{-3}$

The pK_a of the acid = 2.12. The Henderson-Hasselbalch equation will help to determine the ratio of the conjugate pair to make the pH 2.50.

We'll assume that (since the second K_a is approximately 10^{-8}, that we're dealing only with the first ionization. This *is an approximation*.

$$pH = pK_a + \log \frac{[\text{conjugate base}]}{[\text{acid}]} \text{ so } 2.50 = 2.12 + \log \frac{[\text{conjugate base}]}{[\text{acid}]}$$

$$0.38 = \log \frac{[A]}{[HA]} \text{ and the ratio of } \frac{[A]}{[HA]} = 2.37$$

The amount of HA = (0.100 L • 0.230M HA) = 0.0230 mol HA.

Rearranging the ratio: [A] = 2.37[HA]. Knowing that the acid present is *either* the molecular acid, HA, or the conjugate base of the acid, A, we can write:

A + HA = 0.0230 mol and with A = 2.37 HA

2.37 HA + HA = 0.0230mol ; 3.37 HA = 0.0230 mol; and HA =   0.0068 mol.

We need to consume (0.0230 mol HA- 0.0068 mol HA) or 0.0162 mol HA.

This requires 0.0162 mol NaOH, and from 0.150M NaOH we need:

0.0162 mol NaOH = 0.150 mol/L • V and V = 0.108 L or 110 mL of NaOH (2 sf).

103. Separate solutions containing Ag^+, Cu^{2+}, and Pb^{2+} into three separate test tubes:

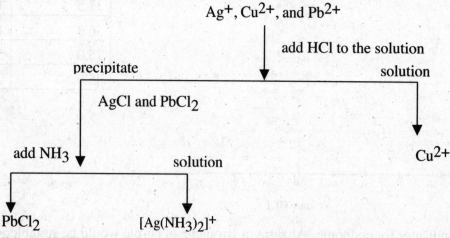

The addition of HCl precipitate both silver and lead(II) ions. Separating the filtrate from the precipitate removes the copper(II) ions.

346

Lead(II) and silver can be separated in one of two ways:

(a) lead(II) chloride is soluble in very hot water, so the addition of hot water to the precipitate will dissolve the lead chloride. Filtering the hot solution will remove the dissolved lead salt from the silver chloride.

(b) Silver forms a complex ion with ammonia, while lead(II) does not. Adding NH_3 to the solution containing both silver and lead(II) will dissolve the AgCl solid—as the silver complex ion forms. Filtering the remaining solid lead(II) chloride will separate the lead salt from the silver complex ion.

SUMMARY AND CONCEPTUAL QUESTIONS

105. To separate CuS and $Cu(OH)_2$:

The K_{sp} for CuS is 6×10^{-37} and that for $Cu(OH)_2$ is 2.2×10^{-20}.

The small size of the K_{sp} for CuS indicates that it is not very soluble. Addition of acid will cause the more soluble hydroxide to dissolve, leaving the CuS in the solid state.

107. Silver phosphate can be more soluble in water than calculated from Ksp data owing to competing reactions. The phosphate anion is also the conjugate base of a weak acid, and will undergo a reaction with water (hydrolysis), which lowers the $[PO_4^{3-}]$, and causes the solubility equilibrium to shift to the right, increasing the solubility of the phosphate salt. This is another example of LeChatelier's Principle in operation.

109. For the equilibrium between acetic acid and its conjugate base (acetate ion).

(a). One can understand the fraction of acetic acid present by viewing the Ka expression for the acid:

$$Ka = \frac{[CH_3CO_2^-][H_3O^+]}{[CH_3CO_2H]}$$, As pH increases, the concentration of H_3O^+ decreases. Since Ka remains constant, the "mathematical" result is that the "numerator term" ($[CH_3CO_2^-]$) must increase at the expense of the "denominator term" ($[CH_3CO_2H]$). Chemically, this means that the fraction of CH_3CO_2H decreases.

(b) Predominant species at pH = 4.0

This is easy to answer by rearranging the Ka expression for the acid.

$$\frac{Ka}{[H_3O^+]} = \frac{[CH_3CO_2^-]}{[CH_3CO_2H]}$$. Substituting values for Ka and the $[H_3O^+]$

$$\frac{1.8 \times 10^{-5}}{1.0 \times 10^{-4}} = \frac{[CH_3CO_2^-]}{[CH_3CO_2H]} = 0.18.$$

This means that the molecular acid is the predominant specie at pH = 4.

Predominant species at pH = 6.0?

Once again, examine the Ka expression, substituting 1×10^{-6} for the $[H_3O^+]$.

$$\frac{1.8 \times 10^{-5}}{1.0 \times 10^{-6}} = \frac{[CH_3CO_2^-]}{[CH_3CO_2H]} = 18.$$ Now the conjugate base is present at a concentration

level almost 20 times that of the molecular acid.

(c) The pH at the equivalence point (when the fraction of acid is equal to the fraction of conjugate base). Let's calculate the $[H_3O^+]$ at this point.

$$1.8 \times 10^{-5} = \frac{[0.5][H_3O^+]}{[0.5]}$$. Note that I simply chose 0.5 as the concentration for both

species—any number here will do. So $[H_3O^+] = 1.8 \times 10^{-5}$ and

pH = -log(1.8 × 10⁻⁵) or 4.74.

111. For salicylic acid:

(a) Approximate values for the bond angles:

Bond	Angle
i	120
ii	120
iii	109
iv	120

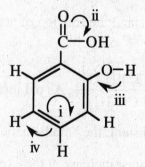

(b) The hybridization of C in the ring is sp^2. With three atoms attached to any one carbon, the three groups separate by an angle of approximately 120°. This is also true of the C in the carboxylate group.

(c) What is Ka of acid?

Using SA to represent the molecular acid, the monoprotic acid would have a Ka expression of:

$Ka = \dfrac{[H^+][A^-]}{[SA]}$. We calculate the concentration of SA:

$$\dfrac{\dfrac{1.00 \text{ g SA}}{122.12 \text{ g/mol SA}}}{0.460 \text{ L}} = 1.78 \times 10^{-2} \text{ mol/L SA}$$

Noting that pH = 2.4, lets us know that $[H^+] = 10^{-2.4}$ or 3.98×10^{-3} M.

Since SA is a monoprotic acid, the concentration of A$^-$ is also 3.98×10^{-3} M.

The Ka is then: $\dfrac{[3.98 \times 10^{-3}][3.98 \times 10^{-3}]}{(1.78 \times 10^{-2}) - (3.98 \times 10^{-3})} = 1.15 \times 10^{-3}$ or 1×10^{-3} (1 sf)

(d) If pH = 2.0, the % of salicylate ion present:

$Ka = \dfrac{[H^+][A^-]}{[SA]} = 1 \times 10^{-3}$; rearranging to isolate the salicylate ion: $\dfrac{[A^-]}{[SA]} = \dfrac{1 \times 10^{-3}}{1 \times 10^{-2}}$

At pH = 2.0, the salicylate ion is 10% of the acid form, with the molecular acid, SA, being 90% of the acid present.

(e) In a titration of 25.0 mL of a 0.014M solution of SA titrated with 0.010 M NaOH:

(i) What is pH at the halfway point?

At the halfway point, the $[A^-]$ = [SA] so $[H^+] = 1 \times 10^{-3}$, so pH = 3.0

(ii) What is pH at the equivalence point?

At the equivalence point, only salt (A$^-$) remains. Begin by asking "What's the concentration of the salt?". The amount of SA present is

(0.0250L)(0.014 mol SA/L) = 0.00035 mol.

The amount of 0.010 M NaOH that contains this number of moles of NaOH is (0.00035mol)/(0.010 mol NaOH/L) = 0.035 L or 35 mL. At the equivalence point, the total volume is then (35 + 25.0) or 60. mL.

The concentration is then: 3.5×10^{-4} mol/0.060 L = 5.8×10^{-3} M.

Substituting into the Kb expression for the salicylate anion reacting:

$$A^- (aq) + H_2O(l) \Leftrightarrow HA(aq) + OH^-(aq)$$

$$Kb = \frac{[HA][OH^-]}{[A^-]} = \frac{1.0 \times 10^{-14}}{1.0 \times 10^{-3}}$$

Since [HA] = [OH$^-$], we can substitute into the Kb expression to get:

$$\frac{[OH^-]^2}{[5.8 \times 10^{-3}]} = 1.0 \times 10^{-11} \text{ and } [OH^-]^2 = 5.8 \times 10^{-14} ; [OH^-] = 2.4 \times 10^{-7} \text{ M},$$

so pOH = 6.62, and pH = 7.38.

Applying Chemical Principles

1. What volume (in mL) of 0.035 mass percent NaCN solution contains 0.10 g of sodium cyanide? Assume the density of the solution is 1.0 g/mL.

$$\frac{0.10 \text{ g NaCN}}{1} \bullet \frac{100 \text{ g solution}}{0.035 \text{ g NaCN}} \bullet \frac{1.0 \text{ mL solution}}{1.0 \text{ g solution}} = 290 \text{ mL} \quad (2 \text{ sf})$$

3. What is the equilibrium concentration of $Au^+(aq)$ in a solution that is 0.0071 M CN^- and 1.1×10^{-4} M $[Au(CN)_2]^-$. From Appendix K, $K = 2.0 \times 10^{38}$.

The equilibrium expression is: $\dfrac{\left[Au(CN)_2^-\right]}{\left[Au^{+1}\right]\left[CN^{-1}\right]^2} = 2.0 \times 10^{38}$

Rearranging the equation to solve for $[Au^{+1}]$: $\dfrac{\left[Au(CN)_2^-\right]}{2.0 \times 10^{38}\left[CN^{-1}\right]^2} = \left[Au^{+1}\right]$

Substituting the given concentrations: $\dfrac{\left[1.1 \times 10^{-4}\right]}{2.0 \times 10^{38}\left[7.1 \times 10^{-3}\right]^2} = \left[Au^{+1}\right] = 1.1 \times 10^{-38}$M

Examining the magnitude of the gold ion concentration above, it is **very** reasonable to conclude that 100% of the gold in solution is present as the $[Au(CN)_2]^-$ complex ion.

5. Write a balanced chemical equation for the reaction of $NaAu(CN)_2(aq)$ and $Zn(s)$:

$$Zn(s) + 2 Na^+(aq) + 2 [Au(CN)_2]^-(aq) \Leftrightarrow 2 Na^+(aq) + 2 Au(s) + [Zn(CN)_4]^{2-} (aq)$$

Chapter 19
Principles of Reactivity: Entropy and Free Energy

Units for thermodynamic processes are typically expressed for the balanced equation given. Hence the equation for the formation of HCl: $H_2 + Cl_2 \rightarrow 2HCl$ has a $\Delta_r G°$, $\Delta_r H°$, and $\Delta_r S°$ that represent the formation of 2 mol of HCl. We can express this as "energy change"/mol-rxn, where "energy change" typically has units of kJ (for G,H) and J/K (for S). In this chapter, we shall omit the "mol-rxn" notation in the interest of brevity. *Unless otherwise noted*, answers should be read: $\Delta_r G°$ and $\Delta_r H°$, units will be **kJ/mol-rxn**, and for $\Delta_r S°$, units will be **J/K• mol-rxn**.

PRACTICING SKILLS

Entropy

1. Compound with the higher entropy:

   (a) CO_2 (s) at –78° vs **CO_2 (g) at 0 °C**: Entropy increases with temperature.

   (b) H_2O (l) at 25 °C vs **H_2O (l) at 50 °C**: Entropy increases with temperature.

   (c) Al_2O_3 (s) (pure) vs **Al_2O_3 (s) (ruby)**: Entropy of a solution (even a solid one) is greater than that of a pure substance.

   (d) **1 mol N_2 (g) at 1bar** vs 1 mol N_2 (g) at 10 bar: With the increased P, molecules have greater order.

3. Entropy changes:

   (a) KOH (s) $\rightarrow$ KOH(aq)
   $$\Delta_r S° = 91.6 \frac{J}{K•mol} (1 \text{ mol}) - 78.9 \frac{J}{K•mol}(1 \text{ mol}) = + 12.7 \frac{J}{K• \text{mol-rxn}}$$

   The increase in entropy reflects the greater disorder of the solution state.

   (b) Na (g) $\rightarrow$ Na (s)
   $$\Delta_r S° = 51.21 \frac{J}{K•mol} (1 \text{ mol}) - 153.765 \frac{J}{K•mol} (1 \text{ mol}) = -102.55 \frac{J}{K• \text{mol-rxn}}$$

   The lower entropy of the solid state is evidenced by the negative sign.

(c) Br_2 (l) $\rightarrow$ Br_2 (g)

$$\Delta_r S° = 245.42 \frac{J}{K\bullet mol} (1\ mol)\quad - 152.2 \frac{J}{K\bullet mol}(1\ mol)\quad = +93.2 \frac{J}{K\bullet mol\text{-rxn}}$$

The increase in entropy is expected with the transition to the disordered state of a gas.

(d) HCl (g) $\rightarrow$ HCl (aq)

$$\Delta_r S° = 56.5 \frac{J}{K\bullet mol} (1\ mol)\quad - 186.2 \frac{J}{K\bullet mol} (1\ mol)\quad = -129.7 \frac{J}{K\bullet mol\text{-rxn}}$$

The lowered entropy reflects the greater order of the solution state over the gaseous state.

5. Standard Entropy change for compound formation from elements:

(a) HCl (g): Cl_2 (g) + H_2 (g) $\rightarrow$ 2 HCl (g)

$$\Delta_r S° = 2 \bullet S° \text{ HCl (g)} - [1 \bullet S° \text{ } Cl_2 \text{ (g)} + 1 \bullet S° \text{ } H_2 \text{ (g)}]$$

$$= (2\ mol)(186.2 \frac{J}{K\bullet mol}) - [(1\ mol)(223.08 \frac{J}{K\bullet mol}) + (1\ mol)(130.7 \frac{J}{K\bullet mol})]$$

$$= +18.6 \frac{J}{K}\quad \text{and} +9.3 \frac{J}{K\bullet mol\text{-rxn}}$$

(b) $Ca(OH)_2$ (s): Ca(s) + O_2 (g) + H_2 (g) $\rightarrow$ $Ca(OH)_2$ (s)

$$\Delta_r S° = 1 \bullet S° \text{ } Ca(OH)_2 \text{ (s)} - [1 \bullet S° \text{ Ca (s)} + 1 \bullet S° \text{ } O_2 \text{ (g)} + 1 \bullet S° \text{ } H_2 \text{ (g)}]$$

$$= (1\ mol)(83.39 \frac{J}{K\bullet mol}) - [(1\ mol)(41.59 \frac{J}{K\bullet mol}) + (1\ mol)(205.07 \frac{J}{K\bullet mol}) +$$

$$(1\ mol)(130.7 \frac{J}{K\bullet mol})]$$

$$= -293.97 \frac{J}{K\bullet mol\text{-rxn}} \text{ (or } -294.0 \frac{J}{K\bullet mol\text{-rxn}} \text{ to 4 sf)}$$

7. Standard molar entropy changes for:

(a) 2 Al (s) + 3 Cl_2 (g) $\rightarrow$ 2 $AlCl_3$ (s)

$$\Delta_r S° = 2 \bullet S° \text{ } AlCl_3 \text{ (s)} - [2 \bullet S° \text{ Al (s)} + 3 \bullet S° \text{ } Cl_2 \text{ (g)}]$$

$$= (2\ mol)(109.29 \frac{J}{K\bullet mol}) - [(2\ mol)(28.3 \frac{J}{K\bullet mol}) + (3\ mol)(223.08 \frac{J}{K\bullet mol})]$$

$$= -507.3 \frac{J}{K\bullet mol\text{-rxn}}$$

(b) 2 CH_3OH (l) + 3 O_2 (g) $\rightarrow$ 2 CO_2 (g) + 4 H_2O (g)

$$\Delta_r S° = [2 \cdot S° \ CO_2 \ (g) + 4 \cdot S° \ H_2O \ (g)] - [2 \cdot S° \ CH_3OH \ (l) + 3 \cdot S° \ O_2 \ (g)]$$

$$= [(2 \ mol)(213.74 \ \frac{J}{K \cdot mol}) + (4 \ mol)(188.84 \ \frac{J}{K \cdot mol})] -$$

$$[(2 \ mol)(127.19 \ \frac{J}{K \cdot mol}) + (3 \ mol)(205.07 \ \frac{J}{K \cdot mol})]$$

$$= + 313.25 \ \frac{J}{K \cdot mol\text{-}rxn}$$

In (a) we see the sign of $\Delta_r S°$ as negative—expected since the reaction results in the decrease in the number of moles of gas. In (b), the sign of is positive—as expected, since the reaction produces a larger number of moles of gaseous products than gaseous reactants.

$\Delta_r S°$ (universe) and Spontaneity

9. Is the reaction: Si (s) + 2 Cl$_2$ (g) → SiCl$_4$ (g) spontaneous?

$$\Delta_r S° \ (system) = 1 \cdot S° \ SiCl_4 \ (g) \ - [1 \cdot S° \ Si \ (s) + 2 \cdot S° \ Cl_2 \ (g)]$$

$$= (1 \ mol)(330.86 \ \frac{J}{K \cdot mol}) - [(1 \ mol)(18.82 \ \frac{J}{K \cdot mol}) + (2 \ mol)(223.08 \ \frac{J}{K \cdot mol})]$$

$$= -134.12 \ \frac{J}{K \cdot mol\text{-}rxn}$$

To calculate $\Delta_r S°$ (surroundings), we calculate $\Delta H°$(system):

$$\Delta H° = 1 \cdot \Delta H° \ SiCl_4 \ (g) \ - [1 \cdot \Delta H° \ Si \ (s) + 2 \cdot \Delta H° \ Cl_2 \ (g)]$$

$$= (1 \ mol)(- 662.75 \frac{kJ}{mol}) - [(1 \ mol)(0 \ \frac{kJ}{mol}) + (2 \ mol)(0 \ \frac{kJ}{mol})] = -662.75 \ \frac{kJ}{mol\text{-}rxn}$$

$$\Delta_r S° \ (surroundings) = - \Delta H°/T = (662.75 \times 10^3 \ J/mol)/298.15 \ K = 2222.9 \ \frac{J}{K \cdot mol\text{-}rxn}$$

$$\Delta_r S° \ (universe) = \Delta_r S° \ (system) + \Delta_r S° \ (surroundings) =$$

$$(-134.12 + 2222.9) = 2088.7 \frac{J}{K \cdot mol\text{-}rxn}$$

11. Is the reaction: 2 H$_2$O(l) → 2 H$_2$ (g) + O$_2$ (g) spontaneous?

$$\Delta_r S° \ (system) = [2 \cdot S° \ H_2 \ (g) + 1 \cdot S° \ O_2 \ (g)] - 2 \cdot S° \ H_2O \ (l)$$

$$= [(2 \text{ mol})(130.7 \tfrac{J}{K\bullet mol}) + (1 \text{ mol})(205.07 \tfrac{J}{K\bullet mol})] - (2 \text{ mol})(69.95 \tfrac{J}{K\bullet mol})$$

$$= +326.57 \tfrac{J}{K} \text{ and for decomposition of 1 mol of water: } +326.57/2 =$$

$$163.3 \tfrac{J}{K\bullet mol\text{-}rxn}$$

To calculate $\Delta_r S°$ (surroundings), we calculate $\Delta H°$(system):

$$\Delta H°\text{(system)} = [2 \bullet \Delta H° \, H_2 \, (g) + 1 \bullet \Delta H° \, O_2 \, (g)] - 2 \bullet \Delta H° \, H_2O \, (l)$$

$$= [(2 \text{ mol})(0 \tfrac{kJ}{mol}) + (1 \text{ mol})(0 \tfrac{kJ}{mol})] - (2 \text{ mol})(-285.83 \tfrac{kJ}{mol})$$

$$= +571.66 \tfrac{kJ}{mol} \text{ and for decomposition of 1 mol of water: } +571.66/2 = 285.83 \tfrac{kJ}{mol\text{-}rxn}$$

$$\Delta_r S° \text{ (surroundings)} = -\Delta H°/T = -(285.83 \times 10^3 \text{ J/mol})/298.15 \text{ K} = -958.68 \tfrac{J}{K\bullet mol\text{-}rxn}$$

$$\Delta_r S° \text{ (universe)} = \Delta_r S° \text{ (system)} + \Delta_r S° \text{ (surroundings)} = 163.3 + -958.68$$

$$\Delta_r S° \text{ (universe)} = -795.4 \tfrac{J}{K\bullet mol\text{-}rxn}.$$

Since this value < zero, the process is not spontaneous.

13. Using Table 19.2 classify each of the reactions:

(a) $\Delta H°_{system}$ = - , $\Delta_r S°_{system}$ = - Product-favored at lower T

(b) $\Delta H°_{system}$ = + , $\Delta_r S°_{system}$ = - Not product-favored under any conditions

Gibbs Free Energy

15. Calculate $\Delta_r G°$ for:

(a) $2 \, Pb \, (s) + O_2 \, (g) \rightarrow 2 \, PbO \, (s)$

$$\Delta_r H° = (2 \text{ mol})(-219 \tfrac{kJ}{mol}) - [0 + 0] = -438 \text{ kJ}$$

$$\Delta_r S° = (2 \text{ mol})(66.5 \tfrac{J}{K\bullet mol}) -$$

$$[(2 \text{ mol})(64.81 \tfrac{J}{K\bullet mol}) + (1 \text{ mol})(205.07 \tfrac{J}{K\bullet mol})] = -201.7 \text{ J/K}$$

$$\Delta_r G° = \Delta_r H° - T\Delta_r S° = -438 \text{ kJ} - (298.15 \text{ K})(-201.7 \text{ J/K})(\frac{1.000 \text{ kJ}}{1000. \text{ J}}) = -378 \text{ kJ}$$

Reaction is product-favored since $\Delta G < 0$. With the very large negative ΔH, the process is enthalpy driven.

(b) $NH_3(g) + HNO_3(aq) \rightarrow NH_4NO_3(aq)$

$$\Delta_r H° = (1 \text{ mol})(-339.87 \frac{kJ}{mol}) - [(1 \text{ mol})(-45.90 \frac{kJ}{mol}) + (1 \text{ mol})(-207.36 \frac{kJ}{mol})] = -86.61 \text{ kJ}$$

$$\Delta_r S° = (1 \text{ mol})(259.8 \frac{J}{K\bullet mol}) -$$

$$[(1 \text{ mol})(192.77 \frac{J}{K\bullet mol}) + (1 \text{ mol})(146.4 \frac{J}{K\bullet mol})] = -79.4 \frac{J}{K\bullet mol\text{-rxn}}$$

$$\Delta_r G° = \Delta_r H° - T\Delta_r S° = -86.61 \text{ kJ} - (298.15 \text{ K})(-79.4 \text{ J/K})(\frac{1.000 \text{ kJ}}{1000. \text{ J}}) = -62.9 \frac{kJ}{mol\text{-rxn}}$$

Reaction is product-favored since $\Delta G < 0$. With the very large negative ΔH, the process is enthalpy driven.

17. Calculate the molar free energies of formation for:
   (a) $CS_2(g)$; The reaction is: C (graphite) + 2 S (s,rhombic) $\rightarrow CS_2(g)$

$$\Delta_r H° = (1 \text{ mol})(116.7 \frac{kJ}{mol}) - [0 + 0] = +116.7 \text{ kJ}$$

$$\Delta_r S° = (1 \text{ mol})(237.8 \frac{J}{K\bullet mol}) - [(1 \text{ mol})(5.6 \frac{J}{K\bullet mol}) + (2 \text{ mol})(32.1 \frac{J}{K\bullet mol})]$$

$$= +168.0 \text{ J/K}$$

$$\Delta_f G° = \Delta_f H° - T\Delta_r S° = (116.7 \text{ kJ}) - (298.15 \text{ K})(168.0 \frac{J}{K})(\frac{1.000 \text{ kJ}}{1000 \text{ J}})$$

$$= +66.6 \text{ kJ} \qquad \text{Appendix value: } 66.61 \text{ kJ/mol}$$

   (b) NaOH (s)   The reaction is: Na (s) + $\frac{1}{2}$ $O_2(g)$ + $\frac{1}{2}$ $H_2(g)$ $\rightarrow$ NaOH (s)

$$\Delta_r H° = (1 \text{ mol})(-425.93 \frac{kJ}{mol}) - [0 + 0 + 0] = -425.93 \frac{kJ}{mol\text{-rxn}}$$

$$\Delta_r S° = (1 \text{ mol})(64.46 \frac{J}{K\bullet mol}) -$$

$$[(1 \text{ mol})(51.21 \frac{J}{K\bullet mol}) + (\frac{1}{2} \text{ mol})(205.07 \frac{J}{K\bullet mol}) + (\frac{1}{2} \text{ mol})(130.7 \frac{J}{K\bullet mol})]$$

$$= -154.6 \; \frac{J}{K \cdot mol\text{-}rxn}$$

$$\Delta_r G° = \Delta_r H° - T\Delta_r S° = (-425.93 \text{ kJ}) - (298.15 \text{ K})(-154.6 \text{ J/K})(\frac{1.000 \text{ kJ}}{1000. \text{ J}})$$

$$= -379.82 \text{ kJ} \qquad \text{Appendix value: } -379.75 \text{ kJ/mol}$$

(c) ICl (g) The reaction is: $\frac{1}{2} I_2 (g) + \frac{1}{2} Cl_2 (g) \rightarrow ICl (g)$

$$\Delta_r H° = (1 \text{ mol})(+17.51 \; \frac{kJ}{mol}) - [0 + 0] = +17.51 \; \frac{kJ}{mol\text{-}rxn}$$

$$\Delta_r S° = (1 \text{ mol})(247.56 \; \frac{J}{K \cdot mol}) -$$

$$[(\frac{1}{2}\text{mol})(116.135 \; \frac{J}{K \cdot mol}) + (\frac{1}{2} \text{ mol})(223.08 \; \frac{J}{K \cdot mol})]$$

$$= +77.95 \; \frac{J}{K \cdot mol\text{-}rxn} \text{ J/K}$$

$$\Delta_r G° = \Delta_r H° - T\Delta_r S° = (+17.51 \text{ kJ}) - (298.15 \text{ K})(+77.95 \text{ J/K})(\frac{1.000 \text{ kJ}}{1000. \text{ J}})$$

$$= -5.73 \text{ kJ} \qquad \text{Appendix value: } -5.73 \text{ kJ/mol}$$

Free Energy of Formation

19. Calculate $\Delta_r G°$ for the following equations. Are they product-favored?

(a) $2 K (s) + Cl_2 (g) \rightarrow 2 KCl (s)$

$$\Delta_r G° = [2 \cdot \Delta_f G° \text{ KCl(s)}] - [1 \cdot \Delta_f G° \text{ Cl}_2\text{(g)} + 2 \cdot \Delta_f G° \text{ K (s)}]$$

$$= [(2 \text{ mol})(-408.77 \; \frac{kJ}{mol}] - [(1 \text{ mol})(0 \; \frac{kJ}{mol}) + (2 \text{ mol})(0 \; \frac{kJ}{mol})]$$

$$= -817.54 \text{ kJ}$$

With a $\Delta G < 0$, the reaction is product-favored.

(b) $2 CuO (s) \rightarrow 2 Cu (s) + O_2 (g)$

$$\Delta_r G° = [2 \cdot \Delta_f G° \text{Cu(s)} + \Delta_f G° O_2 (g)] - [2 \cdot \Delta_f G° \text{ CuO (s)}]$$

$$\Delta_r G° = [(2 \text{ mol})(0 \; \frac{kJ}{mol}) + (1 \text{ mol})(0 \; \frac{kJ}{mol})] - [(2\text{mol})(-128.3 \; \frac{kJ}{mol})]$$

$$= +256.6 \text{ kJ}$$

With a $\Delta G > 0$, the reaction is not product-favored.

(c) $4\,NH_3\,(g) + 7\,O_2\,(g) \rightarrow 4\,NO_2\,(g) + 6\,H_2O\,(g)$

$\Delta_r G° = [4 \cdot \Delta G°_f\,NO_2\,(g) + 6 \cdot \Delta G°_f\,H_2O\,(g)] - [4 \cdot \Delta G°_f\,NH_3\,(g) + 7 \cdot \Delta G°_f\,O_2\,(g)]$

$\Delta_r G° = [(4\,mol)(+51.23\,\frac{kJ}{mol}) + (6\,mol)(-228.59\,\frac{kJ}{mol}] -$

$[(4\,mol)(-16.37\,\frac{kJ}{mol}) + (7\,mol)(0\,\frac{kJ}{mol})] = -1101.14\,kJ$

With a $\Delta G < 0$, the reaction is product-favored.

21. Value for $\Delta_f G°$ of $BaCO_3(s)$:

$\Delta_r G° = [\Delta_f G°\,BaO(s) + \Delta_f G°\,CO_2(g)] - [\Delta_f G°\,BaCO_3(s)]$

$+219.7\,kJ = [(1\,mol)(-520.38\,\frac{kJ}{mol}) + (1\,mol)(-394.359\,\frac{kJ}{mol})] - \Delta G°_f\,BaCO_3(s)$

$+219.7\,kJ = -914.74\,kJ - \Delta_f G°\,BaCO_3(s)$

$-1134.4\,kJ/mol = + \Delta_f G°\,BaCO_3(s)$

Effect of Temperature on ΔG

23. Entropy-favored or Enthalpy-favored reactions?

(a) $N_2\,(g) + 2\,O_2\,(g) \rightarrow 2\,NO_2\,(g)$

$\Delta_r H° = (2\,mol)(+33.1\,\frac{kJ}{mol}) - [0 + 0] = +66.2\,\frac{kJ}{mol\text{-}rxn}$

$\Delta_r S° = (2\,mol)(+240.04\,\frac{J}{K \bullet mol}) -$

$\quad [(1\,mol)(191.56\,\frac{J}{K \bullet mol}) + (2\,mol)(+205.07\,\frac{J}{K \bullet mol})] = -121.62\,\frac{J}{K \bullet mol\text{-}rxn}$

$\Delta_r G° = (2\,mol)(51.23\,\frac{kJ}{mol}) - [(1\,mol)(0\,\frac{kJ}{mol}) + (1\,mol)(0\,\frac{kJ}{mol})] = 102.5\,kJ$

The reaction is **not** entropy OR enthalpy favored. There is **no** T at which $\Delta G < 0$.

(b) $2\,C\,(s) + O_2\,(g) \rightarrow 2\,CO\,(g)$

$\Delta_r H° = (2\,mol)(-110.525\,\frac{kJ}{mol}) - [0 + 0] = -221.05\,\frac{kJ}{mol\text{-}rxn}$

$\Delta_r S° = (2\,mol)(+197.674\,\frac{J}{K \bullet mol}) -$

$$[(2 \text{ mol})(+5.6 \frac{J}{K \bullet mol}) + (1 \text{mol})(+205.07 \frac{J}{K \bullet mol})] = +179.1 \frac{J}{K \bullet \text{ mol-rxn}}$$

$$\Delta_r G° = (2 \text{ mol})(-137.168 \frac{kJ}{mol}) - [(1 \text{ mol})(0 \frac{kJ}{mol}) + (1 \text{ mol })(0 \frac{kJ}{mol})]$$

$$= -274.336 \text{ kJ}$$

This reaction is **both** entropy- and enthalpy-favored *at all temperatures*.

(c) CaO (s) + CO$_2$ (g) → CaCO$_3$ (s)

$$\Delta_r H° = (1 \text{ mol})(-1207.6 \frac{kJ}{mol}) - [(1\text{mol})(-635.0 \frac{kJ}{mol}) + (1\text{mol})(-393.509 \frac{kJ}{mol})]$$

$$= -179.0 \frac{kJ}{\text{mol-rxn}}$$

$$\Delta_r S° = (1 \text{ mol})(+91.7 \frac{J}{K \bullet mol}) -$$

$$[(1\text{mol})(38.2 \frac{J}{K \bullet mol}) + (1\text{mol})(+213.74 \frac{J}{K \bullet mol})] = -160.2 \frac{J}{K \bullet \text{ mol-rxn}}$$

$$\Delta_r G° = (1 \text{ mol})(-1129.16 \frac{kJ}{mol}) -$$

$$[(1\text{mol})(-603.42 \frac{kJ}{mol}) + (1\text{mol})(-394.359 \frac{kJ}{mol})] = -131.4 \frac{kJ}{\text{mol-rxn}}$$

This reaction is *enthalpy-favored* and will be spontaneous at low temperatures.

(d) 2 NaCl (s) → 2 Na (s) + Cl$_2$ (g)

$$\Delta_r H° = [(2 \text{ mol})(0 \frac{kJ}{mol}) + (1\text{mol})(0 \frac{kJ}{mol})] - (2\text{mol})(-411.12 \frac{kJ}{mol})]$$

$$= +822.24 \frac{kJ}{\text{mol-rxn}}$$

$$\Delta_r S° = [(2 \text{ mol})(+51.21 \frac{J}{K \bullet mol}) + (1\text{mol})(+223.08 \frac{J}{K \bullet mol})] -$$

$$(2\text{mol})(+72.11 \frac{J}{K \bullet mol})] = +181.28 \frac{J}{K \bullet \text{ mol-rxn}}$$

$$\Delta_r G° = [(2 \text{ mol})(0 \frac{kJ}{mol}) + (1\text{mol})(0 \frac{kJ}{mol})] - (2\text{mol})(-384.04 \frac{kJ}{mol})] = +768.08 \frac{kJ}{\text{mol-rxn}}$$

This reaction is *entropy-favored* and will be spontaneous at high temperatures.

25. For the decomposition of $MgCO_3$ (s) $\rightarrow$ MgO (s) + CO_2 (g):

(a) $\Delta_r S°$(system) = [1 • S° MgO (s) + 1 • S° CO_2 (g)] - 1 • S° $MgCO_3$ (s)

$$= [(1mol)(26.85 \frac{J}{K•mol}) + (1mol)(213.74 \frac{J}{K•mol})] - (1mol)(65.84 \frac{J}{K•mol})$$

$$= +174.75 \frac{J}{K• mol\text{-}rxn}$$

Calculate $\Delta H°$(system):

$\Delta H°$(system) = [1 • $\Delta H°$ MgO (s) + 1 • $\Delta H°$ CO_2 (g)] - 1 • $\Delta H°$ $MgCO_3$ (s)

$$= [(1\ mol)(-601.24 \frac{kJ}{mol}) + (1\ mol)(-393.509 \frac{kJ}{mol})] - (1\ mol)(-1111.69 \frac{kJ}{mol})$$

$$= +116.94 \frac{kJ}{mol\text{-}rxn}$$

Using the $\Delta H°$(system) and $\Delta_r S°$(system), we can calculate $\Delta_r G°$.

$$\Delta_r G° = \Delta H° - T • S° = +116.94 \frac{kJ}{mol\text{-}rxn} - (298\ K)(+174.75 \frac{J}{K• mol\text{-}rxn})(\frac{1\ kJ}{1000\ J}) =$$

$$= +64.87 \frac{kJ}{mol\text{-}rxn}.$$

(b) The sign of $\Delta_r G°$ is +, so the process is not spontaneous at 298 K.

(c) From Table 19.2 we observe that this type of reaction (ΔH = + and $\Delta_r S$ = +) is

   spontaneous at higher T (product-favored).

Free Energy and Equilibrium Constants

27. Calculate K_p for the reaction:

$$\frac{1}{2} N_2(g) + \frac{1}{2} O_2(g) \Leftrightarrow NO\ (g) \qquad \Delta_r G° = + 86.58\ kJ/mol\ NO$$

$\Delta_r G° = -\ RT \ln K_p$ so 86.58×10^3 J/mol = $- (8.3145 \frac{J}{K • mol})(298.15\ K) \ln K_p$

$- 34.926 = \ln K_p$ and $6.8 \times 10^{-16} = K_p$

Note that the + value of $\Delta_r G°$ results in a value of K_p which is small--reactants are favored.

A negative value would result in a large K_p -- a process in which the products were favored.

29. From the $\Delta G°$ and Kp, determine if the hydrogenation of ethylene is product-favored:

Using $\Delta_f G°$ data from the Appendix: C_2H_4 (g) + H_2 (g) $\Leftrightarrow$ C_2H_6 (g)

$$\Delta_r G° = (1 \text{ mol})(-31.89 \frac{kJ}{mol}) - [(1\text{mol})(68.35 \frac{kJ}{mol}) + (1\text{mol})(0 \frac{kJ}{mol})] = -100.24 \text{ kJ}$$

and since $\Delta_r G° = - RT \ln K_p$

$$-100.24 \times 10^3 \text{ J/mol} = - (8.3145 \frac{J}{K \bullet mol})(298.15 \text{ K}) \ln K_p$$

$$40.436 = \ln K_p \text{ and } K_p = 3.64 \times 10^{17}$$

The *negative value* of ΔG means the reaction is product-favored.

The *large value* of ΔG means that Kp is very large.

General Questions

31. Compound with higher standard entropy:

    (a) HF (g) vs HCl (g) vs **HBr (g)**: Entropy increases with molecular size (mass).

    (b) NH_4Cl (s) vs **NH_4Cl (aq)**: Entropy of solutions is greater than that of the solid.

    (c) N_2 (g) vs **C_2H_4 (g)**: Entropy increases with molecular complexity.

    (d) NaCl(s) vs **NaCl (g)**: Entropy of the gaseous state is very high. The solid state has lower
       entropy.

33. For the reaction C_6H_6 (l) + 3 H_2(g) $\rightarrow$ C_6H_{12}(l), $\Delta_r H° = - 206.7$ kJ, and $\Delta_r S° = -361.5$ J/K

    Is the reaction spontaneous under standard conditions? Calculating $\Delta G°$ will answer this
    question.

$$\Delta_r G° = \Delta_r H° - T\Delta_r S°$$

$$= -206.7 \text{ kJ} - (298.15 \text{ K})(-361.5 \frac{J}{K})(\frac{1.000 \text{ kJ}}{1000 \text{ J}}) = \text{ kJ}$$

$$= -206.7 \text{ kJ} - (-107.8 \text{ kJ}) = -98.9 \frac{kJ}{mol\text{-}rxn}$$

The negative value for $\Delta_r G°$ tells us that the reaction would be spontaneous (product-

favored) under standard conditions. The negative value for $\Delta_r H°$ tells us that the reaction is

enthalpy-driven.

35. Calculate $\Delta_r H°$ and $\Delta_r S°$ for the combustion of ethane:

$$C_2H_6\,(g)\quad+\quad 7/2\,O_2\,(g) \rightarrow 2\,CO_2(g)\quad+\quad 3\,H_2O\,(g)$$

$\Delta_f H°$ (kJ/mol)   -83.85   0   -393.509   -241.83

S° (J/K•mol)   +229.2   +205.07   +213.74   +188.84

$\Delta_r H° = [2 \cdot \Delta H°_f\,CO_2(g) + 3 \cdot \Delta H°_f\,H_2O\,(g)] - [1 \cdot \Delta H°_f\,C_2H_6\,(g) + 7/2 \cdot \Delta H°_f\,O_2\,(g)]$

$$= [(2mol)(-393.509\,\frac{kJ}{mol}) + (3mol)(-241.83\,\frac{kJ}{mol})] - [(1mol)(-83.85\,kJ/mol) + 0]$$

$$= -1428.66\,\frac{kJ}{mol\text{-}rxn}$$

$\Delta_r S° = [2 \cdot S°\,CO_2(g) + 3 \cdot S°\,H_2O\,(g)] - [1 \cdot S°\,C_2H_6\,(g) + 7/2 \cdot S°\,O_2\,(g)]$

$$= [(2\,mol)(213.74\,\frac{J}{K \cdot mol}) + (3\,mol)(188.84\,\frac{J}{K \cdot mol})] -$$

$$[(1\,mol)(229.2\,\frac{J}{K \cdot mol}) + (7/2\,mol)(205.07\,\frac{J}{K \cdot mol})] = +47.1\,\frac{J}{K \cdot mol\text{-}rxn}$$

$$\Delta_r S°(\text{surroundings}) = \frac{-\Delta H_{rxn}}{T} \quad (\text{Assuming we're at 298 K})$$

$$= \frac{1428.66\,kJ}{298.15\,K} \cdot \frac{1000\,J}{1\,kJ} \quad = 4791.7\,J/K$$

so $\Delta_r S°(\text{system}) + \Delta_r S°(\text{surroundings}) = +47.1\,J/K + 4791.7\,J/K$

$$= 4838.8\,J/K \text{ or } +4840\,\frac{J}{K \cdot mol\text{-}rxn}\,(3\,sf)$$

Since $\Delta_r H° = -$ and $\Delta_r S° = +$, the process is product-favored.

This calculation is consistent with our expectations. We know that hydrocarbons burn completely (in the presence of sufficient oxygen) to produce carbon dioxide and water.

37. (a) Calculate $\Delta_r G°$ for $NH_3\,(g) + HCl\,(g) \rightarrow NH_4Cl\,(s)$

S° (J/K•mol)   192.77   186.2   94.85
$\Delta_f H°$ (kJ/mol)   − 45.90   - 92.31   - 314..55

$\Delta_r S° = 1 \cdot S°\,NH_4Cl\,(s) - [1 \cdot S°\,NH_3\,(g) + 1 \cdot S°\,HCl\,(g)]$

$$= (1\,mol)(94.85\,J/K \cdot mol) - [(1\,mol)(192.77\,J/K \cdot mol) +$$

$$(1\,mol)(186.2\,J/K \cdot mol)] = -284.1\,J/K$$

$\Delta_r H° =$ 1 • $\Delta_f H°$ NH4Cl (s) - [1 • $\Delta_f H°$NH3 (g) + 1 • $\Delta_f H°$ HCl (g)]

$\qquad$ = (1 mol)(- 314.55 kJ/mol) - [(1 mol)(- 45.90 kJ/mol) + (1 mol)(- 92.31 kJ/mol)]

$\qquad$ = - 176.34 kJ

$\Delta_r G° = \Delta_r H° - T \Delta_r S°$

$\qquad$ = - 176.34 kJ - (298.15 K)(- 284.1 J/K)($\dfrac{1.000 \text{ kJ}}{1000 \text{ J}}$)

$\qquad$ = - 176.34 + 84.67 = - 91.64 $\dfrac{\text{kJ}}{\text{mol-rxn}}$

$\Delta_r S°$(surroundings) $\quad = \dfrac{- \Delta_r H}{T}$ (Assuming we're at 298 K)

$\qquad = \dfrac{176.34 \text{ kJ}}{298.15 \text{ K}} • \dfrac{1000 \text{ J}}{1 \text{ kJ}} \quad = 591.45$ J/K

so $\Delta_r S°$(system) + $\Delta_r S°$(surroundings) = -284.12 J/K + 591.45 J/K = +307.3 $\dfrac{\text{J}}{\text{K• mol-rxn}}$

The value for $\Delta_r G°$ for the equation is negative, indicating that is product-favored. The reaction is enthalpy driven (Δ_r H < 0).

(b) Calculate K_p for the reaction:

$\Delta_r G° = $ - RT ln K_p so -91.64 x 10^3 J/mol = - (8.3145 $\dfrac{\text{J}}{\text{K • mol}}$)(298.15 K) ln K_p

36.97 = ln K_p and 1.13 x 10^{16} = K_p

39. Calculate K_p for the formation of methanol from its elements:

Begin by calculating a $\Delta_r G°$

$\Delta_r G° = \Delta_f G_{CH3OH(l)} - [\Delta_f G_{C(graphite)} + 1/2 • \Delta_f G_{O2(g)} + 2 • \Delta_f G_{H2(g)}]$

$\Delta_r G° = $ -166.14 kJ – (0 kJ + 0 kJ + 0 kJ) = -166.14 kJ

$\Delta_r G° = $ - RT ln K_p

-166.1 x 10^3 J/mol = - (8.3145 $\dfrac{\text{J}}{\text{K • mol}}$)(298.15 K) ln K_p

67.00 = ln K_p and 1.3 x 10^{29} = K_p

The large value of K_p indicates that this process is product-favored at 298 K. Judging by the relative numbers of gaseous particles, (without doing a calculation), one can see that ΔrS for the reaction is < 0, so higher temperatures would reduce the value of K_p.

Regarding the connection between $\Delta G°$ and K, the more negative the value of $\Delta G°$, the larger the value of K.

41. Calculate the $\Delta_r S°$ for the vaporization of ethanol at 78.0 °C.

The $\Delta_r S = \dfrac{\Delta_{vap} H}{T} = \dfrac{39.3 \times 10^3 \text{ J}}{351 \text{ K}} = 112 \dfrac{\text{J}}{\text{K} \cdot \text{mol-rxn}}$

43. For the decomposition of phosgene:

$\Delta_r H° = [\Delta_f H° \text{CO (g)} + \Delta_f H° \text{ Cl}_2 \text{ (g)}] - [\Delta_f H° \text{ COCl}_2\text{(g)}]$

$\Delta_r H° = [(1 \text{ mol.})(-110.525 \dfrac{\text{kJ}}{\text{mol}}) + (1\text{mol})(0 \dfrac{\text{kJ}}{\text{mol}})] - [(1 \text{ mol})(-218.8 \dfrac{\text{kJ}}{\text{mol}})] = 108.275 \dfrac{\text{kJ}}{\text{mol}}$

$\Delta_r S° = [S° \text{ CO (g)} + S° \text{ Cl}_2 \text{ (g)}] - [S° \text{ COCl}_2\text{(g)}]$

$\Delta_r S° = [(1 \text{ mol})(197.674 \dfrac{\text{J}}{\text{K} \bullet \text{mol}}) + (1 \text{ mol})(223.07 \dfrac{\text{J}}{\text{K} \bullet \text{mol}})] -$

$[(1 \text{ mol})(283.53 \dfrac{\text{J}}{\text{K} \bullet \text{mol}})] = 137.2 \dfrac{\text{J}}{\text{K} \bullet \text{mol-rxn}}$

Using the $\Delta_r S$ data, we can see that raising the temperature will favor the endothermic decomposition of this substance.

45. For the reaction of sodium with water: Na (s) + H_2O(l) $\rightarrow$ NaOH(aq) + $\dfrac{1}{2}$ H_2 (g)

Predict signs for $\Delta_r H°$ and $\Delta_r S°$:

This one seems easy! The reaction of sodium with water gives off heat, and the heat frequently ignites the hydrogen gas that is concomitantly evolved. $\Delta_r H° = $ -.

Regarding entropy, the system changes from one with a solid (low entropy) and a liquid (higher entropy) to a solution (*frequently* higher entropy than liquid) and a gas (high entropy). So we would predict that the entropy would increase, i.e. $\Delta_r S° = $ +.

Now for the calculation:

$\Delta_r H° = [1 \bullet \Delta_f H° \text{ NaOH(aq)} + \dfrac{1}{2} \bullet \Delta_f H° H_2\text{(g)}] - [1 \bullet \Delta_f H° \text{ Na(s)} + 1 \bullet \Delta_f H° H_2O\text{(l)}]$

$= [(1 \text{ mol})(-469.15 \dfrac{\text{kJ}}{\text{mol}}) + (\dfrac{1}{2}\text{mol})(0)] - [(1 \text{ mol})(0) + (1 \text{ mol})(-285.83 \dfrac{\text{kJ}}{\text{mol}})]$

$= -183.32 \dfrac{\text{kJ}}{\text{mol-rxn}}$

$$\Delta_r S° = [1 \bullet S° \ NaOH(aq) + \frac{1}{2} \bullet S° \ H_2(g)] - [1 \bullet S° \ Na(s) + 1 \bullet S° \ H_2O(l)]$$

$$= [(1 \ mol)(48.1 \ \frac{J}{K \bullet mol}) + (\frac{1}{2} \ mol)(130.7 \ \frac{J}{K \bullet mol})] -$$

$$[(1 \ mol)(51.21 \ \frac{J}{K \bullet mol}) + (1 \ mol)(69.95 \ \frac{J}{K \bullet mol})]$$

$$= -7.7 \ \frac{J}{K \bullet mol\text{-}rxn}$$

As expected, the $\Delta_r H°$ for the reaction is negative! The surprise comes in the calculation for

$\Delta_r S°$. While we anticipate the sign to be positive, we find a slightly negative number—

reflecting the order (hence a decrease in entropy) that can occur as solutions occur.

47. For the reaction : $BCl_3(g) + 3/2 \ H_2(g) \rightarrow B(s) + \ 3HCl(g)$

$S° (\frac{J}{K \bullet mol})$	290.17	130.7	5.86	186.2
$\Delta_f H° (\frac{kJ}{mol})$	-402.96	0	0	-92.31

$$\Delta_r H° = [3 \bullet \Delta_f H° \ HCl(g) + 1 \bullet \Delta_f H° \ B(s)] - [1 \bullet \Delta_f H° \ BCl_3(g) + 3/2 \ \Delta_f H° \ H_2(g)]$$

$$\Delta_r H° = [(3 \ mol)(-92.31 \frac{kJ}{mol}) + (1 \ mol)(0)] - [(1 \ mol)(-402.96 \frac{kJ}{mol}) + (3/2 \ mol)(0)]$$

$$= 126.03 \ \frac{kJ}{mol\text{-}rxn}$$

$$\Delta_r S° = [3 \bullet S° \ HCl(g) + 1 \bullet S° \ B(s)] - [1 \bullet S° \ BCl_3(g) + 3/2 \ S° \ H_2(g)]$$

$$\Delta_r S° = [(3 \ mol)(186.2 \ \frac{J}{K \bullet mol}) + (1 \ mol)(5.86 \ \frac{J}{K \bullet mol})] - [(1 mol)(290.17 \frac{J}{K \bullet mol})$$

$$+ (3/2 \ mol)(130.7 \frac{J}{K \bullet mol})] = \ 78.2 \frac{J}{K \bullet mol\text{-}rxn}$$

$$\Delta_r G° = \ \Delta_r H° - T \ \Delta_r S° = 126.03 \ kJ - (298.15 \ K)(78.2 \frac{J}{K})(\frac{1.000 \ kJ}{1000 \ J}) = 103 \ \frac{kJ}{mol\text{-}rxn}$$

The reaction is not product-favored.

49. Calculate $\Delta_r G°$ for conversion of N_2O_4 to NO_2:

$\Delta_r G° = - RTln \ K = - (8.3145 \ x \ 10^{-3} \ kJ/K \bullet mol)(298 \ K) \ ln \ 0.14 = 4.87 \ kJ$

compare with the calculated $\Delta_f G°$ values:

$\Delta_r G° = 2 \ \Delta_f G° \ [NO_2 \ (g)] - \Delta_f G° [\ N_2O_4 \ (g)] = (2 \bullet 51.23 \ kJ/mol) - (1 \bullet 97.73 \ kJ/mol) = 4.73 \ kJ$

51. Calculate $\Delta_r G°$ for conversion of butane to isobutane, given K = 2.50:

$$\Delta_r G° = -RT\ln K = -(8.3145 \times 10^{-3} \text{ kJ/K} \cdot \text{mol})(298 \text{ K}) \ln 2.50 = -2.27 \frac{\text{kJ}}{\text{mol-rxn}}$$

53. For the reaction: $2 \text{ SO}_3 \text{ (g)} \Leftrightarrow 2 \text{ SO}_2 \text{ (g)} + \text{O}_2 \text{ (g)}$

$$\Delta_r H° = [2 \cdot \Delta_f H° \text{ SO}_2 \text{ (g)} + 1 \cdot \Delta_f H° \text{ O}_2 \text{ (g)}] - [2 \cdot \Delta_f H° \text{ SO}_3 \text{ (g)}]$$

$$= [(2 \text{ mol})(-296.84 \frac{\text{kJ}}{\text{mol}}) + 0] - [(2 \text{ mol})(-395.77 \frac{\text{kJ}}{\text{mol}})] = 197.86 \text{ kJ}$$

$$\Delta_r S° = [2 \cdot S° \text{ SO}_2 \text{ (g)} + 1 \cdot S° \text{ O}_2 \text{ (g)}] - [2 \cdot S° \text{ SO}_3 \text{ (g)}]$$

$$= [(2 \text{ mol})(248.21 \frac{\text{J}}{\text{K} \cdot \text{mol}}) + (1 \text{ mol})(205.07 \frac{\text{J}}{\text{K} \cdot \text{mol}})] - [(2 \text{ mol})(256.77 \frac{\text{J}}{\text{K} \cdot \text{mol}})]$$

$$= 187.95 \frac{\text{J}}{\text{K} \cdot \text{mol-rxn}}$$

(a) Is the reaction product-favored at 25 °C?:

$$\Delta_r G° = \Delta_r H° - T\Delta_r S° = 197.86 \text{ kJ} - (298.15 \text{ K})(187.95 \frac{\text{J}}{\text{K}})(\frac{1.000 \text{ kJ}}{1000 \text{ J}}) = 141.82 \frac{\text{kJ}}{\text{mol-rxn}}$$

The reaction is not product-favored.

(b) The reaction can become product-favored if there is some T at which $\Delta r G° < 0$. To see if such a T is feasible, let's set $\Delta r G° = 0$ and solve for T! *Remember that the units of energy must be the same, so let's convert units of J (for the entropy term) into units of kJ*

$$\Delta_r G° = \Delta_r H° - T\Delta_r S°$$

$$0 = 197.86 \text{ kJ} - T(0.18795 \frac{\text{kJ}}{\text{K}})$$

$$T = \frac{197.86 \text{ kJ}}{0.18795 \frac{\text{kJ}}{\text{K}}} = 1052.7 \text{ K} \quad \text{or} \quad (1052.7 - 273.1) = 779.6 \text{ °C}$$

(c) The equilibrium constant for the reaction at 1500 °C. Since we know that

$$\Delta_r G° = \Delta_r H° - T\Delta_r S° = -RT\ln K, \text{ we can solve for K if we know } \Delta_r G° \text{ at} 1500 \text{ °C}$$

$$\Delta_r G° = \Delta_r H° - T\Delta_r S° = -RT\ln K$$

$$= 197.86 \text{ kJ} - (1773 \text{ K})(187.95 \text{ J/K})(\frac{1.000 \text{ kJ}}{1000 \text{ J}}) = -135.4 \text{ kJ}$$

Substitute into the equation $(\Delta_r G^\circ = - RT \ln K)$:

$$- 135.4 \text{ kJ} = - 8.314 \frac{J}{K \cdot mol} \cdot \frac{1 \text{ kJ}}{1000 \text{ J}} \cdot 1773 \text{ K} \cdot \ln K \text{ and } K = 9.7 \times 10^3 \text{ or } 1 \times 10^4 \text{ (1 sf)}$$

55. Reaction: H_2S (g) + 2 O_2 (g) → H_2SO_4 (l)

    $\Delta_f H^\circ$(kJ/mol)   -20.63 0   - 814

    S° (J/K • mol)   205.79   205.07   156.9

$$\Delta_r H^\circ = [(1 \text{ mol})(- 814 \frac{kJ}{mol})] - [(1 \text{ mol})(-20.63 \frac{kJ}{mol}) + 0] = -793 \frac{kJ}{mol\text{-rxn}}$$

$$\Delta_r S^\circ = [(1 \text{ mol})(156.9 \frac{J}{K \cdot mol})] - [(1 \text{ mol})(205.79 \frac{J}{K \cdot mol}) + (2 \text{ mol})(205.07 \frac{J}{K \cdot mol})]$$

$$= -459.0 \frac{J}{K \cdot mol\text{-rxn}}$$

$$\Delta_r G^\circ = \Delta_f H^\circ - T \Delta_r S^\circ$$

$$= - 793 \text{ kJ} - (298.15 \text{ K})(- 459.0 \frac{J}{K})(\frac{1.000 \text{ kJ}}{1000 \text{ J}}) = - 657 \frac{kJ}{mol\text{-rxn}}$$

The reaction is product-favored at 25 °C $(\Delta G^\circ < 0)$ and enthalpy-driven $(\Delta_r H^\circ < 0)$.

57. Calculate the $\Delta_r G^\circ$ for the transition of S_8 (rhombic) → S_8 (monoclinic)

    (a) At 80 °C $\Delta_r G^\circ = \Delta_r H^\circ - T \Delta_r S^\circ$

$$\Delta_r G^\circ = 3.213 \text{ kJ} - (353 \text{ K})(0.0087 \frac{kJ}{K}) = 0.14 \frac{kJ}{mol\text{-rxn}}$$

    At 110 °C $\Delta_r G^\circ = \Delta_r H^\circ - T \Delta_r S^\circ$

$$\Delta_r G^\circ = 3.213 \text{ kJ} - (383 \text{ K})(0.0087 \frac{kJ}{K}) = - 0.12 \frac{kJ}{mol\text{-rxn}}$$

The rhombic form of sulfur is the more stable at lower temperature, while the monoclinic form is the more stable at higher temperature. The transition to monoclinic form is product-favored at temperatures above 110 degrees C.

(b) The temperature at which $\Delta_r G^\circ = 0$:

$$\Delta_r G^\circ = 3.213 \text{ kJ} - (T)(0.0087 \frac{kJ}{K}) \text{ substituting: } 0 = 3.213 \text{ kJ} - (T)(0.0087 \frac{kJ}{K})$$

$$T = \frac{3.213 \text{ kJ}}{0.0087 \frac{kJ}{K}} = 370 \text{ K} = 96 \text{ °C}$$

96 °C is the temperature at which the phase transition begins.

IN THE LABORATORY

59. Is decomposition of silver(I) oxide product-favored at 25 °C ?

Calculate: $\Delta_r H°$ and $\Delta_r S°$:

$$\Delta_r H° = ([4 \bullet \Delta_f H° \; Ag(s)] + [1 \bullet \Delta_f H° \; O_2(g)]) - [2 \bullet \Delta_f H° \; [Ag_2O(s)]$$

$$\Delta_r H° = \; 0 \; kJ - [2 \; mol \bullet -31.1 \; kJ/mol] = 62.2 \frac{kJ}{mol\text{-}rxn}$$

and for $\Delta_r S°$:

$$\Delta_r S° = ([4 \bullet S° \; Ag(s)] + [1 \bullet S° \; O_2(g)]) - [2 \bullet S° \; [Ag_2O(s)]$$

$$\Delta_r S° = ([4mol \bullet 42.55 \; J/K \bullet mol] + [1mol \bullet 205.07 \; J/K \bullet mol]) - [2mol \bullet 121.3 \; J/K \bullet mol]$$

$$= [170.2 \; J/K + 205.07 \; J/K] - [242.6 \; J/K] = +132.7 \; J/K$$

While enthalpic considerations do **not** favor product fomation, entropic considerations **do**. The Gibbs Free Energy change would be:

$$\Delta_r G° = ([4 \bullet \Delta_f G° \; Ag(s)] + [1 \bullet \Delta_f G° \; O_2(g)]) - [2 \bullet \Delta_f G° \; [Ag_2O(s)]$$

$$= (0 \; kJ) - (2 \; mol \bullet -11.32 \; kJ/mol) = \; 22.64 \; kJ, \text{ so this change } \textbf{does not favor} \text{ product}$$

formation.

The signs of $\Delta_r H°$ and $\Delta_r S°$ indicate that there **may be** some T at which the reaction is product-favored. So let's calculate the temperature at which $\Delta_r G° = 0$:

$$\Delta_r G° = \; 62.2 \; kJ \; - (T)(0.1327 \frac{kJ}{K}) \quad \text{Note the conversion of } \Delta S \text{ units to kJ!}$$

$$0 = 62.2 \; kJ \; - (T)(0.1327 \frac{kJ}{K})$$

$$T = \frac{62.2 \; kJ}{0.1327 \frac{kJ}{K}} = 469 \; K = 196°C$$

At temperatures greater than 196 °C, the reaction would be product-favored.

61. Calculate $\Delta_f G°$ for HI(g) at 350 °C, given the following equilibrium partial pressures:

P(H_2) = 0.132 bar, P(I_2) = 0.295 bar, and P(HI) = 1.61 bar. At 350 °C, 1 bar, I_2 is a gas.

$$1/2 \; H_2(g) + 1/2 \; I_2(g) \Leftrightarrow HI(g)$$

Calculate Kp: $\dfrac{P_{HI}}{P_{H_2}^{1/2} \bullet P_{I_2}^{1/2}} = \dfrac{1.61}{0.363 \bullet 0.543} = 8.16$

Knowing that $\Delta_rG° = - RT\ln K$, we can solve:

$\Delta_rG° = - RT\ln K = - (8.3145 \dfrac{J}{K \bullet mol})(623.15K)\ln 8.16 = -10{,}873J$ or -10.9 kJ/mol

63. (a) Calculate $\Delta_rG°$ and K for the reaction at 727 °C:

$\Delta_rG° = ([2 \bullet \Delta_fG° \, CO(g)] + [1 \bullet \Delta_fG° \, TiC(s)]) - ([1 \bullet \Delta_fG° \, [TiO_2 (s)] + [3 \bullet \Delta_fG° \, [C (s)])$

$\Delta_rG° = ([2mol \bullet -200.2 \text{ kJ/mol}] + [1 \text{ mol} \bullet -162.6 \text{ kJ/mol}]) - ([1 \text{ mol} \bullet -757.8 \text{ kJ/mol}] + [0])$

$= (-400.4 \text{ kJ} + -162.6 \text{ kJ}) – (-757.8 \text{ kJ}) = -563.0 \text{ kJ} + 757.8 \text{ kJ} = 194.8 \dfrac{kJ}{mol\text{-}rxn}$

K would equal:

$\Delta_rG° = - RT\ln K$ and $194.8 \times 10^3 J = - (8.3145 \dfrac{J}{K \bullet mol})(1000K)\ln K$

[Note the conversion of the energy units of ΔG to accommodate J in the value of R!]

$\dfrac{194.8 \times 10^3 J}{-(8.3145 \dfrac{J}{K \bullet mol})(1000 \text{ K})} = \ln K = -23.43$ and $K = 6.68 \times 10^{-11}$

(b) The value of K indicates that the reaction is **not product-favored at this T**.

(c) Three of the four substances in the equilibrium are solids, hence do not appear in the K expression. The K expression would have the composition: $K = P^2(CO)$. According to LeChatelier's principle, reducing the concentration (and the pressure) of CO would tend to shift the equilibrium to the right, favoring product formation.

SUMMARY AND CONCEPTUAL QUESTIONS

65. An examination of the equation $Hg(l) \Leftrightarrow Hg(g)$ shows that the equilibrium constant expression would be $Kp = P_{Hg(g)}$. So to find the temperature at which Kp = 1.00 bar and 1/760 bar, we need only to find the temperature at which the vapor pressure of mercury is 1.00 bar and 1/760 bar, respectively.

We can calculate the T for Kp = 1.00 bar easily. At the equilibrium point, we can calculate T at which Kp = 1.00bar if we know $\Delta G°$. Since at equilibrium, $\Delta G° = 0$, we can rewrite the equation: $\Delta G° = \Delta H° - T\Delta_r S°$ to read: $\Delta H° / \Delta_r S = T$

$$\Delta_r H° = (1 \text{ mol})(61.38 \frac{kJ}{mol}) - (1 \text{ mol})(0) = 61.38 \text{ kJ} \text{ and for entropy:}$$

$$\Delta_r S° = (1 \text{ mol})(174.97 \frac{J}{K \cdot mol}) - (1 \text{ mol})(76.02 \frac{J}{K \cdot mol}) = 98.95 \frac{J}{K}$$

Substituting into the equation:

$$\Delta_r H° / \Delta_r S = T = (61.38 \text{ kJ} \cdot 1000 \text{ J/kJ}) / 98.95 \frac{J}{K} = 620.3 \text{ K. or } 347.2 °C.$$

(b) Temperature at which Kp = 1/760 Using the Clausius-Clapeyron equation:

$$\ln\left(\frac{P_2}{P_1}\right) = \frac{\Delta H}{R}\left(\frac{1}{T_1} - \frac{1}{T_2}\right) \text{ and } \ln\left(\frac{1}{760}\right) = \frac{61.38 \times 10^3 \frac{J}{mol}}{8.3145 \frac{J}{K \cdot mol}}\left(\frac{1}{620.3K} - \frac{1}{T_2}\right)$$

$$\frac{8.3145 \frac{J}{K \cdot mol} \cdot -6.633}{61.38 \times 10^3 \frac{J}{mol}} = \left(\frac{1}{620.3K} - \frac{1}{T_2}\right) \text{ so } -8.98 \times 10^{-4} \frac{1}{K} = \left(\frac{1}{620.3K} - \frac{1}{T_2}\right)$$

$-1/T_2 = -8.98 \times 10^{-4} \text{ 1/K} - 1.61 \times 10^{-3} = -2.51 \times 10^{-3}$ and $T_2 = 398.3$ K or 125.2 °C.

So in summary, Kp = 1 at 347.2 °C and is 1/760 at 125.2 °C.

67. Following statements false or true?

(a) The entropy of a liquid increases on going from the liquid to the vapor state at any temperature. **True**. For a given substance, the entropy of the vapor state of that substance is greater than for the liquid state.

(b) An exothermic reaction will always be spontaneous. **False**. While exothermic reactions are *almost always spontaneous*, the entropy does play a role, and,should the entropy increase greatly enough, cause the reaction to be non-spontaneous (i.e. reactant-favored).

(c) Reactions with a + $\Delta_r H$ and a +$\Delta_r S$ can never be product-favored. **False**. At very high temperatures, such reactions can be product-favored ($\Delta_r G < 0$).

(d) If $\Delta_r G$ is < 0, the reaction will have an equilibrium constant greater than 1. **True**. Since $\Delta_r G$ and K are related by the expression $\Delta_r G = -RT \ln K$, if $\Delta G < 0$, then mathematically K will be greater than 1.

69. If we dissolve a solid (e.g. table salt), the process proceeds spontaneously ($\Delta_r G° < 0$).

If $\Delta_r H° = 0$, we can write: $\Delta_r G° = \Delta_r H° - T\Delta_r S°$ and $(-) = 0 - (+)\Delta_r S°$.

The only mathematical condition for which this equation is true is if $\Delta_r S° = +$, hence the

process is entropy driven.

71. For the reaction: $2\, C_2H_6(g) + 7\, O_2(g) \rightarrow 4\, CO_2(g) + 6\, H_2O(g)$
   (a) Predict whether signs of $\Delta_r S°$(system), $\Delta_r S°$(surroundings), $\Delta_r S°$(universe) are greater

   than, equal to, or less than 0.

   $\Delta_r S°$(system) will be > 0, since 9 mol of gas form 10 mol of gas as the reaction proceeds.

   $\Delta_r S°$(surroundings) will be > 0, since the reaction liberates heat, and would increase the

   entropy of the surroundings. With both $\Delta_r S°$(system) and $\Delta_r S°$(surroundings) increasing,

   $\Delta_r S°$(universe) would also increase.

   (b) Predict signs of $\Delta_r H°$, and $\Delta_r G°$: Since the reaction is exothermic, $\Delta_r H°$ would be "-".

   With a negative $\Delta_r H°$ and an increasing entropy, $\Delta_r G°$ would be "-" as well.

   (c) Will value of Kp be very large, very small, or nearly 1? With the relatively large number

   of moles of carbon dioxide and water being formed, $\Delta_r H°$ will be *large and negative*, and

   with the increasing entropy, $\Delta_r G°$ will also be relatively *large and negative*.

   Since $\Delta_r G = -RT \ln K$, we anticipate Kp being **very large**.

   Will Kp be larger or smaller at temperatures greater than 298 K?

   Rearrange the expression:

   $\dfrac{\Delta G}{RT} = -\ln K$. As T increases, the term on the left will decrease, resulting in a larger value

   of K ($-\ln K$ decreases-> so K increases).

73. Calculate the $\Delta_r S$ for (1) $C(s) + 2\, H_2(g) \rightarrow CH_4(g)$

   $$\Delta_r S_1° = (1\ \text{mol})(+186.26\ \tfrac{J}{K \cdot mol}) -$$

   $$[(1\,\text{mol})(+5.6\ \tfrac{J}{K \cdot mol}) + (2\,\text{mol})(+130.7\ \tfrac{J}{K \cdot mol})] = -80.7\ \tfrac{J}{K \cdot mol\text{-rxn}}$$

   Calculate the $\Delta_r S$ for (2) $CH_4(g) + \tfrac{1}{2}\, O_2(g) \rightarrow CH_3OH(l)$

$$\Delta_r S_2^\circ = (1\ mol)(+127.19\ \frac{J}{K \cdot mol}) -$$

$$[(1mol)(+186.26\ \frac{J}{K \cdot mol}) + (\frac{1}{2}mol)(+205.07\ \frac{J}{K \cdot mol})] = -$$

$$161.60\ \frac{J}{K \cdot mol\text{-}rxn}$$

Calculate the $\Delta_r S$ for (3) $C(s) + 2\ H_2\ (g) + \frac{1}{2}\ O_2\ (g) \rightarrow CH_3OH(l)$

$$\Delta_r S_3^\circ = (1\ mol)(+127.19\ \frac{J}{K \cdot mol}) -$$

$$[(1mol)(+5.6\ \frac{J}{K \cdot mol}) + (2mol)(+130.7\ \frac{J}{K \cdot mol}) + (\frac{1}{2}mol)(+205.07\ \frac{J}{K \cdot mol})]$$

$$= -\ 242.3\ \frac{J}{K \cdot mol\text{-}rxn}$$

So $\Delta_r S_1^\circ + \Delta_r S_2^\circ = (-80.7\ \frac{J}{K}) + (-161.60\ \frac{J}{K}) = -242.3\ \frac{J}{K \cdot mol\text{-}rxn}$

75. (a) Confirm that $Mg(s) + 2\ H_2O(l) \rightarrow Mg(OH)_2\ (s) + H_2(g)$ is a spontaneous reaction.

$\Delta_f H^\circ$ (kJ/mol) 0   -285.83   -924.54   0

S° (J/K $\cdot$ mol) 32.67   69.95   63.18   130.7

$$\Delta_r H^\circ = [(1\ mol)(-924.54\ \frac{kJ}{mol}) + (1mol)(0)] - [(1mol)0 + (2\ mol)(-285.83\ \frac{kJ}{mol})]$$

$$= -352.88\ \frac{kJ}{mol\text{-}rxn}$$

$$\Delta_r S^\circ = [(1\ mol)(63.18\ \frac{J}{K \cdot mol}) + (1mol)(130.7\ \frac{J}{K \cdot mol})] -$$

$$[(1mol)(32.67\ \frac{J}{K \cdot mol}) + (2\ mol)(69.95\ \frac{J}{K \cdot mol})]$$

$$= 21.31\ \frac{J}{K \cdot mol\text{-}rxn}$$

$$\Delta_r G^\circ = \Delta_f H^\circ - T\ \Delta_r S^\circ$$

$$= -\ 352.88\ kJ - (298.15\ K)(21.31\ \frac{J}{K})(\frac{1.000\ kJ}{1000\ J}) = -359.23\ \frac{kJ}{mol\text{-}rxn}$$

With a negative $\Delta_r G^\circ$, we anticipate the reaction to be spontaneous.

(b) Mass of Mg to produce sufficient energy to heat 225 mL of water (D = 0.996 g/mL) from

25°C to the boiling point (100 °C)? [100-25 = 75 °C or 75K]

Heat required: 225 mL • 0.996 g/mL • 4.184 J/g•K • 75 K. = 70,322.58 J or 70.3 kJ

The $\Delta_r H^\circ$ = -352.88 kJ for 1 mol of Mg.

$$\frac{70.3\,kJ}{1} \cdot \frac{1\,mol\,Mg}{352.88\,kJ} = 0.2\,mol\,Mg \text{ or } 24.3\,g/mol \cdot 0.2\,mol = 4.84\,g\,Mg$$

77. (a) Equation for the reaction of hydrazine and oxygen.

$$N_2H_4\,(l) + O_2(g) \rightarrow 2\,H_2O(l) + N_2(g)$$

Oxygen is the **oxidizing agent,** and hydrazine is the **reducing agent.** There are several ways to assess this. Note that the oxidation state for O_2 is 0 (as reactant) and –2 (as product)—it has been reduced (by hydrazine). Note that hydrazine **loses H**—in going from reactant to product—a definition for being oxidized.

(b) Calculate $\Delta_rH°$, $\Delta_rS°$, and $\Delta_rG°$:

$$\Delta_rH° = [2 \cdot \Delta_fH°\,H_2O\,(l) + 1 \cdot \Delta_fH°\,N_2\,(g)] - [1 \cdot \Delta_fH°\,N_2H_4\,(l) + 1 \cdot \Delta_fH°\,O_2\,(g)]$$

$$= [(2\,mol)(-285.830\,\tfrac{kJ}{mol}) + 0] - [(1\,mol)(50.63\,\tfrac{kJ}{mol}) + 0] = -622.29\,\tfrac{kJ}{mol\text{-}rxn}$$

$$\Delta_rS° = [2 \cdot S°H_2O\,(l) + 1 \cdot S°\,N_2\,(g)] - [1 \cdot S°\,N_2H_4\,(l) + 1 \cdot S°\,O_2\,(g)]$$

$$= [(2\,mol)(69.95\,\tfrac{J}{K\cdot mol}) + (1\,mol)(191.56\,\tfrac{J}{K\cdot mol})] -$$

$$[(1\,mol)(121.52\,\tfrac{J}{K\cdot mol}) + (1\,mol)(205.07\,\tfrac{J}{K\cdot mol})] = 4.87\,\tfrac{J}{K\cdot mol\text{-}rxn}$$

$$\Delta_rG° = \Delta_rH° - T\Delta_rS°$$

$$= -622.29\,kJ - (298\,K)(4.87\,\tfrac{J}{K})(\tfrac{1.000\,kJ}{1000.\,J}) = -623.74\,\tfrac{kJ}{mol\text{-}rxn}$$

(c) T change of 5.5×10^4 L of water (assuming 1 mole of N_2H_4 reacts):

1 mol of hydrazine releases –622.29 kJ,

Heat = $m \cdot c \cdot \Delta t$ [Assume D of water = 0.996 g/mL]

622.29×10^3 J = 5.5×10^4 L $\cdot$ 996 g/L $\cdot$ 4.184 J/g $\cdot$K $\cdot \Delta t$.

$$\frac{6.2229 \times 10^5\,J}{\left(5.5 \times 10^4\,L \cdot 996\,\tfrac{g}{L} \cdot 4.184\,\tfrac{J}{g\cdot K}\right)} = \Delta t. \text{ Solving for } \Delta t \text{ gives: } 2.7 \times 10^{-3}\,K.$$

(d) Solubility of $O_2 = 0.000434$ g O_2/100g water.

5.5×10^4 L • 996 g/L • 4.34×10^{-4} g O_2/100g water • 1 mol O_2/32.00 g O_2

7.5 mol O_2 (or approximately 240 g)

(e) If hydrazine is present in 5% solution, what mass of hydrazine solution is needed to consume the O_2 present?

$$\frac{7.5 \text{ mol } O_2}{1} \cdot \frac{1 \text{ mol } N_2H_4}{1 \text{ mol } O_2} \cdot \frac{32.05 \text{ g } N_2H_4}{1 \text{ mol } N_2H_4} \cdot \frac{100 \text{ g solution}}{5.00 \text{ g } N_2H_4} = 4.8 \times 10^3 \text{ g solution}$$

(2 sf)

(f) Assuming N_2 escapes as gas, calculate V of N_2 at STP:

The balanced equation tells us that 7.5 mol of O_2 will liberate 7.5 mol of N_2.

At STP, 7.5 mol of this gas will occupy (7.5 mol • 22.4 L/mol) or 170 L. (2 sf)

79. The key phrase needed to answer the question: "What is the sign......" is "Iodine dissolves readily....". This phrase tell us that $\Delta_r G°$ is negative.

Enthalpy-driven processes are exothermic. The "neutrality" of the ΔH for this reaction tells us that the process is NOT enthalpy-driven. Since the iodine goes from the solid state to the "solution" state, we anticipate an increase in entropy, and would therefore state that the process is entropy-driven.

81. Equation for decomposition of 1 mol of $CH_3OH(g)$ to elements:

$CH_3OH(g) \rightarrow C(s) + 2 H_2(g) + \frac{1}{2} O_2(g)$

(a) According to Appendix L, the $\Delta_f H°$ formation for methanol is -201.0 kJ/mol. The equation above is the reverse of that process, so we anticipate that the $\Delta_r H°$ for this process will be positive (endothermic). Additionally the decomposition has a positive $\Delta_r S°$, since the number of moles of gas increase during the process. So spontaneity increases as T increases.

(b) There is no T between 400K and 1000K at which the reaction is spontaneous.

83. (a) Calculate $\Delta_r G°$ at 298K, 800 K, and 1300 K for the reaction: $N_2(g) + 3 H_2(g) \Leftrightarrow 2 NH_3(g)$

At 298K: $\Delta_r H° = (2 \text{ mol} \bullet -45.90 \text{ kJ/mol}) - (0 + 0) = -91.80$ kJ

and $\Delta_r S° = (2 \text{ mol} \bullet 192.77 \text{ J/K} \bullet \text{mol}) - (1 \text{mol} \bullet 191.56 \text{ J/K} \bullet \text{mol} + 3 \text{mol} \bullet 130.7 \text{ J/K} \bullet \text{mol})$
$= -198.12$ J/K   and

$$\Delta_r G^\circ = \Delta_f H^\circ - T\Delta_r S^\circ = -91.80\ \text{kJ} - (298\ \text{K})(-0.19812\ \text{kJ/K}) = -32.74\ \frac{\text{kJ}}{\text{mol-rxn}}$$

At 800K:

$$\Delta_r G^\circ = \Delta_f H^\circ - T\Delta_r S^\circ = -107.4\ \text{kJ} - (800\text{K})(-0.2254\ \text{kJ//K}) = 72.92\ \frac{\text{kJ}}{\text{mol-rxn}}$$

At 1300 K:

$$\Delta_r G^\circ = \Delta_f H^\circ - T\Delta_r S^\circ = -112.4\ \text{kJ} - (1300\text{K})(-0.2280\ \text{kJ/K}) = 184.0\ \frac{\text{kJ}}{\text{mol-rxn}}$$

A quick examination of the values of $\Delta_r G^\circ$ indicates that the free energy change becomes more positive as T increases.

(b) Calculate K for the reaction at 298K, 800K, 1300K:

At 298K:

$$\frac{-32.74 \times 10^3 \text{J}}{-(8.3145\ \frac{\text{J}}{\text{K}\cdot\text{mol}})(298\ \text{K})} = \ln K = 13.21 \text{ and } K = 5.48 \times 10^5$$

At 800K:

$$\frac{72.92 \times 10^3 \text{J}}{-(8.3145\ \frac{\text{J}}{\text{K}\cdot\text{mol}})(800\ \text{K})} = \ln K = -10.96 \text{ and } K = 1.73 \times 10^{-5}$$

At 1300 K:

$$\frac{184 \times 10^3 \text{J}}{-(8.3145\ \frac{\text{J}}{\text{K}\cdot\text{mol}})(1300\ \text{K})} = \ln K = -17.02 \text{ and } K = 4.05 \times 10^{-8}$$

(c) For which T will mole fraction of NH_3 be largest?

The "bottom line" is easy to assess. The partial pressure of ammonia (and hence the mol fraction) will be greatest for the temperature at which K is greatest (298 in this case).

Applying Chemical Principles

1. (a) Use $\Delta_f G°$ values from Appendix L to calculate $\Delta G°$ and K_{eq} for the reaction under standard conditions and 298.15 K.

$\Delta_r G° = \Delta_f G°$ (graphite) - $\Delta_f G°$ (diamond)

$\Delta_r G° = 0.0$ kJ/mol(1mol) $- 2.900$ kJ/mol(1mol) $= -2.9$ kJ   (sf limited to 2 sf arbitrarily)

$\Delta_r G° = - RTlnK_{eq}$ and $-2.9 \times 10^3 J = - (8.3145 \frac{J}{K \bullet mol})(298.15 \text{ K})\ln K_{eq}$

Solving for ln K_{eq}: $\dfrac{-2.9 x 10^3 J}{\left(-8.3145 \dfrac{J}{K \bullet mol}\right)(298.15 K)} = \ln K_{eq}$ and $1.1698 = \ln K_{eq}$

and $K_{eq} = 3.22$

(b) Use $\Delta_f H°$ and S° values from Appendix L to calculate $\Delta G°$ and K_{eq} for the reaction at 1000K. Assume that enthalpy and entropy values are valid at these temperatures. Does heating shift the equilibrium toward the formation of diamond or graphite?

$\Delta_r H° = \Delta_f H°$ (graphite) - $\Delta_f H°$ (diamond)

$\Delta_r H° = 0.0$ kJ/mol (1mol) $- 1.8$ kJ/mol(1mol) $= -1.8$ kJ

$\Delta_r S° = S°$ (graphite) - S° (diamond)

$\Delta_r S° = 5.6 \dfrac{J}{K \bullet mol}$ (1mol) $- 2.377 \dfrac{J}{K \bullet mol}$ (1mol) $= 3.2 \dfrac{J}{K}$

$\Delta_r G° = \Delta_f H° - T\Delta_r S° = -1.8$ kJ$(\dfrac{1000 J}{1 kJ}) - (1000 \text{ K})(3.2 \dfrac{J}{K}) = - 5000 J$ or -5.0 kJ

$\Delta_r G° = - RTlnK_{eq}$

-5000 J $= - (8.3145 \dfrac{J}{K \bullet mol})(1000 \text{ K})\ln K_{eq}$ and $\dfrac{-5000 J}{\left(-8.3145 \dfrac{J}{K \bullet mol}\right)(1000 K)} = \ln K_{eq}$

$0.6013 = \ln K_{eq}$ and $K_{eq} = 1.8$

(c) Why is the formation of diamond favored at high pressures?

Examine the phase diagram and you'll note that as P increases, diamond is favored for any given T.

(d) Why is the conversion done at much higher T and P? Higher temperatures will increase the RATE (kinetics, right?) for the process!

Chapter 20
Principles of Reactivity: Electron Transfer Reactions

PRACTICING SKILLS
Balancing Equations for Oxidation-Reduction Reactions

1. Balance the following:

	reactant is	overall process is
(a) $Cr\,(s) \rightarrow Cr^{3+}\,(aq) + 3\,e^-$	reducing agent	oxidation
(b) $AsH_3\,(g) \rightarrow As\,(s) + 3\,H^+\,(aq) + 3\,e^-$	reducing agent	oxidation
(c) $VO_3^-\,(aq) + 6\,H^+\,(aq) + 3\,e^- \rightarrow$ $\qquad V^{2+}\,(aq) + 3\,H_2O\,(l)$	oxidizing agent	reduction
(d) $2\,Ag\,(s) + 2\,OH^-\,(aq) \rightarrow Ag_2O(s)$ $\qquad + H_2O(l) + 2e^-$	reducing agent	oxidation

Note: e^- are used to balance charge; H^+ balances only H atoms; H_2O (or OH^- in base) balances both H and O atoms.

3. Balance the equations (in acidic solutions):

Balancing redox equations in neutral or acidic solutions may be accomplished in several steps. They are:

1. Separating the equation into two equations which represent reduction and oxidation
2. Balancing mass of elements (other than H or O)
3. Balancing mass of O by adding H_2O
4. Balancing mass of H by adding H^+
5. Balancing charge by adding electrons
6. Balancing electron gain (in the reduction half-equation) with electron loss (in the oxidation half-equation)
7. Combining the two half equations

For the parts of this problem, each step will be identified with a number corresponding to the list above. In addition, the physical states of all species will be omitted in all but the final step. While this omission is <u>not generally recommended,</u> it should increase the clarity of the steps involved. In addition when a step leaves a half equation unchanged from the previous step, we have omitted the half equation.

(a) $Ag(s) + NO_3^-(aq) \rightarrow NO_2(g) + Ag^+(aq)$

Oxidation half-equation	Reduction half-equation	Step
$Ag \rightarrow Ag^+$	$NO_3^- \rightarrow NO_2$	1 & 2
	$NO_3^- \rightarrow NO_2 + H_2O$	3
	$2 H^+ + NO_3^- \rightarrow NO_2 + H_2O$	4
$Ag \rightarrow Ag^+ + 1e^-$	$2 H^+ + NO_3^- + 1e^- \rightarrow NO_2 + H_2O$	5 & 6
$2 H^+(aq) + NO_3^-(aq) + Ag(s) \rightarrow Ag^+(aq) + NO_2(g) + H_2O(l)$		7

(b) $MnO_4^-(aq) + HSO_3^-(aq) \rightarrow Mn^{2+}(aq) + SO_4^{2-}(aq)$

Oxidation half-equation	Reduction half-equation	Step
$HSO_3^- \rightarrow SO_4^{2-}$	$MnO_4^- \rightarrow Mn^{2+}$	1 & 2
$H_2O + HSO_3^- \rightarrow SO_4^{2-}$	$MnO_4^- \rightarrow Mn^{2+} + 4H_2O$	3
$H_2O + HSO_3^- \rightarrow SO_4^{2-} + 3 H^+$	$8 H^+ + MnO_4^- \rightarrow Mn^{2+} + 4H_2O$	4
$H_2O + HSO_3^- \rightarrow SO_4^{2-} + 3 H^+ + 2e^-$	$8 H^+ + MnO_4^- + 5e^- \rightarrow Mn^{2+} + 4H_2O$	5
$5 H_2O + 5 HSO_3^- \rightarrow 5 SO_4^{2-} + 15 H^+ + 10e^-$	$16 H^+ + 2MnO_4^- + 10e^- \rightarrow 2 Mn^{2+} + 8H_2O$	6
$5 HSO_3^-(aq) + H^+(aq) + 2 MnO_4^-(aq) \rightarrow 5 SO_4^{2-}(aq) + 2 Mn^{2+}(aq) + 3 H_2O(l)$		7

(c) $Zn(s) + NO_3^-(aq) \rightarrow Zn^{2+}(aq) + N_2O(g)$

Oxidation half-equation	Reduction half-equation	Step
$Zn \rightarrow Zn^{2+}$	$NO_3^- \rightarrow N_2O$	1
	$2 NO_3^- \rightarrow N_2O$	2
	$2 NO_3^- \rightarrow N_2O + 5 H_2O$	3
	$10 H^+ + 2 NO_3^- \rightarrow N_2O + 5 H_2O$	4
$Zn \rightarrow Zn^{2+} + 2 e^-$	$10 H^+ + 2 NO_3^- + 8 e^- \rightarrow N_2O + 5 H_2O$	5
$4 Zn \rightarrow 4 Zn^{2+} + 8 e^-$		6
$10 H^+(aq) + 2 NO_3^-(aq) + 4 Zn(s) \rightarrow 4 Zn^{2+}(aq) + N_2O(g) + 5 H_2O(l)$		7

(d) $Cr(s) + NO_3^-(aq) \rightarrow Cr^{3+}(aq) + NO(g)$

Oxidation half-equation	Reduction half-equation	Step
$Cr \rightarrow Cr^{3+}$	$NO_3^- \rightarrow NO$	1 & 2
	$NO_3^- \rightarrow NO + 2 H_2O$	3
	$4 H^+ + NO_3^- \rightarrow NO + 2 H_2O$	4
$Cr \rightarrow Cr^{3+} + 3 e^-$	$4 H^+ + NO_3^- + 3 e^- \rightarrow NO + 2 H_2O$	5 & 6
$Cr(s) + 4 H^+(aq) + NO_3^-(aq) \rightarrow NO(g) + 2 H_2O(l) + Cr^{3+}(aq)$		7

5. Balancing redox equations in basic solutions may be accomplished in several steps. There is only a *slight change* from the "acidic solution".

1. Separating the equation into two equations which represent reduction and oxidation
2. Balancing mass of elements (other than H or O)
3. Balancing mass of O by adding H_2O
4. Balancing mass of H by adding H^+
5. Balancing charge by adding electrons
6. Balancing electron gain (in the reduction half-equation) with electron loss (in the oxidation half-equation)
7. Combine the two half equations, removing any redundancies.
8. *Add as many OH$^-$ to both sides of the equation as there are H$^+$ ions, to form water.*
9. *Remove any redundancies in H$_2$O molecules.*

As before, each step will be identified with a number corresponding to the list above, and physical states of all species will be omitted in all but the final step.

(a) $Al(s) + H_2O(l) \rightarrow Al(OH)_4^-(aq) + H_2(g)$

Oxidation half-equation	Reduction half-equation	Step
$Al \rightarrow Al(OH)_4^-$	$H_2O \rightarrow H_2$	1 & 2
$Al + 4 H_2O \rightarrow Al(OH)_4^-$	$H_2O \rightarrow H_2 + H_2O$	3
$Al + 4 H_2O \rightarrow Al(OH)_4^- + 4H^+$	$2 H^+ + H_2O \rightarrow H_2 + H_2O$	4
$Al + 4 H_2O \rightarrow Al(OH)_4^- + 4 H^+ + 3 e^-$	$2 H^+ + H_2O + 2 e^- \rightarrow H_2 + H_2O$	5
$2Al + 8 H_2O \rightarrow 2 Al(OH)_4^- + 8 H^+ + 6 e^-$	$6 H^+ + 3 H_2O + 6 e^- \rightarrow 3H_2 + 3 H_2O$	6
$2 Al + 8 H_2O \rightarrow 2 Al(OH)_4^- + 3 H_2 + 2 H^+$		7
$2 Al + 8 H_2O + 2 OH^- \rightarrow 2 Al(OH)_4^- + 3 H_2 + 2 H^+ + 2 OH^-$		8
$2Al(s) + 6 H_2O(l) + 2 OH^-(aq) \rightarrow 2 Al(OH)_4^-(aq) + 3 H_2(g)$		9

(b) $CrO_4^{2-}(aq) + SO_3^{2-}(aq) \rightarrow Cr(OH)_3(s) + SO_4^{2-}(aq)$

Oxidation half-equation	Reduction half-equation	Step
$SO_3^{2-} \rightarrow SO_4^{2-}$	$CrO_4^{2-} \rightarrow Cr(OH)_3$	1 & 2
$H_2O + SO_3^{2-} \rightarrow SO_4^{2-}$	$CrO_4^{2-} \rightarrow Cr(OH)_3 + H_2O$	3
$H_2O + SO_3^{2-} \rightarrow SO_4^{2-} + 2 H^+$	$5 H^+ + CrO_4^{2-} \rightarrow Cr(OH)_3 + H_2O$	4
$H_2O + SO_3^{2-} \rightarrow SO_4^{2-} + 2 H^+ + 2 e^-$	$5 H^+ + CrO_4^{2-} + 3 e^- \rightarrow Cr(OH)_3 + H_2O$	5
$3 H_2O + 3 SO_3^{2-} \rightarrow 3 SO_4^{2-} + 6 H^+ + 6 e^-$	$10 H^+ + 2 CrO_4^{2-} + 6 e^- \rightarrow 2 Cr(OH)_3 + 2 H_2O$	6
$H_2O(l) + 3 SO_3^{2-} + 4 H^+ + 2 CrO_4^{2-} \rightarrow 2 Cr(OH)_3 + 3 SO_4^{2-}$		7
$H_2O(l) + 3 SO_3^{2-} + 4 H^+ + 4\,OH^- + 2 CrO_4^{2-} \rightarrow 2 Cr(OH)_3 + 3 SO_4^{2-} + 4\,OH^-$		8
$5 H_2O(l) + 3 SO_3^{2-}(aq) + 2 CrO_4^{2-}(aq) \rightarrow 2 Cr(OH)_3(s) + 3 SO_4^{2-}(aq) + 4\,OH^-(aq)$		9

(c) $Zn(s) + Cu(OH)_2(s) \rightarrow Zn(OH)_4^{2-}(aq) + Cu(s)$

Oxidation half-equation	Reduction half-equation	Step
$Zn \rightarrow Zn(OH)_4^{2-}$	$Cu(OH)_2 \rightarrow Cu$	1 & 2
$4 H_2O + Zn \rightarrow Zn(OH)_4^{2-}$	$Cu(OH)_2 \rightarrow Cu + 2 H_2O$	3
$4 H_2O + Zn \rightarrow Zn(OH)_4^{2-} + 4 H^+$	$2 H^+ + Cu(OH)_2 \rightarrow Cu + 2 H_2O$	4
$4 H_2O + Zn \rightarrow Zn(OH)_4^{2-} + 4 H^+ + 2 e^-$	$2 H^+ + Cu(OH)_2 + 2 e^- \rightarrow Cu + 2 H_2O$	5 & 6
$2 H_2O(l) + Zn(s) + Cu(OH)_2(s) \rightarrow Zn(OH)_4^{2-}(aq) + 2 H^+(aq) + Cu(s)$		7
$2 H_2O(l) + Zn(s) + Cu(OH)_2(s) + 2OH^-(aq) \rightarrow Zn(OH)_4^{2-}(aq) + 2 H^+(aq) + 2OH^-(aq) + Cu(s)$		8
$Zn(s) + Cu(OH)_2(s) + 2\,OH^-(aq) \rightarrow Zn(OH)_4^{2-}(aq) + Cu(s)$		9

(d) $HS^-(aq) + ClO_3^-(aq) \rightarrow S(s) + Cl^-(aq)$

Oxidation half-equation	Reduction half-equation	Step
$HS^- \rightarrow S$	$ClO_3^- \rightarrow Cl^-$	1 & 2
	$ClO_3^- \rightarrow Cl^- + 3\,H_2O$	3
$HS^- \rightarrow S + H^+$	$6\,H^+ + ClO_3^- \rightarrow Cl^- + 3\,H_2O$	4
$HS^- \rightarrow S + H^+ + 2\,e^-$	$6\,e^- + 6\,H^+ + ClO_3^- \rightarrow Cl^- + 3\,H_2O$	5
$3\,HS^- \rightarrow 3\,S + 3\,H^+ + 6\,e^-$		6
$3\,HS^- + 3\,H^+ + ClO_3^- \rightarrow Cl^- + 3\,H_2O + 3\,S$		7
$3\,HS^- + 3\,H^+ + 3\,OH^- + ClO_3^- \rightarrow Cl^- + 3\,H_2O + 3\,S + 3\,OH^-$		8
$3\,HS^-(aq) + ClO_3^-(aq) \rightarrow Cl^-(aq) + 3\,S(s) + 3\,OH^-(aq)$		9

Constructing Voltaic Cells

7. For the reaction: $2\,Cr(s) + 3\,Fe^{2+}(aq) \rightarrow 2\,Cr^{3+}(aq) + 3\,Fe(s)$:

Electrons in the external circuit flow from the <u>Cr</u> electrode to the <u>Fe</u> electrode. Negative ions move in the salt bridge from the <u>iron</u> half-cell to the <u>chromium</u> half-cell. The half-reaction at the anode is <u>$Cr(s) \rightarrow Cr^{3+}(aq) + 3\,e^-$</u> and that at the cathode is <u>$Fe^{2+}(aq) + 2\,e^- \rightarrow Fe(s)$</u>.

Note that the reaction shows that Cr is being oxidized (to Cr^{3+}) by Fe^{2+} which is being reduced (to Fe). The electrons leave the anode and head to the cathode via the external circuit (wire). With reduction occurring in the cathode half-cell, a net deficit of positive ions accumulates, necessitating the assistance of + ions from the salt bridge. See SQ20.2 and 20.3 for additional questions of this type.

9. Like SQ20.7, we can complete the paragraph in part (c), by deciding on the spontaneous or product-favored reaction between the iron and oxygen half cells. The reduction potentials are:

$$O_2(g) + 4\,H^+(aq) + 4e^-(aq) \rightarrow 2\,H_2O(l) \qquad E° = 1.229V$$

$$Fe^{2+}(aq) + 2\,e^- \rightarrow Fe(s) \qquad E° = -0.44V$$

(a) Oxidation half-reaction: $2\,Fe(s) \rightarrow 2\,Fe^{2+}(aq) + 4\,e^-$

Reduction half-reaction: $O_2(g) + 4\,H^+(aq) + 4e^-(aq) \rightarrow 2\,H_2O(l)$

Net cell reaction: $2\,Fe(s) + O_2(g) + 4\,H^+(aq) \rightarrow 2\,H_2O(l) + 2\,Fe^{2+}(aq)$

How to decide which half-reaction occurs as oxidation? An examination of Table 20.1 (or Appendix M) shows O_2 as a stronger oxidizing agent than Fe^{2+}.

Alternatively, Fe is a stronger reducing agent than H_2O. Either of these conclusions points to the direction of reaction, which is product-favored.

(b) The anode half-reaction is the oxidation half-reaction: $2\ Fe(s) \rightarrow 2\ Fe^{2+}(aq) + 4\ e^-$

At the cathode, the reduction half-reaction: $O_2(g) + 4\ H^+(aq) + 4\ e^-(aq) \rightarrow 2\ H_2O(l)$

(c) Electrons in the external circuit flow from the <u>Fe</u> electrode to the <u>O_2</u> electrode. Negative ions move in the salt bridge from the <u>oxygen</u> half-cell to the <u>iron</u> half-cell.

Commercial Cells

11. Similarities and differences between dry cells, alkaline batteries, and Ni-cad batteries:
The first two types of cells are non-rechargeable batteries—also called primary batteries. They also share the common anode, zinc. Ni-cad batteries are rechargeable. Alkaline and Ni-cad batteries are in a basic environment, whereas dry cells are in an acidic environment.

Standard Electrochemical Potentials

13. Calculate E° for each of the following, and decided if it is product-favored as written:
$$E°_{cell} = E°_{cathode} - E°_{anode}$$

(a) $2\ I^-(aq) + Zn^{2+}(aq) \rightarrow I_2(s) + Zn(s)$

Cathode reaction: $Zn^{2+}(aq) + 2\ e^- \rightarrow Zn(s)$	E° = -0.763 V
Anode reaction: $2\ I^-(aq) \rightarrow I_2(s) + 2\ e^-$	<u>E° = +0.535 V</u>
Cell voltage:	E° = -1.298 V(not product-favored)

(b) $Zn^{2+}(aq) + Ni(s) \rightarrow Zn(s) + Ni^{2+}(aq)$

Cathode reaction: $Zn^{2+}(aq) + 2\ e^- \rightarrow Zn(s)$	E° = -0.763 V
Anode reaction: $Ni(s) \rightarrow Ni^{2+}(aq) + 2\ e^-$	<u>E° = -0.25 V</u>
Cell voltage:	E° = -0.51 V(not product-favored)

(c) $2\ Cl^-(aq) + Cu^{2+}(aq) \rightarrow Cu(s) + Cl_2(g)$

Cathode reaction: $Cu^{2+}(aq) + 2\ e^- \rightarrow Cu(s)$	E° = +0.337 V
Anode reaction: $2\ Cl^-(aq) \rightarrow Cl_2(g) + 2\ e^-$	<u>E° = +1.360 V</u>
Cell voltage:	E° = -1.023 V(not product-favored)

(d) $Fe^{2+}(aq) + Ag^+(aq) \rightarrow Fe^{3+}(aq) + Ag(s)$

　　　Cathode reaction: $Ag^+(aq) + e^- \rightarrow Ag(s)$ 　　　　$E° = +0.7994$ V

　　　Anode reaction: $Fe^{2+}(aq) \rightarrow Fe^{3+}(aq) + e^-$ 　　<u>$E° = +0.771$ V</u>

　　　　　　　　　　　Cell voltage: 　　　　　$E° = 0.028$ V (product-favored)

15. (a) $Sn^{2+}(aq) + 2 Ag(s) \rightarrow Sn(s) + 2 Ag^+(aq)$

　　　Cathode reaction: $Sn^{2+}(aq) + 2e^- \rightarrow Sn (s)$ 　　　$E° = -0.14$ V

　　　Anode reaction: $2 Ag (s) \rightarrow 2 Ag^+ (aq) + 2e^-$ 　　<u>$E° = +0.7994$ V</u>

　　　　　　　　　　Cell voltage: 　　　　　$E° = -0.94$ V (not product-favored)

(b) $2 Al(s) + 3 Sn^{4+}(aq) \rightarrow 3 Sn^{2+}(aq) + 2 Al^{3+}(aq)$

　　　Cathode reaction: $3 Sn^{4+}(aq) + 6e^- \rightarrow 3 Sn^{2+}(aq)$ 　$E° = +0.15$ V

　　　Anode reaction: $2 Al(s) \rightarrow 2 Al^{3+}(aq) + 6e^-$ 　　<u>$E° = -1.66$ V</u>

　　　　　　　　　　Cell voltage: 　　　　　$E° = 1.81$ V (product-favored)

(c) $2 ClO_3^-(aq) + 10 Ce^{3+}(aq) + 12 H^+(aq) \rightarrow Cl_2(g) + 10 Ce^{4+}(aq) + 6 H_2O(l)$

　　　Cathode reaction: $2 ClO_3^-(aq) + 10 e^- + 12 H^+(aq)$

　　　　　　　　　$\rightarrow Cl_2(g) + 6 H_2O(l)$ 　　　$E° = +1.47$ V

　　　Anode reaction: $10 Ce^{3+} (aq) \rightarrow 10 Ce^{4+}(aq)$

　　　　　　　　　　　　　　$+ 10 e^-$ 　　　<u>$E° = +1.61$ V</u>

　　　　　　　　　　Cell voltage: 　　　　　$E° = -0.14$ V (not product-favored)

(d) $3 Cu(s) + 2 NO_3^-(aq) + 8 H^+(aq) \rightarrow 3 Cu^{2+}(aq) + 2 NO(g) + 4 H_2O(l)$

　　　Cathode reaction: $2 NO_3^-(aq) + 8 H^+(aq) + 6 e^-$ 　$E° = +0.96$ V

　　　　　　　　　$\rightarrow 2 NO (g) + 4 H_2O (l)$

　　　Anode reaction: $3 Cu(s) \rightarrow 3 Cu^{2+}(aq) + 6e^-$ 　　<u>$E° = +0.337$ V</u>

　　　　　　　　　　Cell voltage: 　　　　　$E° = 0.62$ V 　　　(product-favored)

Ranking Oxidizing and Reducing Agents

17. From the following half-reactions:

(a) The metal most easily oxidized:

From the list **Al** is the most easily oxidized metal. Having the most negative reduction

potential, Al is the strongest reducing agent of the group, and reducing agents are

oxidized as they perform their task.

(b) Metals on the list capable of reducing Fe^{2+} to Fe:

Zn and **Al** both have more negative reduction potentials than Fe, hence are stronger

reducing agents, and can reduce Fe^{2+} to Fe.

(c) A balanced equation for the reaction of Fe^{2+} with Sn. Is the reaction product-favored?

$Fe^{2+}(aq) + Sn(s) \rightarrow Fe(s) + Sn^{2+}(aq)$; Since Fe is a stronger reducing agent than Sn, this reaction would not have an $E° > 0$, and the reaction would be reactant-favored.

(d) A balanced equation for the reaction of Zn^{2+} with Sn. Is the reaction product-favored?

$Zn^{2+}(aq) + Sn(s) \rightarrow Zn(s) + Sn^{2+}(aq)$; Zn is a stronger reducing agent than Sn, this reaction would not have an $E° > 0$, and the reaction would be reactant-favored.

19. Element from the group that is the best reducing agent?

Cr has the most negative standard reduction potential of the group. Recall that the more negative the reduction potential, the stronger a substance is as a reducing agent.

21. Ion from the group that is most easily reduced?

The specie with the most positive reduction potential is the strongest oxidizing agent of the list, and with that role, becomes the most easily reduced. **Ag^+** fits that role from this list.

23. Regarding the halogens:

(a) The halogen most easily reduced is the one with the most positive reduction potential.

F_2 has the most reduction potential of the halogens.

(b) MnO_2 has a reduction potential of 1.23V. Both F_2 and Cl_2 have more positive reduction potentials than MnO_2, and are better oxidizing agents than MnO_2.

Electrochemical Cells Under Nonstandard Conditions

25. The voltage of a cell that has dissolved species at 0.025 M:

Calculate the standard voltage of the cell:

(1) $2 H_2O(l) + 2e^- \rightarrow H_2(g) + 2 OH^-(aq)$ $E° = -0.8277$ V

(2) $[Zn(OH)_4]^{2-}(aq) + 2 e^- \rightarrow Zn(s) + 4 OH^-(aq)$ $E° = -1.22$ V

The net equation is given as: $Zn(s) + 2 H_2O(l) + 2 OH^-(aq) \rightarrow Zn(OH)_4^-(aq) + H_2(g)$

The equilibrium expression (Q) would be: $\dfrac{\left[Zn(OH)_4^-\right]P_{H2}}{[OH^-]^2}$

The hydrogen pressure is 1 bar, and we know the concentrations of the other terms: 0.025 M.

Note that n (in the Nernst equation) corresponds to 2, since the balanced overall equation indicates that 2 moles of electrons are lost and 2 moles of electrons are gained.

The cell equation indicates Zn as a reactant, meaning that reaction (2) runs in the reverse direction, making that process an oxidation, so the $E°$anode = -1.22 V.

Calculate $E°_{cell}$ = $E°$cathode − $E°$anode = -0.8277 − (-1.22) = 0.39 V

The Nernst equation $\qquad E_{cell} = E°_{cell} - \dfrac{0.0257}{n} \ln \dfrac{\left[Zn(OH)_4^-\right]P_{H2}}{[OH^-]^2}$

$E_{cell} = 0.39 - \dfrac{0.0257}{2}\ln \dfrac{[0.025]\cdot 1}{[0.025]^2} = 0.39 - (0.0257\cdot 3.69/2) = 0.34\ V$

27. The voltage of a cell that has Ag in a 0.25 M solution of Ag^+ and Zn electrode in 0.010 M Zn^{2+}: Calculate the standard voltage of the cell:

   (1) Zn^{2+} (aq) + 2e$^-$ → Zn (s) $\qquad\qquad\qquad$ E° = -0.763 V

   (2) Ag^+ (aq) + e$^-$ → Ag(s) $\qquad\qquad\qquad$ E° = +0.7994 V

The net equation is given as: 2 Ag^+(aq) + Zn (s) → Zn^{2+} (aq) + 2 Ag(s).

The equilibrium expression (Q) would be: $\dfrac{[Zn^{2+}]}{[Ag^+]^2}$

The cell will run in the direction that is product favored, so we can calculate

$\qquad$ E° cell = $E°$cathode − $E°$anode = +0.7994 − (-0.763) = 1.562 V

We also note from the balanced equation that 2 moles(n) of electrons are transferred.

Using the Nernst equation $\quad E_{cell} = E°_{cell} - \dfrac{0.0257}{n} \ln \dfrac{[Zn^{2+}]}{[Ag^+]^2}$

$E_{cell} = 1.562 - \dfrac{0.0257}{2}\ln \dfrac{[0.010]}{[0.25]^2} = 1.562 - (0.0257/2)\cdot (-1.83) = 1.562 + 0.0235 = 1.59\ V$

29. The voltage of a cell that has Ag in a ? M solution of Ag^+ and Zn electrode in 1.0 M Zn^{2+}:

   Calculate the standard voltage of the cell:

   (1) Zn^{2+} (aq) + 2e$^-$ → Zn (s) $\qquad\qquad\qquad$ E° = -0.763 V

   (2) Ag^+ (aq) + e$^-$ → Ag(s) $\qquad\qquad\qquad$ E° = +0.7994 V

The net equation is given as: $2Ag^+(aq) + Zn(s) \rightarrow Zn^{2+}(aq) + 2Ag(s)$.

The equilibrium expression (Q) would be: $\dfrac{[Zn^{2+}]}{[Ag^+]^2}$

The cell will run in the direction that is product favored, so we can calculate

$E^\circ{}_{cell} = E^\circ{}_{cathode} - E^\circ{}_{anode} = +0.7994 - (-0.763) = 1.562\ V$

We also note from the balanced equation that 2 moles(n) of electrons are transferred.

Using the Nernst equation $\quad E_{cell} = E^\circ{}_{cell} - \dfrac{0.0257}{n}\ln\dfrac{[Zn^{2+}]}{[Ag^+]^2}$

Given the Ecell = 1.48V, we should be able to calculate the value of the "ln term".

Recall that we **know** the concentration of $[Zn^{2+}]$, but **don't know** the $[Ag^+]$.

$1.48 = 1.562 - \dfrac{0.0257}{2}\ln\dfrac{[Zn^{2+}]}{[Ag^+]^2}$ or $1.48 = 1.562 - (0.0257/2)\bullet\ln\dfrac{[Zn^{2+}]}{[Ag^+]^2}$

$\dfrac{-(1.48-1.562)2}{0.0257} = \ln\dfrac{[Zn^{2+}]}{[Ag^+]^2}$. or $6.381 = \ln\dfrac{[Zn^{2+}]}{[Ag^+]^2}$; and $\dfrac{[Zn^{2+}]}{[Ag^+]^2} = 590.7$

So $\dfrac{[1.0]}{590.7} = [Ag^+]^2$ and $[Ag^+] = 0.040\ M$

Electrochemistry, Thermodynamics, and Equilibrium

31. Calculate $\Delta_r G^\circ$ and K for the reactions:

(a) $2Fe^{3+}(aq) + 2I^-(aq) \Leftrightarrow 2Fe^{2+}(aq) + I_2(aq)$

Using the potentials: $Fe^{3+}(aq) + e^- \rightarrow 2Fe^{2+}\qquad E^\circ = 0.771V$

$\qquad\qquad\qquad\qquad 2I^-(aq) \rightarrow I_2(aq) + 2e^-\qquad E^\circ = 0.621V$

$E^\circ{}_{cell} = (0.771V - 0.621V) = 0.150\ V$

The relationship between ΔG° and E° is: $\Delta G^\circ = -nFE^\circ$:

$\Delta_r G^\circ = -(2\ mol\ e)(96{,}500\ C/mol\ e)(0.150\ V)$ and $1V = 1J/C$ so

$\Delta_r G^\circ = -(2\ mol\ e)(96{,}500\ C/mol\ e)(0.150\ J/C) = -28950\ J$ or $-29.0\ kJ$

$\Delta_r G^\circ = -RT\ln K$ so $-\Delta G^\circ/RT = \ln K$

$$\frac{28950 \text{ J}}{8.314 \dfrac{\text{J}}{\text{K} \bullet \text{mol}} \bullet 298.15\text{K}} = \ln K \text{ so } 11.68 = \ln K \text{ and } K = 1 \times 10^5$$

(b) I_2 (aq) + 2Br⁻ (aq) → 2 I⁻ (aq) + Br_2 (l)

Using the potentials: Br_2 (l) + 2 e⁻ → 2Br⁻ (aq)   E° = 1.08 V   and

I_2 (aq) + 2 e⁻ → 2I⁻ (aq)   E° = 0.621V

We calculate an E°$_{cell}$, noting that in the cell reaction given, molecular iodine is being reduced (the cathode) and bromide ion is being oxidized (the anode).

E°$_{cell}$ = (0.621V – 1.08V) = - 0.459 V

The relationship between $\Delta_r G°$ and E° is: $\Delta_r G°$ = -nFE° so.

$\Delta_r G°$ = -(2 mol e)(96,500 C/mol e)(-0.459 V) and 1V = 1J/C so

$\Delta_r G°$ = -(2 mol e)(96,500 C/mol e)(-0.459 J/C) = + 88,587 J or +88.6 kJ (to 3 sf)

$\Delta_r G°$ = - RTlnK so - ΔG°/RT = ln K

$$\frac{(-)+ 88587 \text{ J}}{8.314 \dfrac{\text{J}}{\text{K} \bullet \text{mol}} \bullet 298.15 \text{ K}} = \ln K \text{ so } -35.74 = \ln K \text{ and } K = 3 \times 10^{-16}$$

33. Calculate K$_{sp}$ for AgBr using the following reactions:

   (1) AgBr(s) + 1 e⁻ → Ag(s) + Br⁻(aq)   E° = 0.0713 V

   (2) Ag⁺(aq) + 1e⁻ → Ag(s)   E° = 0.7994 V

Write the Ksp expression for AgBr; AgBr(s) ⇔ Ag⁺ (aq) + Br⁻ (aq)

Note that we can accomplish this as an overall reaction, by reversing equation (2) and adding that to equation (1). Equation (1) is presently written as a reduction (naturally) and the *reverse* of Equation (2) would be an oxidation—so the roles for our "cell" are defined.

E°cell = E°cathode = E°anode = 0.0713 V – 0.7994 V = -0.7281V

Using our $\Delta_r G°$ relationships: $\Delta_r G°$ = -nFE°= - RTlnK so lnK = $\dfrac{\text{nFE}^{\cdot}}{\text{RT}}$

$$\text{So } \ln K = \frac{1 \text{ mol e} \bullet 96,500 \dfrac{\text{C}}{\text{mol e}} \bullet -0.7281 \dfrac{\text{J}}{\text{C}}}{8.314 \dfrac{\text{J}}{\text{K} \bullet \text{mol}} \bullet 298.15 \text{ K}} = -28.36 \text{ and } K = 4.9 \times 10^{-13}$$

35. Calculate the $K_{formation}$ for $AuCl_4^-$ (aq)

(1) $AuCl_4^-(aq) + 3e^- \rightarrow Au(s) + 4 Cl^-(aq)$ $E° = 1.00$ V

(2) $Au^{3+}(aq) + 3e^- \rightarrow Au(s)$ $E° = 1.50$ V

The formation reaction for the complex is: $Au^{3+} (aq) + 4 Cl^- (aq) \Leftrightarrow AuCl_4^- (aq)$

To achieve this reaction as a net reaction, we need to reverse equation (1) and add it to equation (2).

The $E°_{cell}$ for that process would be: $E°_{cathode} - E°_{anode} = 1.50 - 1.00 = 0.50$ V

Using our $\Delta_r G°$ relationships: $\Delta_r G° = -nFE° = -RT\ln K$ so $\ln K = \dfrac{nFE°}{RT}$

So $\ln K = \dfrac{3 \text{ mol e} \cdot 96,500 \dfrac{C}{\text{mol e}} \cdot 0.50 \dfrac{J}{C}}{8.314 \dfrac{J}{K \cdot mol} \cdot 298.15 \text{ K}} = 58.42$ and $K_{formation} = 2 \times 10^{25}$

Electrolysis

37. Diagram of an electrolysis apparatus
for molten NaCl:

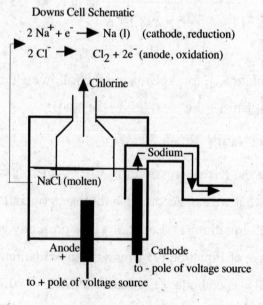

Downs Cell Schematic

$2 Na^+ + e^- \rightarrow Na$ (l)   (cathode, reduction)

$2 Cl^- \rightarrow Cl_2 + 2e^-$ (anode, oxidation)

Chlorine

Sodium

NaCl (molten)

Anode
+

to + pole of voltage source

Cathode
-
to - pole of voltage source

39. For the electrolysis of a solution of KF(aq), what product is expected at the anode: O_2 or F_2

Example 20.10 provides additional help with this concept. The **bottom line** is that the process that occurs in an electrolysis is the one requiring the smaller applied potential. For an electrolysis: $E°cell = E°cathode - E°anode$. For aqueous KF, those voltages are:

$2 F^- (aq) \rightarrow F_2 (g) + 2e^-$ $E° = +2.87$ V

$2 H_2O (l) \rightarrow O_2 (g) + 4 H^+ (aq) + 4 e^-$ $E° = +1.23$ V

At the cathode: $2 H_2O(l) + 2 e^- \rightarrow H_2(g) + 2 OH^-(aq)$ $E° = -0.83$ V

[K has such a large negative reduction potential, that it *will not be reduced*.]

So the two choices are: $E°cell = E°cathode - E°anode$

     For fluorine oxidation: $E°cell = (-0.83 - +2.87) = -3.7$ V

     For oxygen oxidation: $E°cell = (-0.83 - +1.23) = -2.06$ V

Oxygen oxidation will require the lower applied potential, and will be produced at the anode.

41. For the electrolysis of KBr (aq):

(a) The reaction occurring at the cathode: $2 H_2O(l) + 2 e^- \rightarrow H_2(g) + 2 OH^-(aq)$

(b) The reaction occurring at the anode: $2 Br^-(aq) \rightarrow Br_2(l) + 2e^-$

Counting Electrons

43. Solutions to problems of this sort are best solved by beginning with a factor containing the desired units. Connecting this factor to data provided usually gives a direct path to the answer.

units desired
$\downarrow$

$$\frac{58.69 \text{ g Ni}}{1 \text{ mol Ni}} \cdot \frac{1 \text{ mol Ni}}{2 \text{ mol e}^-} \cdot \frac{1 \text{ mol e}^-}{9.65 \times 10^4 \text{ C}} \cdot \frac{1 \text{ C}}{1 \text{ amp} \cdot \text{s}}$$

$$\cdot \frac{0.150 \text{amps}}{1} \cdot \frac{60s}{1 \text{ min}} \cdot \frac{12.2 \text{ min}}{1} = 0.0334 \text{ g Ni}$$

The second factor ($\frac{1 \text{ mol Ni}}{2 \text{ mol e}^-}$) is arrived at by looking at the reduction half-reaction:

$Ni^{2+} (aq) + 2 e^- \rightarrow Ni(s)$

All other factors are either data or common unity factors (e.g. $\dfrac{60s}{1\ min}$).

45. Follow the pattern from SQ20.43, in starting with a unit that has the units of the desired answer—in this case units of *time*.

$$\dfrac{1\ amp \cdot s}{1\ C} \cdot \dfrac{1}{0.66\ amp} \cdot \dfrac{96500\ C}{1\ mol\ e} \cdot \dfrac{2\ mol\ e}{1\ mol\ Cu} \cdot \dfrac{1\ mol\ Cu}{63.546\ g\ Cu} \cdot \dfrac{0.50\ g\ Cu}{1} = 2300\ s.$$

47. Once again, since the requested answer has units of *hours*, start with a unit that has hours in it.

$$\dfrac{1\ hr}{3600\ s} \cdot \dfrac{1\ amp \cdot s}{1\ C} \cdot \dfrac{1}{1.0\ amp} \cdot \dfrac{96500\ C}{1\ mol\ e} \cdot \dfrac{3\ mol\ e}{1\ mol\ Al} \cdot \dfrac{1\ mol\ Al}{26.98\ g\ Al} \cdot \dfrac{84\ g\ Al}{1} = 250\ hr$$

GENERAL QUESTIONS

49. Balanced equations for the following half-reactions:

(a) $UO_2^+(aq) \rightarrow U^{4+}(aq)$

Reduction half-equation	
$UO_2^+ (aq) \rightarrow U^{4+}(aq)$	Balance all non H,O
$UO_2^+ (aq) \rightarrow U^{4+}(aq) + 2H_2O(l)$	Balance O with H_2O
$4\ H^+(aq) + UO_2^+(aq) \rightarrow U^{4+}(aq) + 2H_2O\ (l)$	Balance H with H^+
$4\ H^+(aq) + UO_2^+(aq) + 1\ e^- \rightarrow U^{4+}(aq) + 2\ H_2O(l)$	Balance charge with e^-

(b) $ClO_3^- (aq) \rightarrow Cl^-(aq)$

Reduction half-equation	
$ClO_3^- (aq) \rightarrow Cl^-(aq)$	Balance all non H,O
$ClO_3^- (aq) \rightarrow Cl^-(aq) + 3H_2O(l)$	Balance O with H_2O
$6\ H^+(aq) + ClO_3^- (aq) \rightarrow Cl^-(aq) + 3H_2O(l)$	Balance H with H^+
$6\ H^+(aq) + ClO_3^-(aq) + 6\ e^- \rightarrow Cl^-(aq) + 3H_2O(l)$	Balance charge with e^-

(c) $N_2H_4 (aq) \rightarrow N_2 (g)$

Oxidation half-equation	
$N_2H_4 (aq) \rightarrow N_2 (g)$	Balance all non H,O
$N_2H_4 (aq) \rightarrow N_2 (g) + 4\ H_2O(l)$	Balance H with H_2O
$4\ OH^-(aq) + N_2H_4 (aq) \rightarrow N_2 (g) + 4\ H_2O(l)$	Balance O with OH^-
$4\ OH^-(aq) + N_2H_4(aq) \rightarrow N_2(g) + 4\ H_2O(l) + 4e^-$	Balance charge with e^-

(d) $ClO^-(aq) \rightarrow Cl^-(aq)$

Reduction half-equation	
$ClO^- (aq) \rightarrow Cl^-(aq)$	Balance all non H,O
$ClO^- (aq) \rightarrow Cl^-(aq) + OH^- (aq)$	Balance O with OH^-
$H_2O(l) + ClO^- (aq) \rightarrow Cl^-(aq) + OH^- (aq) + OH^- (aq)$	Balance H with H_2O rebalance H with OH^-
$H_2O(l) + ClO^- (aq) + 2 e^- \rightarrow Cl^-(aq) + 2 OH^-(aq)$	Balance charge with e^-

51. For the electrochemical cell involving Mg and Ag:

(a) Parts of the cell:

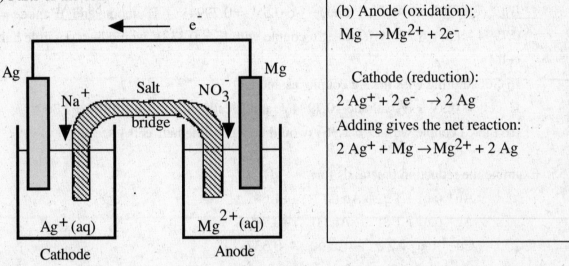

(b) Anode (oxidation):

$Mg \rightarrow Mg^{2+} + 2e^-$

Cathode (reduction):

$2 Ag^+ + 2 e^- \rightarrow 2 Ag$

Adding gives the net reaction:

$2 Ag^+ + Mg \rightarrow Mg^{2+} + 2 Ag$

(c) The flow of electrons in the outer circuit, as described on the diagram above is from the Mg electrode to the Ag electrode. The ion flow in the salt bridge is also shown on the diagram above. The salt bridge is necessary to negate the charge differential that would grow as the cell operates. For example, in the Ag half-cell, the compartment that originally contained equal amounts of + and – charges (from Ag^+ and NO_3^- ions respectively) would accrue a net "-" charge, as the "+" silver ions are reduced. The Na^+ ions from the salt bridge would flow into the Ag compartment, neutralizing that growing negative charge.

53. (a) Half-cells that might be used in conjunction with the Ag+/Ag half-cell to produce a cell with voltage close to 1.7V. Consider cells in which the silver half-cell could function either as cathode or anode. $E° Ag^+ = 0.7994V$

Recall that $E°$ cell = $E°$cathode – $E°$ anode.

With $E°$ cell desired to be 1.7V, we can substitute the Ag/Ag^+ half cell first as cathode, then as anode.

E° cell = E°cathode – E° anode   1.7V = 0.7994V – E° anode, and E°anode

= 0.7994-1.7 = -0.90 V (so the Chromium half-cell, with E° = -0.91 V would be an

appropriate half-cell.)

And E° cell = E°cathode – E° anode 1.7V = E°cathode – 0.7994V and E°cathode =

0.7994 +1.7 = 2.5V so fluorine (E° = 2.87V) is a reasonable suggestion as the other

half-cell.

(b) Half-cells that might be used in conjunction with the Ag^+/Ag half-cell to produce a cell

with voltage close to 0.5V. Consider cells in which the silver half-cell could function

either as cathode or anode.   E° Ag^+ = 0.7994V. As in (a) above, we can substitute the

Ag/Ag^+ half cell first as cathode, then as anode, for a E° cell = 0.5V

(i)E° cell = E°cathode – E° anode so 0.5V = 0.7994V – E° anode, and E°anode =

0.7994 - 0.5 = 0.299 V (so the Cu couple with E° = 0.337V would be a possible half-

cell.)

(ii)Substituting with the Ag couple as anode:

E° cell = 0.5V = E° cathode - 0.7994V, and E°cathode = 0.5 + 0.7994 = 1.30 V

(so the Cl couple with E° = 1.36V would be a possible half-cell.)

55. Examine the reduction potentials for: E°

Au^+ (aq) + 1 e^- → Au (s)   +1.68

Ag^+ (aq) + 1 e^- → Ag (s)   +0.7994

Cu^{2+} (aq) + 2 e^- → Cu (s)   +0.337

Sn^{2+} (aq) + 2e^- → Sn (s)   -0.14

Co^{2+} (aq) + 2 e^- → Co (s)   -0.28

Zn^{2+} (aq) + 2 e^- → Zn (s)   -0.763

To clarify the trends, I have listed them in the descending reduction potential typical of

Reduction Potential Charts, as in Table 20.1.

(a) Metal ion that is the weakest oxidizing agent? — Zn^{2+} Strength of oxidizing agents

increase with the *increasing* + E° ·

(b) Metal ion that is the strongest oxidizing agent? –Au^+

(c) Metal that is the strongest reducing agent? –Zn(s) (Since its oxidized partner is the

weakest oxidizing agent, the metal is the strongest reducing agent.)

(d) Metal that is the weakest reducing agent?—Au(s)

(e) Will Sn(s) reduce Cu^{2+} to Cu(s)? – Yes. Use the "NW to SE rule to see that this will give

a positive E°cell.

(f) Will Ag(s) reduce Co^{2+}(aq) to Co(s)? No. See part (e)

(g) Which metal ions from the list can be reduced by Sn(s). Using the logic as in part (e)

Au^+, Ag^+, and Cu^{2+} can be reduced by Sn.

(h) Metals that can be oxidized by Ag^+(aq) –Any metal "below" Ag: Cu, Sn, Co, Zn

57. Examine the following reductions $\underline{E°}$

Cu^{2+} (aq) + 2 e$^-$ →Cu (s)   +0.337

Fe^{2+} (aq) + 2 e$^-$ →Fe (s)   -0.44

Cr^{3+} (aq) + 3 e$^-$ →Cr (s)   -0.74

Mg^{2+} (aq) + 2 e$^-$ →Mg (s)   -2.37

(a) In which of the voltaic cells would the S.H.E. be the cathode?

$E°_{cell}$ must be positive, and the E° S.H.E.= 0.00V, so for $E°_{cell}$ = 0.00V – (E°anode) to be

positive, the E°anode would have to be *negative*, so the iron, chromium, and magnesium

half-cells would fit this description.

(b) Voltaic cell with the highest and lowest potentials?

For $E°_{cell}$ = 0.00V – (E°anode) to have the largest value the E°anode would have to be

the most negative (Magnesium)

For $E°_{cell}$ = 0.00V – (E°anode) to have the smallest value the E°anode would have to be

the least negative (Iron). While Cu^{+2} is a tempting choice, note that the Cu couple, when

paired with the S.H.E. would give a negative voltage!

59. Mass of Al metal produced from electrolysis of Al^{3+} salt with 5.0V and 1.0 x 10^{5}A in 24 hr.

$$\frac{26.98 \text{ g Al}}{1 \text{ mol Al}} \cdot \frac{1 \text{mol Al}}{3 \text{ mol e}^-} \cdot \frac{1 \text{ mol e}^-}{9.65 \text{ x } 10^4 \text{ C}} \cdot \frac{1 \text{ C}}{1 \text{ amp} \cdot \text{s}} \cdot \frac{1.0 \text{ x } 10^5 \text{ amp}}{1} \cdot \frac{3600 \text{ s}}{1 \text{ hr}} \cdot \frac{24 \text{ hr}}{1}$$

$$= 8.1 \text{ x } 10^5 \text{ g Al}$$

61. (a) $E°_{cell}$ = 0.142V = E°cathode – E° anode, so 0.142V = -0.126 – E° anode

[Note that the value, -0.126V, is arrived at by examining the Reduction Potential Table]

Rearranging to solve: E° anode = -0.126 - 0.142 = - 0.268V

(b) Given these data, estimate Ksp for $PbCl_2$.

The expression for this salt is: Ksp =$[Pb^{2+}][Cl^-]^2$

From equation 20.3, we can calculate the equilibrium constant (Q):

$$Ecell = E°cell - \frac{0.0257}{n}\ln Q$$

If we are at equilibrium, Ecell = 0, and Q = Ksp.

$$0.0 = -0.142 - \frac{0.0257}{2}\ln Ksp \text{ and rearranging: } \frac{-(0.0+0.142)\cdot 2}{0.0257} = \ln Ksp$$

$$-11.1 = \ln Ksp \text{ and } Ksp = 2 \times 10^{-5}$$

63. What is $\Delta_r G°$ for the reaction, given that E°cell = +2.12V

Note that the # of electrons (n) would be 2 (Zn -> Zn^{2+})

$$\Delta_r G° = -nFE° = -(2 \text{ mole})(96500 \frac{C}{\text{mol e-}})(+2.12V)\left(\frac{1 J}{1 V\cdot C}\right)\left(\frac{1 kJ}{1000 J}\right) = -409 \text{ kJ}$$

65. Mass of Cl_2 and Na produced by 7.0V with a current of 4.0×10^4A, flowing for 1 day?

$$\frac{70.91 \text{ g } Cl_2}{1 \text{ mol } Cl_2} \cdot \frac{1 \text{ mol } Cl_2}{2 \text{ mol e-}} \cdot \frac{1 \text{mole}^-}{9.65 \times 10^4 C} \cdot$$

$$\frac{1C}{1\text{amp}\cdot s} \cdot \frac{4.0 \times 10^4 \text{ amp}}{1} \cdot \frac{3600s}{1hr} \cdot \frac{24 \text{ hr}}{1 \text{ day}} = 1.3 \times 10^6 \text{ g } Cl_2$$

and for sodium: $\frac{23.0 \text{gNa}}{1\text{molNa}} \cdot \frac{1\text{molNa}}{1\text{mole}^-} \cdot \frac{1 \text{ mol e}^-}{9.65 \times 10^4 C} \cdot$

$$\frac{1 C}{1 \text{ amp} \cdot s} \cdot \frac{4.0 \times 10^4 \text{ amp}}{1} \cdot \frac{3600 s}{1 \text{ hr}} \cdot \frac{24 \text{ hr}}{1 \text{ day}} = 8.2 \times 10^5 \text{ g Na}$$

The energy consumed:

$$\frac{1 \text{kwh}}{3.6 \times 10^6 J} \cdot \frac{1 J}{1 V \cdot C} \cdot \frac{7.0 V}{1} \cdot \frac{1C}{1\text{amp s}} \cdot \frac{4.0 \times 10^4 \text{ amp}}{1} \cdot \frac{3600s}{1hr} \cdot \frac{24hr}{1 \text{ day}} = 6720 \text{ kwh}$$

(6700 khw to 2 sf)

67. To calculate the charge on the Ru^{n+} ion, we need to know two things:

1. How many moles of elemental ruthenium are reduced?

2. How many moles of electrons caused that reduction?

Moles of ruthenium: $\frac{0.345 \text{ gRu}}{1} \cdot \frac{1\text{molRu}}{101.07\text{gRu}} = 3.41 \times 10^{-3} \text{ mol Ru}$

Moles of electrons:

$$0.44 \text{ amp} \cdot \frac{1 C}{1 \text{ amp} \cdot s} \cdot \frac{3600 s}{1 \text{ hr}} \cdot \frac{(25/60) \text{ hr}}{1} \cdot \frac{1 \text{ mol e}^-}{9.65 \times 10^4 C} = 6.8 \times 10^{-3} \text{ mol e}^-$$

Recall that our general reduction reactions are written: $M^{+x} + x\ e^- \rightarrow M$

If we know the number of $\dfrac{\text{moles of electrons}}{\text{mol of metal}}$, we know the charge on the cation, hence

for the Rh^{x+} ion we have $\dfrac{6.8 \times 10^{-3}\text{mol e}^-}{3.41 \times 10^{-3}\text{mol Ru}} = 2.0\ \dfrac{\text{mole}^-}{\text{molRu}}$.

The ion is therefore the Ru^{2+} ion! The formula for the nitrate salt would be $Ru(NO_3)_2$.

69. Mass of Cl_2 produced by electrolysis with 3.0×10^5 amperes at 4.6V in a 24-hr day.

$$\dfrac{70.91\ \text{g Cl}_2}{1\ \text{mol Cl}_2} \cdot \dfrac{1\ \text{mol Cl}_2}{2\ \text{mol e}^-} \cdot \dfrac{1\ \text{mol e}^-}{96500\ \text{C}} \cdot \dfrac{1\ \text{C}}{1\ \text{amp} \bullet \text{s}} \cdot \dfrac{3.0 \times 10^5\ \text{amp}}{1} \cdot \dfrac{3600\ \text{s}}{1\ \text{hr}} \cdot \dfrac{24\ \text{hr}}{1} =$$

$$9.5 \times 10^6\ \text{g Cl}_2$$

71. The products formed in the electrolysis of aqueous $CuSO_4$ are Cu(s) and O_2(g).

Write equations for the anode and cathode reactions.

Anode: $2\ H_2O(l) \rightarrow O_2(g) + 4\ H^+(aq) + 4\ e^-$

Cathode: $Cu^{2+}(aq) + 2\ e^- \rightarrow Cu(s)$

73. The fact that H_2 is formed at the cathode rather than NO reflects the relative **rates** of the two

reactions. If water is reduced faster than nitric acid, we would expect hydrogen gas to be

preferentially produced. At the anode, we anticipate the formation of oxygen gas, according

to the equation: $2\ H_2O(l) \rightarrow O_2(g) + 4\ H^+(aq) + 4\ e^-$

75. The half-cells: $Pt\ |\ Fe^{3+}(aq)$, (0.50 M) and $Fe^{2+}(aq)$,$(1.0 \times 10^{-5}\ M)$ and

$Hg^{2+}(0.020\ M)\ |\ Hg$. Which electrode is the anode?

Examine the two half-reactions:

$Fe^{3+}(aq) + e^- \rightarrow Fe^{2+}(aq)$ $E° = +0.771\ V$

$Hg^{2+}(aq) + 2e^- \rightarrow Hg(l)$ $E° = +0.855\ V$

Under standard conditions, the mercury electrode would be the cathode, so the iron electrode

would serve as the anode!

The cell reaction would be:

$$Hg^{2+}(aq) + 2Fe^{2+}(aq) \rightarrow Hg(l) + 2Fe^{3+}(aq)$$

So $E°_{cell}$ = E°cathode – E° anode = (+ 0.855 V) – (+ 0.771 V) = 0.084 V

What will be the potential of the voltaic cell?

Using the Nernst equation $\quad E_{cell} = E°_{cell} - \dfrac{0.0257}{n} \ln \dfrac{\left[Fe^{3+}\right]^2}{\left[Hg^{2+}\right]\left[Fe^{2+}\right]^2}$

Recall that the "ln" term has "right-hand species over left-hand species"!

$$E_{cell} = 0.084 - \dfrac{0.0257}{2} \ln \dfrac{[0.50]^2}{[0.020]\left[1.0x10^{-5}\right]^2} = 0.084 - (0.0257/2) \bullet (25.55)$$

$$= 0.084 + (-0.328) = -0.244 \text{ V}$$

Note that cells *do not* operate at "negative" voltages, so the E_{cell} will be +0.244V. This also

tells us that the reaction *will not occur* as we've predicted. So the spontaneous process will

be $Hg(l) + 2Fe^{3+}(aq) \rightarrow Hg^{2+}(aq) + 2Fe^{2+}(aq)$, with mercury functioning as the anode!

77. Cell potential for: $Pt|H_2$ (P = 1 bar)|H⁺ (aq, 1.0 M)||Fe³⁺ (aq, 1.0M), Fe²⁺ (aq, 1.0M)|Pt

The cell convention indicates that the spontaneous reaction would be:

H_2 (g) + 2 Fe^{3+}(aq) → 2 H⁺(aq) + 2 Fe^{2+} (aq). Note that we can simply read the convention

"from left to right". Note also that the equation is NOT balanced automatically, so we check

the number of electrons transferred—2. With the conditions noted, the cell is in it's

"standard" state, so calculate E°cell:

$E°_{cell}$ = E° (Fe^{3+}) – E° (H_2) = (+ 0.771 V) – (+ 0.0 V) = +0.771 V

Now what about the cell at "lower" pHs? If you note in the equation above that hydrogen

ions will be *produced* as the reaction occurs, a lower pH [a greater concentration of H⁺]

would favor the "left side" of this equation, and the reaction would be less favored!

To demonstrate the change in the reaction with pH quantitatively, let's calculate the cell

potential for a reaction in which [H⁺(aq)] is 1.0×10^{-7} M—the pH = 7!

Using the Nernst equation $\quad E_{cell} = E°_{cell} - \dfrac{0.0257}{n} \ln \dfrac{\left[H^+\right]^2\left[Fe^{2+}\right]^2}{\left[Fe^{3+}\right]^2}$

$$E_{cell} = +0.771\ V - \frac{0.0257}{2}\ ln\frac{[H^+]^2[Fe^{2+}]^2}{[Fe^{3+}]^2} = +0.771 - \frac{0.0257}{2}\ ln\frac{[1x10^{-7}]^2[1.0]^2}{[1.0]^2}$$

$$E_{cell} = +0.771\ V - (0.0128)(-32.236) = 0.771\ V + 0.414V = 1.185\ V$$

So the potential has gone UP as the pH has decreased---in accordance with our prediction.

79. Two $Ag^+(aq)|Ag$ half-cells are constructed. The first has $[Ag^+] = 1.0\ M$, the second has

$[Ag^+] = 1.0 \times 10^{-5}\ M$. When linked together with a salt bridge and external circuit, a cell

potential is observed. (This kind of voltaic cell is referred to as a concentration cell.)

(a) Draw a picture of this cell consisting of one half-cell with $[Ag^+] = 1.0\ M$ and the second with

$[Ag^+] = 1.0 \times 10^{-5}\ M$.

For this example, we have filled the
salt bridge with $NaNO_3$ as the
electrolyte. Many such electrolytes
would be possible.

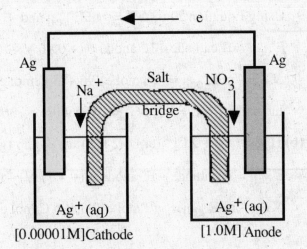

(b) Calculate the cell potential.

Since the silver ion concentrations are *not* 1.0 M, we need to use the Nernst equation.

$$E_{cell} = E°_{cell} - \frac{0.0257}{1}\ ln\frac{[Ag^+]}{[Ag^+]}$$

As *both* half-cells are Ag, the $E°_{cell} = 0.0V$. Substituting the concentrations into the Nernst:

$$E_{cell} = 0.00 - 0.0257 \bullet ln\frac{[Ag^+]}{[Ag^+]} = -0.0257 \bullet ln\frac{[1.0x10^{-5}]}{[1.0]} = -0.0257(-11.51) = 0.30\ V$$

81. Equilibrium constants for the following reations. Is the equilibrium, as written, reactant- or
product-favored?

(a) $2\ Cl^-\ (aq) + Br_2\ (aq) \Leftrightarrow Cl_2\ (aq) + 2\ Br^-(aq)$

$E°_{cell} = E°_{cathode} - E°\ _{anode} = (+1.08\ V) - (+1.36\ V) = -0.28\ V$

Using Equation 20.7, $\ln K = \dfrac{nE^0}{0.0257}$ so $\ln K = \dfrac{2(-0.28)}{0.0257} = -21.79$ and $K = 3.4 \times 10^{-10}$, and the equilibrium as written is reactant-favored.

(b) $Fe^{2+} (aq) + Ag^+ (aq) \Leftrightarrow Fe^{3+} (aq) + Ag (s)$

$E^\circ_{cell} = E^\circ cathode - E^\circ\ anode = (+ 0.799\ V) - (+ 0.771\ V) = + 0.028\ V$

and $\ln K = \dfrac{nE^0}{0.0257}$ so $\ln K = \dfrac{1(+0.028)}{0.0257} = +1.09$ and $K = 3.0$, and the equilibrium as written is product-favored.

83. Calculate ΔrG° for the following reactions:

(a) $3\ Cu (s) + 2\ NO_3^- (aq) + 8\ H^+ (aq) \rightarrow 3\ Cu^{2+} (aq) + 2\ NO (g) + 4\ H_2O (l)$

Using Equation 20.6, $\Delta G^\circ = - nFE^\circ_{cell}$ and

$E^\circ_{cell} = E^\circ cathode - E^\circ\ anode = (+ 0.96\ V) - (+0.337\ V) = + 0.62\ V$

$\Delta G^\circ = - nFE^\circ_{cell} = - (6\ mol\ e)(96{,}485\ C/mol\ e)(+ 0.62\ V) = -360660.93\ VC$

and since 1 VC = 1 J, $\Delta G^\circ = - 360660 J$ or - 360 kJ

(b) $H_2O_2 (aq) + 2\ Cl^- (aq) + 2\ H^+ (aq) \rightarrow Cl_2 (g) + 2\ H_2O (l)$

$E^\circ_{cell} = E^\circ cathode - E^\circ\ anode = (+ 1.77\ V) - (+ 1.36\ V) = + 0.41\ V$

$\Delta G^\circ = - nFE^\circ_{cell} = - (2\ mol\ e)(96{,}485\ C/mol\ e)(+ 0.41\ V) = - 79117.7\ J$ and $\Delta G^\circ = - 79$ kJ

85. Balance the equations:

(a) Separating the equation into an oxidation and reduction half-reaction we get:

$Ag^+ + 1e^- \rightarrow Ag$ (reduction)

$C_6H_5CHO \rightarrow C_6H_5CO_2H$ (oxidation)

Using the method outlined earlier in this chapter , we balance the oxidation half-equation: (Numbers in parentheses correspond to the steps listed in SQ20.3)

$C_6H_5CHO + H_2O \rightarrow C_6H_5CO_2H$ (3)

$C_6H_5CHO + H_2O \rightarrow C_6H_5CO_2H + 2\ H^+$ (4)

$C_6H_5CHO + H_2O \rightarrow C_6H_5CO_2H + 2\ H^+ + 2\ e^-$ (5)

Note that the reduction half equation gains 1 electron/ silver ion. To balance electron gain with electron loss, we multiply the reduction equation by 2.

$$2Ag^+ + 2e^- \rightarrow 2Ag$$

$$\underline{C_6H_5CHO + H_2O \rightarrow C_6H_5CO_2H + 2\,H^+ + 2\,e^-}$$

$$2Ag^+ + C_6H_5CHO + H_2O \rightarrow C_6H_5CO_2H + 2\,H^+ + 2Ag$$

(b) Separating the equation into an oxidation and reduction half-equation we get:

$$Cr_2O_7{}^{2-} \rightarrow Cr^{3+} \qquad\qquad \text{(reduction)}$$

$$C_2H_5OH \rightarrow CH_3CO_2H \qquad \text{(oxidation)}$$

Performing the "steps" on each half-equation

$Cr_2O_7{}^{2-} \rightarrow Cr^{3+}$	$C_2H_5OH \rightarrow CH_3CO_2H$	
$Cr_2O_7{}^{2-} \rightarrow 2\,Cr^{3+}$	$C_2H_5OH \rightarrow CH_3CO_2H$	(2)
$Cr_2O_7{}^{2-} \rightarrow 2\,Cr^{3+} + 7H_2O$	$C_2H_5OH + H_2O \rightarrow CH_3CO_2H$	(3)
$14H^+ + Cr_2O_7{}^{2-} \rightarrow 2\,Cr^{3+} + 7H_2O$	$C_2H_5OH + H_2O \rightarrow CH_3CO_2H + 4H^+$	(4)
$14H^+ + Cr_2O_7{}^{2-} + 6e^- \rightarrow 2\,Cr^{3+} + 7H_2O$		

$$C_2H_5OH + H_2O \rightarrow CH_3CO_2H + 4H^+ + 4e^- \quad (5)$$

Multiplying the reduction half-equation by 2, and the oxidation half-equation by 3 will equalize electron gain with electron loss.

$$28H^+ + 2\,Cr_2O_7{}^{2-} + 12e^- \rightarrow 4\,Cr^{3+} + 14H_2O$$

$$3C_2H_5OH + 3H_2O \rightarrow 3CH_3CO_2H + 12H^+ + 12e^-$$

Adding the two equations, and removing any duplications:

$$28H^+ + 2\,Cr_2O_7{}^{2-} + 12e^- \rightarrow 4\,Cr^{3+} + 14H_2O$$

$$\underline{3C_2H_5OH + 3H_2O \rightarrow 3CH_3CO_2H + 12H^+ + 12e^-}$$

$$16H^+ + 2Cr_2O_7{}^{2-} + 3C_2H_5OH \rightarrow 3CH_3CO_2H + 11H_2O + 4\,Cr^{3+}$$

87. Comparing the silver/zinc battery with the lead storage battery:

(a) The stoichiometry of the silver/zinc battery indicates a reaction of one mole each of silver oxide, zinc, and water. The mass of one mole of each of these three substances is:

1 mol Ag_2O	231.7 g
1 mol Zn	65.4 g
1 mol H_2O	18.0 g
	315.1 g

The energy associated with the battery is:

$$\frac{1.59\text{ V}}{1} \cdot \frac{96500\text{ C}}{1\text{ mol e}} \cdot \frac{1\text{ J}}{1\text{ V} \cdot \text{C}} \cdot \frac{2\text{ mol e}}{1\text{ mol reactant}} \cdot \frac{1\text{ mol reactant}}{315.1\text{ g}} = 973.88\text{ J/g or } 0.974\text{ kJ/g}$$

(b) Performing the same calculations for the lead storage battery, using a stoichiometric amount for the overall battery reaction:

$$\begin{array}{ll}
1\text{ mol Pb} & 207.2\text{ g} \\
1\text{ mol PbO}_2 & 239.2\text{ g} \\
2\text{ mol H}_2\text{SO}_4 & \underline{196.2\text{ g}} \\
& 642.6\text{ g}
\end{array}$$

$$\frac{2.0\text{ V}}{1} \cdot \frac{96500\text{ C}}{1\text{ mol e}} \cdot \frac{1\text{ J}}{1\text{ V} \cdot \text{C}} \cdot \frac{2\text{ mol e}}{1\text{ mol reactant}} \cdot \frac{1\text{ mol reactant}}{642.6\text{ g}} = 600.\text{ J/g or } 0.60\text{ kJ/g (2sf)}$$

(c) The silver/zinc battery produces more energy/gram.

89. Using the procedure outlined in SQ20.3

(a1) $Mn^{2+} + NO_3^- \rightarrow NO + MnO_2$

Oxidation half-equation	Reduction half-equation	Step
$Mn^{2+} \rightarrow MnO_2$	$NO_3^- \rightarrow NO$	1 & 2
$2\,H_2O + Mn^{2+} \rightarrow MnO_2$	$NO_3^- \rightarrow NO + 2\,H_2O$	3
$2\,H_2O + Mn^{2+} \rightarrow MnO_2 + 4\,H^+$	$4\,H^+ + NO_3^- \rightarrow NO + 2\,H_2O$	4
$2\,H_2O + Mn^{2+} \rightarrow MnO_2 + 4\,H^+ + 2\,e^-$	$4\,H^+ + NO_3^- + 3\,e^- \rightarrow NO + 2\,H_2O$	5
$6\,H_2O + 3\,Mn^{2+} \rightarrow 3\,MnO_2 + 12\,H^+ + 6\,e^-$	$8\,H^+ + 2\,NO_3^- + 6\,e^- \rightarrow 2\,NO + 4\,H_2O$	6
$2\,H_2O(l) + 3\,Mn^{2+}(aq) + 2\,NO_3^-(aq) \rightarrow 2\,NO(g) + 3\,MnO_2(s) + 4\,H^+(aq)$		7

(a2) $NH_4^+ + MnO_2 \rightarrow N_2 + Mn^{2+}$

Oxidation half-equation	Reduction half-equation	Step
$2\,NH_4^+ \rightarrow N_2$	$MnO_2 \rightarrow Mn^{2+}$	1 & 2
$2\,NH_4^+ \rightarrow N_2$	$MnO_2 \rightarrow Mn^{2+} + 2\,H_2O$	3
$2\,NH_4^+ \rightarrow N_2 + 8\,H^+$	$4\,H^+ + MnO_2 \rightarrow Mn^{2+} + 2\,H_2O$	4
$2\,NH_4^+ \rightarrow N_2 + 8\,H^+ + 6\,e^-$	$4\,H^+ + MnO_2 + 2\,e^- \rightarrow Mn^{2+} + 2\,H_2O$	5
$2\,NH_4^+ \rightarrow N_2 + 8\,H^+ + 6\,e^-$	$12\,H^+ + 3\,MnO_2 + 6\,e^- \rightarrow 3\,Mn^{2+} + 6\,H_2O$	6
$2\,NH_4^+(aq) + 4\,H^+(aq) + 3\,MnO_2(s) \rightarrow N_2(g) + 3\,Mn^{2+}(aq) + 6\,H_2O(l)$		7

(b) E°cell for the two reactions:

For a1: E°cell = E°cathode - E°anode = 0.96 V – 1.23V = -0.27 V

For a2: E°cell = E°cathode - E°anode = 1.23 V – (-0.272)V = 1.50 V

91. For the disproportionation of Iron(II):

(a) The half-reactions that make up the disproportionation reaction:

$$Fe^{2+}(aq) + 2\ e^- \rightarrow Fe(s) \qquad\qquad E^{\circ} = -0.44\ V \quad (cathode)$$

$$\underline{2\ Fe^{2+}(aq) \rightarrow 2\ Fe^{3+}(aq) + 2\ e^- \qquad\qquad E^{\circ} = 0.771\ V\ (anode)}$$

$$3\ Fe^{2+}(aq) \rightarrow 2\ Fe^{3+}(aq) + Fe(s) \qquad E^{\circ} = -0.44\ V - 0.771\ V = -1.211V$$

(b) Given the voltage calculated, the disproportionation reaction is **not product-favored**.

(c) $\ln K = \dfrac{2\ mol\ e \bullet 96{,}500\dfrac{C}{mol\ e} \bullet -1.211\dfrac{J}{C}}{8.314\ \dfrac{J}{K\bullet mol} \bullet 298\ K} = -94.34$ and $K_{formation} = 1 \times 10^{-41}$

IN THE LABORATORY

93. (a) Cell diagram:

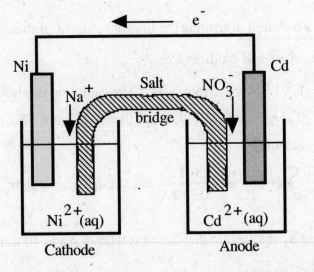

(b) As shown in part (a), Cd serves as the anode, so it must be oxidized to the 2^+ cation.

Likewise Ni ions would be reduced to elemental Ni.

The balanced equation is: $Ni^{2+}(aq) + Cd(s) \rightarrow Cd^{2+}(aq) + Ni(s)$

(c) The anode (Cd) serves as the source of electrons to the external circuit, and we label it "-". The cathode is labeled "+".

(d) The E°cell = E°cathode - E°anode = (-0.25) – (-0.403) = +0.15 V

(e) As shown on the diagram, electrons flow from Cd to Ni compartments.

(f) The direction of travel for the sodium and nitrate ions are shown on the diagram.

(g) The K for the reaction: $\Delta_rG° = -nFE°cell$

$$= -(2 \text{ mol e})(96500 \text{ C/mol e})(+0.15 \text{ J/C}) = 28950J$$

$$\Delta_rG° = -RT\ln K \text{ and } \ln K = \frac{nFE}{RT}$$

$$\text{So } \ln K = \frac{2 \text{ mol e} \cdot 96,500 \frac{C}{\text{mol e}} \cdot 0.15 \frac{J}{C}}{8.314 \frac{J}{K \cdot mol} \cdot 298 \text{ K}} = 11.68 \text{ and } K = 1 \times 10^5$$

(h) If $[Cd^{2+}] = 0.010M$ and $[Ni^{2+}] = 1.0M$, the value for Ecell = ?

$$\text{Using the Nernst equation } E_{cell} = E°_{cell} - \frac{0.0257}{n} \ln \frac{[Cd^{2+}]}{[Ni^{2+}]}$$

$$E_{cell} = 0.15 - \frac{0.0257}{2} \ln \frac{[0.010]}{[1.0]} = 0.15 - (0.0257/2) \cdot (-4.605) = 0.15 + 0.0592 = 0.21 \text{ V}$$

(i) Lifetime use of battery?

We have 1.0 L of each solution, we should determine the limiting reagent, if there is one.

The spontaneous reaction reduces Ni^{2+} and oxidizes Cd.

The Nickel solution contains: $1.0 \text{ L} \cdot 1.0M = 1.0 \text{ mol } Ni^{2+}$. The cadmium electrode weighs 50.0 g, so we have: $50.0 \text{ g Cd} \cdot (1 \text{ mol Cd}/112.41 \text{ g Cd}) = 0.445 \text{ mol Cd}$, so Cd is the limiting reagent. Now we can calculate:

$$0.445 \text{ mol Cd} \cdot \frac{2 \text{ mol e}}{1 \text{ mol Cd}} \cdot \frac{96500 \text{ C}}{1 \text{ mol e}} \cdot \frac{1 \text{ amp} \cdot s}{1C} \cdot \frac{1}{0.050 \text{ A}} = 1.7 \times 10^6 \text{ s or 480 hr.}$$

95. What amount of Au will be deposited by the current that deposits 0.089g Ag in 10 minutes?

$$\frac{0.089 \text{ g Ag}}{1} \cdot \frac{1 \text{ mol Ag}}{107.9 \text{ g Ag}} \cdot \frac{1 \text{ mol e}^-}{1 \text{ mol Ag}} = 8.2 \times 10^{-4} \text{ mol e}^-$$

$$\frac{8.2 \times 10^{-4} \text{ mol e}^-}{1} \cdot \frac{1 \text{ mol Au}}{3 \text{ mol e}^-} \cdot \frac{197.0 \text{ g Au}}{1 \text{ mol Au}} = 0.054 \text{ g Au}$$

97. Explain the reaction of Cu^{2+} with I^-:

The appropriate reduction potentials:

$$Br_2 + 2e^- \rightarrow 2Br^-   1.08V$$

$$Cl_2 + 2e^- \rightarrow Cl^-   1.36V$$

$$I_2 + 2e^- \rightarrow 2I^-   0.535V$$

The relative values for the reduction potentials tell us that I^- is the strongest reducing agent, capable of reducing the Cu(II) ion, to Cu(I). The Cu^{+1} ion would react with the I^- to form the insoluble CuI. The equation: $2\, Cu^{2+}(aq) + 4\, I^-\, (aq) \rightarrow 2\, CuI(s) + I_2(aq)$

SUMMARY AND CONCEPTUAL QUESTIONS

99. In the electrolysis of 150 g of CH_3SO_2F:

(a) The mass of HF required to electrolyze 150 g of CH_3SO_2F:

$$\frac{150\ g\ CH_3SO_2F}{1} \cdot \frac{1\ mol\ CH_3SO_2F}{98.10\ g\ CH_3SO_2F} \cdot \frac{3\ mol\ HF}{1\ mol\ CH_3SO_2F} \cdot \frac{20.01\ g\ HF}{1\ mol\ HF} = 92\ g\ HF$$

$$\frac{150\ g\ CH_3SO_2F}{1} \cdot \frac{1\ mol\ CH_3SO_2F}{98.10\ g\ CH_3SO_2F} \cdot \frac{1\ mol\ CF_3SO_2F}{1\ mol\ CH_3SO_2F} \cdot \frac{152.07\ g\ CF_3SO_2F}{1\ mol\ CF_3SO_2F} =$$

$$230\ gCF_3SO_2F$$

$$\frac{150\ g\ CH_3SO_2F}{1} \cdot \frac{1\ mol\ CH_3SO_2F}{98.10\ g\ CH_3SO_2F} \cdot \frac{3\ mol\ H_2}{1\ mol\ CH_3SO_2F} \cdot \frac{2.02\ g\ H_2}{1\ mol\ H_2} = 9.3\ g\ H_2$$

(b) H_2 produced at the anode or cathode? Since H is being reduced from +1 to 0, it will be produced at the cathode.

(c) Energy consumed: $\dfrac{1\ kwh}{3.60 \times 10^6\ J} \cdot \dfrac{1\ J}{1\ V \bullet C} \cdot \dfrac{8.0\ V}{1} \cdot \dfrac{250\ C}{1\ s} \cdot \dfrac{3600\ s}{1\ hr} \cdot \dfrac{24\ hr}{1} = 48\ kwh$

101. Since the reaction depends on the oxidation of elemental hydrogen to water

(2 mol e^- per mol H_2), we must determine the amount of H_2 present:

$$n = \frac{(200.\ atm)(1.0\ L)}{(0.0821 \dfrac{L \bullet atm}{K \bullet mol})(298\ K)} = 8.2\ mol\ H_2$$

The amount of time this cell can produce current:

$$8.2 \text{ mol } H_2 \cdot \frac{2 \text{ mol } e^-}{1 \text{ mol } H_2} \cdot \frac{9.65 \times 10^4 \text{ C}}{1 \text{ mol } e^-} \cdot \frac{1 \text{ A} \cdot \text{s}}{1 \text{ C}} \cdot \frac{1}{1.5 \text{ A}} = 1.1 \times 10^6 \text{ s} \quad (290 \text{ hrs})$$

103. (a) Amount of glucose(mole) needed to furnish 2400 kcal per 24 hours:

$$\frac{2400 \text{ kcal}}{1} \cdot \frac{4.184 \text{ kJ}}{1 \text{ kcal}} \cdot \frac{1 \text{ mol glucose}}{2800 \text{ kJ}} = 3.6 \text{ mol glucose}$$

Amount of oxygen consumed in the process:

From the balanced equation, note that 1 mol glucose requires 6 mol oxygen, so

3.6 mol glucose • 6 mol O_2/1 mol glucose = 22 mol O_2

(b) moles of electrons to reduce 22 mol O_2:

From Appendix M, note that 1 mol of oxygen requires 4 mol of electrons, so

22 mol oxygen (actually 21.5 mol to 3sf) would require 4 x 21.5 or 86 mol electrons.

(c) Current from the combustion of 3.6 mol glucose:

$$\frac{3.6 \text{ mol glucose}}{24 \text{ hr}} \cdot \frac{86 \text{ mol } e^-}{3.6 \text{ mol glucose}} \cdot \frac{1 \text{ hr}}{3600 \text{ s}} \cdot \frac{96500 \text{ C}}{1 \text{ mol } e^-} \cdot \frac{1 \text{ ampere} \cdot \text{s}}{1 \text{ C}} = 96 \text{ amperes}$$

(d) Watts expended for 1.0V:

$$\frac{1 \text{ watt} \cdot \text{s}}{1 \text{ J}} \cdot \frac{1 \text{ J}}{1 \text{ V} \cdot \text{C}} \cdot \frac{1.0 \text{ V}}{1} \cdot \frac{1 \text{ C}}{1 \text{ ampere} \cdot \text{s}} \cdot \frac{96 \text{ amperes}}{1} = 96 \text{ watts}$$

Applying Chemical Principles

1. To be effective as a sacrificial anode, the zinc and the copper must be in contact.

3. Which of the following metals could serve as a sacrificial anode on a steel hull, assuming that
   the reduction potential for steel is equivalent to that of iron?
   The reduction potentials are:

$$Sn^{2+} + 2e^- \rightarrow Sn \qquad -0.14 \text{ V}$$
$$Ag^+ + 1e^- \rightarrow Ag \qquad +0.799 \text{ V}$$
$$Fe^{2+} + 2e^- \rightarrow Fe \qquad -0.44 \text{ V}$$
$$Ni^{2+} + 2e^- \rightarrow Ni \qquad -0.25 \text{ V}$$
$$Cr^{3+} + 3e^- \rightarrow Cr \qquad -0.74 \text{ V}$$

The only reduction potential that is more negative than that of iron is that belonging to
chromium, making chromium the only metal in this list capable of replacing Zn as a sacrifical
anode. Compare the reduction potential for zinc ($Zn^{2+} + 2e^- \rightarrow Zn \quad E° = -0.763$ V).

5. For the cell: $Zn \mid Zn(OH)_2(s) \mid OH^-(aq) \mid Cu(OH)_2(s) \mid Cu(s)$

   (a) The balanced equation for the cathode reaction: $Cu(OH)_2(s) + 2 e^- \rightarrow Cu(s) + 2 OH^-(aq)$

   (b) The balanced equation for the anode reaction: $Zn(s) + 2 OH^-(aq) \rightarrow Zn(OH)_2(s) + 2 e^-$

   (c) The balanced equation for the overall reaction:
   $$Cu(OH)_2(s) + Zn(s) \rightarrow Zn(OH)_2(s) + Cu(s)$$

   (d) What is the potential (in volts) of the cell at pH = 7.90 and 25 °C?
   Before you drag out Walther's equation, note that all reactants and products are solids, so
   in the "ln" term, all those quantities would have a value of 1, so the ln(1) = 0.
   The Ecell = E°cathode - E°anode = (- 0.36V) – (- 1.245) = 0.89 V.
   NOTE: The reduction potentials are those *in base*:
   $$Cu(OH)_2(s) + 2 e^- \rightarrow Cu(s) \quad E° = - 0.36 \text{ V}$$
   $$Zn(OH)_2(s) + 2 e^- \rightarrow Zn(s) \quad E° = - 1.245 \text{ V}$$

Chapter 21
The Chemistry of the Main Group Elements

PRACTICING SKILLS

Properties of the Elements

1. Examples of two basic oxides, and equations showing the oxide formation from its elements:

$$4 \, Li(s) + O_2(g) \rightarrow 2 \, Li_2O(s) \, ; \text{ is basic owing to the reaction:}$$

$$Li_2O(s) + H_2O(l) \rightarrow 2LiOH(aq)$$

$$2 \, Ca(s) + O_2(g) \rightarrow 2 \, CaO(s) \, ; \text{ is basic owing to the reaction:}$$

$$CaO(s) + H_2O(l) \rightarrow Ca(OH)_2(s)$$

3. Name and symbol with a valence configuration $ns^2 \, np^1$:
   The configuration shown is characteristic of the elements in Group 3A (13)

B:Boron	Al:Aluminum	Ga:Gallium	In:Indium	Tl:Thallium

5. Select an alkali metal; Write the balanced equation with Cl_2:

$$2 \, Na \; + Cl_2 \rightarrow 2 \, NaCl$$

Reactions of alkali metals with halogens are exothermic—given the excellent ability of these metals to function as reducing agents, and the ability of Cl_2 to function as oxidizing agents. Typically compounds formed from elements of greatly different electronegativities (a metal and a nonmetal) are ionic in nature.

7. Predict color, state of matter, water solubility of compound in SQ21.5:
   Such compounds are typically colorless solids; are high melting (a solid), and are water soluble.

9. Calcium is **not** expected to be found free in the earth's crust. Its great reactivity with oxygen and water would dispose the metal to exist as an oxide.

11. The oxides listed in order of increasing basicity:

Metal oxides are basic, nonmetal oxides are acidic. The three listed range from a nonmetal oxide (CO_2) to a metal oxide, SnO_2. So the order is: $CO_2 < SiO_2 < SnO_2$

13. Balanced equations for the following reactions:

(a) $2\,Na(s) + Br_2(l) \rightarrow 2\,NaBr(s)$ (c) $2\,Al(s) + 3\,F_2(g) \rightarrow 2\,AlF_3(s)$

(b) $2\,Mg(s) + O_2(g) \rightarrow 2\,MgO(s)$ (d) $C(s) + O_2(g) \rightarrow CO_2(g)$

Hydrogen

15. Balanced chemical equation for hydrogen gas reacting with oxygen, chlorine, and nitrogen.

$2\,H_2(g) + O_2(g) \rightarrow 2\,H_2O(g)$

$H_2(g) + Cl_2(g) \rightarrow 2\,HCl(g)$

$3\,H_2(g) + N_2(g) \rightarrow 2\,NH_3(g)$

17. The reaction: $CH_4(g) + H_2O(g) \rightarrow CO(g) + 3\,H_2(g)$

Data from Appendix L:

$\Delta_f H°$(kJ/mol)	-74.87	-241.83	-110.525	0
$S°$ (J/K•mol)	+186.26	+188.84	+197.674	+130.7
$\Delta_f G°$ (kJ/mol)	-50.8	-228.59	-137.168	0

$\Delta_r H° = [(1\,mol•-110.525\ kJ/mol) + (3\,mol• 0)] - [(1\,mol • -74.87\ kJ/mol)+(1\,mol•-241.83\ kJ/mol)]$

$= 206.18\ kJ$

$\Delta_r S° = [(1\,mol• +197.674\ J/K•mol) + (3\,mol• +130.7J/K•mol)] -$

$[(1\,mol • +186.26J/K•mol)+(1\,mol• +188.84J/K•mol)]$

$= +214.7\ J/K$

$\Delta_r G° = [(1\,mol•-137.168\ kJ/mol) + (3\,mol• 0)] - [(1\,mol • -50.8\ kJ/mol)+(1\,mol•-228.59\ kJ/mol)]$

$= 142.2\ kJ$

19. Prepare a balanced equation for each of the 3 steps, and show that the sum is the decomposition of water to form hydrogen and oxygen.

(a) $2\ SO_2\ (g) + 4\ H_2O\ (l) + 2\ I_2\ (s) \rightarrow 2\ H_2SO_4\ (l) + 4\ HI\ (g)$

(b) $2\ H_2SO_4\ (l) \rightarrow 2\ H_2O\ (l) + 2\ SO_2\ (g) + O_2\ (g)$

(c) $4\ HI\ (g) \rightarrow 2\ H_2\ (g) + 2\ I_2\ (g)$

Net: $2\ H_2O\ (l) \rightarrow 2\ H_2\ (g) + O_2\ (g)$

Alkali Metals

21. Equations for the reaction of sodium with the halogens:

$2\ Na(s)\ +\ F_2(g) \rightarrow 2\ NaF(s)$

$2\ Na(s)\ +\ Cl_2(g) \rightarrow 2\ NaCl(s)$

$2\ Na(s)\ +\ Br_2(l) \rightarrow 2\ NaBr(s)$

$2\ Na(s)\ +\ I_2(s) \rightarrow\ 2\ NaI(s)$

> Physical properties of the alkaline metal halides: (1) ionic solids (2) high melting and boiling points (3) white color (4) water soluble

23. In the electrolysis of aqueous NaCl:

(a) The balanced equation for the process:

$2\ NaCl(aq)\ +\ 2\ H_2O(l) \rightarrow 2\ NaOH(aq) + Cl_2(g) + H_2(g)$

(b) Anticipated masses ratios:

$$\frac{1\ mol\ Cl_2}{2\ mol\ NaOH} = \frac{70.9\ g\ Cl_2}{80.0\ g\ NaOH} = 0.88\ g\ Cl_2/g\ NaOH$$

$$\text{Actual:}\ \frac{1.14 \times 10^{10}\ kg\ Cl_2}{1.19 \times 10^{10}\ kg\ NaOH} = 0.96\ \frac{kg\ Cl_2}{kg\ NaOH}$$

The difference in ratios means that alternative methods of producing chlorine are used. One of these is the Kel-Chlor process which uses HCl, NOCl, and O_2. Other products are NO and H_2O.

Alkaline Earth Elements

25. Balanced equations for the reaction of magnesium with nitrogen and oxygen:

$3\ Mg(s)\ +\ N_2(g) \rightarrow\ \ \ \ Mg_3N_2(s)$

$2\ Mg(s)\ +\ O_2(g) \rightarrow\ \ \ \ 2\ MgO(s)$

27. Uses of limestone:

agricultural: to furnish Ca^{2+} to plants and neutralize acidic soils

building: lime (CaO) is used in mortar and absorbs CO_2 to form $CaCO_3$

steel-making: $CaCO_3$ furnishes lime (CaO) in the basic oxygen process. The lime reacts with gangue (SiO_2) to form calcium silicate.

The balanced equation for the reaction of limestone with carbon dioxide in water:

$$CaCO_3(s) + H_2O(l) + CO_2(g) \rightarrow Ca^{2+}(aq) + 2\ HCO_3^-(aq)$$

This reaction is important in the formation of "hard water" (not particularly a great happening for plumbing) and stalagmites and stalactites (aesthetically pleasing in caves).

29. The amount of SO_2 that could be removed by 1200 kg of CaO by the reaction:

$$CaO\ (s) + SO_2\ (g) \rightarrow CaSO_3\ (s)$$

$$1.2 \times 10^6\ g\ CaO \cdot \frac{1\ mol\ CaO}{56.079\ g\ CaO} \cdot \frac{1\ mol\ SO_2}{1\ mol\ CaO} \cdot \frac{64.059\ g\ SO_2}{1\ mol\ SO_2} = 1.4 \times 10^6\ g\ SO_2$$

Boron and Aluminum

31. Structure for the cyclic anion in the salt $K_3B_3O_6$, and the chain anion in $Ca_2B_2O_5$

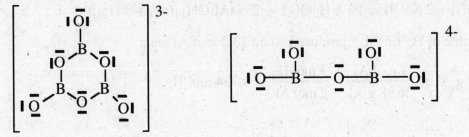

33. For the reactions of boron hydrides:

(a) Balanced Equation: $2\ B_5H_9(g) + 12\ O_2(g) \rightarrow 5\ B_2O_3(s) + 9\ H_2O(g)$

(b) Heat of combustion for $B_5H_9(g)$:

$$\Delta_{comb}H^\circ = [9 \bullet \Delta_fH^\circ\ (H_2O(g)) + 5 \bullet \Delta_fH^\circ(B_2O_3\ (s))]$$

$$- [2 \bullet \Delta_fH^\circ(B_5H_9(g)) + 12 \bullet \Delta_fH^\circ(O_2(g))]$$

$$\Delta_{comb}H^\circ = [9mol \bullet (-241.83kJ/mol) + 5 \bullet (-1271.9\ kJ/mol)]$$

$$- [2 \bullet (73.2\ kJ/mol)) + 12 \bullet (0)] = -8682.37\ kJ$$

So in kJ/mol: $-8682.37 kJ/2 mol\ B_5H_9(g) = -4341.2\ kJ/mol$

The text reports the Heat of Combustion for B_2H_6 is $-2038\ kJ/mol$, so the Heat of Combustion for B_5H_9 is more than twice as great than that of diborane's combustion.

(c) Compare Heats of Combustion for $C_2H_6(g)$ with $B_2H_6(g)$:

For B: $B_2H_6(g) + 3\ O_2(g) \rightarrow B_2O_3(s) + 3\ H_2O(g)$ $\Delta H°_{comb} = -2038\ kJ/mol$

So the energy/g is $-2038\ kJ/mol$ or, for $27.67\ g\ B_2H_6$ $(-73.7\ kJ/g\ B_2H_6)$

For C: $2C_2H_6(g) + 7\ O_2(g) \rightarrow 4\ CO_2(g) + 6\ H_2O(g)$

$\Delta_{comb}H° = [6\bullet(-241.83 kJ/mol) + 4\bullet(-393.509\ kJ/mol)] - [2\bullet(-83.85 kJ/mol) + 7\bullet 0)]$

$\Delta_{comb}H° = -2857.32\ kJ$ and $-1,428.66\ kJ/mol\ C_2H_6$ or for $30.07\ g\ C_2H_6$ $(47.5\ kJ/g)$

35. The equations for the reaction of aluminum with HCl, Cl_2 and O_2:

$2\ Al(s) + 6\ HCl(aq) \rightarrow 2\ Al^{3+}(aq) + 6\ Cl^-(aq) + 3\ H_2(g)$

$2\ Al(s) + 3\ Cl_2(g) \rightarrow 2\ AlCl_3(s)$

$4\ Al(s) + 3\ O_2(g) \rightarrow 2\ Al_2O_3(s)$

37. The equation for the reaction of aluminum dissolving in aqueous NaOH:

$2\ Al(s) + 2\ NaOH(aq) + 6\ H_2O(l) \rightarrow 2\ NaAl(OH)_4(aq) + 3\ H_2(g)$

Volume of H_2 (in mL) produced when 13.2 g of Al react:

$$13.2\ g\ Al \bullet \frac{1\ mol\ Al}{26.98\ g\ Al} \bullet \frac{3\ mol\ H_2}{2\ mol\ Al} = 0.734\ mol\ H_2$$

$$V = \frac{(0.734\ mol\ H_2)(0.082057\ \frac{L \bullet atm}{K \bullet mol})(295.7\ K)}{735\ mm\ Hg \bullet \frac{1\ atm}{760\ mm\ Hg}} = 18.4\ L$$

39. The equation for the reaction of aluminum oxide with sulfuric acid:

$Al_2O_3(s) + 3\ H_2SO_4(aq) \rightarrow Al_2(SO_4)_3(s) + 3\ H_2O(l)$

$$1.00 \times 10^3\ g\ Al_2(SO_4)_3 \bullet \frac{1\ mol\ Al_2(SO_4)_3}{342.1\ g\ Al_2(SO_4)_3} \bullet \frac{1\ mol\ Al_2O_3}{1\ mol\ Al_2(SO_4)_3} \bullet$$

$$\frac{102.1\ g\ Al_2O_3}{1\ mol\ Al_2O_3} \bullet \frac{1\ kg}{1\times10^3\ g} = 0.298\ kg\ Al_2O_3$$

$$1.00 \times 10^3 \text{ g Al}_2(\text{SO}_4)_3 \cdot \frac{1 \text{ mol Al}_2(\text{SO}_4)_3}{342.1 \text{ g Al}_2(\text{SO}_4)_3} \cdot \frac{3 \text{ mol H}_2\text{SO}_4}{1 \text{ mol Al}_2(\text{SO}_4)_3} \cdot$$

$$\frac{98.07 \text{ g H}_2\text{SO}_4}{1 \text{ mol H}_2\text{SO}_4} \cdot \frac{1 \text{ kg}}{1 \times 10^3 \text{ g}} = 0.860 \text{ kg H}_2\text{SO}_4$$

Silicon

41. The structure of pyroxenes is tetrahedral with the silicon atom being surrounded by four oxygen atoms, as shown in the graphic on page 986. The linkage of tetrahedra results in each silicon atom "owning" 2 oxygen atoms, and "sharing"1/2 of 2 atoms (net 1 O atom) between adjacent tetrahedra giving each Si 3 O atoms.

43. The structure for the silicate anion is shown to the right: Like the pyroxenes (SQ21.41), the silicon atoms have an average of 3 O atoms, and a net formula of SiO_3^{2-}

    With six of these in the ring, the **net charge** on the anion (n) is 12-. Charges on the O atoms have been omitted for clarity.

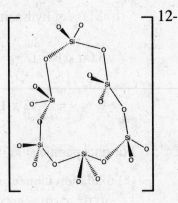

Nitrogen and Phosphorus

45. The $\Delta_f G°$ data from Appendix L are shown below:

compound:	NO(g)	NO$_2$(g)	N$_2$O(g)	N$_2$O$_4$(g)
$\Delta_f G° \left(\dfrac{\text{kJ}}{\text{mol}}\right)$	+86.58	+51.23	+104.20	+97.73

To ask if the oxide is stable with respect to decomposition is to ask about the $\Delta_r G°$ for the

process: $N_xO_y \;\rightarrow\; \dfrac{x}{2} N_2 \text{ (g)} + \dfrac{y}{2} O_2(g)$

This reaction is *the reverse* of the $\Delta_f G°$ for each of these oxides. The $\Delta_r G°$ for such a process would have an *opposite sign* from that data given above.

Since that sign would be negative, the process is product-favored. Hence **all** of the oxides shown above are unstable with respect to decomposition.

47. Calculate $\Delta_r H°$ for the reaction: $2\,NO(g)\ +\ O_2(g)\ \rightarrow\ 2\,NO_2(g)$

$$\Delta_f H°\ (\frac{kJ}{mol})\quad +90.29\qquad\quad 0\qquad\qquad +33.1$$

$$\Delta_f G°\ (\frac{kJ}{mol})\quad +86.58\qquad\quad 0\qquad\qquad +51.23$$

$$\Delta_r H°\ =\ [(2\ mol)(+33.1\ \frac{kJ}{mol})]\ -\ [(2\ mol)(+90.29\ \frac{kJ}{mol})]\ =\ -114.4\ kJ$$

$$\Delta_r G°\ =\ [(2\ mol)(+51.23\ \frac{kJ}{mol})]\ -\ [(2\ mol)(+86.58\ \frac{kJ}{mol})]\ =\ -70.7\ kJ:$$

The reaction is exothermic [negative($\Delta_r H°$) and product-favored (negative $\Delta_r G°$)].

49. (a) The reaction of hydrazine with dissolved oxygen:

   $N_2H_4(aq) + O_2(g)\ \rightarrow N_2(g) + 2\ H_2O(l)$

   (b) Mass of hydrazine to consume the oxygen in 3.00×10^4 L of water:

$$3.00 \times 10^4\ L\ \cdot\ \frac{0.0044\ g\ O_2}{0.100\ L}\ \cdot\ \frac{1 mol\ O_2}{32.0\ g\ O_2}\ \cdot\ \frac{1\ mol\ N_2H_4}{1\ mol\ O_2}\ \cdot\ \frac{32.05 g N_2H_4}{1\ mol\ N_2H_4}$$

$$=\ 1.32 \times 10^3\ g\ N_2H_4$$

51. For phosphonic acid, $H_4P_2O_5$:

   (a) oxidation number of P atom:

   $4(H) + 2(P) + 5(O) = 0$ where the symbol represents the oxidation number of the atom.

   $4(+1) + 2(P) + 5(-2) = 0$; $2(P) -6 = 0$; $2(P) =+6$ so P $=+3$

   (b) The sructure of diphosphonic acid, $H_4P_2O_5$, and the maximum number of protons that can dissociate in this acid:

   The "acidic" H atoms are those attached to the O atoms, and NOT those directly bonded to the P atoms—so 2 protons can dissociate.

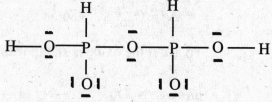

Oxygen and Sulfur

53. (a) Allowable release of SO_2: (0.30 %)

$$1.80 \times 10^6 \text{ kg } H_2SO_4 \cdot \frac{1 \text{ mol } H_2SO_4}{98.08 \text{ g } H_2SO_4} \cdot \frac{1 \text{ mol } SO_2}{1 \text{ mol } H_2SO_4} \cdot \frac{64.06 \text{ g } SO_2}{1 \text{ mol } SO_2}$$

$$\cdot \frac{0.0030 \text{ kg } SO_2 \text{ released}}{1.00 \text{ kg } SO_2 \text{ produced}} = 3.5 \times 10^3 \text{ kg } SO_2$$

(b) Mass of $Ca(OH)_2$ to remove 3.53×10^3 kg SO_2:

$$3.53 \times 10^3 \text{ kg } SO_2 \cdot \frac{1 \text{ mol } SO_2}{64.06 \text{ g } SO_2} \cdot \frac{1 \text{ mol } Ca(OH)_2}{1 \text{ mol } SO_2} \cdot \frac{74.09 \text{ g } Ca(OH)_2}{1 \text{ mol } Ca(OH)_2}$$

$$= 4.1 \times 10^3 \text{ kg } Ca(OH)_2$$

55. The disulfide ion S_2^{2-} can be pictured as:

$$\left[\; \ddot{\underset{\displaystyle\cdot\cdot}{\ddot{S}}} \; \ddot{\underset{\displaystyle\cdot\cdot}{\ddot{S}}} \; \right]^{2-}$$

Fluorine and Chlorine

57. Calculate the equivalent net cell potential for the oxidation:

$2[Mn^{2+}(aq) + 4H_2O(l) \rightarrow MnO_4^-(aq) + 8H^+(aq) + 5e^-]$ -1.51V

$2BrO_3^-(aq) + 12H^+(aq) + 10e^- \rightarrow Br_2(aq) + 6H_2O(l)]$ +1.44V

---

net $2Mn^{2+}(aq) + 2H_2O(l) + 2BrO_3^-(aq) \rightarrow 2MnO_4^-(aq) + 4H^+(aq) + Br_2(aq)$ -0.07V

NOTE: The first equation has been reversed from its usual "reduction form", indicating

that we want to have this process occur as an oxidation. Recall that the task is to ask

if BrO_3^- can oxidize Mn^{2+}! So we arrange the two half-equations to give us a

specific net equation, and then use that format to ask, "Will it go this way?" If E°_{cell}

is > 0, the answer is **Yes**. If E°cell is < 0, the answer is **No**. Given that we calculate

E°_{cell} by *subtracting* the E°_{anode} from the $E^\circ_{cathode}$, I have given the reduction

potential for the manganese half-reaction a sign that is *opposite* the normal sign. If

we add the values, this is the mathematical equivalent of our usual procedure.

The negative net potential indicates that this process doesn't favor products with

1.0 M bromate ion.

59. The balanced equation for the reaction of Cl_2 with Br^-

$$2\ Br^-\ (aq)\ \longrightarrow\ Br_2\ (l) + 2\ e^- \qquad\qquad -1.08\ V$$

$$Cl_2\ (g) + 2\ e^-\ \longrightarrow\ 2\ Cl^-\ (aq) \qquad\qquad +1.36\ V$$

net: $Cl_2\ (g)\ + 2\ Br^-\ (aq)\ \longrightarrow\ 2\ Cl^-\ (aq) + Br_2\ (l) \quad + 0.28\ V$

- Note that bromide ions are losing electrons (and donating them to chlorine), causing chlorine to be reduced—so bromide is the reducing agent and chlorine, removing the electrons from bromide ions, is causing the bromide ions to be oxidized—so chlorine is the oxidizing agent.

- Note also that the voltage for the cell we "constructed" is positive, making this process a "product-favored" reaction.

61. Mass of F_2 that can be produced per 24hr by a current of 5.00×10^3 amps (at 10.0V):

$$\frac{38.00\ g\ F_2}{1\ mol\ F_2} \cdot \frac{1\ mol\ F_2}{2\ mol\ e^-} \cdot \frac{1\ mol\ e^-}{9.65 \times 10^4\ C} \cdot \frac{1\ C}{1\ amp \cdot s}$$

$$\cdot\ \frac{5.00 \times 10^3\ amps}{1} \cdot \frac{3600\ s}{1\ hr} \cdot \frac{24\ hr}{1} = 8.51 \times 10^4\ g\ F_2$$

General Questions

63. Describe the elements in the third period:

Atomic No.	Element	(a) Type	(b) Color	(c) State
11	Sodium	metal	silvery	solid
12	Magnesium	metal	silvery	solid
13	Aluminum	metal	silvery	solid
14	Silicon	metalloid	black, shiny	solid
15	Phosphorus	nonmetal	red, white, black	solid

	(a)	(b)	(c)
Atomic No.	Element	Type	Color	State
16	Sulfur	nonmetal	yellow	solid
17	Chlorine	nonmetal	pale green	gas
18	Argon	nonmetal	colorless	gas

65. Reactions of K, Ca, Ga, Ge, As

(a) Balanced equations of the elements with elemental chlorine:

$2 K(s) + Cl_2(g) \rightarrow 2 KCl(s)$ $Ge(s) + 2 Cl_2(g) \rightarrow GeCl_4(l)$

$Ca(s) + Cl_2(g) \rightarrow CaCl_2(s)$ $2As(s) + 3 Cl_2(g) \rightarrow 2 AsCl_3(l)$

$2 Ga(s) + 3 Cl_2(g) \rightarrow 2 GaCl_3(s)$

(b) Bonding in the products:

KCl and $CaCl_2$ are ionic; all others are covalent. This bonding is attributable—to a first approximation--to the differing electronegativities of the atoms bonded to each other.

Compound	Electron-Pair geometry	Molecular geometry
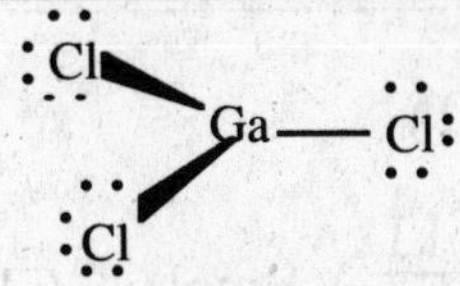	Planar Trigonal	Planar Trigonal
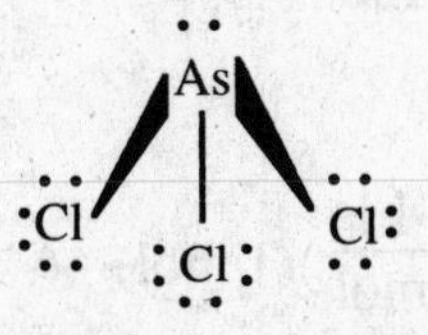	Tetrahedral	Pyramidal

(c) Electron dot structures for the products; electron-pair geometry; molecular geometry

67. Complete and balance the equations:

   (a) $2 KClO_3(s) + heat \rightarrow 2 KCl(s) + 3 O_2(g)$

   (b) $2 H_2S(g) + 3 O_2(g) \rightarrow 2 H_2O(g) + 2 SO_2(g)$

   (c) $2 Na(s) + O_2(g) \rightarrow Na_2O_2(s)$

   (d) $P_4(s) + 3 KOH(aq) + 3 H_2O(l) \rightarrow PH_3(g) + 3 KH_2PO_4(aq)$

   (e) $NH_4NO_3(s) + heat \rightarrow N_2O(g) + 2 H_2O(g)$

   (f) $2 In(s) + 3 Br_2(l) \rightarrow 2 InBr_3(s)$

   (g) $SnCl_4(l) + 2 H_2O(l) \rightarrow 4 HCl(aq) + SnO_2(s)$

69. $\dfrac{1.0 \times 10^5 \, MT \, SiC}{1} \cdot \dfrac{100 \, MT \, sand}{70 \, MT \, SiC} = 1.4 \times 10^5 \, MT \, sand$

71. Calculate $\Delta_r G°$ for the decomposition of the metal carbonates for Mg, Ca, and Ba

$MCO_3(s)$	$\rightarrow$	$MO(s)$	$+ CO_2(g)$		M
-1028.2		- 568.93	- 394.359	(Mg) $\Delta G°$ ($\dfrac{kJ}{mol}$)	0
- 1129.16		- 603.42	- 394.359	(Ca) $\Delta G°$	0
- 1134.41		- 520.38	- 394.359	(Ba) $\Delta G°$	0

$\Delta_r G° = \Delta_f G° \, MO + \Delta_f G° \, CO_2 - \Delta_f G° \, MCO_3$

$MgCO_3 = (- 568.93 \dfrac{kJ}{mol})(1 \, mol) + (-394.359 \dfrac{kJ}{mol})(1 \, mol) - (-1028.2 \dfrac{kJ}{mol})(1 \, mol)$

$\qquad = 64.9 \, kJ$

$CaCO_3 = (- 603.42 \dfrac{kJ}{mol})(1 \, mol) + (-394.359 \dfrac{kJ}{mol})(1 \, mol) - (- 1129.16 \dfrac{kJ}{mol})(1 \, mol)$

$\qquad = 131.38 \, kJ$

$BaCO_3 = (- 520.38 \dfrac{kJ}{mol})(1 \, mol) + (-394.359 \dfrac{kJ}{mol})(1 \, mol) - (-1134.41 \dfrac{kJ}{mol})(1 \, mol)$

$\qquad = 219.67 \, kJ$

The relative tendency for decomposition is then $MgCO_3 > CaCO_3 > BaCO_3$.

73. (a) Since $\Delta_rG° < 0$ for the reaction to be product-favored, calculate the value for

$\Delta_fG°$ MX that will make the $\Delta_fG°$ zero.

$$\Delta_rG° = \Delta G°_f(MX_n) - n\,\Delta G°_f(HX)$$

for HCl $= \Delta_fG°(MX_n) - n(-95.1\,\frac{kJ}{mol})$

so if $n(-95.1\ kJ) = \Delta_fG°(MX_n)$ then $\Delta G° = 0$,

and if $n\,(-95.1\ kJ) > \Delta G_f°(MX_n)$ then $\Delta_rG° < 0$.

(b) Examine $\Delta_fG°$ MX values for

metal:	Ba	Pb	Hg	Ti
$\Delta_fG°$ MX:	-810.4	-314.10	-178.6	-737.2
n:	2	2	2	4
n(-95.1):	-190.2	-190.2	-190.2	-380.4

For Barium, Lead, and Titanium, $n\,(-95.1) > \Delta_rG°(MX)$; We expect these reactions to be

spontaneous(product-favored).

75. The average O-F bond energy in OF_2, given that the $\Delta_fH° = +24.5$ kJ/mol:

The reaction is represented as: $O_2\,(g) + 2F_2\,(g) \rightarrow 2OF_2\,(g)$

ΔH = Energy input – Energy released

Energy input: 1 mol O=O = 1 mol • 498 kJ/mol =   498 kJ
              2 mol F-F = 2 mol • 155 kJ/mol =   310 kJ
                        Total input   =   808 kJ

Energy released : 4 mol O-F = 4 x (where x = O-F bond energy)
                        Total input   = 4x kJ

ΔH = Energy input – Energy released

+49.0 kJ = 808 kJ – 4x kJ so +49.0 kJ - 808 kJ = -4x and

-759 kJ = -4x   and 190 kJ = x. So the O-F bond energy is 190 kJ/mol

77. For the equation: $H_2NN(CH_3)_2(l) + 2\ N_2O_4(l) \rightarrow 3\ N_2(g) + 4\ H_2O(g) + 2\ CO_2(g)$

(a) The oxidizing and reducing agents: Noting that $N_2O_4(l)$ loses O as it reacts, the oxide

is reduced—making N_2O_4 the **oxidizing agent.** This means that $H_2NN(CH_3)_2(l)$

serves as the **reducing agent**.

417

(b) What mass of $N_2O_4(l)$ was consumed if 4100kg of $H_2NN(CH_3)_2$ reacts? Noting that the mass of the substituted hydrazine was given in kg, we can express the answer in kg, and treat the mass *as if it were in grams*.

$$\left(\frac{4100 \text{ kg } H_2NN(CH_3)_2}{1} \cdot \frac{1 \text{ mol } H_2NN(CH_3)_2}{60.10 \text{ g } H_2NN(CH_3)_2}\right) \cdot \frac{2 \text{ mol } N_2O_4}{1 \text{ mol } H_2NN(CH_3)_2} \cdot \frac{92.01 \text{ g } N_2O_4}{1 \text{ mol } N_2O_4}$$

$$= 1.3 \times 10^4 \text{ kg } N_2O_4 \text{ (2 sf)}$$

What mass of N_2, H_2O, CO_2 will be formed?

$$\left(\frac{68.2 \text{ mol } H_2NN(CH_3)_2}{1}\right) \cdot \frac{3 \text{ mol } N_2}{1 \text{ mol } H_2NN(CH_3)_2} \cdot \frac{28.01 \text{ g } N_2O_4}{1 \text{ mol } N_2} = 5.7 \times 10^3 \text{ kg } N_2 \text{ (2sf)}$$

$$\left(\frac{68.2 \text{ mol } H_2NN(CH_3)_2}{1}\right) \cdot \frac{4 \text{ mol } H_2O}{1 \text{ mol } H_2NN(CH_3)_2} \cdot \frac{18.02 \text{ g } H_2O}{1 \text{ mol } H_2O} = 4.9 \times 10^3 \text{ kg } H_2O \text{ (2sf)}$$

$$\left(\frac{68.2 \text{ mol } H_2NN(CH_3)_2}{1}\right) \cdot \frac{2 \text{ mol } CO_2}{1 \text{ mol } H_2NN(CH_3)_2} \cdot \frac{44.01 \text{ g } CO_2}{1 \text{ mol } CO_2} = 6.0 \times 10^3 \text{ kg } CO_2 \text{ (2sf)}$$

79. Examine the enthalpy change for the reaction:

$$2 N_2 (g) + 5 O_2 (g) + 2 H_2O (l) \rightarrow 4 HNO_3 (aq)$$

$$\Delta_r H° = [(4 \text{ mol})(-207.36 \tfrac{kJ}{mol})] - [(2 \text{ mol})(0 \tfrac{kJ}{mol}) + (2 \text{ mol})(0 \tfrac{kJ}{mol}) +$$

$$(2 \text{ mol})(-285.83 \tfrac{kJ}{mol})] = -257.78 \text{ kJ}$$

The reaction is **exothermic** so it is a reasonable "first-guess" that this might be a way to "fix" nitrogen. The only way to be certain is to calculate the $\Delta_r G°$.

$$\Delta_r G° = [(4 \text{ mol})(-111.25 \tfrac{kJ}{mol})] - [(2 \text{ mol})(0 \tfrac{kJ}{mol}) + (2 \text{ mol})(0 \tfrac{kJ}{mol}) +$$

$$(2 \text{ mol})(-237.15 \tfrac{kJ}{mol})] = 29.19 \text{ kJ}$$

The positive value for $\Delta_r G°$ tells us that the **reaction is not likely at 25 °C**. The decrease in the number of moles of gas (entropy decrease) is also a factor that works against this process at 25 °C. The favorable $\Delta H°$ does indicate that a lower temperature might be feasible, and below 268 K, the $\Delta_r G$ is favorable, but at this temperature water is a solid—not a good "sign".

81. This problem requires one to calculate empirical formulas for the boron hydrides.
Using the %B (and %H) as masses, calculate #mol of each element, determine the ratio
of hydrogen/boron atoms, and determine an empirical formula. Using the #B and #H
atoms, and their respective atomic masses creates the column entitled (MM calc).
Using the molar masses provided with the data (MM known) divided by the calculated
"empirical mass", we obtain the number of empirical formulas per molecular formula.
Multiplying that ratio by the #B and #H in the empirical formulas gives the molecular
formula (final column). The calculations are straightforward, with perhaps the exception
of the columns of (#B) and (#H). These were obtained by expressing the data in the
column (Ratio of H/B atoms) as a fraction. So 2.973 is 3, 2.5 becomes 2+1/2, 2.2 is
2+1/5, 1.790 is 1+4/5, 1.4 is 1+2/5. Converting these fractions to integers gives the
numbers in the column (#H atoms).

	% B	% H	mol B	Mol H	Ratio of H/B atom	# B atom	#H atom	MM calc	MM known	Emp./molec. formula	Molec. Formula
A	78.3	21.7	7.243	21.53	3.0	1	3	13.83	27.7	2.00221	B_2H_6
B	81.2	18.8	7.511	18.65	2.5	2	5	26.66	53.3	1.99914	B_4H_{10}
C	83.1	16.9	7.687	16.77	2.2	5	11	65.14	65.1	0.99936	B_5H_{11}
D	85.7	14.3	7.927	14.19	1.8	5	9	63.13	63.1	0.99959	B_5H_9
E	88.5	11.5	8.186	11.41	1.4	5	7	61.11	122.2	1.99966	$B_{10}H_{14}$

83. What current must be used in a Downs cell operating at 7.0 V to produce 1.00 metric ton
(exactly 1000 kg) of sodium per day? Assume 100% efficiency.

$$\frac{1 \text{ A·s}}{1 \text{ C}} \cdot \frac{1 \text{ hr}}{3600 \text{ s}} \cdot \frac{1 \text{ day}}{24 \text{ hr}} \cdot \frac{96485 \text{ C}}{1 \text{ mol e}} \cdot \frac{1 \text{ mol e}}{1 \text{ mol Na}} \cdot \frac{1 \text{ mol Na}}{22.99 \text{ g Na}} \cdot \frac{1000 \text{ g Na}}{1 \text{ kg Na}} \cdot \frac{1000 \text{ kg Na}}{1 \text{ day}}$$

$$= 4.86 \times 10^4 \text{ A}$$

85. The structure for the anion $[Si_3O_9]^{6-}$:

The ring will be puckered since each Si atom will
have tetrahedral group geometry around it.

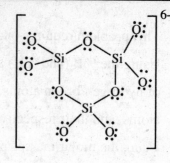

87. Is the following reaction product or reactant-favored?

$$2\ ZnS\ (s) + 3\ O_2\ (g) \rightarrow 2\ ZnO\ (s) + 2\ SO_2\ (g)$$

Calculate the $\Delta G°$ for the reaction:

$\Delta_r G° = [2\Delta_f G°(ZnO) + 2\ \Delta_f G°\ (SO2)] - 2\Delta_f G°\ (ZnS)$

$\Delta_r G° = [2\ mol(-318.30\ kJ/mol) + 2\ mol\ (-300.13\ kJ/mol)] - 2\ mol\ (-201.29\ kJ/mol)$

$= (-636.60\ kJ + -600.26\ kJ) - (-402.58\ kJ)$

$= -1236.86\ kJ + 402.58\ kJ = -834.28\ kJ$

With a negative $\Delta_r G°$, the reaction is product-favored.

A decrease in entropy (loss of moles of gas (in the products side)), would favor the
"reactant-side" with an increase in T.

IN THE LABORATORY

89. In the synthesis of dichlorodimethylsilane:

(a) The balanced equation: $Si(s) + 2\ CH_3Cl(g) \rightarrow (CH_3)_2SiCl_2(l)$

(b) Stoichiometric amount of CH_3Cl to react with 2.65 g silicon:

$$2.65\ g\ Si \cdot \frac{1\ mol\ Si}{28.09\ g\ Si} \cdot \frac{2\ mol\ CH_3Cl}{1\ mol\ Si} = 0.189\ mol\ CH_3Cl$$

$$P = \frac{(0.189\ mol\ CH_3Cl)\left(0.082057\ \dfrac{L \cdot atm}{K \cdot mol}\right)(297.7\ K)}{5.60\ L} = 0.823\ atm$$

(c) Mass of $(CH_3)_2SiCl_2$ produced assuming 100 % yield:

$$0.0943\ mol\ Si \cdot \frac{1\ mol\ (CH_3)_2SiCl_2}{1\ mol\ Si} \cdot \frac{129.1\ g\ (CH_3)_2SiCl_2}{1\ mol\ (CH_3)_2SiCl_2} = 12.2\ g\ (CH_3)_2SiCl_2$$

91. The half equations:

$$N_2H_5^+(aq) \rightarrow N_2(g) + 5\ H^+(aq) + 4\ e^-\ and\ IO_3^-\ (aq) \rightarrow I_2(s)$$

are balanced according to the procedure in Chapter 5 to give:

$$5\ N_2H_5^+(aq) + 4\ IO_3^-(aq) \rightarrow 5\ N_2(g) + 2\ I_2(aq) + 12\ H_2O(l) + H^+(aq)$$

E° for the reaction is: E° for the hydrazine equation (oxidation) = - 0.23 V

$$E° \text{ for the iodate equation (reduction)} = \underline{+ 1.195 \text{ V}}$$

$$E°_{net} = + 1.43 \text{ V}$$

93. The problem indicates that oxide **A** is less volatile than oxide **B**. This tells us that oxide **A** has a *greater molar mass than that of oxide* **B**.

Other data can be schematically illustrated:

(a) **A** + I⁻ → I_2; the resulting I_2 reacts with thiosulfate: $I_2 + 2 S_2O_3^{2-} \rightarrow 2 I^- + S_4O_6^{2-}$

(b) **A** + $AgNO_3$ → AgBr(s)

The two reactions let us know that (a) oxide **A** has a Br⁻ ion (to form AgBr). We also know that the oxide is capable of oxidizing I⁻ to I_2. The literature [Pascal, J.L. *Compt.Rend. Hebd. Seances, Acad. Sci, Section C*, 1974,279,43; and Seppelt, Konrad, *Acc. Chem.Res*. 30(3), 111-113,1997] indicates that the ozonolysis of bromine results in an oxide which upon decomposition forms Br_2O and Br_2O_3. Seppelt reports a terminal OBr attached to a pyramidal O-BrO$_2$. The pyramidal BrO_3 reacts with the iodide ion to produce iodine. The Br attached to the hypobromite reacts with the silver nitrate to form AgBr. The formula of the oxide is then Br_2O_3, and the structure of the oxide, **A**, is (as determined by Pascal, et al.) to be: :B̈r⃛ ⟍Ö̈⟋ B̈r⃛: the structure of Br_2O_3 is:

:B̈r: ⟍Ö̈⟋ B̈r ⟍ :Ö̈: ⟋ Ö̈:

SUMMARY AND CONCEPTUAL QUESTIONS

95. For N_2O_3:

(a) Explain differences in bond lengths for the three N-O bonds:

The Lewis dot picture at right indicates that the "left" N has a N=O, while the "right" N has two oxygens, each connected with a N-O "one and a half" bond, owing to resonance.

Naturally, the N-O bonds (121 pm) are expected to be equal to one other and longer than the N=O double bond (114.2pm).

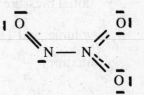

(b) $\Delta H° = +40.5$ kJ/mol N_2O_3; $\Delta G° = -1.59$ kJ/mol;

What are $\Delta S°$ and K for the reaction at 298K?

Given that $\Delta G° = \Delta H° - T\Delta S°$, $\Delta S° = -(\Delta G° - \Delta H°)/T = (\Delta H° - \Delta G°)/T$

$(+40.5$ kJ/mol $+1.59$ kJ/mol$)/298K = (42.09 \times 10^3$ J/mol$)/298K = 141$ J/K•mol

and K can be calculated via the relationship:

$\Delta G° = -RT\ln K$ so -1.59 kJ/mol $= -(8.3145$ J/K•mol$)298K$•$\ln K$

$$\frac{-1.59 \times 10^3 \dfrac{J}{mol}}{-(8.3145 \text{ J/K} \cdot \text{mol}) \cdot 298K} = \ln K = 6.417 \times 10^{-1} \text{ and } K = 1.90$$

(c) Calculate $\Delta_f H°$ for $N_2O_3(g)$

The reaction: $N_2O_3(g) \rightarrow NO(g) + NO_2(g)$ has $\Delta H = +40.5$ kJ/mol N_2O_3.

$\Delta_r H = [\Delta_f H° (NO(g)) + \Delta_f H° (NO_2(g))] - \Delta_f H°(N_2O_3(g)) = +40.5$ kJ

$\Delta_r H = [(+90.29$ kJ/mol$)$•1mol$+(+33.1$ kJ/mol$)$•1mol$]-(1$mol$)$•$\Delta H_f°(N_2O_3(g))=+40.5$kJ

$\Delta_r H = 123.39$kJ$- \Delta_f H°(N_2O_3(g)) = +40.5$ kJ

123.39kJ $- 40.5$ kJ $= \Delta_f H°(N_2O_3(g)) = +82.9$ kJ/mol

97. A procedure to determine the % hydrogen in a mixture of argon and hydrogen in a 1.0-L flask.

A simple solution is to convert the hydrogen to water (burn the mixture). Argon doesn't react (noble gas, right?). Quantitatively collecting the gaseous mixture, and chilling it until water liquefies, one can separate the water from the argon. The reduced pressure of the gas that remains is due to argon. The difference between the reduced pressure and the initial pressure (745 mmHg) would be attributable to the original H_2. Given the pressure, volume, and T, one could calculate the amount of hydrogen gas originally present in the mixture.

99. To extinguish a Na fire, addition of water is NOT ADVISABLE, since the reaction of the water with the Na would produce HEAT and HYDROGEN GAS—which would also burn!! The optimal solution would be to smother the fire—perhaps with sand.

101. Stoppered flask contains either H_2, N_2, or O_2. What experiment would identify the gas?

Several possibilities exist. Chilling a small sample of the gas with liquid N_2(bp $-196°C$), would cause O_2 to liquefy—but not H_2, and only N_2 slowly. If the gas liquefies—it's Oxygen!

If the gas doesn't liquefy at all, the gas is either H_2 (bp $-253°C$) or N_2(bp $-196°C$).

Allowing a small sample of the gas to escape through a narrow opening (perhaps the tip of a plastic medicine dropper) while holding a smoldering splint at the tip of the opening would confirm the presence of H_2 (the escaping gas would burn), or N_2 (the escaping gas would **not** burn). *CAUTION: Burning H_2 in glass vessels is TO BE AVOIDED. Serious physical damage frequently results!*

103. $1.00 \text{ kg Cl}_2 \cdot \dfrac{1 \times 10^3 \text{ g Cl}_2}{1.0 \text{ kg Cl}_2} \cdot \dfrac{1 \text{ mol Cl}_2}{70.906 \text{ g Cl}_2} \cdot \dfrac{2 \text{ mol e}^-}{1 \text{ mol Cl}_2} \cdot \dfrac{9.6485 \times 10^4 \text{ C}}{1 \text{ mol e}^-}$

= 2,720 Coulombs

The power is then:

$\dfrac{2720 \text{ C}}{1} \cdot \dfrac{4.6 \text{ V}}{1} \cdot \dfrac{1 \text{ J}}{1 \text{ V} \cdot \text{C}} \cdot \dfrac{1 \text{ kwh}}{3.6 \times 10^6 \text{ J}} = 3.5 \text{ kwh}.$ (2 sf)

The factor, $\dfrac{1 \text{ kwh}}{3.6 \times 10^6 \text{ J}}$, is derived from the fact that a kwh = 1000 watts $\cdot$ 1 hr

and the relationship that 1 J = 1 watt $\cdot$ s or 1 J/s = 1 watt

$1 \text{ kwh} = 1000 \text{ watts} \cdot 1 \text{ hr} \cdot \dfrac{1 \text{J/s}}{1 \text{watt}} \cdot \dfrac{3600 \text{ s}}{1 \text{ hr}} = \dfrac{3.60 \times 10^6 \text{ J}}{1 \text{ kwh}}$

105. The change to more positive values in **Reduction Potentials** as one descends the group (from Al to Tl) indicates a diminishing ability for the metals to act as a reducing agent. Another noticeable trend is the large **differences** between the reduction potentials of Al and Ga, and between In and Tl. The trend toward more positive reduction potentials also indicates the increased stability of +1 oxidation states (over higher oxidation states) for the heavier elements in the family.

Applying Chemical Principles

1. Classify the compounds: GaAs, SBr_2, Mg_3N_2, BP, C_3N_4, CuZn, and $SrBr_2$ on the

   van Arkel diagram

Element (E'neg)	Element (E'neg)	Difference (E'net)	Average E'neg
Ga(1.8)	As (2.2)	0.4	2.0
S(2.6)	Br (3.0)	0.4	2.8
Mg(1.3)	N(3.0)	1.7	2.15
B(2.0)	P(2.2)	0.2	2.1
C(2.5)	N(3.0)	0.5	2.75
Cu(1.9)	Zn(1.6)	0.3	1.75
Sr(1.0)	Br(3.0)	2.0	2.0

   (a) Which of the compounds are metallic? **CuZn**

   (b) Which of the compounds are semiconductors? **GaAs, BP**

       Are either (or both) element(s) in these compounds metalloid(s)? **For GaAs, Ga
       is a metal while As is a metalloid. In BP, B is a metalloid, and P is a
       nonmetal**.

   (c) Which of the compounds are ionic? **Mg_3N_2, $SrBr_2$**

       Are the compounds composed of a metal and a nonmetal? **Yes**

   (d) Carbon nitride (C_3N_4) is predicted to be harder than diamond (which is currently

       the hardest known substance), but too little has been synthesized to enable a

       comparison. What type of bonding is predicted for C_3N_4? **Covalent**

   (e) Which of the compounds are covalent? **SBr_2, C_3N_4**

       Are both elements in these compounds nonmetallic? **Yes**

Chapter 22
The Chemistry of the Transition Elements

PRACTICING SKILLS

Properties of Transition Elements

1. (a) Cr^{3+} 3d [↑][↑][↑][ ][ ] 4s [ ] paramagnetic

(b) V^{2+} 3d [↑][↑][↑][ ][ ] 4s [ ] paramagnetic

(c) Ni^{2+} 3d [↑↓][↑↓][↑↓][↑][↑] 4s [ ] paramagnetic

(d) Cu^{+} 3d [↑↓][↑↓][↑↓][↑↓][↑↓] 4s [ ] diamagnetic

3. Ions from first series transition metals that are isoelectronic with:

(a) Fe^{3+} has 5 3d electrons so it is isoelectronic with Mn^{2+}

(b) Zn^{2+} has 10 3d electrons so it is isoelectronic with Cu^{+}(see 1(d) above)

(c) Fe^{2+} has 6 3d electrons so it is isoelectronic with Co^{3+}

(d) Cr^{3+} has 3 3d electrons so it is isoelectronic with V^{2+} (see 1(b) above)

5. Balance:
Each is balanced by the procedure shown in Chapter 5 of the Solutions Manual. Refer to that procedure. While each of these is a reduction-oxidation reaction, one can also use the "inspection method" to obtain the balanced equations:

(a) $Cr_2O_3(s) + 2 Al(s) \rightarrow Al_2O_3(s) + 2 Cr(s)$

(b) $TiCl_4(l) + 2 Mg(s) \rightarrow Ti(s) + 2 MgCl_2(s)$

(c) $2 [Ag(CN)_2]^-(aq) + Zn(s) \rightarrow 2 Ag(s) + [Zn(CN)_4]^{2-}(aq)$

(d) $3 Mn_3O_4(s) + 8 Al(s) \rightarrow 9 Mn(s) + 4 Al_2O_3(s)$

Formulas of Coordination Compounds

7. Classify each of the following as monodentate or multidentate:

    (a) CH_3NH_2 monodentate (lone pair on N)

    (b) $CH_3C{\equiv}N$ monodentate (lone pair on N)

    (c) N_3^- monodentate (lone pair on a N atom)

    (d) ethylenediamine   multidentate (lone pairs on terminal N atoms)

    (e) Br^- monodentate (lone pair on Br)

    (f) phenanthroline   multidentate (lone pairs on N atoms)

9.

Compound	Metal	Oxidation Number
(a) $[Mn(NH_3)_6]SO_4$	Mn	+2

Ammonia is a neutral ligand. SO_4^{2-} means that Mn has to have a +2 oxidation number

(b) $K_3[Co(CN)_6]$	Co	+3

The 3 K^+ ions mean that the complex ion must have a -3 charge. Each CN^- has a –1 charge, so Co must have a +3 charge.

(c) $[Co(NH_3)_4Cl_2]Cl$	Co	+3

The chloride anion means that the complex ion must have a net +1 charge. While the ammonia ligand is neutral, each Cl has a –1 charge, so Co must have +3.

(d) $Cr(en)_2Cl_2$	Cr	+2

The ethylenediamine ligand is neutral, so with each Cl having a –1 charge,

Cr must be +2.

11. Formula of complex: $[Ni(NH_3)_3(H_2O)en]^{2+}$. The complex has to have a 2+ charge, since each of the three types of ligands is neutral.

Naming Coordination Compounds

13. Formulas for:

    (a) dichlorobis(ethylenediamine)nickel(II) $Ni(en)_2Cl_2$

    (b) potassium tetrachloroplatinate(II) $K_2[PtCl_4]$

    (c) potassium dicyanocuprate(I) $K[Cu(CN)_2]$

(d) tetraamminediaquairon(II) $[Fe(NH_3)_4(H_2O)_2]^{2+}$

15. <u>Formula</u>   <u>Name</u>

(a) $[Ni(C_2O_4)_2(H_2O)_2]^{2-}$ diaquabis(oxalato)nickelate(II) ion

(b) $[Co(en)_2Br_2]^+$ dibromobis(ethylenediamine)cobalt(III) ion

(c) $[Co(en)_2(NH_3)Cl]^{2+}$ amminechlorobis(ethylenediamine)cobalt(III) ion

(d) $Pt(NH_3)_2(C_2O_4)$ diammineoxalatoplatinum(II)

17. The name or formula for the ions or compounds shown below:

(a) $[Fe(H_2O)_5(OH)]^{2+}$ Pentaaquahydroxoiron(III) ion

(b) $K_2[Ni(CN)_4]$ Potassium tetracyanonickelate(II)

(c) $K[Cr(C_2O_4)_2(H_2O)_2]$ Potassium diaquabis(oxalato)chromate(III)

(d) $(NH_4)_2[PtCl_4]$ Ammonium tetrachloroplatinate(IV)

Isomerism

19. Geometric Isomers of

(a) $Fe(NH_3)_4Cl_2$

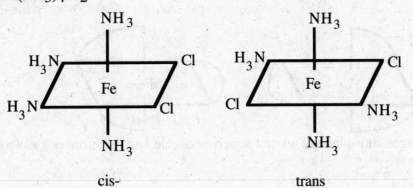

cis-   trans

(b) $Pt(NH_3)_2(SCN)(Br)$

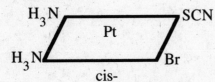
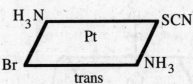

cis-   trans

427

(c) Co(NH₃)₃(NO₂)₃

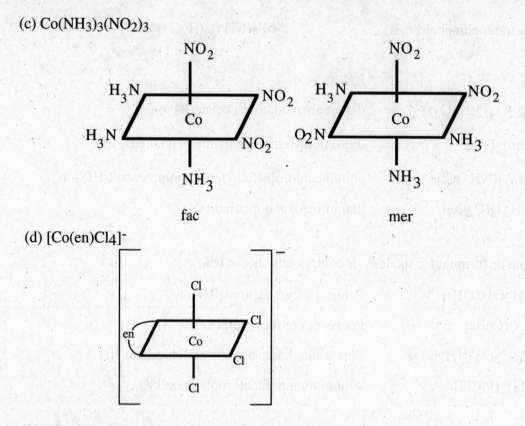

fac   mer

(d) [Co(en)Cl₄]⁻

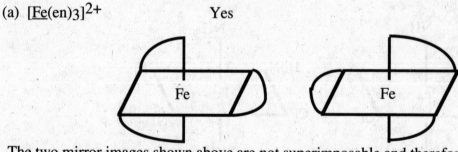

21. Which of the following species has a chiral center?

In these diagrams the curved lines represent the H₂N-CH₂CH₂-NH₂ ligand.

(a) [Fe(en)₃]²⁺   Yes

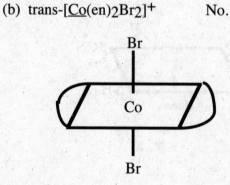

The two mirror images shown above are not superimposable and therefore possess a chiral center.

(b) trans-[Co(en)₂Br₂]⁺   No.

(c) fac-[Co(en)(H₂O)Cl₃]   No

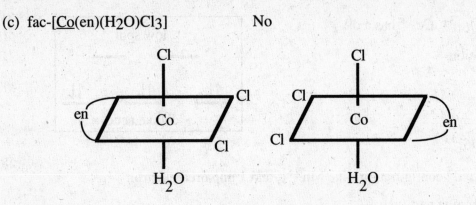

As you can see by the two structures above, a 180° rotation along the Cl-Co-H₂O axis, would make these two mirror images superimposable. The *mer* complex would also be superimposable, and therefore possesses no chiral center.

(d) Pt(NH₃)(H₂O)Cl(NO₂)

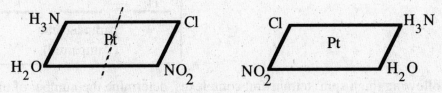

Above are two mirror images of the complex. Rotation of the first complex along the dotted axis by 180° results in the second complex. There are no nonsuperimposable isomers.

Magnetic Properties of Complexes

23. In the name of clarity, the counterion has been omitted. The counterions do not affect the magnetic behavior of the complex ion.

(a) [Mn(CN)₆]⁴⁻

Mn²⁺ has a d⁵ configuration

low spin strong field ligand

paramagnetic

(b) $[Co(NH_3)_6]^{3+}$ Co^{3+} has a d^6

configuration

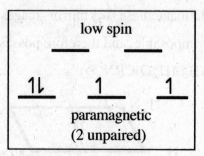

(c) $[Fe(H_2O)_6]^{3+}$

Fe^{3+} has a d^5 configuration (like Mn^{2+}), and 1 unpaired electron

(see diagram in part (a)).

(d) $[Cr(en)_3]^{2+}$

Cr^{2+} has the d^4

configuration

25. For the following (high spin) tetrahedral complexes, determine the number of unpaired electrons:

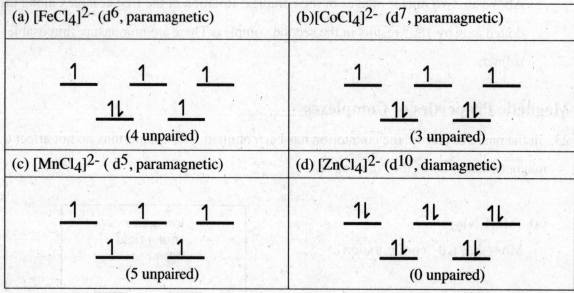

(a) $[FeCl_4]^{2-}$ (d^6, paramagnetic)	(b)$[CoCl_4]^{2-}$ (d^7, paramagnetic)
(4 unpaired)	(3 unpaired)
(c) $[MnCl_4]^{2-}$ (d^5, paramagnetic)	(d) $[ZnCl_4]^{2-}$ (d^{10}, diamagnetic)
(5 unpaired)	(0 unpaired)

27. For $[Fe(H_2O)_6]^{2+}$

(a) The coordination number of iron is 6. 6 monodentate ligands are attached.

(b) The coordination geometry is octahedral. (Six groups attached to the central metal ion)

(c) The oxidation state of iron is 2+. The charge on the complex is 2+.

Water is a neutral ligand.

(d) Fe^{2+} is a d^6 case. Water is a weak-field ligand (high spin complex); there are 4 unpaired electrons.

(e) The complex would be paramagnetic

29. The weak-field ligand, Cl^-, permits either tetrahedral or square-planar geometries. With the d^8 Ni^{2+} ion, such weak-field splitting provides 2 unpaired electrons in the tetrahedral $NiCl_4^{2-}$. The strong-field ligand, CN^-, results in a square-planar complex with no unpaired electrons—a diamagnetic complex.

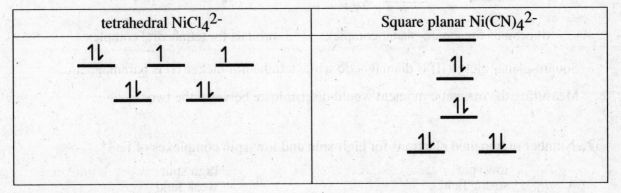

tetrahedral $NiCl_4^{2-}$	Square planar $Ni(CN)_4^{2-}$

Spectroscopy of Complexes

31. 500nm light is in the blue region of the spectrum (page 1045 of your text). The transmitted light, and the color of the solution--is yellow.

Organometallic Compounds

33. Do the following satisfy the EAN rule?

Complex	(a) $[Mn(CO)_6]^+$	(b) $Co(C_5H_5^-)(CO)(PR_3)$	(c) $Mn(C_5H_5^-)(CO)_3$
Electrons from:Metal	Mn^+ d^6	Co^+ d^8	Mn^+ d^6
from Ligand 1	6 x CO: 2	1 x $C_5H_5^-$:6	1 x $C_5H_5^-$:6
from Ligand 2		1 x CO:2	3 x CO:2
from Ligand 3		1 x PR_3:2	
Total electrons	6 + 12 =18	8+6+2+2 = 18	6 + 6 + 6 = 18

GENERAL QUESTIONS

35. Describe an experiment to determine if:

Nickel in $K_2[NiCl_4]$ is in a square planar or tetrahedral environment.

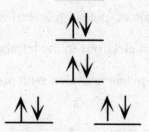   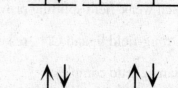

d- orbitals for square-planar complex   d- orbitals for tetrahedral complex

Square-planar nickel(II) is diamagnetic while tetrahedral nickel(II) is paramagnetic.
Measuring the magnetic moment would discriminate between the two.

37. Number of unpaired electrons for high-spin and low-spin complexes of Fe^{2+}:

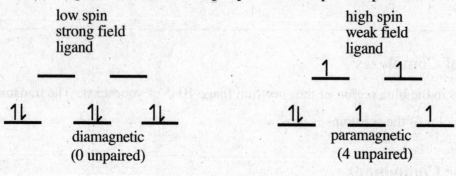

        low spin   high spin
        strong field   weak field
        ligand   ligand

        diamagnetic   paramagnetic
        (0 unpaired)   (4 unpaired)

39. Which of the following complexes are square planar?

Of the four complexes, $[Ni(CN)_4]^{2-}$ and $[Pt(CN)_4]^{2-}$ are square planar, since such

complexes have the geometry assumed by d^8 metal ions. The other complexes are

tetrahedral.

41. Geometric isomers of $Pt(NH_3)(CN)Cl_2$:

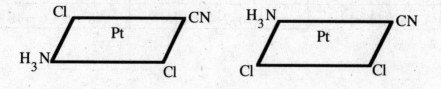

43. Complex absorbs 425-nm light, the blue-violet end of the visible spectrum. So red and green are transmitted, and the complex appears to be yellow.

45. For the high-spin complex $Mn(NH_3)_4Cl_2$ determine:

(a) The oxidation number of manganese is +2. With a neutral overall complex, and 2 chloride ions (each with a –1 charge) and 4 neutral ammonia ligands, manganese has to be +2.

(b) With six monodentate ligands, the coordination number for manganese is 6.

(c) With six monodentate ligands, the coordination geometry for manganese is octahedral.

(d) Number of unpaired electrons: 5 The ligands attached are weak-field ligands resulting in five unpaired electrons.

(e) With unpaired electrons, the complex is paramagnetic.

(f) Cis- and trans- geometric isomers are possible.

47. Structures for cis- and trans- isomers of $CoCl_3 \cdot 4NH_3$

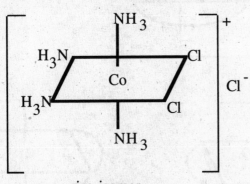

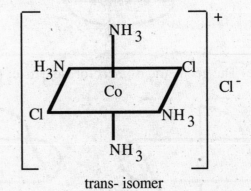

cis- isomer

trans- isomer

cis-tetraamminedichlorocobalt(III) chloride   *trans*-tetraamminedichlorocobalt(III) chloride

49. Formula of a complex containing a Co^{3+} ion, two ethylenediamine molecules, one water molecule, and one chloride ion: $[Co(en)_2(H_2O)Cl]^{2+}$

The ethylenediamine molecules and the water molecule are neutral, so with a (3+) and (1-) charge, the net charge on the ion is 2+.

51. (a) Structures for the fac- and mer- isomers of Cr(dien)Cl$_3$. In these diagrams the curved

lines represent the H$_2$N-CH$_2$CH$_2$-NH-CH$_2$CH$_2$-NH$_2$ ligand--with attachments to the

metal ion through the electron pair on the N atoms.

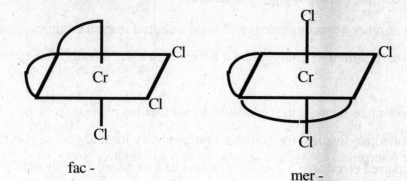

fac -   mer -

(b) Two different isomers of mer-Cr(dien)BrCl$_2$:

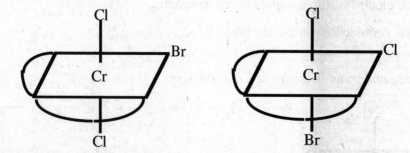

(c) The geometric isomers for isomers of [Cr(dien)$_2$]$^{3+}$

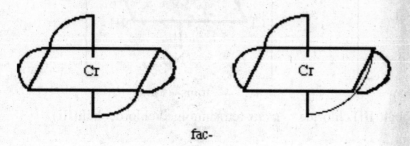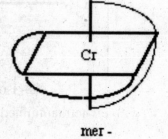

fac-   mer -

53. Three geometric isomers of [Co(en)(NH$_3$)$_2$(H$_2$O)$_2$]$^{3+}$:

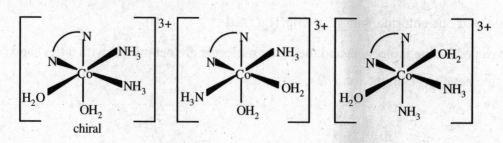

chiral

55. For the two Mn^{2+} complexes:

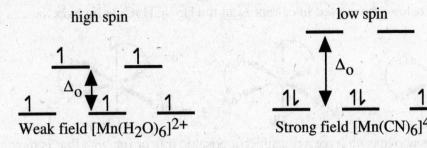

high spin   low spin

Weak field $[Mn(H_2O)_6]^{2+}$ Strong field $[Mn(CN)_6]^{4-}$

The cyano ligand is a *strong field* ligand, so the difference in energies of the doubly vs triply

degenerate d orbitals is greater than with the *weak field* aqua ligand. The differences in these

energy levels are referred to as Δ_O, and are shown in the diagram above.

57. For the complexes, a systematic name or formula:
   (a) $(NH_4)_2[CuCl_4]$ ammonium tetrachlorocuprate(II)
   (b) $Mo(CO)_6$ hexacarbonylmolybdenum(0)
   (c) tetraaquadichlorochromium(III) chloride $[Cr(H_2O)_4Cl_2]Cl$

   (d) aquabis(ethylenediamine)thiocyanatocobalt(III) nitrate
                          $[Co(H_2O)(H_2NCH_2CH_2NH_2)_2(SCN)](NO_3)_2$

59. For the complex ion $[Co(CO_3)_3]^{3-}$:

   (a) Color of complex: blue or cyan—absorbing the red light (at 640 nm), the blue or cyan

      color would be observed.

   (b) Place the CO_3^{2-} ion in the spectrochemical series:

      $F^- < CO_3^{2-} < C_2O_4^{2-} < H_2O < NH_3, en < phen < CN^-$

   (c) Is the complex para- or diamagnetic?

      If the carbonate ion gives rise to a small Δ, the electrons would tend to occupy as many d

      orbitals as possible—since the pairing energy would be greater than the Δ. With the d

      electrons occupying as many orbitals as possible—unpaired electrons would result and

      the complex would be paramagnetic.

61. The five geometric isomers of Cu(H2NCH2CO2)2(H2O)2:

   In the diagrams below, the curved lines represent the H2NCH2CO2 ligands.

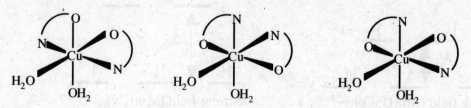

   The three isomers above each have a nonsuperimposable mirror image—that is they contain a chiral center. The two isomers shown below have no chiral center.

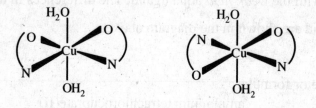

63. For Ni^{2+} and Pd^{2+}, the configuration is d^8, with each losing 2 s electrons in the formation of the ion. (See the solution to SQ22.35 as you consider this solution.)

   (a) A paramagnetic complex would have to be tetrahedral, whereas a diamagnetic complex would be square-planar.

   (b) The tetrahedral complex will have no isomers. The square planar Pd(II) complex will have two isomers:

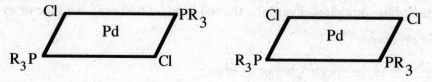

IN THE LABORATORY

65.   A + BaCl2 → ppt (BaSO4)   ⇒ A = [Co(NH3)5Br]SO4

        B + BaCl2 → no ppt   ⇒ B = [Co(NH3)5SO4]Br

   Complex A has the sulfate ion in the outer sphere of the transition metal compound. As such, this compound (like many ionic compounds) dissolves in water—and dissociates, liberating sulfate ions. Barium ions react with the sulfate ions to produce the precipitate.

   Complex B has the sulfate ion as a part of the inner sphere of the compound. This ion is bound tightly to the cobalt ion, and not available to the barium ions—hence no precipitate.

The reaction between A and $BaCl_2$:

$$[Co(NH_3)_5Br]SO_4(aq) + BaCl_2(aq) \rightarrow BaSO_4\ (s) + [Co(NH_3)_5Br]^{2+}(aq) + 2\ Cl^-(aq)$$

67. A 0.213 g sample of $UO_2(NO_3)_2$ contains xxxx mol of $UO_2(NO_3)_2$.

$$\frac{0.213\ g\ UO_2(NO_3)_2}{1} \bullet \frac{1\ mol\ g\ UO_2(NO_3)_2}{394.038\ g\ UO_2(NO_3)_2} = 5.41 \times 10^{-4}\ mol\ of\ UO_2(NO_3)_2$$

(a) If MnO_4^- is reduced to Mn^{2+}, each mol of MnO_4^- gains 5 mol of electrons (as Mn^{7+} is reduced to Mn^{2+}. The number of mol of MnO_4^- is:

$$\frac{0.0173\ mol\ MnO_4^-}{1\ L} \bullet \frac{0.01247\ L}{1} = 2.15731 \times 10^{-4}\ mol\ MnO_4^-.\ With\ this\ information,$$

we know that 5 mol e^-/mol MnO_4^- • 2.15731 x 10^{-4} mol MnO_4^-. = 1.08 x10^{-3} mol e^- Knowing that we have 5.41 x 10^{-4} mol of $UO_2(NO_3)_2$ and 1.08 x10^{-3} mol e^-, we

calculate the number of electrons per mol of uranium salt.

$$\frac{1.08 \times 10^{-3}\ mol\ electrons}{5.41 \times 10^{-4}\ mol\ UO_2(NO_3)_2} = 2\ mol\ electrons/mol\ uranium\ salt.$$

So with U having an oxidation state of +6 (in UO_2^{2+}), and each mol of U gaining 2 mol

electrons, the U would be reduced to +4. So n = 4.

(b) Balanced net ionic equation for reduction of UO_2^{2+} by zinc:

We know that Zn will be oxidized to 2+ (the ion that Zn forms):

$Zn \rightarrow Zn^{2+}$ and the reduction equation: $UO_2^{2+} \rightarrow U^{4+}$ (with a 2 electron change)

The net reduction equation: $UO_2^{2+} + 4\ H^+ + 2e^- \rightarrow U^{4+} + 2\ H_2O$, with an overall

equation: $UO_2^{2+}(aq) + 4H^+(aq) + Zn(s) \rightarrow U^{4+}(aq) + 2\ H_2O(l) + Zn^{2+}(aq)$

(c) The net ionic equation for the oxidation of U^{4+} to UO_2^{2+} by MnO_4^-.

The oxidation equation is: $U^{4+} \rightarrow UO_2^{2+}$. The reduction equation is: $MnO_4^- \rightarrow Mn^{2+}$

In acid, the oxidation equation is: $U^{4+} + 2\ H_2O \rightarrow UO_2^{2+} + 4H^+ + 2e^-$.

The reduction equation (in acid): $MnO_4^- + 8\ H^+ + 5\ e^- \rightarrow Mn^{2+} + 4\ H_2O$

To equalize electron gain and loss, multiply the oxidation equation by 5 and the reduction equation by 2 to obtain:

$$5\ U^{4+} + 10\ H_2O \rightarrow 5\ UO_2{}^{2+} + 20\ H^+ + 10\ e^-$$

$$2\ MnO_4{}^- + 16\ H^+ + 10\ e^- \rightarrow 2\ Mn^{2+} + 8\ H_2O$$

Adding the two equations, and removing any redundancies:

$$2\ MnO_4{}^-(aq) + 5\ U^{4+}(aq) + 2\ H_2O\ (l) \rightarrow 5\ UO_2{}^{2+}(aq) + 4\ H^+(aq) + 2\ Mn^{2+}(aq).$$

Summary and Conceptual Questions

69. The relative stabilities of the hexaammine complexes with Co^{2+}, Ni^{2+}, Cu^{2+}, and Zn^{2+}:

From Appendix K, the data are:

$Co^{2+}(aq) + 6\ NH_3\ (aq) \Leftrightarrow [Co(NH_3)_6]^{2+}$ $K_f = 1.3 \times 10^5$

$Ni^{2+}(aq) + 6\ NH_3\ (aq) \Leftrightarrow [Ni(NH_3)_6]^{2+}$ $K_f = 5.5 \times 10^8$

$Cu^{2+}(aq) + 4\ NH_3\ (aq) \Leftrightarrow [Cu(NH_3)_4]^{2+}$ $K_f = 2.1 \times 10^{13}$

$Zn^{2+}(aq) + 4\ NH_3\ (aq) \Leftrightarrow [Zn(NH_3)_4]^{2+}$ $K_f = 2.9 \times 10^9$

While the *general order is followed*, the copper tetraammine complex has the largest K_f of this series, a trend that is generally followed for many transition metal complexes.

71. Indicate the number of electrons associated with each Rh compound:

		Electrons:
Step 1 reactant **Total electrons:** **8 + 2 + 6 = 16**		Rh^+ has 8 d electrons Cl^- ligand contributes 2 e^- *Each* PR_3 ligand contributes 2
Step 1 product **Total electrons:** **6 + 2 + 6 + 4 =** **18**		Rh^{3+} has 6 d electrons Cl^- ligand contributes 2 e^- *Each* PR_3 ligand contributes 2 *Each* H^- ligand contributes 2
Step 2 product **Total electrons:** **6 + 2 + 4 + 4 =** **16**		Much like the complex described above, with the loss of one PR_3 (2 e^- donor) ligand.

Step 3 product **Total electrons:** **6 + 2 + 4 + 4 + 2** **= 18**		Note that the ethylene ligand (2 e⁻) has been added to the previous compound.
Step 4 product **Total electrons:** **6 + 2 + 4 + 2 =** **16**		Change from previous complex: Replaced one 2 e⁻ donor (H⁻) with the ethyl (2 e⁻ donor) ligand, and lost the ethylene ligand.
Step 5 product **Total electrons:** **8 + 2 + 4 = 14**		Change from previous complex: Lost an H⁻ (2 e⁻ donor), a PR₃ (2 e⁻ donor) and the ethyl (2 e donor). Note that Rh is now a 3+ (d^8) ion.

Applying Chemical Principles

1. Coordination number of the metal in the Fe-TAML complex is 5. Note 4 atoms in the plane with the Fe, and a water molecule coordinated above the plane.

Prototype TAML activator (Model from the Institute for Green Science: www.chem.cmu.edu/groups/collins

3. –dentate is the word used to indicate the number of "teeth" a chelating agent uses.

5. As one can see from the diagram the five coordinated atoms form a square pyramid. When the water molecule (above the plane of the Fe and N atoms) is removed, the geometry is best described as square planar.

Chapter 23
Nuclear Chemistry

PRACTICING SKILLS
Important Concepts

1. Discovery, contributors, significance:

   (a) 1896, discovery of radioactivity—H. Becquerel—indicated the complexity of the atom, and spurred research for the fundamental particles.

   (b) 1898, discovery of Ra and Po—M. Curie—identified Ra and Po as trace components of pitchblende, linking Ra and Po as members of the decay series of uranium.

   (c) 1918, first artificial nuclear reaction—E. Rutherford bombarded N with alpha particles; protons ejected and a new element was formed (O).

   (d) 1932, (n,γ) reactions—E. Fermi discovered the use of low-energy neutrons in bombarding nuclei. The nucleus, having absorbed the neutron, would emit γ radiation. Many modern medical isotopes are produced using this technique.

   (e) 1939, fission reactions; O. Hahn and F. Strassman discovered the process of nuclear fission upon discovering Ba in a sample of U that had been bombarded with n.

3. Data for graph of binding energy /nucleon: Once the mass defect of an isotope, Einstein's equation allows one to calculate the binding energy associated with that mass defect. Dividing the binding energy with the number of nucleons provides the desired data.

5. Nuclear reactions are carried out by bombarding one nucleus with, typically, another particle—neutron, alpha particle,etc. Resulting products are characterized. To make an atom of an element with atomic number greater than 92, one can bombard U-238 with a neutron. As the products emit β particles, atoms of Neptunium-239 and Plutonium-239 can be formed.

7. Carbon-14 can be used for dating of old objects by use of the ratio of C-14 to C-12 in the object under examination. Living items exchange (through respiration) C-14 and C-12, so that a living specie contains a ratio of C-14:C-12 which is the same as species around it. When the specie dies, the exchange ceases, and the C-14 in the dead specie decays. An assumption is that the amount of C-14 in the atmosphere remains constant. A limitation of radiocarbon dating is a result of the long half-life for C-14. Objects that aren't very old have very small changes in the amount of C-14 from current living species, and estimation by this technique is subject to large error.

9. A radioactive decay series is a set of sequential nuclear reactions showing the formation of isotopes, with the eventual formation of a non-radioactive isotope. As noted in SQ23.1b, as Uranium decays, both radium and polonium are formed.

Nuclear Reactions

11. Balance the following nuclear equations, supplying the missing particle.

[The missing particle is emboldened.]

(a) $^{54}_{26}\text{Fe} + ^{4}_{2}\text{He} \rightarrow 2\,^{1}_{0}\text{n} + \mathbf{^{56}_{28}\text{Ni}}$

(b) $^{27}_{13}\text{Al} + ^{4}_{2}\text{He} \rightarrow\ ^{30}_{15}\text{P} + \mathbf{^{1}_{0}\text{n}}$

(c) $^{32}_{16}\text{S} + ^{1}_{0}\text{n} \rightarrow\ ^{1}_{1}\text{H} + \mathbf{^{32}_{15}\text{P}}$

(d) $^{96}_{42}\text{Mo} + ^{2}_{1}\text{H} \rightarrow\ ^{1}_{0}\text{n} + \mathbf{^{97}_{43}\text{Tc}}$

(e) $^{98}_{42}\text{Mo} + ^{1}_{0}\text{n} \rightarrow\ ^{99}_{43}\text{Tc} + \mathbf{^{0}_{-1}\text{e}}$

(f) $^{18}_{9}\text{F} \rightarrow\ ^{18}_{8}\text{O} + ^{0}_{+1}\beta$

13. Balance the following nuclear equations, supplying the missing particle.

[The missing particle is emboldened.]

(a) $^{111}_{47}\text{Ag} \rightarrow\ ^{111}_{48}\text{Cd} + \mathbf{^{0}_{-1}\text{e}}$

(b) $^{87}_{36}\text{Kr} \rightarrow\ ^{0}_{-1}\text{e} + \mathbf{^{87}_{37}\text{Rb}}$

(c) $^{231}_{91}\text{Pa} \rightarrow\ ^{227}_{89}\text{Ac} + \mathbf{^{4}_{2}\text{He}}$

(d) $^{230}_{90}\text{Th} \rightarrow\ ^{4}_{2}\text{He} + \mathbf{^{226}_{88}\text{Ra}}$

(e) $^{82}_{35}\text{Br} \rightarrow\ ^{82}_{36}\text{Kr} + \mathbf{^{0}_{-1}\text{e}}$

(f) $\mathbf{^{24}_{11}\text{Na}} \rightarrow\ ^{24}_{12}\text{Mg} + ^{0}_{-1}\text{e}$

15. $^{235}_{92}U \rightarrow ^{231}_{90}Th \rightarrow ^{231}_{91}Pa \rightarrow ^{227}_{89}Ac \rightarrow ^{227}_{90}Th \rightarrow ^{223}_{88}Ra \rightarrow ^{219}_{86}Rn \rightarrow ^{215}_{84}Po$

$\quad\quad\quad\quad + \quad\quad\quad + \quad\quad\quad + \quad\quad\quad + \quad\quad\quad + \quad\quad\quad + \quad\quad\quad +$

$\quad\quad ^{4}_{2}He \quad\quad ^{0}_{-1}e \quad\quad ^{4}_{2}He \quad\quad ^{0}_{-1}e \quad\quad ^{4}_{2}He \quad\quad ^{4}_{2}He \quad\quad ^{4}_{2}He$

and continuing (from Po-215) we have:

$^{215}_{84}Po \rightarrow ^{211}_{82}Pb \rightarrow ^{211}_{83}Bi \rightarrow ^{211}_{84}Po \rightarrow ^{207}_{82}Pb$

$\quad\quad\quad + \quad\quad\quad + \quad\quad\quad + \quad\quad\quad +$

$\quad ^{4}_{2}He \quad\quad ^{0}_{-1}e \quad\quad ^{0}_{-1}e \quad\quad ^{4}_{2}He$

Nuclear Stability and Nuclear Decay

17. The particle emitted in the following reactions: [The missing particle is emboldened.]

(a) $^{198}_{79}Au \rightarrow ^{198}_{80}Hg + \mathbf{^{0}_{-1}e}$

(b) $^{222}_{86}Rn \rightarrow ^{218}_{84}Po + \mathbf{^{4}_{2}He}$

(c) $^{137}_{55}Cs \rightarrow ^{137}_{56}Ba + \mathbf{^{0}_{-1}e}$

(d) $^{110}_{49}In \rightarrow ^{110}_{48}Cd + \mathbf{^{0}_{1}e}$

19. Predict the probable mode of decay for each of the following:

(a) $^{80}_{35}Br$ (large number of neutrons /proton—beta emission) $^{80}_{35}Br \rightarrow ^{80}_{36}Kr + ^{0}_{-1}e$

(b) $^{240}_{98}Cf$ (large isotope- alpha emission) $^{240}_{98}Cf \rightarrow ^{236}_{96}Cm + ^{4}_{2}He$

(c) $^{61}_{27}Co$ (mass # > atomic number—beta emission) $^{61}_{27}Co \rightarrow ^{61}_{28}Ni + ^{0}_{-1}e$

(d) $^{11}_{6}C$ (more protons than neutrons—positron emission or K-capture)

$\quad\quad ^{11}_{6}C \rightarrow ^{11}_{5}B + ^{0}_{1}e \quad$ or $\quad ^{11}_{6}C + ^{0}_{-1}e \rightarrow ^{11}_{5}B$

21. Beta particle and positron emission:

(a) Beta particle emission occurs (usually) when the ratio of neutrons/protons is high

Hydrogen-3 has 1 proton and 2 neutrons—**beta particle emission (forms $^{3}_{2}He$)**

Oxygen-16 has 8 protons and 8 neutrons—not expected

Fluorine-20 has 9 protons and 11 neutrons-- **beta particle emission (forms $^{20}_{10}Ne$)**

Nitrogen-13 has 7 protons and 6 neutrons – not expected

(b) Position emission occurs when the neutron/proton ratio is too low:

Uranium-238 has 92 protons and 146 neutrons—not expected

Fluorine-19 has 9 protons and 10 neutrons—not expected

Sodium-22 has 11 protons and 11 neutrons—positron emission expected (**forms $^{22}_{10}Ne$**)

Sodium-24 has 11 protons and 13 neutrons—not expected

23. The change in mass (Δ m) for ^{10} B is:

Δ m $\quad = 10.01294 - [5(1.00783) + 5(1.00867)]$

$\quad\quad = 10.01294 - 10.0825 \quad = -0.06956$ g/mol

Binding energy is: $\Delta mc^2 = (6.956 \times 10^{-5}$ kg/mol$)(3.00 \times 10^8$ m/s$)^2 \left(\dfrac{1\ J}{1\ kg \bullet m^2 \bullet s^{-2}} \right)$

$\quad\quad\quad\quad\quad\quad = 6.26 \times 10^{12}$ J/mol

The **binding energy per nucleon**: $\dfrac{6.26 \times 10^9\ kJ}{10\ mol\ nucleons} = 6.26 \times 10^8\ \dfrac{kJ}{nucleon}$

The mass change for ^{11} B is:

Δ m $\quad = 11.00931 - [5(1.00783) + 6(1.00867)]$

$\quad\quad = 11.00931 - 11.09117 \quad = -0.08186$ g/mol

Binding energy is: $\Delta mc^2 \quad = (8.186 \times 10^{-5}$ kg/mol$)(3.00 \times 10^8$ m/s$)^2 \left(\dfrac{1\ J}{1\ kg \bullet m^2 \bullet s^{-2}} \right)$

$\quad\quad\quad\quad\quad\quad = 7.367 \times 10^{12}$ J/mol

The **binding energy per nucleon**: $\dfrac{7.37 \times 10^9\ kJ}{11\ mol\ nucleons} = 6.70 \times 10^8\ \dfrac{kJ}{nucleon}$

25. The binding energy per nucleon for calcium-40:

The change in mass (Δ m) for ^{40} Ca is:

Δ m $\quad = 39.96259 - [20(1.00783) + 20(1.00867)]$

$\quad\quad = 39.96259 - 40.3300$

$\quad\quad = -0.3674$ g/mol

The energy change is then:

$$\Delta E = (3.674 \times 10^{-4} \text{ kg/mol})(3.00 \times 10^8 \text{ m/s})^2 \left(\frac{1 \text{ J}}{1 \text{ kg} \cdot \text{m}^2 \cdot \text{s}^{-2}} \right)$$

$$= 3.307 \times 10^{13} \text{ J/mol or } 3.307 \times 10^{10} \text{ kJ/mol}$$

This energy can be converted into the **binding energy per nucleon:**

$$\frac{3.307 \times 10^{10} \text{ kJ}}{40 \text{ mol nucleons}} = 8.26 \times 10^8 \frac{\text{kJ}}{\text{nucleon}}$$

27. Binding energy per nucleon for Oxygen-16

The change in mass (Δ m) for ^{16}O is:

$$\Delta m = 15.99492 - [8(1.00783) + 8(1.00867)] = -0.13708 \text{ g/mol}$$

The energy change is then:

$$\Delta E = (1.3708 \times 10^{-4} \text{ kg/mol})(3.00 \times 10^8 \text{ m/s})^2 \left(\frac{1 \text{ J}}{1 \text{ kg} \cdot \text{m}^2 \cdot \text{s}^{-2}} \right)$$

$$= 1.234 \times 10^{13} \text{ J/mol or } 1.234 \times 10^{10} \text{ kJ/mol}$$

This energy can be converted into the **binding energy per nucleon:**

$$\frac{1.234 \times 10^{10} \text{ kJ/mol}}{16 \text{ mol nucleons}} = 7.70 \times 10^8 \text{ kJ/nucleon}$$

Rates of Radioactive Decay

29. For ^{64}Cu, $t_{1/2} = 12.7$ hr

The fraction remaining as ^{64}Cu following n half-lives is equal to $\left(\frac{1}{2} \right)^n$.

Note that 63.5 hours corresponds to **exactly five** half-lives.

The <u>fraction</u> remaining as ^{64}Cu is $\left(\frac{1}{2} \right)^5$ or $\frac{1}{32}$ or 0.03125.

The mass remaining is: $(0.03125)(25.0 \text{ }\mu g) = 0.781 \text{ }\mu g$.

31. (a) The equation for β–decay of ^{131}I is: $^{131}_{53}\text{I} \rightarrow ^{\ \ 0}_{-1}e + ^{131}_{54}\text{Xe}$

(b) The amount of ^{131}I remaining after 40.2 days:

For ^{131}I, $t_{1/2}$ is 8.04 days--so 40.2 days is exactly **five** half-lives:

444

The fraction of ^{131}I remaining is $\left(\frac{1}{2}\right)^5$ or $\frac{1}{32}$ or 0.03125.

The amount of the original 2.4 µg remaining will be : $(0.03125)(2.4 \text{ µg}) = 0.075 \text{ µg}$

33. To determine the mass of Gallium-67 left after 13 days, determine the number of half-lives corresponding to 13 days.

$$\frac{13 \text{ days}}{1} \cdot \frac{24 \text{ hrs}}{1 \text{ day}} = 312 \text{ hours}$$

The rate constant is: $k = \dfrac{0.693}{t_{\frac{1}{2}}} = \dfrac{0.693}{78.25 \text{ hr}} = 0.00886 \text{ hrs}^{-1}$

The fraction remaining can be calculated: $\ln(x) = -(0.00886 \text{ hrs}^{-1}) \cdot (312 \text{ hrs})$ and solving for x yields 0.06309 (where x represents the fraction of Ga-67 remaining).

The amount of Gallium-67 remaining is then $(0.06309)(0.015 \text{ mg}) = 9.5 \times 10^{-4} \text{ mg}$.

35. For the decomposition of Radon-222:
(a) The balanced equation for the decomposition of Rn-222 with α particle emission.

$$^{222}_{86}\text{Rn} \rightarrow {}^{4}_{2}\text{He} + {}^{218}_{84}\text{Po}$$

(b) Time required for the sample to decrease to 20.0 % of its original activity:

Since this decay follows 1st order kinetics, we can calculate a rate constant:

$$k = \frac{0.693}{t_{\frac{1}{2}}} = \frac{0.693}{3.82 \text{ days}} = 0.181 \text{ days}^{-1}$$

With this rate constant, using the 1st order integrated rate equation, we can calculate the time required:

$$\ln\left(\frac{20.0}{100}\right) = -(0.181 \text{ days}^{-1}) \cdot t \quad \text{and} \quad \frac{-1.609}{-0.181 \text{ days}^{-1}} = t = 8.87 \text{ days}$$

37. For the decay of Cobalt-60, $t_{1/2}$ is 5.27 yrs:
(a) Time for Co-60 to decrease to 1/8 of its original activity:

Following the methodology of questions 34,36 and 38, determine the rate constant:

$$k = \frac{0.693}{t_{\frac{1}{2}}} = \frac{0.693}{5.27 \text{ yr}} = 0.131 \text{ yr}^{-1}$$

Substituting into the equation:

$$\ln(\tfrac{1}{8}) = -0.131\,\text{yr}^{-1} \bullet t \quad \text{and}$$

$$\ln(0.125) = -0.131\ \text{yr}^{-1} \bullet t \quad \text{and solving for t} = 15.8\ \text{yrs}$$

A " short-cut" is available here if you notice that 1/8 corresponds to $(\tfrac{1}{2})^3$. Said another

way, one-eighth of the Co-60 will remain after **three half-lives** have passed, so

$3 \bullet 5.27\ \text{yrs} = 15.8\ \text{years} \ !!$

(b) Fraction of Co-60 remaining as Co-60 after 1.0 years:

Now we can solve for the fraction on the "left-hand side" of the rate equation:

$$\ln\,(\text{fraction remaining}) = -\,k \bullet t = -\,0.131\ \text{yr}^{-1} \bullet 1.0\ \text{yr}$$

$\ln\,(\text{fraction remaining}) = -0.131$ and $e^{-0.131}$ = fraction remaining

fraction remaining = 0.877 , so 88% remains after 1.0 years.

Nuclear Reactions

39. For the decay of Plutonium-239: $\ _{94}^{239}\text{Pu} + \ _{2}^{4}\text{He} \rightarrow \ _{95}^{240}\text{Am} + \ _{1}^{1}\text{H} + 2\ _{0}^{1}\textbf{n}$

41. Synthesis of Element 114: $\ _{94}^{242}\text{Pu} + \ _{20}^{48}\text{Ca} \rightarrow \ _{114}^{287}\text{Uuq} + 3\ _{0}^{1}\text{n}$

43. Complete the following equations using deuterium bombardment:

   [The missing particle is emboldened.]

   (a) $\ _{48}^{114}\text{Cd} + \ _{1}^{2}\text{D} \rightarrow \ _{\textbf{48}}^{\textbf{115}}\textbf{Cd} + \ _{1}^{1}\text{H}$

   (c) $\ _{20}^{40}\text{Ca} + \ _{1}^{2}\text{D} \rightarrow \ _{19}^{38}\text{K} + \ _{\textbf{2}}^{\textbf{4}}\textbf{He}$

   (b) $\ _{3}^{6}\text{Li} + \ _{1}^{2}\text{D} \rightarrow \ _{\textbf{4}}^{\textbf{7}}\textbf{Be} + \ _{0}^{1}\text{n}$

   (d) $\ _{\textbf{29}}^{\textbf{63}}\textbf{Cu} + \ _{1}^{2}\text{D} \rightarrow \ _{30}^{65}\text{Zn} + \gamma$

45. The equation for the bombardment of Boron-10 with a neutron, and the subsequent release of

   an alpha particle: $\ _{5}^{10}\text{B} + \ _{0}^{1}\text{n} \rightarrow \ _{3}^{7}\text{Li} + \ _{2}^{4}\text{He}$

GENERAL QUESTIONS

47. The rate constant, $k = \dfrac{0.693}{t_{1/2}} = \dfrac{0.693}{4.8 \times 10^{10} \text{ yr}} = 1.44375 \times 10^{-11} \text{ yr}^{-1}$

At some time, t, we have 1.8×10^{-3} mol Rb. We also have 1.6×10^{-3} mol Sr, which resulted from the decay of an equal amount of Rb. So the initial amount of Rb = $(1.6 + 1.8) \times 10^{-3}$.

Substituting into the first-order equation we get: $\ln\dfrac{1.8}{3.4} = -(1.44 \times 10^{-11} \text{ yr}^{-1})t$

$\dfrac{\ln(0.529)}{-1.44 \times 10^{-11} \text{yr}^{-1}} = \dfrac{-0.636}{-1.44 \times 10^{-11} \text{yr}^{-1}} = t = 4.4 \times 10^{10} \text{ yr}$

49. Graph for P-31 of disintegrations per minute as a function of time for a period of 1 year.

Calculate rate constant: $k = \dfrac{0.693}{t_{1/2}} = \dfrac{0.693}{14.28 \text{days}}$

$= 0.04854 \text{ days}^{-1}$ On the graph is plotted the results of the calculation:

$\ln\dfrac{x}{3.2 \times 10^6} = -(0.04853 \text{ days}^{-1})t$

I have substituted multiples of 14.28 days, so each data point corresponds to a half-life.

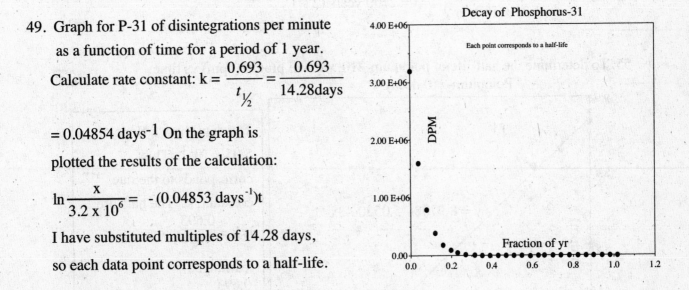

51. The decay of Uranium-238 to produce Plutonium-239:

(a) $^{238}_{92}\text{U} + ^{1}_{0}\text{n} \rightarrow ^{239}_{92}\text{U} + \gamma$

(b) $^{239}_{92}\text{U} \rightarrow ^{239}_{93}\text{Np} + ^{0}_{-1}\text{e}$

(c) $^{239}_{93}\text{Np} \rightarrow ^{239}_{94}\text{Pu} + ^{0}_{-1}\text{e}$

(d) $^{239}_{94}\text{Pu} + ^{1}_{0}\text{n} \rightarrow 2^{1}_{0}\text{n} + \text{other nuclei} + \text{energy}$

IN THE LABORATORY

53. The age of the fragment can be determined if: (1st) we calculate a rate constant and, (2nd) we use the 1st order integrated rate equation

(1st) $k = \dfrac{0.693}{t_{1/2}} = \dfrac{0.693}{5730 \text{ yr}} = 1.21 \times 10^{-4} \text{ yr}^{-1}$

(2nd) Now we can calculate the time required for the Carbon-14:Carbon-12 to decay to 72% of that ratio in living organisms.

$$\ln\left(\frac{72}{100}\right) = -(1.21 \times 10^{-4} \text{ yr}^{-1}) \bullet t$$

$$\frac{-0.3285}{-1.21 \times 10^{-4} \text{ yr}^{-1}} = t = 2700 \text{ years (2 sf)}$$

55. To determine the half-life of polonium-210, we will plot ln (dpm) vs time.

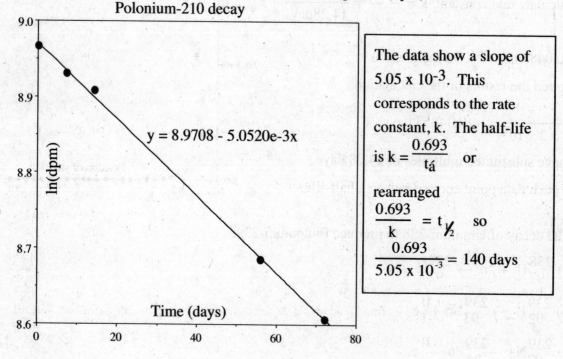

Polonium-210 decay

$y = 8.9708 - 5.0520\text{e-}3x$

The data show a slope of 5.05×10^{-3}. This corresponds to the rate constant, k. The half-life is $k = \dfrac{0.693}{t_{á}}$ or rearranged $\dfrac{0.693}{k} = t_{1/2}$ so

$\dfrac{0.693}{5.05 \times 10^{-3}} = 140$ days

57. If the ratio of $\dfrac{Pb\text{-}206}{U\text{-}238}$ is 0.33, a ratio of 0.25 of Pb-206 to 0.75 of U-238 provides an appropriate ratio. Calculate a rate constant and use the 1st order rate equation to calculate the time required for this to occur:

$$k = \dfrac{0.693}{4.5 \times 10^9 \text{ yr}} = 1.54 \times 10^{-10} \text{ yr}^{-1} \text{ and } \ln(\dfrac{0.75}{1.00}) = -1.54 \times 10^{-10} \text{ yr}^{-1} \bullet t$$

$$-0.288 = -1.54 \times 10^{-10} \text{ yr}^{-1} \bullet t \text{ and } t = \dfrac{-0.288}{-1.54 \times 10^{-10} \text{yr}^{-1}} = 1.9 \times 10^9 \text{ yr}$$

SUMMARY AND CONCEPTUAL QUESTIONS

59. The energy liberated by one pound of ^{235}U:

$$\dfrac{2.1 \times 10^{10} \text{ kJ}}{1 \text{ mol } ^{235}U} \bullet \dfrac{1 \text{ mol } ^{235}U}{235 \text{ g U}} \bullet \dfrac{453.6 \text{ g U}}{1.000 \text{ lb}} = \dfrac{4.1 \times 10^{10} \text{ kJ}}{1.000 \text{ lb U}}$$

comparing this amount of energy to the energy per ton of coal yields:

$$\dfrac{4.1 \times 10^{10} \text{ kJ}}{1.000 \text{ lb U}} \bullet \dfrac{1 \text{ ton coal}}{2.6 \times 10^7 \text{ kJ}} = 1.6 \times 10^3 \text{ tons coal/lb U}$$

61. If we assume that the catch represents a homogeneous sample of the tagged fish, the problem is rather simple. The percentage of tagged fish in the sample is $\dfrac{27}{5250} = 0.00514$ or 0.51%. If our 1000 fish represent 0.51% of the fish in the lake, the number is approximately:

$$\dfrac{1000}{0.0051} = 190,000 \text{ fish.}$$

63. For the radioactive decay series of U-238:

(a) Why can the masses be expressed as m = 4n +2?

These masses correlate in this fashion since the **principal** mode of decay in the U-238 series is **alpha particle** emission. The mass of an alpha particle is **4** (since it is basically 2 neutrons + 2 protons). Hence the loss of an alpha particle (say from U-238 to Th-234) results in a **mass loss of 4 units**. The series results in a total loss of 8 alpha particles and 6 beta particles. Since the beta particles (electrons) have an insignificant mass (compared to a neutron or a proton), the loss of a beta particle--or for that matter 6 of them--does not

significantly affect the masses of the daughter products. These masses correspond to n
values: n = (51 -> 59).

(b) Equations corresponding to the decay series for U-235 and Th-232 :

The U-235 series corresponds to the equation **m = 4n+3** with n values: (51 -> 58). The

Th-232 series corresponds to the equation: **m = 4n**

From an empirical standpoint, the masses of the most massive isotopes in the three series

differ by 3 (U-238 $\rightarrow$ U–235) and (U–235 $\rightarrow$ Th-232). So given that the algorithm for the

U-238 series is 4n+2, subtracting 3 gave 4n-1 (which also equals 4n+3)—with n reduced

by 1; subtracting 3 more gives 4n.

The isotopic masses for these 3 series are summarized in the table below:

n	= 4n+2	= 4n+3	= 4n
51	206		
52	210	207	208
53	214	211	212
54	218	215	216
55	222	219	220
56	226	223	224
57	230	227	228
58	234	231	232
59	238	235	

(c) Identify the series to which each of the following isotopes belong:

Isotope	series
226-Ra:	U-238 (4n + 2)
215-At:	U-235 (4n + 3)
228-Th:	Th-232(4n)
210- Bi:	U-238 (4n + 2)

(d) Why is the series "4n+1" missing in the earth's crust?

To occur in the earth's crust, an element must be very stable—that is have a very long
half-life or be non-radioactive! From hydrogen to lawrencium, with the exception of two
isotopes of hydrogen (protium and tritium), every isotope of every element has a nucleus
containing at least one neutron for every proton. With the mass of the neutron and proton
being 1, the change in mass number would have to change by a factor of 2—so 4n+1
would not lead to a long-lived isotope and would not be found in the earth's crust.

65. For Protactinium:

(a) The series containing Protactinium-231 is the U-235 series. It corresponds to the (4n+3) series—See question 66.

(b) A series of reaction to produce Pa-231:

$$^{235}_{92}U \rightarrow {}^{231}_{90}Th + {}^{4}_{2}He \text{ followed by the decay } {}^{231}_{90}Th \rightarrow {}^{0}_{-1}e + {}^{231}_{91}Pa.$$

(c) Quantity of ore to provide 1.0 g of Pa-231 assuming 100% yield:

Stow your calculators! If the ore is 1 part per million, and you want 1.0 g, then you need 1,000,000 g of ore, check?

(d) Decay for Pa-231: $^{231}_{91}Pa \rightarrow {}^{227}_{89}Ac + {}^{4}_{2}He$

67. Which of the isotopes are anticipated in uranium ore? Since both radium and polonium are decay products of uranium, they must belong to either the 4n+2(U-238) or 4n+3 (U-235) decay series. The 4n+3 series would give rise to isotopes with mass numbers: 235→231→227→223→219→215→211; The 4n+2 series would give rise to isotopes with mass numbers: 238→234→230→226→222→218→214→210. Of the 5 isotopes given, Ra-226, and Po-210 are both members of the 4n+2 decay series, and have a sufficient half-life to be detected.

Applying Chemical Principles

1. A balanced equation for the radioactive decomposition of ^{87}Rb: $^{87}_{37}$Rb $\rightarrow$ $^{87}_{38}$Sr $+$ $^{0}_{-1}$e

   With the mass number constant (87), the mass of the second product has to be 0. As the atom number *increases* from 37 to 38, the second product has to have a mass of -1, so the electron fills the bill.

3. What is the rate constant for the decay of the ^{87}Rb, given the half-life is 4.88 x 10^{10} yrs?

   $$k = \frac{0.693}{4.88 x 10^{10} \, yr} = 1.42 \times 10^{-11} \, yr^{-1}$$

5. A strontium-rubidium isochron plot of the data provided:

Sample #	Sr-86	Sr-87	Rb-87	Sr-87/Sr-86	Rb-87/Sr-86
1	1.000	0.819	0.839	0.819	0.839
2	1.063	0.855	0.506	0.804	0.476
3	0.950	0.824	1.929	0.867	2.031
4	1.011	0.809	0.379	0.800	0.375

The plot is as follows:

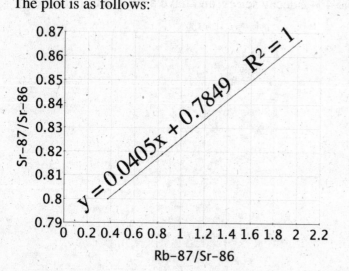

The slope of the line is 0.0405.

The slope (0.0405) = e^{kt} -1. So 1.0405 = e^{kt} (adding 1 to both sides of the equation).

So kt = 0.0397. In question 3 above, we determined that k = 1.42 x 10^{-11} yr^{-1}

Dividing 0.0397 by k gives a t = 2.80 x 10^{9}yr, the age of the meteorite.

Interchapter
The Chemistry of Fuels and Energy Resources

1. Mass of H_2 expected from the reaction of steam with 100. g of methane (CH_4), petroleum(CH_2) and coal (C):

$$CH_4(g) + H_2O(g) \rightarrow CO(g) + 3H_2(g)$$

$$CH_2(l) + H_2O(g) \rightarrow CO(g) + 2H_2(g)$$

$$C(s) + H_2O(g) \rightarrow CO(g) + H_2(g)$$

Based on the equations above, we convert the mass to moles of the C containing specie, use the stoichiometric ratio indicated by the balanced equation, and convert to mass of H_2, using the molar mass of H_2.

$$\frac{100.\text{ g } CH_4}{1} \cdot \frac{1 \text{ mol } CH_4}{16.04 \text{ g } CH_4} \cdot \frac{3 \text{ mol } H_2}{1 \text{ mol } CH_4} \cdot \frac{2.02 \text{g } H_2}{1 \text{mol } H_2} = 37.7 \text{ g } H_2$$

$$\frac{100.\text{ g } CH_2}{1} \cdot \frac{1 \text{ mol } CH_2}{14.03 \text{ g } CH_2} \cdot \frac{2 \text{ mol } H_2}{1 \text{ mol } CH_2} \cdot \frac{2.02 \text{g } H_2}{1 \text{mol } H_2} = 28.7 \text{ g } H_2$$

$$\frac{100.\text{ g } C}{1} \cdot \frac{1 \text{ mol } C}{12.01 \text{ g } C} \cdot \frac{1 \text{ mol } H_2}{1 \text{ mol } C} \cdot \frac{2.02 \text{g } H_2}{1 \text{mol } H_2} = 16.8 \text{ g } H_2$$

3. Calculate the energy evolved (in kJ) when 70. lb of coal is burned. We make the assumption that coal can be represented as carbon (C).

$$\frac{70.\text{ lb } C}{1} \cdot \frac{454 \text{ g } C}{1 \text{ lb } C} \cdot \frac{33 \text{ kJ}}{1 \text{ g } C} = 1.0 \times 10^6 \text{ kJ}$$

5. Energy consumption in U.S. is: 7.0 gallons of oil (or 70. lb coal) per person per day. Compare the energy produced by 7.0 gallons of oil with that of 70. lb of coal (see problem 3 above). From Table 1, the energy released per gram of crude petroleum (45 kJ/g) is found.

$$\frac{7.0 \text{ gal oil}}{1} \cdot \frac{4 \text{ qts}}{1 \text{ gal}} \cdot \frac{1000 \text{ mL}}{1.0567 \text{ qt}} \cdot \frac{0.8 \text{ g oil}}{1 \text{ mL oil}} \cdot \frac{45 \text{ kJ}}{1 \text{ g oil}} = 9 \times 10^5 \text{ kJ}.$$

The energy released is slightly less for petroleum than for coal, but within the same order of magnitude.

7. (a) Enthalpy change for combustion of 1.00 kg of ethanol and of 1.00 kg of isooctane:

$$C_2H_5OH \ (l) + 3 \ O_2(g) \ \rightarrow \ 2 \ CO_2(g) + 3 \ H_2O(l)$$

$\Delta H°_f(kJ/mol) \quad -277.0 \qquad\qquad 0 \qquad\qquad\quad -393.509 \quad -285.83$

$\Delta H°_{rxn} \ = \ \sum \Delta H°_f \text{ products} \ - \ \sum \Delta H°_f \text{ reactants}$

$\Delta H°_{rxn} \ = \ [(2 \text{ mol})(-393.509 \ \frac{kJ}{mol}) + (3 \text{ mol})(-285.83 \ \frac{kJ}{mol})]$

$\qquad\qquad - \ [(1 \text{ mol})(-277.0 \ \frac{kJ}{mol}) + (3 \text{ mol})(0 \ \frac{kJ}{mol})]$

$\Delta H°_{rxn} \ = \ (-1644.508 \text{ kJ}) - (-277 \text{ kJ}) = -1367.508 \text{ kJ/mol ethanol.}$

And on a *per gram basis*: $\dfrac{-1.37 \times 10^3 \text{ kJ}}{1 \text{ mol C}_2\text{H}_5\text{OH}} \bullet \dfrac{1 \text{ mol C}_2\text{H}_5\text{OH}}{46.0688 \text{ g C}_2\text{H}_5\text{OH}} = -29.7 \text{ kJ/g ethanol and}$

for 1.00 kg of ethanol, the enthalpy change is -29.7×10^3 kJ or -2.97×10^4 kJ

For isooctane:

$$2 \ C_8H_{18} \ (l) + 25 \ O_2(g) \ \rightarrow \ 16 \ CO_2(g) + 18 \ H_2O(l)$$

$\Delta H°_f(kJ/mol) \quad -259.3 \qquad\qquad 0 \qquad\qquad\quad -393.509 \quad -285.83$

$\Delta H°_{rxn} \ = \ \sum \Delta H°_f \text{ products} \ - \ \sum \Delta H°_f \text{ reactants}$

$\Delta H°_{rxn} \ = \ [(16 \text{ mol})(-393.509 \ \frac{kJ}{mol}) + (18 \text{ mol})(-285.83 \ \frac{kJ}{mol})]$

$\qquad\qquad - \ [(2 \text{ mol})(-259.3 \ \frac{kJ}{mol}) + (0 \ \frac{kJ}{mol})]$

$\Delta H°_{rxn} \ = \ (-11441.084 \text{ kJ}) - (-518.6 \text{ kJ}) = -10922.484 \text{ kJ for 2 mol of isooctane!}$

$\qquad\quad = \ -5461.242 \text{ kJ/mol isooctane.}$

On a *per gram basis*: $\dfrac{-5.461 \times 10^3 \text{ kJ}}{1 \text{ mol C}_8\text{H}_{18}} \bullet \dfrac{1 \text{ mol C}_8\text{H}_{18}}{114.23 \text{ g C}_8\text{H}_{18}} = -47.8 \text{ kJ/g isooctane and}$

for 1.00 kg of isooctane, the enthalpy change is -47.8×10^3 kJ or -4.78×10^4 kJ

Comparing the two fuels, isooctane delivers more energy than ethanol.

(b) Which fuel produces more CO_2 per kg?

Consider the balanced equations for the two fuels. From the balanced equations, 1 mole of ethanol produces 2 moles of carbon dioxide, while 1 mole of isooctane produces 8. (ratios of coefficients!)

1 mol of ethanol has a mass of 46.1 grams, and 1 mol of isooctane a mass of 114 grams.

Divide the mass of each into the # of moles of carbon dioxide:

$$\frac{2 \text{ mol } CO_2}{46.1 \text{g } C_2H_5OH} = 0.0434 \text{ mol } CO_2/\text{g } C_2H_5OH \text{ and for isooctane:}$$

$$\frac{8 \text{ mol } CO_2}{114 \text{g } C_8H_{18}} = 0.0702 \text{ mol } CO_2/\text{g } C_8H_{18}$$

Note that the amount of CO_2 (as well as the mass) is greater per gram (and per kg) for isooctane.

(c) Which is better fuel in terms of energy production and greenhouse gases?

Isooctane produces more energy per kg of fuel than ethanol, but also produces more greenhouse gas (carbon dioxide) per kg!

9. If a washer uses 940 kwh/year, what is this energy in units of kJ?

$$\frac{940 \text{ kwh}}{1 \text{ yr}} \cdot \frac{1000 \text{ watt} \cdot \text{hours}}{1 \text{ kwh}} \cdot \frac{1 \text{ J/s}}{1 \text{ watt}} \cdot \frac{3600 \text{ s}}{1 \text{ hour}} \cdot \frac{1 \text{ kJ}}{1000 \text{ J}} = 3.4 \times 10^6 \text{ kJ/yr}$$

Cost per month to operate the machine if energy costs $0.08/kwh

$$\frac{\$0.08}{1 \text{ kwh}} \cdot \frac{940 \text{ kwh}}{1 \text{ yr}} \cdot \frac{1 \text{ yr}}{12 \text{ mo}} = \$6.27 / \text{mo}$$

11. Confirm that oxidation of 1.0 L of CH_3OH to form $CO_2(g)$ and $H_2O(l)$ in a fuel cell provides at least 5.0 kwh of energy (D of $CH_3OH = 0.787$ g/mL).

We need to convert 1.0 L of methanol into mass and moles:

$$\frac{1.0 \text{ L } CH_3OH}{1} \cdot \frac{787 \text{ g } CH_3OH}{1.0 \text{ L}} \cdot \frac{1 \text{ mol } CH_3OH}{32.04 \text{ g } CH_3OH} = 25 \text{ mol } CH_3OH$$

Now the question is what energy change occurs when methanol burns. That question is answered by using thermodynamic data (from Appendix L)

2 CH_3OH (l) + 3 O_2 (g) → 2 $CO_2(g)$ + 4 $H_2O(l)$

$\Delta_rH = [2 \ \Delta_fH°CO_2(g) + 4 \ \Delta_fH°H_2O(l)] - [2 \ \Delta_fH°CH_3OH(l) + 3 \ \Delta_fH°O_2(g)]$

$\Delta_rH = [2\text{mol} \cdot (-393.509 \text{ kJ/mol}) + 4\text{mol} \cdot (-285.83 \text{ kJ/mol})]$

$- [2\text{mol} \cdot (-238.4 \text{ kJ/mol}) + 3 \text{ mol} \cdot 0)]$

$\Delta_rH = -1,453.5 \text{ kJ}$

The energy change per mol is then multiplied by the # of moles of methanol associated with 1.0 L of CH_3OH

$$\frac{-1,453.5 \text{ kJ}}{2 \text{ mol CH}_3\text{OH}} \cdot \frac{25 \text{ mol CH}_3\text{OH}}{1} = -1.8 \times 10^4 \text{ kJ (2 sf)}$$

Converting that energy change to units of kwh:

$$\frac{1.8 \times 10^4 \text{ kJ}}{1} \cdot \frac{1000 \text{ J}}{1 \text{ kJ}} \cdot \frac{1 \text{ watt}}{1 \text{ J/s}} \cdot \frac{1 \text{ hr}}{3600 \text{ s}} \cdot \frac{1 \text{ kw}}{1000 \text{ watt}} = 5.0 \text{ kwh (2sf)}$$

13. The parking lot is 325 m long and 50.0 m wide, or a surface area of 16250 m^2.

If the solar radiation is 2.6×10^7 J/m^2 (per day), the parking lot would receive:

2.6×10^7 J/m$^2 \cdot 16250$ m$^2 = 4.3 \times 10^{11}$ J (per day).

15. Energy consumed to drive 1.00 mile by a car rated at 55.0 mpg. The density of gasoline is

0.737 g/cm^3; gasoline produces 48.0 kJ/g.

$$\frac{48.0 \text{ kJ}}{1 \text{ g gasoline}} \cdot \frac{0.737 \text{ g gasoline}}{1 \text{ cm}^3 \text{ gasoline}} \cdot \frac{1000 \text{ cm}^3}{1 \text{ L}} \cdot \frac{1 \text{ L}}{1.0567 \text{ qt}} \cdot \frac{4 \text{ qt}}{1 \text{ gal}} \cdot \frac{1 \text{ gal}}{55.0 \text{ mile}} \cdot \frac{1.00 \text{ mile}}{1} =$$

$$2.43 \times 10^3 \text{ kJ}$$

Interchapter
Milestones in the Development of Chemistry and the Modern View of Atoms and Molecules

1. Critique Dalton's hypothesis of an atom as a "solid, massy hard, impenetrable, moveable particle."

   Several experimentalists—Rutherford among them—showed that atoms could be penetrated by having high energy particles (alpha particles in Rutherford's case) fired toward atoms (again in Rutherford's case, gold). Also were atoms "solid, hard impenetrable" when bonds were made between two atoms, the internuclear distance would be a simple summation of atomic radii, which is not the case. What is true about Dalton's hypothesis is that atoms can move—translation—from one place to another. Diffusion and molecular motion, in general, are examples of this atomic/molecular motion.

3. Ratio of mass of electron to the mass of a proton:

   The currently accepted mass of the electron is 9.10983×10^{-28} g.

   The currently accepted mass of the proton is 9.10983×10^{-28} g.

   The ratio is then: $\dfrac{9.109383 \times 10^{-28} \text{ g/electron}}{1.672622 \times 10^{-24} \text{ g/proton}} = 5.446170 \times 10^{-4}$ g electron/g proton

   Given the small ratio, let's take the reciprocal of this number:

   $\dfrac{5.446170 \times 10^{-4} \text{ g electron/g proton}}{1} = 1836$. So that proton has a mass that is 1836 times that of the electron.

Interchapter
The Chemistry of Life:Biochemistry

1. Lewis structures for:

(a) Valine with amino and carboxyl groups in un-ionized form

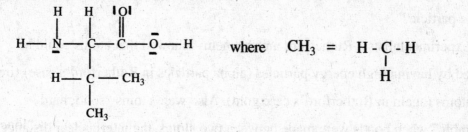

(b) Valine in zwitterionic form

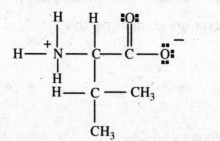

(c) At physiological pH, the zwitterion predominates since the carboxylic acid group and the amino group are ionized.

3. Two different ways in which glycine and alanine may be combined:

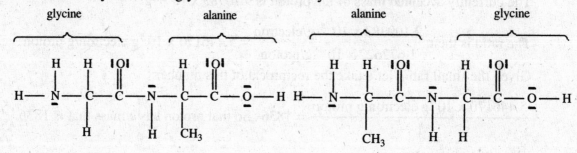

The differences result from varying which "amino group" is bonded to which "carboxyl group".

5. Two Lewis structures for the dipeptide alanine-isoleucine that show the resonance structures of the amide linkage. Resonance structures are a result of shifting pairs of electrons, in this case, the lone pair of electrons from N to create the C=N bond.

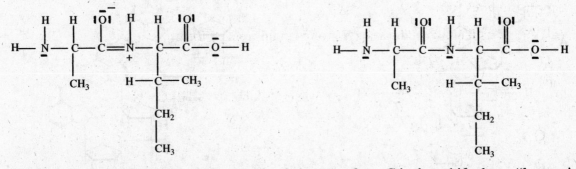

The first structure shows the shifts. A pair of electrons from C is then shifted to a "lone pair" position. The resulting O bears a "-" charge, while the N bears a "+" charge.

7. (a) Lewis structure for the sugar ribose:

    In the Lewis structure, the solid lines indicating covalent bonds are above the plane of the ribose ring, while the dotted lines are below the plane of the ribose ring.

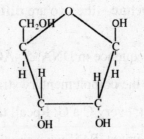

   (b) Lewis structure for adenosine:
       Note that for the nucleoside the adenine (base) is attached to the ribose ring. The attachment occurs via the formation of water (H- from the adenine moiety, and OH from the ribose molecule.)

   (c) Lewis structure for nucleotide: adenosine 5'-monophosphate. Note that the phosphate group is attached to the 5' C.

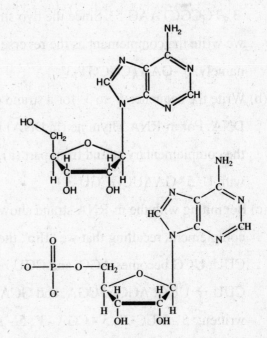

9. Do DNA sequences ATGC and CGTA represent the same molecule?

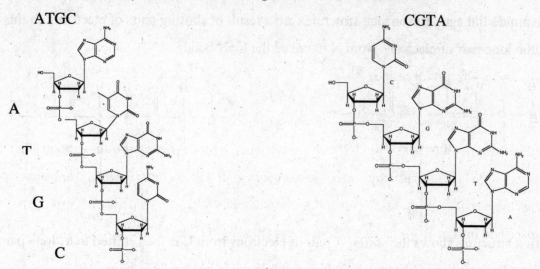

ATGC   CGTA

As one can see, while the individual pieces are the same, the assembly does not result in the same molecular structure—the two are **different**.

11. For the nucleotide sequence in DNA: 5'-ACGCGATTC-3':

(a) The sequence of the complementary strand would be found by noting that every A has a T as a complement; every C a G. Recall that the 5' end of a DNA strand pairs with the 3' end of its complement. So if we write the complement to the sequence above, we'd get: 3'-TGCGCTAAG-5'. Since the two strands of DNA have 5' ends paired with 3' ends, we write the complement as the reverse of the listing above, namely: 5'-GAATCGCGT-3'.

(b) Write the sequence 5' to 3' for a strand of m-RNA to complement the original strand of DNA. For m-RNA Thymine (in DNA) is replaced with Uracil (in RNA). Hence taking the complementary strand from part (a), we replace every T with U: 5'-GAAUCGCGU-3'

(c) Beginning with the m-RNA strand shown in part (b), we now code each "letter" with its complement, recalling that we "flip" the 5' and 3' ends of the segment: so GAA becomes CUU, UCG becomes AGC, and CGU becomes GCA---and swapping the ends yields: CUU → UUC; AGC →CGA; and GCA →ACG so the three anticodons would be written: 5'- UUC-3'; 5'- CGA -3'; 5'- ACG -3'

(d) The sequence of amino acids coded by the three codons on m-RNA (part(b)):

GAA codes for glu (glutamic acid)

UCG codes for ser (serine)

CGU codes for arg (arginine)

13. (a) Describe what occurs in the process of transcription:

The information contained in DNA is "transcribed" by the process of transcription. The process is NOT a straight copy, but "an exchange", in which complementary nitrogen bases appear in the "product"---m-RNA. So when DNA has an "A", that "A" is transcribed into m-RNA as U, a "C" is transcribed as "G", a "G" as "C":

So a sequence in DNA that is CGCAA is transcribed into m-RNA as GCGUU.

(b) Describe what occurs in the process of translation: The information that m-RNA contains is "decoded" by t-RNA . That decoding results in the formation of an amino acid sequence (much as we did in 11(d) above). See Table 1 in the Biochemistry Interchapter (p 499) for the amino acids that result from this translation.

15. The structure that all steroids have in common is:

There are three six-membered rings and a five-member ring. As is fairly common in larger organic molecules, these geometric structures are assumed to have C atoms at each vertex.

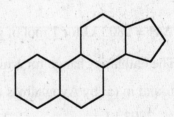

17. Which of the following statements are true?

(a) Breaking the P-O bond in ATP is exothermic. FALSE. The cleavage of P-O bonds in ATP requires energy—so the process is ENDOTHERMIC, not exothermic.

(b) Making a new bond between the P atoms in the phosphate group being cleaved off and the OH group of water is an exothermic process. TRUE- The formation of ADP from ATP releases energy. While the cleavage of a P-O bond requires energy, as does the cleavage of the O-H bond, the energy released upon formation of the P-OH bond releases more energy that is input in the cleavage of the P-O and O-H bonds, resulting in a net EXOTHERMIC process.

(c) Breaking bonds is an ENDOTHERMIC process. TRUE—To cleave bonds, energy must be put in to the process.

(d) Energy released in hydrolysis of ATP may be used to run endothermic reactions in a cell. TRUE. As noted on p 504 of the text (and in part (b) above), the hydrolysis of ATP requires an input of energy (breaking a P-O bond and a O-H bond). It also releases energy as the P-OH bond is formed—more energy than is required for the breaking—with the NET RELEASE of energy that can be channeled into driving otherwise endothermic processes in a cell.

19. (a) The enthalpy change for the production of one mole of glucose by photosynthesis at 25°C:

The process may be represented: $6 CO_2(g) + 6 H_2O(l) \rightarrow C_6H_{12}O_6(s) + 6 O_2(g)$

The energy change is:

$\Delta_rH = (\Delta_fH° \ C_6H_{12}O_6(s) + 6 \cdot \Delta_fH° \ O_2(g)) - (6 \cdot \Delta_fH° \ CO_2(g) + 6 \cdot \Delta_fH° \ H_2O(l))$

$\Delta_rH = (1mol \cdot (-1273.3 \ kJ/mol) + 6 \cdot (0))$

$- \quad (6mol \cdot (-393.509 \ kJ/mol) + 6 \ mol \cdot (-285.83 \ kJ/mol))$

$\Delta_rH = 2803 \ kJ$ for 1 mol of glucose.

(b) The enthalpy change for 1 molecule of glucose is found by dividing the energy change found in (a) by Avogadro's number:

$$\frac{2803 \ kJ}{1 \ mol \ glucose} \cdot \frac{1 \ mol \ glucose}{6.022 \ x \ 10^{23} \ molecules \ glucose} = 4.655 \ x \ 10^{-21} \ kJ/molecule$$

(c) What is the energy of a photon of light with a wavelength of 650 nm?

To calculate the energy of one photon of light with 650 nm wavelength, we need to first calculate the frequency of the radiation:

$$frequency = \frac{speed \ of \ light}{wavelength} = \frac{2.9979 x 10^8 \ m/s}{6.5 \ x 10^2 \ nm} \cdot \frac{1.00 \ x \ 10^9 \ nm}{1.00 \ m} = 4.61 \ x \ 10^{14} \ s^{-1}$$

The energy is $E = h\upsilon$ or $(6.626 \ x \ 10^{-34} \ J \cdot s \cdot photons^{-1})(4.61 \ x \ 10^{14} \ s^{-1})$
$= 3.06 \ x \ 10^{-19} \ J \cdot photons^{-1}$

(d) One molecule of glucose requires $4.655 \ x \ 10^{-21}$ kJ or $4.655 \ x \ 10^{-18}$ kJ. One photon with wavelength 650 nm has $3.06 \ x \ 10^{-19}$ J of energy, one photon will be **insufficient** to cause the production of one molecule of glucose.

Interchapter
The Chemistry of Modern Materials

1. Degrees of magnification in Figure 14(b,c,d):

   With the length of the scale bar in each figure being 5.0 mm, we can simply divide the explicit length provided in the text into 5.0mm, to obtain the magnification:

   For (b): 5.0 mm/1 mm = 5 times magnification

   For (c): 5.0 mm/20 μm = 5.0 mm/0.020 mm = 250 times magnification

   For (d): 5.0 mm/500 nm = 5.0 mm/500 x 10^{-9} m

   $\qquad\qquad\qquad\quad$ = 5.0 mm/5.00 x 10^{-6} mm = 10000 time magnification

3. Energy produced/minute by a 1.0 cm^2 solar cell operating at 25% efficiency, if 925 watts/m^2 are striking the earth's surface.

   The energy striking the cell would be: $\dfrac{925 \text{ watts}}{1 \text{ m}^2} \cdot \dfrac{(1m)^2}{(100 \text{ cm})^2} \cdot \dfrac{1.0 \text{ cm}^2}{1} = 0.0925$ watts

   Since the cell operates at 25% efficiency, the power absorbed would be 0.023125 watts (or 1/4 of 0.0925 watts).

   Now we convert between power (watts) and energy (Joules), with the factor 1 watt = 1J/s

   $0.023125 \text{ watts} \cdot \dfrac{1 \frac{J}{s}}{1 \text{ watt}} \cdot \dfrac{60 \text{ s}}{1 \text{ min}} = 1.4$ J/minute (2 sf)

5. How would one calculate the density of pewter?

   Pewter is an alloy of antimony (7.5%), copper (1.5 %), and tin (91%).

   The densities of these three elements from www.webelements.com:

Element	Density (g/cm^3)
Antimony	6.697
Copper	8.920
Tin	7.310

The density is the sum of the masses contributed by the three components:

$$(0.075)(6.697 \text{g/cm}^3) + (0.015)(8.920 \text{g/cm}^3) + (0.91)(7.310 \text{ g/cm}^3) = 7.3 \text{ g/cm}^3 \text{ (2 sf)}$$

7. We need to calculate the volume of the space between the two sheets of glass.

   The area is 180 cm x 150 cm with a 2.0 mm gap. We have the density of aerogel in units of mg/cm³, so calculate the volume using the metrics provided *with units of cm*.

   $$180 \text{ cm} \cdot 150 \text{ cm} \cdot \frac{1.0 \text{ cm}}{10 \text{ mm}} \cdot \frac{2.0 \text{ mm}}{1} = 5400 \text{ cm}^3$$

   As Density = M/V, we multiply V x D to get the mass of aerogel.

   5400 cm^3 x $1.9 \text{ mg/cm}^3 = 10260$ mg (or 10. g to 2sf).

Interchapter
The Chemistry of the Environment

1. Molar concentrations of Na^+ and Cl^- in seawater:

   From Table 2: Na^+ (in mmol/L) = 460   and Cl^- = 550

   The concentration in mmol/L differs from the molar concentrations by a factor of 1000, so

   460 mmol Na^+/L = 0.460 mol Na^+/L   and 550 mmol Cl^-/L = 0.550 mol Cl^- /L.

3. Mass of NaCl could be obtained by evaporating 1.0 L of seawater :

   1.0 L of seawater could produce (at most) 0.460 mol NaCl.

   $$1.0 \text{ L} \cdot \frac{0.460 \text{ mol NaCl}}{1 \text{ L}} \cdot \frac{58.44 \text{ g NaCl}}{1 \text{ mol NaCl}} = 26.9 \text{ g NaCl (3 sf)}$$

5. Pressure exerted by water vapor that is 40,000 ppm.

   Note that 40,000 ppm corresponds to a concentration of 4 parts (water vapor) in 100 parts of
   a gaseous mixture. [Easily seen by expressing the two concentrations as ratios—one fraction
   with a denominator of 1,000,000 and the other with a denominator of 100.] **Assume** that the
   total pressure of the gaseous mixture is 760 mm Hg (or 1 atm). We know that the pressure
   exerted by a gas is related to the amount of gas present. So in a mixture of (e.g.) 100
   molecules of gas, 4 molecules are water vapor. So 4/100 of the total pressure is due to water
   vapor, or (0.04 • 760 mm Hg) = 30.4 mm Hg. Reference to Appendix G shows that a
   temperature between 29 and 30 °C will support a vapor pressure of approximately
   30.4 mmHg.

7. Mass of Mg obtained by precipitation of $Mg(OH)_2$, formation of $MgCl_2$, and subsequent
   electrolysis from 1.0 L of seawater.

   $$1.0 \text{ L} \cdot \frac{0.052 \text{ mol Mg}^{2+}}{1 \text{ L}} \cdot \frac{24.3 \text{ g Mg}^{2+}}{1 \text{ mol Mg}^{2+}} = 1.26 \text{ g/L or 1.3g Mg/L (2 sf)}$$

Preparation of 100. kg of Mg requires what volume of seawater (D = 1.025 g/cm^3)

$$\frac{100.\ \text{kg Mg}}{1} \cdot \frac{1 \times 10^3 \text{g Mg}}{1\ \text{kg Mg}} \cdot \frac{1\ \text{L seawater}}{1.26\ \text{g Mg}} = 7.9 \times 10^4\ \text{L seawater.}$$

Remember that rounding numbers (in this case 1.26 to 1.3) will give slightly differing answers.

9. Graduated cylinder contains 100. mL of liquid water at 0 °C. An ice cube of volume 25 cm^3 is dropped into the cylinder.

(a) If 92% of the ice is below water level, then the corresponding mass of water is displaced by the ice, resulting in an apparent gain of volume. The volume increase is 0.92 x 25 cm^3 or 23 cm^3. So the volume in the graduated cylinder will increase to 123 mL.

(b) When the ice melts, the volume will be equal to the volume of the water **plus** the volume of the ice, which is now liquid water. So the question is, "What mass of water is contained in the original 25 cm^3 of ice? With a density of 0.92, the mass of the ice will be 23 g. This mass plus the mass of the original water (now **both** with a density of 1.0 g/cm^3) will be 123 g and with a volume of 123 mL.